The Fossil Vertebrates of Florida

Florida A&M University, Tallahassee
Florida Atlantic University, Boca Raton
Florida Gulf Coast University, Ft. Myers
Florida International University, Miami
Florida State University, Tallahassee
University of Central Florida, Orlando
University of Florida, Gainesville
University of North Florida, Jacksonville
University of South Florida, Tampa
University of West Florida, Pensacola

The Fossil Vertebrates of Florida

Edited by Richard C. Hulbert, Jr.

University Press of Florida

Gainesville · Tallahassee · Tampa · Boca Raton

Pensacola · Orlando · Miami · Jacksonville · Ft. Myers

06 05 04 03 02 01 6 5 4 3 2 1

LIBRARY OF CONGRESS CATALOGING-IN-PUBLICATION DATA
The fossil vertebrates of Florida / edited by Richard C. Hulbert, Jr.
p. cm.
Includes bibliographical references and index.
ISBN 0-8130-1822-6 (alk. paper)
1. Vertebrates, Fossil—Florida. 2. Animals, Fossil—Florida. I. Hulbert, Richard C.
QE841.F655 2001
566'.09759—dc21   00-030258

The University Press of Florida is the scholarly publishing agency
for the State University System of Florida, comprising Florida A&M
University, Florida Atlantic University, Florida Gulf Coast Univer-
sity, Florida International University, Florida State University, Uni-
versity of Central Florida, University of Florida, University of North
Florida, University of South Florida, and University of West Florida.

University Press of Florida
15 Northwest 15th Street
Gainesville, FL 32611-2079
http://www.upf.com

Dedication

To Walter Auffenberg, Thomas H. Barbour, Pierce Brodkorb, J. W. Gidley,
H. J. Gut, Oliver P. Hay, Joseph Leidy, Stanley J. Olsen, Clayton E. Ray,
E. H. Sellards, George G. Simpson, S. David Webb, and Theodore E. White,
for laying the foundation of Florida vertebrate paleontology.

Contents

Preface

The Plaster Jacket series of pamphlets began in 1967 as an effort to communicate information about the fossil vertebrates of Florida between professional paleontologists and the public. For the most part, scholarly papers and monographs written by professionals are filled with jargon, and their subject matter is presented in great detail. They are also often published by obscure technical journals or expensive university or specialty publishers and are not readily available. On the other hand, less technical books on paleontology are often inaccurate or out-of-date and have little information specific to Florida. The Plaster Jacket series tried to bridge the gap between these extremes, and judging by its popularity it succeeded.

Perhaps in no other state is vertebrate paleontology such a popular subject of serious interest to the general public than Florida. The reason is simple: Florida abounds in fossils. And not in hostile deserts and inaccessible badlands, but seemingly everywhere—on beaches and river bottoms, at construction sites, and in quarries. Many of the original Plaster Jackets are now long out of print, and recent discoveries and studies have outdated the others. The Florida Paleontological Society recognized that the original purpose of the Plaster Jacket series was still viable but wanted to put the information in book form—thus the genesis of this volume.

First, what this book is not: It is not a guide to where to find fossils, except in the most general sense. Three good sources of this information are readily available: Robin Brown's *Florida's Fossils* (1988, Pineapple Press, reprinted 1996); Margaret Thomas's *Fossil Vertebrates—Beach and Bank Collecting for Amateurs* (1968, currently available from the Florida Paleontological Society); and Mark Renz's *Fossiling in Florida: A Guide for Diggers and Divers* (1999, University Press of Florida). Nor is it an exhaustive guide to identifying fossil specimens, although it may help in some cases. The best sources for identifying fossils are comparative skeletons of modern animals, other fossils that have already been properly identified, and experience. Novices are often amazed at how rapidly experienced avocational and professional paleontologists can identify fossil bones and teeth. This is the result of many hours of practice and is for the most part something you cannot get out of a book (although helpful references are available; see references at the end of chapter 1).

This book describes the fossil vertebrates found in Florida. It tries to answer several questions. What animals lived in Florida? How are they related to one another and to living animals? When did they first appear, and (for many) when did they become extinct? What did they eat? What do they tell us about past environments? Most of these animals lived throughout the southeastern United States and, in many instances, across much of North America. Thus, much of the presented information is applicable beyond the state boundaries of Florida. References at the end of each chapter provide the interested reader with a guide to more information, both in printed form and on the Internet. The addresses (URLs) provided for Internet sites were correct and functioning as of 2000, but over time they may change or be deleted. Updates, corrections, and new discoveries will be posted on an Internet site that can be reached through http://www.flmnh.ufl.edu/natsci/vertpaleo/vertpaleo.htm.

No special technical background is assumed, only an interest in learning about fossil vertebrates. All terms are explained in the introductory chapter or the glossary. The first chapter provides basic information on vertebrate anatomy, biologic classification, geologic time scales, and the geologic history of Florida.

Funds for updating the Plaster Jacket series and producing this book were provided by the Florida Paleontological Society. Computer and photographic facilities, supplies, and access to fossil specimens were provided by the Florida Museum of Natural History, Gainesville, and the Department of Geology and Geography, Georgia

Southern University, Statesboro. The G. G. Simpson Library of Paleontology at the Florida Museum of Natural History was a valuable resource.

Numerous individuals have greatly contributed to the completion of this work. I received abundant assistance over the years from the staff at the Florida Museum of Natural History, including Marc Frank, Russell McCarty, Gary Morgan, Roger Portell, Art Poyer, and Erika Simons. Vickie Kersey DuBois assisted with the photographic images. The following reviewed all or portions of the book and greatly improved its content: J. A. Baskin, R. C. Brown, D. P. Domning, S. D. Emslie, R. Franz, G. Hubbell, W. D. Lambert, H. G. McDonald, B. J. MacFadden, P. A. Meylan, G. S. Morgan, R. Purdy, B. J. Shockey, E. H. Simons, E. Taylor, S. P. Vives, S. D. Webb, and D. P. Whistler. Many of the illustrated specimens were collected by avocational paleontologists and donated to the Florida Museum of Natural History.

State and Regional Fossil Clubs and Organizations

As Florida's population boomed through the twentieth century, mining and construction activities and the advent of scuba diving opened the door to the state's buried treasure of vertebrate fossils. Fossil collecting by both professional and dedicated avocational paleontologists increased over the years. For the most part, these two groups interacted in a symbiotic, mutually beneficial partnership. This informal association culminated in October of 1976 with a meeting on the University of Florida campus organized by David Webb and Howard Converse. Talks on a variety of paleontologic and geologic topics were presented by both professionals and experienced amateurs.

Following a second successful fall meeting the next year, it was apparent that a more formal arrangement was needed to build upon the cooperation and goodwill generated over the previous years. This was the genesis of the Florida Paleontological Society, Inc. (FPS). The FPS is a nonprofit corporation whose goals are to advance the science of paleontology within the state of Florida, to disseminate knowledge of paleontology, and to foster cooperation with all persons interested in the state's ancient fauna and flora. Toward these goals, the FPS annually sponsors the Gary S. Morgan Student Research Award to assist a university student (in Florida) in conducting paleontologic research. FPS publishes a newsletter, short articles, and books on fossil collecting and preparation. It also organizes annual meetings and fossil collecting field trips.

Membership in the FPS is open to anyone interested in paleontology. For information about joining the society, write to Secretary, Florida Paleontological Society, Florida Museum of Natural History, P.O. Box 117800, University of Florida, Gainesville, FL 32611-7800, or e-mail fps@flmnh.ufl.edu.

In addition to the statewide FPS, there are many regional organizations and clubs comprised of avocational paleontologists and fossil enthusiasts. Most hold regular meetings, sponsor fossil fairs, and organize field trips to collect fossils and visit museums. A comprehensive list of Florida's fossil clubs is available on the Internet at www. flmnh.ufl.edu/orgs/club.htm. Among them, these are the most active:

Bone Valley Fossil Society, 2704 Dixie Rd., Lakeland, FL 33801

Florida Fossil Hunters, 320 W. Rich Ave., Deland, FL 32720-4128

Fossil Club of Miami, 12540 SW 37th Street, Miami, FL 33175

Manasota Fossil Club, P.O. Box 1601, Tallevast, FL 34270

Paleontological Society of Lee County, P.O. Box 151651, Cape Coral, FL 33915-6111

Southwest Florida Fossil Club, c/o Steve Wilson, P.O. Box 1308, Arcadia, FL 34265

Space Coast Fossil Hunters, 2125 N. Indian River Drive, Cocoa, FL 32922

Tampa Bay Fossil Club, c/o Terry Sellari, P.O. Box 290561, Temple Terrace, FL 33687

Authorship

All of the original authors of the Plaster Jacket series gave the Florida Paleontological Society permission to reissue their work in revised form. A balanced, cohesive, up-to-date work was produced using their original text. For some chapters this required almost total rewriting; others needed only minor editing and updating. Four chapters were written specifically for this book (1, 2, 3, and 9), as were sections of others when certain taxonomic groups had not been covered by any of the Plaster Jackets (such as many of the rodents). The authors of the Plaster Jackets who dealt with vertebrate paleontology of Florida and their respective topics are as follows:

	Original authors	Topic	Present chapter
PJ #1	Norman Tessman	sharks	4
PJ #2	Elizabeth S. Wing	skates and rays	4
PJ #3	Walter Auffenberg	snakes	7
PJ #4	S. David Webb	proboscideans	15
PJ #5	Walter Auffenberg	crocodilians	7
PJ #7	Camm Swift and Elizabeth S. Wing	bony fish	4
PJ #8	Robert A. Martin	aquatic rodents	12
PJ #9	John Waldrop	horses	14
PJ #11	Norman Tessman	carnivores	11
PJ #12	Jesse S. Robertson	bison	13
PJ #13	S. David Webb	edentates	10
PJ #15	Roy H. Reinhart	sea cows	16
PJ #16	Walter Auffenberg	turtles	6
PJ #17	S. David Webb	peccaries	13
PJ #22	Steve P. Christman	rattlesnakes	7
PJ #27	Clayton E. Ray	seals and walruses	11
PJ #29	Gary S. Morgan	whales	17
PJ #30	Bruce J. MacFadden	rhinoceroses	14
PJ #32	Jon A. Baskin	saber-tooth carnivores	11
PJ #36	Annalisa Berta	dogs	11
PJ #39	S. David Webb	camels	13
PJ #42	Jonathan J. Becker	birds	8
PJ #43	Gary S. Morgan and Ann E. Pratt	marine mammals	16, 17
PJ #44	Peter A. Meylan	amphibians and reptiles	5–7
PJ #45–46	S. David Webb	ruminants	13

SUGGESTED MANNER OF CITATION

For most chapters:
Hulbert, R. C., [and name(s) of original author(s), if any]. 2001. Chapter title. Pp. 00–00 *in* R. C. Hulbert (ed.), *The Fossil Vertebrates of Florida*. University Press of Florida, Gainesville.

For chapters 5–7 only:
Meylan, P. A., W. A. Auffenberg, and R. C. Hulbert. 2001. Chapter title. Pp. 00–00 *in* R. C. Hulbert (ed.), *The Fossil Vertebrates of Florida*. University Press of Florida, Gainesville.

Abbreviations

ka   thousands of years ago.

Ma   millions of years ago.

UF   Vertebrate paleontology collection, Florida Museum of Natural History, Gainesville.

UF/FGS   Florida Geological Survey collection, now housed with UF collection.

UF/PB   Pierce Brodkorb collection, now housed with UF collection.

AMNH   American Museum of Natural History, New York, New York.

LACM   Natural History Museum of Los Angeles County, Los Angeles, California.

MCZ   Museum of Comparative Zoology, Harvard University, Cambridge, Massachusetts.

TRO   Timberlane Research Organization, Lake Wales, Florida.

UCMP   Museum of Paleontology, University of California, Berkeley.

UNSM   University of Nebraska State Museum, Lincoln.

USNM   Department of Paleobiology, National Museum of Natural History, Smithsonian Institution, Washington, D.C.

Introduction

FOSSILS AND PALEONTOLOGY

A fossil is any indication of past life. For animals, the most common fossils are remains of the inorganic skeletal system, such as bones, teeth, and shells. Under special circumstances, when bacterial decay is prevented, organic body tissues can be preserved. Examples of this are the famous frozen mammoths of Alaska and Siberia and insects embedded in amber. Also considered to be true fossils, even though they are not directly part of the animal's body, are trace fossils, features such as tracks, burrows, nests, and coprolites (fossilized excrement). Vertebrate fossils in Florida are most often limited to skeletal material and coprolites because of the nature of the state's geology. Examples of different types of fossils are shown in figure 1.1.

Paleontology is the scientific study of fossils. It is often confused with archaeology, which is the study of past humans, their artifacts, and cultures. An argument could be made that archaeology is only a specialized branch of paleontology because humans are biological organisms. However these two scientific fields have long, independent histories and different academic affiliations. Paleontology is traditionally associated with the natural sciences, particularly biology and geology, whereas archaeology is grouped with the social sciences, such as anthropology, psychology, and sociology. There are two areas where the interests of paleontologists and archae-

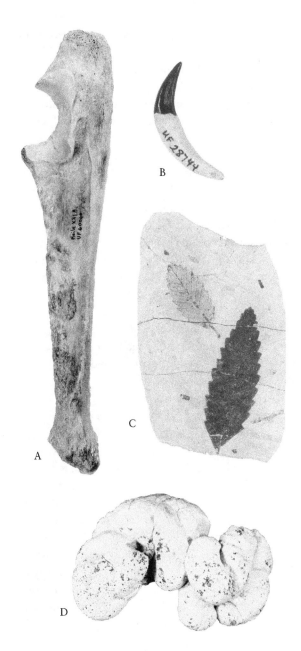

Fig. 1.1. Examples of different types of fossils. *A,* a fossil bone (the right ulna) of the saber-toothed cat *Xenosmilus* from Haile 21A, Alachua County, late Pliocene, UF 60000; *B,* UF 28744, a fossil tooth of the long-snouted dolphin *Pomatodelphis* from the Phosphoria Mine, Polk County, middle Miocene. The enamel-covered crown is more shiny and preserved a much darker color than the root; *C,* impressions of tree leaves preserved in a fine-grained sedimentary rock (age and locality unknown); *D,* a coprolite (fossilized excrement), probably of an alligator, from Haile 7C, Alachua County, late Pliocene. *A* about 0.5×; *B–D* about 1×.

ologists overlap (sometimes cooperatively, sometimes contentiously). The first is the study of the fossilized remains of the group of mammals that includes humans, the primates. The second is the remains of prey hunted by prehistoric man. The latter are the subject of a subdivision of archaeology, zooarchaeology. Zooarchaeology is the study of animal (that is, nonhuman) fossils from archaeological sites, and in particular uses them to reconstruct diet, behavior, and environments of ancient humans. In North America, most accumulations of vertebrate fossils less than 10,000 years old (= 10 ka), are associated with human artifacts and are therefore studied by archaeologists. Older fossils are studied by paleontologists. This leaves a relatively short interval between when humans first entered North America from Asia (sometime between 14 ka and 30 ka) and the end of the Pleistocene Epoch (10 ka) when the interests of paleontologists and archaeologists strongly coincide.

The primary subdivisions of the science of paleontology are based on the kinds of fossil organisms being studied. The major fields include vertebrate paleontology (the study of vertebrates, those animals with backbones); invertebrate paleontology (the study of animals without backbones); micropaleontology (the study of microscopic organisms, mostly kinds of aquatic plankton); paleobotany (the study of fossil plants); and ichnology (the study of trace fossils). Paleoecology is the paleontologic equivalent to the modern biologic field of ecology, the study of the relationships between organisms and their interactions with the environment. A paleoecologist studies the same topics but deals with fossils instead of living organisms. Taphonomy is the study of the processes by which plants and animals become buried in sediment and are turned into fossils. The subject matter of this book is the paleontology and paleoecology of the vertebrate animals that lived in the region now called Florida.

A basic knowledge of what bones and teeth look like is essential training for both professional and avocational paleontologists. To the untrained eye, a variety of oddly shaped natural objects can easily be mistaken for fossils. In Florida, these "pseudofossils" commonly are chert nodules or phosphate pebbles. These can take on shapes that appear to be bones or teeth. Careful comparison shows that true bone possesses detailed features that these lack. The most revealing is a definite internal structure. Most bones are either hollow with supporting struts or have an internal spongelike texture. Even the more solid bones of sea cows show growth rings. In contrast, pseudofossils usually lack internal structure and have homogeneous, solid centers.

Even when an object found in the ground or on a riverbed truly is a bone, there is no guarantee that it is a fossil. It could belong to an animal that died within the last few hundred years (or even to last week's barbecue dinner!). These items are rarely of any interest to scientists. There are several tests to determine if a bone is or is not a fossil. None are 100 percent reliable, so an expert opinion should be obtained before discarding a potentially important find. Fossil bones are often permineralized, which means that the open spaces within the bone have been filled by mineral deposits. This makes the fossil more dense (heavier) than a modern bone. There are several problems with this test. A true fossil that is not permineralized may be just as lightweight as a bone from a modern animal. Fossils from dry caves are often not permineralized. Another problem with this test is that the bones of living manatees (see chap. 16) are naturally very dense to act as ballast. Thus this test is invalid for these animals, which are very common in Florida. Finally, bones deposited in mineral-rich waters of springs can become completely permineralized in only a few decades (Neill, 1957).

Another test is to gently tap on or *carefully* drop the suspected fossil from a height of about an inch or so onto concrete or a similar hard surface. A true fossil will most often make a sound similar to dropped porcelain or rock (a "ping"); recent bone makes a duller sound, more like a piece of wood (a "thud"). This works best when known fossils are on hand to make comparisons with, and of course should not be used with fragile specimens. Suspected fossils can also be burned with a match or lighter. Recent bone may retain organic material and give off a pungent odor similar to burnt hair. A true fossil should not produce such a foul smell. Color is a *very poor* indicator of fossilization. Many true fossils are white or tan, resembling fresh bone, while modern bones can rapidly take on a brown color after immersion in one of Florida's stained rivers.

The identification of a bone may indirectly demonstrate whether or not it is a fossil. If it can confidently be identified as a member of a long-extinct group, then it must be a fossil. Likewise, if it belongs to a group recently introduced to this continent by Europeans, then it is not a fossil. Examples of these in North America are chickens, pigs, cattle (although often difficult to distinguish from native bison), house mice, and the Norway rat.

Paleontology uses terminology and knowledge from the fields of biology and geology. The most basic biological disciplines applicable to vertebrate paleontology are anatomy of the skeleton and systematics (the study of the evolutionary relationships among organisms and their classification). Equally important are geological topics. The remainder of this chapter provides an introduction to these subjects and introduces much of the terminology required to fully comprehend the succeeding chapters. For further information, the reference section lists several basic textbooks and Internet Web sites covering each of these particular fields.

THE VERTEBRATE SKELETON

Introduction

Table 1.1 presents the definitions of directional terms used in vertebrate anatomy and paleontology. They are most often used comparatively—for example, the heart is anterior to the kidneys. During descriptions of features on bones and teeth, the subject is oriented as it was in the living animal. Reference can then be made to a bone's proximal or distal end or its lateral surface, for example (fig. 1.2).

Vertebrate skeletons are made of several different substances, which greatly differ in physical properties such as strength and elasticity. The primary skeletal materials are cartilage, bone, collagen, dentine, and enamel. Cartilage is typically made entirely of organic materials and lacks minerals. It usually decays completely and is very rarely preserved in fossils. The cartilaginous vertebrae of some sharks and related fish are embedded with small crystals of calcium-rich minerals to produce what is called calcified cartilage. This material can fossilize. Bone and other mineralized skeletal materials in verte-brates contain microscopic crystals of the mineral hydroxyapatite that are formed by specialized cells of the body. Hydroxyapatite has a complex chemical formula but is one of a group of minerals that can simply be called calcium phosphate because they are mostly composed of the elements calcium, phosphorus, and oxygen. Bone and dentine are hybrid substances; both are formed from a mix of hydroxyapatite crystals and fibers of the protein collagen. The proportions of the two vary, but most bone is made of two-thirds to three-fourths hydroxyapatite, and one-third to one-fourth collagen and other organic compounds. The basic type of dentine is similar in composition to bone, but the two differ in the arrangement of the internal structure of the crystals. In contrast to bone and dentine, enamel is composed almost entirely (about 97 percent) of hydroxyapatite crystals, with only a minor organic component. Dentine and enamel are only found in teeth and the plates or scales of external armor of some fishes. Most fish lack true enamel but have specialized, hardened forms of dentine in its place. Only detailed microscopic investigation can tell true enamel apart from these specialized types of dentine, sometimes called enameloid. Enamel has a glossy, smooth surface (fig. 1.1B), which the experienced fossil hunter soon learns to detect and distinguish from the usually duller surfaces of bone and dentine. Teeth, particularly those with a thick coating of enamel, are more durable than bone and therefore tend to fossilize more often than skeletal elements made of bone.

The vertebrate skeleton has a number of functions and must work in concert with the muscular and circula-

Table 1.1. Commonly used anatomical directional terms

Term	Definition
anterior	located toward the front or the head
posterior	located toward the rear or tail (opposite of anterior)
dorsal	located toward the back or top of the animal
ventral	located toward the ground (opposite of dorsal)
lateral	located toward the side, away from the middle
medial	located toward the middle or interior of the animal (opposite of lateral)
proximal	the end of a bone closer to the middle of the body
distal	the end of a bone farther away from the middle of the body (opposite of proximal)
lingual	the side of a tooth closer to the tongue, usually its medial side[a]
labial	the side of a tooth farther away from the tongue, usually its lateral side (*buccal* is a synonym of *labial*)
occlusal	the side of a tooth facing directly out of the gum—i.e., the ventral surface of an upper tooth or the dorsal surface of a lower tooth. During chewing, the opposing occlusal surfaces of the upper and lower teeth contact each other.

Note: These terms are applied as if to a living, four-legged animal (not to an upright, bipedal organism like a human). See figure 1.2 for examples.

a. On an incisor, the lingual side is often the posterior side.

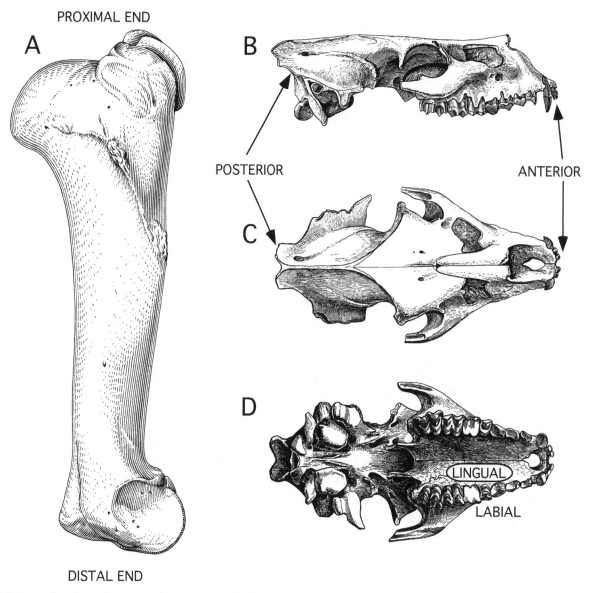

PROXIMAL END

A

B

POSTERIOR

ANTERIOR

C

D

LINGUAL

LABIAL

DISTAL END

Fig. 1.2. Examples of use of anatomic direction terms for skeletal elements (see table 1.1). *A*, lateral view of a humerus (upper-arm bone) of the camel *Camelops*, with indicated proximal and distal ends; *B*, right lateral, *C*, dorsal, and *D*, ventral views of the skull of an oreodont (chap. 13). *A* after Webb (1965), courtesy S. David Webb, Florida Museum of Natural History. *B–D* after Peterson (1906); reproduced by permission of the Carnegie Museum of Natural History Museum, Pittsburgh.

tory systems. Protection against injury or predators is an obvious function of the skeletal system. The skull protects the brain and vital sense organs of the head, while the rib cage surrounds the heart and lungs. Some vertebrates also embed pieces of protective bone in their skin (turtles and armadillos being two familiar examples). The skeletal system also supports the body against the force of gravity, keeping it upright, and provides attachment points for muscles and tendons. Bone is a dynamic tissue and is constantly being broken down and reformed. As part of this process, it functions as a storehouse for calcium and phosphorous, two of the body's

essential elements. Some scientists have argued that this was the original function of bone in ancestral vertebrates and that it only later took on its other roles. The jaws and their associated teeth may be used for protection, but primarily function in obtaining and chewing food. Depending on the particular vertebrate, the skeletal elements can also be sites of manufacture of blood cells, can store fat, and, in birds, are part of a specialized respiratory system.

Each individual piece or element of the skeletal system has one or more functions, and these determine its shape, size, and internal structure. Skeletal elements can be long, slender cylinders, flat plates, cubic blocks, or an irregular shape. The skeleton of each individual contains hundreds of separate elements (for example, a human normally has 206 bones and 32 teeth; other vertebrates can have far more). It is very uncommon for these to remain together after death, joined as they were in life. The rare fossils that are found like this, with the skeletal elements joined as in life, are described as being articulated and are highly prized for their scientific value. Only a handful of Florida's fossil localities have produced articulated specimens (fig. 1.3A). Somewhat more common, but still unusual, is to find all or a portion of the skeleton of a single individual in which the elements are no longer attached as they were in life but which are nevertheless close to each other and analysis can prove they once fit together. Such a fossil is described as being an associated skeleton. However, by far the majority of fossils are neither articulated nor associated skeletons (fig. 1.3B). Instead, they are isolated skeletal elements or a broken part of one.

Fossils of complete elements are usually easily identified as to which part they formed in the original skeleton, after a moderate amount of experience. Some of the books listed at the end of this chapter illustrate many skeletal elements of different types of vertebrates. Most paleontologists use skeletons from modern animals with each element labeled to aid them in identifying partial or nondescript fossils. While some fragments many contain enough information to identify them, others may be so incomplete as to prevent exact identification.

As the function(s) of a skeletal element influence its shape and size, paleoecologists can use this relationship to deduce information concerning the animal when it was alive. For example, the elements of the limbs differ greatly depending on how the animal moved—swimming, fast running, digging, tree climbing, etc. Similarly, the teeth and jaws of carnivores and herbivores can easily be distinguished. More careful study can reveal even greater detail about the diet of the fossil animal; for example, the bone-crushing habit of certain hyenas is evident in their teeth. A fairly new technique using sophisticated equipment determines the precise composition of the enamel from a fossil tooth, such as the percentages of the different isotopes of carbon atoms (carbon-12 vs. carbon-13). This provides more clues as to the animal's diet.

The Axial Skeleton

The vertebrate skeleton is divided into two groups of elements that are collectively called the axial and appendicular skeletons. The axial skeleton consists of those elements forming the central core (or axis) of the body: the elements of the head, vertebrae, ribs, and sternebrae. The appendicular skeleton includes the elements of the paired lateral fins in fish or the front and hind limbs in land vertebrates (tetrapods). The bones of the shoulder (or pectoral) and pelvic regions, which serve to structurally unite the limbs with the axial skeleton, are also considered parts of the appendicular skeleton.

A   B

Fig. 1.3. Florida vertebrate fossil localities. *A,* Moss Acres Racetrack site, Marion County, late Miocene. This locality produced articulated skeletons, a very rare occurrence in Florida. The fossil in front of the collector, Gary Morgan, is an articulated limb of the rhinoceros *Aphelops. B,* Leisey Shell Pit 1A, Hillsborough County, early Pleistocene. Although an extremely rich site, most of the fossils collected here were isolated (a few were associated individuals, none were articulated).

The skeletal elements found in the vertebrate head are surprisingly numerous and complexly integrated with each other. In the oldest vertebrates, such as the jawless agnathan fish of the early Paleozoic Era, the skeletal elements of the head were not organized into a true skull (Forey and Janvier, 1994). Instead, they consisted of more discrete groups separated by their function—bony plates in the skin provided protection, rodlike structures supported the gill openings, and a boxlike structure enclosed the brain. Later vertebrates added a fourth functional group of cranial elements, the upper and lower jawbones with their associated teeth. Originally these four skeletal complexes were only loosely integrated and operated more or less independently. This arrangement is still found in modern sharks and other chondrichthyans (chap. 4). Only bony fish and their descendants have fully integrated these cranial elements into a true skull. Fossil skulls, or portions of them, are generally the most prized part of the body for paleontologists, because they provide the most scientific information about the animal. Examples of representative vertebrate skulls are shown in figures 1.4–1.6.

Vertebrae form the individual elements of the "backbone," one of the characteristic features of vertebrate animals (and where the name comes from). An individual vertebra consists of a main body called the centrum, from which project a number of processes (fig. 1.7A, B). Two of these typically combine to form an arch on top of the centrum (the neural arch) through which passes the spinal cord. Paired right and left ribs articulate with a variable number of the vertebrae. In mammals, only a specific section of about 10 to 20 consecutive vertebrae bear ribs. Proportionally many more of the vertebrae have ribs in vertebrates that are not mammals. One common way to classify vertebrae is by the shapes of the front and back surfaces of their centra (fig. 1.7C–F). The vertebrae of fish do not strongly articulate with each other, allowing great flexibility. Land vertebrates must support their bodies against the force of gravity, so their vertebrae strongly interlock using projections from the neural arch known as zygapophyses. Fossil vertebrae and ribs are common because each individual animal has a greater number of them than the elements of the head or limbs. With some practice, vertebrae can easily be sorted into such basic groups as shark, bony fish, amphibian, snake, alligator, bird, and mammal. More complete identification usually takes a good collection of comparative specimens and a fair degree of expertise. For some vertebrates, most notably snakes, fossil vertebrae are the most extensively studied part of the skeleton. Most of the time, however, vertebrae receive only

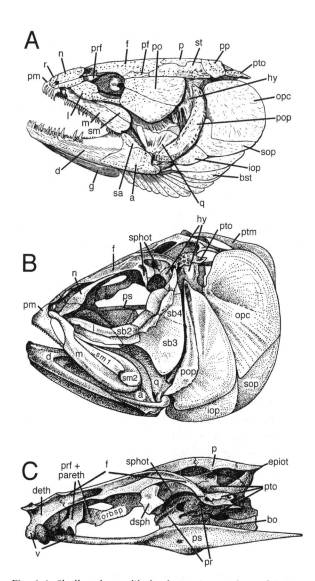

Fig. 1.4. Skull and mandibular bones in two bony fish (Osteichthyes). *A*, left lateral view of *Amia calva*, the bowfin; *B*, left lateral view of *Clupea*, a herring; *C*, left lateral view of the braincase (chondrocranium) of *Clupea*, showing the interior bones of the skull. Note the great complexity of fish skulls, with many more bones than in those of other vertebrates. *Amia* is a much less specialized fish than *Clupea* (see chap. 4). Abbreviations: *a*, angular; *bo*, basioccipital; *bst*, branchiostegal rays; *d*, dentary; *deth*, dermethmoid; *dsph*, dermosphenotic; *epiot*, epiotic; *g*, gular plate; *f*, frontal; *hy*, hyomandibular; *iop*, interopercular; *l*, lacrimal; *m*, maxilla; *n*, nasal; *opc*, opercular; *orbsp*, orbitosphenoid; *p*, parietal; *pareth*, parethmoid; *pm*, premaxilla; *po*, postorbital; *pop*, preopercular; *pp*, postparietal; *pr*, prootic; *prf*, prefrontal; *ps*, parasphenoid; *pt*, pterygoid; *ptm*, posttemporal; *pto*, pterotic; *q*, quadrate; *r*, rostral; *sa*, surangular; *sb*, suborbital (numbered in *B*); *sm*, supramaxilla (numbered in *B*); *sop*, subopercular; *sphot*, sphenotic; *st*, supratemporal; *v*, vomer. Figures from *The Vertebrate Body*, 4th ed., by Alfred Sherwood Romer, copyright © 1970 by Saunders College Publishing, reproduced by permission of the publisher.

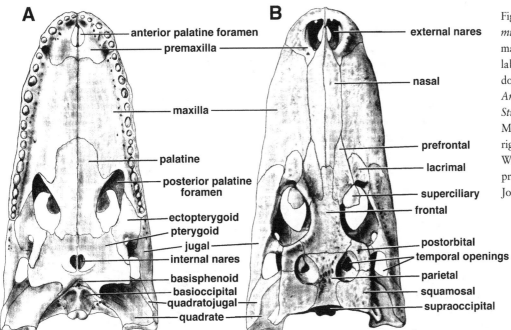

A

- anterior palatine foramen
- premaxilla
- maxilla
- palatine
- posterior palatine foramen
- ectopterygoid
- pterygoid
- jugal
- internal nares
- basisphenoid
- basioccipital
- quadratojugal
- quadrate

B

- external nares
- nasal
- prefrontal
- lacrimal
- superciliary
- frontal
- postorbital
- temporal openings
- parietal
- squamosal
- supraoccipital

Fig. 1.5. Skull of *Alligator mississippiensis*, with major bones and openings labeled. *A*, ventral, and *B*, dorsal views. Figures from *Analysis of Vertebrate Structure*, 4th ed., by Milton Hildebrand, copyright © 1994 by John Wiley & Sons, Inc. Reprinted by permission of John Wiley & Sons, Inc.

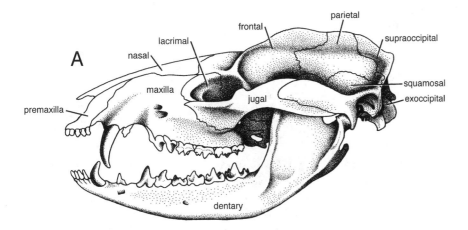

A

- parietal
- frontal
- supraoccipital
- lacrimal
- nasal
- squamosal
- maxilla
- exoccipital
- premaxilla
- jugal
- dentary

Fig. 1.6. Skull and mandibular bones of three mammals. *A*, left lateral view of the opossum, *Didelphis*; *B*, dorsal view of the domestic cat, *Felis*; *C*, ventral views of the domestic dog, *Canis*. Figures from *Morphogenesis of the Vertebrates*, 2d ed., by Theodore W. Torrey, copyright © 1967 by John Wiley & Sons, Inc. Reprinted by permission of John Wiley & Sons, Inc.

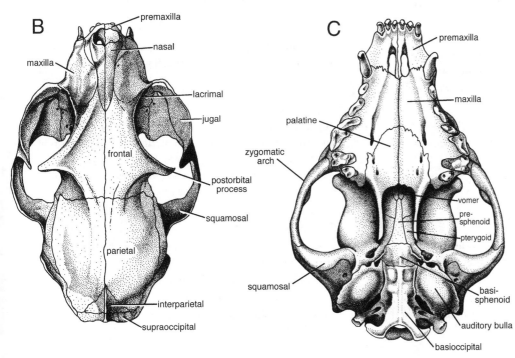

B

- premaxilla
- nasal
- maxilla
- lacrimal
- jugal
- frontal
- postorbital process
- squamosal
- parietal
- interparietal
- supraoccipital

C

- premaxilla
- maxilla
- palatine
- zygomatic arch
- vomer
- pre-sphenoid
- pterygoid
- squamosal
- basi-sphenoid
- auditory bulla
- basioccipital

cursory treatment from paleontologists compared to skull and limb bones. An exception to this generalization is during studies of the locomotion of fossil animals (for example, Zhou et al., 1992; Giffin, 1992). Except when completing a skeleton in a museum exhibit, ribs usually get even less attention from paleontologists than vertebrae.

The Appendicular Skeleton

The appendicular skeleton of most fish is relatively weak, because most swim using muscles attached to axial skeletal elements (vertebrae and ribs). Land-dwelling tetrapods, which use their limbs for locomotion and support, have large appendicular skeletal elements. The basic plan of the tetrapod front and hind limb are the same (fig. 1.8), although the corresponding elements differ in form and are usually easy to distinguish. The joint between the most proximal limb element (humerus or femur) and the supporting pectoral or pelvis element is like a ball and socket, permitting a wide range of motion. In contrast, the joint between the proximal limb element and the paired second elements (fig. 1.8), the elbow and knee joints in humans, are more hingelike, with a more limited range of motion. The six major limb elements, the humerus, radius, and ulna of the forelimb, and the femur, tibia, and fibula of the hind limb, are usually sturdy and fossilize well.

Articulating with the distal ends of the radius and ulna are two or three rows of much smaller bones called carpals. There can be up to twelve individual carpals, but many tetrapods have fewer (most mammals, including humans, have six). Carpals have irregular shapes and, in mammals, intricate surfaces where they articulate with each other. Corresponding to the carpals of the front limb are the tarsal elements of the hind limb. In mammals, two of the tarsal elements, the astragalus and calcaneum, are larger than the others and fairly distinctive. Their shape determines the range of motion of the ankle joint.

The distal end of the tetrapod limb consists of the digits; each digit contains a series of elongate bones. The first, or most proximal, of these elements, the one that articulates with the carpals (or tarsals), is called a metacarpal (or metatarsal; fig. 1.8). Metapodial is a term used to refer to either metacarpals or metatarsals. The metapodial is often the longest element of the digit. The remaining elements of the digit are called phalanges. The number of phalanges per digit varies among tetrapods, although frequently there are three. The digits in a tetrapod are numbered starting from the inner (medial) side, so the human thumb and big toe are digit 1, the index

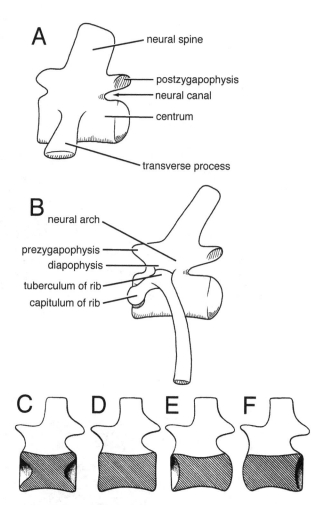

Fig. 1.7. Basic structures of tetrapod vertebrae, shown in lateral view, anterior to left. *A,* lumbar vertebra, which lacks a true rib but has instead a laterally projecting transverse process; *B,* a dorsal or thoracic vertebra that articulates with a rib, in this case a "two-headed" rib, one with both a capitulum and tuberculum. *C–F,* the basic types of vertebrae, shown somewhat diagrammatically, anterior to left. *C,* typical amphicoelus type, with both ends of centrum concave; *D,* acoelus type, with both ends of the centrum flat; *E,* procoelus type, anterior end of centrum concave, posterior end convex; *F,* opisthocoelus type, anterior end of centrum convex, posterior end concave. *A–B* Figures from *Analysis of Vertebrate Structure* by Milton Hildebrand, copyright © 1974 by John Wiley & Sons, Inc. Reprinted by permission of John Wiley & Sons, Inc. *C–F* Modified from Romer (1956); copyright © 1956 by The University of Chicago Press and reproduced by permission of the publisher.

finger is digit 2, and so on. Although the oldest tetrapods had more than five digits, all living and most fossil tetrapods have five or fewer digits per limb. When a tetrapod group has lost digits over the course of evolution, the remaining digits are still referred to using the number of the equivalent digits in a five-toed tetrapod. For example, an animal which has lost digit 1 is said to have

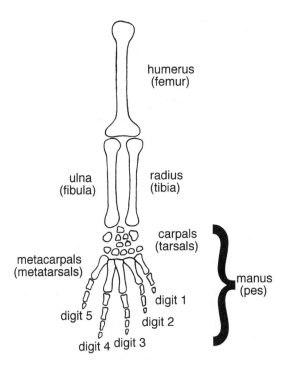

Fig. 1.8. Basic arrangement and names of the limb bones in tetrapod vertebrates. Names in parentheses refer to the hind limb, those which are not are forelimb elements, with the exception of the phalanges (which are used for both). Figures from *Morphogenesis of the Vertebrates*, 2d ed., by Theodore W. Torrey, copyright © 1967 by John Wiley & Sons, Inc. Reprinted by permission of John Wiley & Sons, Inc.

digits 2 through 5 (they are not renumbered 1 through 4). Reference to a metapodial for a specific digit is done in this form: the metacarpal of digit 2 is called metacarpal 2 (this number is often written with roman numerals).

Sesamoids are bony elements that form in tendons at places where they are stressed. The largest sesamoid is the patella (kneecap), which inserts over the joint between the femur and tibia. Smaller, usually paired sesamoids are found at the distal ends of some metapodials. For example, humans usually have very small sesamoids with metacarpal 1 and metatarsal 1.

Teeth and Dental Terminology

Teeth are technically not part of the internal skeleton but are instead highly modified scales that over the course of evolution have become attached to bones lining the mouth. Fossils of teeth are common, and they are studied in great detail, especially those of mammals and sharks. Some fossil species are known only from their teeth. A large number of terms are used to describe teeth; the basic ones will be described in this section. Most of these are only applicable to mammals. Comprehension of

these terms is needed to understand the scientific literature of mammalian paleontology. Terminology related to shark teeth is discussed in chapter 4. See table 1.1 for directional terms used with teeth.

Mammalian teeth consist of an outer layer of enamel over a dentine core, with an inner pulp cavity. The pulp cavity narrows at the tooth base to form one or more root canals (fig. 1.9). Nutrients are supplied to the living portions of the tooth by blood vessels that pass through the root canals, which also carry nerves. The crown of a tooth is defined as the portion that is covered by enamel. The root is that portion of the tooth that is embedded in bone and where the dentine is surrounded by cement. Fibrous tissue and cement bind the tooth to the socket(s) in the jaw that are called alveoli.

In most mammals, an individual's teeth are specialized into four general types (fig. 1.10) that differ in their appearance and have different functions. This condition is known as heterodonty. The opposite condition, found in vertebrates whose teeth are all basically the same

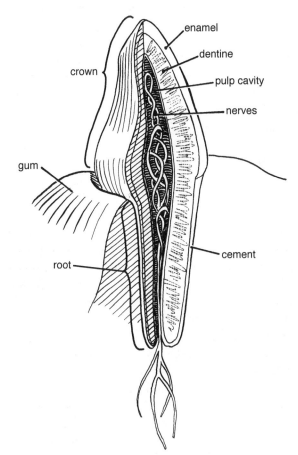

Fig. 1.9. Cross section through a simple mammalian tooth, showing its interior structures. Modified from Shipman et al. (1985); reproduced by permission of Harvard University Press, Cambridge.

shape, is homodonty. The anteriormost mammalian teeth are the incisors, which are typically flattened or conical, and function in obtaining food. Primitively, mammals have five upper incisors and four lowers (on each side), for a total of eighteen. This number has been substantially decreased in many groups. The second type of tooth is the canine, of which there is only one in each of the four jaw quadrants. It is usually conical and larger than the preceding incisors, greatly enlarged in carnivorous mammals that use the canine to puncture and tear their prey. Large canines are also found in males of some groups, where they are used in displays and fighting for dominance (for example, baboons). In other groups, like rodents, the canine tooth is lost altogether. The teeth posterior to the canine are collectively referred to as cheek teeth. Cheek teeth usually have two or more roots, while incisors and canines tend to have one. In some groups, typically herbivores, a wide gap called the diastema separates the canine (or the posteriormost incisor if the canine is lost) and the cheek teeth. The cheek teeth chew the food procured by the incisors and canines. Cheek teeth are divided into two groups: the more anterior premolars and the more posterior molars (fig. 1.10). Primitive mammals had up to eleven cheek teeth (five premolars and six molars), but this number is reduced in all living groups. Premolars tend to have simpler crowns than do molars. In many herbivores, however, one or more of the premolars may take on the appearance and function of true molars. This process of molarization produces more surface area for chewing.

Premolars and molars differ in another aspect, whether or not during the life of the animal they are replaced by a second tooth. In most vertebrates, excluding mammals, teeth are continually shed and replaced in an orderly fashion throughout the life of the animal. In mammals, there are instead only two sets of teeth that are formed during the life of the animal. The deciduous series (commonly called the "milk" teeth) begins formation before birth and consists of a set of incisors, a canine, premolars, and molars. The sequence of eruption of these teeth varies among mammal groups, but usually all teeth are in place before the individual becomes a full adult. As the individual matures, the adult or permanent series erupts, replacing some of the deciduous teeth. The adult series consists only of incisors, canine, and premolars; the molars of the deciduous series are not replaced. Other members of the deciduous tooth series can be retained and function along with those of the adult series in certain species. For example, in rodents the incisors are never replaced and so are actually members of the deciduous series. Deciduous teeth can usually (but

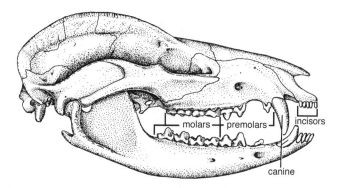

Fig. 1.10. Right lateral view of the skull and mandible of the opossum, *Didelphis virginiana*, showing the relationship between the skull bones and tooth types in mammals. The upper incisors are held in the premaxilla, the alveolus for the canine is located at or just posterior to the suture between the maxilla and premaxilla, and the premolars and molars are held in the maxilla. From *Vertebrate Paleontology and Evolution* by Carroll © 1988 by W. H. Freeman and Company. Used with permission.

not always) be distinguished from their adult successors by differences in size and shape.

The numbering system used by scientists to refer to mammalian teeth is based on their relative position in the mouth. The anteriormost tooth of each of the four types is assigned the number 1; the tooth directly posterior to it, the number 2; and so on. By tradition, this system uses as its starting point the ancestral pattern shared by all placental mammals, three incisors, one canine, four premolars, and three molars. Similar terminology is used for marsupials and other groups of fossil mammals, but with different numbering systems. In species that have reduced the number of teeth, the remaining teeth are numbered to correspond to the matching teeth in the ancestral pattern, and *not* to the actual number present in the mouth. This can be confusing when teeth have been lost from the anterior side of a series, or from the middle. The latter is rare, but the former is usually the case with premolars. Thus a placental mammal with three premolars is generally assumed to possess the second, third, and fourth premolars and to have lost the first. Similarly, in a placental with two premolars (like most humans), unless demonstrated otherwise, they are the third and fourth premolars. A code for referring to teeth is made up of one or two letters referring to tooth type (I for incisor, C for canine, P for premolar, M for molar, D added if it belongs to the deciduous series) and a number referring to the tooth position within the series. For example, M3 is a third molar and DI2 is a deciduous second incisor. Uppercase letters indicate upper teeth (DP2 is an upper second deciduous premolar) and lowercase letters lower teeth (i3 is a lower third incisor). In some publications, the numeral is written as a superscript to indicate an upper tooth or as a subscript to indicate a lower tooth. As there is only one canine tooth, the letter C alone denotes the canine.

The crowns of mammalian molars (and some premolars) are generally much more complicated than those of incisors and canines. Their occlusal surfaces contain a number of topographical high features, called cusps, with intervening lower areas. The arrangement of the cusps is the subject of much study, as their formation is generally similar within a species but different from other species. The terminology used to describe these features is complex, and only the basics will be presented here. Molar terminology is best learned by starting with a mammal that has the simple, primitive pattern; for example, the common opossum, *Didelphis* (fig. 1.11A, B). Terms for lower teeth are distinguished from corresponding terms of the upper teeth by addition of the suffix "-id." The three primary cusps on upper molars are called the protocone, paracone, and metacone (fig. 1.11A), while secondary features are called conules (for example, a paraconule). Surrounding the center of the molar is a shelf or ridge called the cingulum, from which other secondary cusps can form. These are called styles (for example, a parastyle or metastyle). Primitively, the cusps of mammalian molars are separate and distinct from one another. Such a tooth with low, rounded cusps is described as being bunodont. Human molars are examples of bunodont teeth. Many herbivorous mammals connect their cusps with ridges (called lophs), blurring the distinction of the individual cusps. This type of tooth is called lophodont.

The three upper cusps form a triangle, with a lingual protocone, an anterior and labial paracone, and a posterior and labial metacone (fig. 1.11A). Triangular upper molars are the primitive pattern in mammals and are found in many carnivorous and insectivorous species. Herbivorous and some insectivorous mammals often have a fourth major cusp, the hypocone, in the posterior lingual corner of the tooth (fig. 1.11C). This produces a squarer occlusal surface, with more area for chewing. The hypocone evolved independently in many groups of mammals.

The primitive condition for lower molars consists of a raised, triangular anterior section called the trigonid, and a posterior basin, called the talonid (fig. 1.11B). The trigonid bears the three principle cusps, a labial protoconid, an anterior-lingual paraconid, and a posterior-lingual metaconid. Note that the apex of the triangle formed by the three primary cusps points labially in

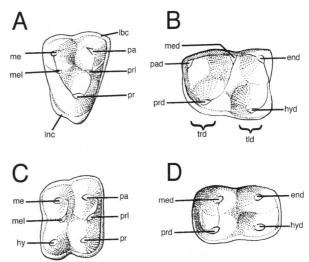

Fig. 1.11. Basic patterns and names of the major cusps found on the molar teeth of mammals. A–B, the primitive, unspecialized pattern, as seen in the opossum; C–D, the more advanced pattern evolved in many herbivorous groups of mammals in which the molars are "squared-up" by addition of a hypocone to the upper molars and loss of the paraconid in the lowers; A, C, right upper molars (anterior to right); B, D, left lower molars (anterior to left). Abbreviations: *end*, entoconid; *hy*, hypocone; *hyd*, hypoconid; *lbc*, labial cingulum; *lnc*, lingual cingulum; *me*, metacone; *med*, metaconid; *mel*, metaconule; *pa*, paracone; *pad*, paraconid; *pr*, protocone; *prd*, protoconid; *prl*, protoconule; *tl*, talonid; *tr*, trigonid. Figures from *Mammalogy* by Terry A. Vaughn, copyright © 1972 by Saunders College Publishing, reproduced by permission of the publisher.

lower teeth and lingually in upper teeth. The talonid bears two major cusps, a labial hypoconid and a lingual entoconid. When upper molars square up by adding a hypocone, the lowers usually lose the paracone to become correspondingly more rectangular (fig. 1.11D). When premolars evolve to become more molarlike, the same cusp terms are used to describe their features, even though they have evolved independently.

Another primitive condition for mammal teeth is for the height of the crown to be shorter than or about equal to the depth of the root(s) (fig. 1.9). Such teeth are brachyodont. Mammals with brachyodont teeth include opossums, pigs, squirrels, and most primates including humans. Mammals that eat abrasive food tend to evolve higher-crowned teeth. When the crown height of a tooth exceeds that of the roots, the tooth is described as being hypsodont. Hypsodont cheek teeth are almost exclusively found in herbivorous groups, especially those with a high proportion of grass in their diets (grazers). Taking hypsodonty to its ultimate limit are mammals that suppress the formation of roots and keep forming enamel and dentine at the base of the crown as fast as it is worn off the occlusal surface (hypselodont teeth). Hypselodont teeth have evolved independently in many groups, especially among rodents.

ANIMAL CLASSIFICATION AND SYSTEMATICS

All human cultures have names they apply to plants and animals in their everyday language. Scientists call these common or vernacular names. Although informal, there is often a loose hierarchical structure to common names. For example, the English word *tree* is a very general term referring to a large number of plants, among them oak trees and pine trees. Each of these also has its subgroups; for example, loblolly pine, slash pine, and Douglas fir are all types of pine trees. Different regions or subcultures within a country often have different common names for what is actually the same animal or plant (for example, the cougar, mountain lion, catamount, and Florida panther are different common names for the same animal). Also, different organisms can have the same common name; for example, the word *dolphin* means both a type of fish and several kinds of small, toothed whales. Cultures speaking different languages will usually have completely different common names for identical creatures. Thus a widespread European bird or tree can easily have a dozen or more common names.

Such a chaotic situation does not sit well with scientists, who wish to communicate precisely with colleagues from around the world. Also, microscopic, fossil, or obscure organisms usually do not even have common names. To solve all of these problems, biologists devised an orderly method of applying names to all organisms, names that are used by scientists regardless of their nationality. This method dates back to the middle of the eighteenth century and the Swedish naturalist Linnaeus, and it has endured with some modification to the present. It is not a perfect system, but its use is so deeply ingrained that it will not be abandoned any time soon. (An alternative method to Linnaean taxonomy for applying names to organisms is known as *phylogenetic nomenclature*; its rules and procedures are still being formed. See Cantino et al., 1999.) The Linnaean system is more than a method to devise names for plants and animals; it also organizes all living (and formerly living) organisms into a natural hierarchical framework. After the widespread acceptance of the concept of organic evolution in the mid-nineteenth century, it was realized that the "naturalness" of the Linnaean system was its concordance with the evolutionary history of biologic organisms. Ever since, biologists and paleontologists have agreed that the formal classification of organisms should

be based on their evolutionary relationships with each other. There have been, however, serious disagreements on how to actually go about determining this.

The Linnaean system is a hierarchical series of ranks (fig. 1.12). The basic unit of the system is the rank of species. Surprisingly, as this *is* the basic unit for the entire system, there is no agreed-upon definition for what a species truly is. There are several different "species concepts." One that is widely used considers species to be "groups of actually or potentially interbreeding natural populations, which are reproductively isolated from other [species]" (Mayr, 1942, 120). This and other theoretical species concepts all have their strong points, but in the museums and laboratories of the world more practical considerations are used to distinguish and identify species. The biologist (or paleontologist) places into a single species those individuals which resemble each other more than they do individuals of different species.

The remaining ranks in the Linnaean hierarchy, from genus up to domain (fig. 1.12), can be viewed as a means to organize species on the basis of their evolutionary relationships. Two or more species placed in the same genus are very closely related (by definition they are more closely related to each other than they are to any species placed in another genus). For example, dogs, wolves, and coyotes are three closely related species that are placed in the same genus, *Canis*. Two species placed in different kingdoms are very distantly related. A cow in a

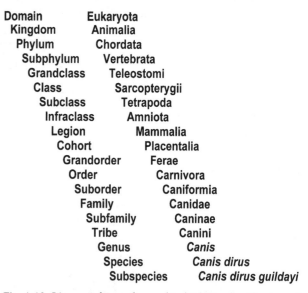

Domain	Eukaryota
Kingdom	Animalia
Phylum	Chordata
Subphylum	Vertebrata
Grandclass	Teleostomi
Class	Sarcopterygii
Subclass	Tetrapoda
Infraclass	Amniota
Legion	Mammalia
Cohort	Placentalia
Grandorder	Ferae
Order	Carnivora
Suborder	Caniformia
Family	Canidae
Subfamily	Caninae
Tribe	Canini
Genus	*Canis*
Species	*Canis dirus*
Subspecies	*Canis dirus guildayi*

Fig. 1.12. Linnaean hierarchy used in biologic classification of animals. The ranks of the hierarchy are shown on the left side and follow McKenna and Bell (1997). On the right are the formal names used in the classification of the dire wolf, an extinct mammalian carnivore known from Florida.

meadow and the grass it is eating are placed in two separate kingdoms, while the bacteria in the cow's stomach are in a third kingdom. Figure 1.12 shows the basic ranks of the hierarchy. Many ranks are necessary to completely show the evolutionary relationships among species. Prefixes are also added to the basic ranks to create new ones. The most commonly used prefixes are *super-* and *sub-* (for example, to create the ranks of superfamily and subfamily), but others are needed. Examples are shown in the checklist in chapter 3.

There are two concepts worth emphasizing at this point. First, while similarity or overall resemblance is a practical criterion for recognition of species, it is *not* for the other levels of the hierarchy. For example, dogs and cows are not placed in the same group (Mammalia) to the exclusion of lizards because they are more similar to each other than they are to lizards (even though that is true). This principle can be observed in figure 1.13, which shows that resemblance and relatedness do *not* always have a one-to-one correspondence. For the purposes of classification, chickens (and other birds) are grouped together with alligators (in a group called the Archosauria), while the lizards go in a separate group (Lepidosauria). The second point, and one not well understood outside the scientific community, is that all determinations of evolutionary relationships among organisms are not definitive, but are instead provisional, what scientists call hypotheses. *All* scientific hypotheses can potentially be proven wrong and replaced with new hypotheses. Thus, with reference to figure 1.13, the lines showing that alligators are more closely related to chickens than they are to lizards is a hypothesis, not a fact (although it is a hypothesis currently accepted by most evolutionary biologists; see Hedges, 1994). It is possible that new evidence will be discovered that disproves this hypothesis and replaces it with something else. As biologists and paleontologists analyze the evolutionary relationships of plants and animals, they are constantly disproving old hypotheses and suggesting new ones. This means that the classification of organisms is not a static collection of names, but a dynamic structure that is always changing. This unfortunate but necessary result means that the formal names of biologic organisms often change, which sometimes hampers their role in communication.

Basic Rules of Scientific Names

The rules and procedures for naming biologic organisms are regulated by international commissions. The International Commission of Zoological Nomenclature (ICZN) has authority over the naming of animals, both modern

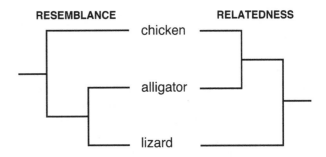

Fig. 1.13. Diagram contrasting the concepts of evolutionary relatedness and anatomical resemblance. The alligator and chicken are regarded as more closely related to each other because they share a closer common ancestor, even though in terms of overall morphologic similarity the alligator is closer to the lizard. Biological classification is based on relatedness and not similarity.

and fossil. The most recent version of the rules was published late in 1999 and went into effect January 1, 2000. The rules of the ICZN have several basic objectives or principles. First is promotion of universality in the scientific names of animals (ideally everyone uses the same name for the same biologic group or taxon). The second objective is to ensure that each name is unique and distinct. The third is use of priority when choosing between two or more possible names (choose the oldest name), unless this leads to great instability.

The formal names of animals in the hierarchy above the species level consist of one word which is always capitalized if used in its formal sense (fig. 1.12). Examples are Animalia, Reptilia, Rodentia, and *Canis*. Names of genera (and subgenera) are always italicized, underlined, or otherwise distinguished from regular text by the use of a different font. The scientific name of a species always consists of two words, not one. The first word is the generic name (the name of the genus in which the species is classified). The second is the trivial name. The trivial name is *not* capitalized, but like the generic name is italicized or underlined. When referring to a species, the two names are used together, for example *Homo sapiens, Tyrannosaurus rex,* and *Rattus rattus.* Note from the latter example that generic and trivial names can have the same spelling (except for capitalization). When the generic name has been used once in written text, it can subsequently be abbreviated to its first letter, if that does not result in confusion or ambiguity (for example, *T. rex*). The letter representing the generic name is still capitalized and italicized. Trivial names are sometimes mistakenly used by themselves. A common example from Florida paleontology is the extinct shark *Carcharodon megalodon,* which is sometimes incorrectly referred to as *Megalodon*. In this case, the trivial name has in effect become a common name and should not be italicized or capitalized when used in this context (for example, "Timmy found two five-inch megalodon teeth last Saturday"). Subspecies names, uncommonly used in vertebrate paleontology, are a three-word combination. An example is *Trachemys scripta petrolei*, which is an extinct subspecies of the living pond turtle *Trachemys scripta*. If a species is divided into two or more subspecies, one of them must have the same subspecific name as the species' trivial name. Thus there is not only a subspecies named *T. scripta petrolei*, but another called *T. scripta scripta*. Similarly, for a genus that is divided into subgenera, one of the subgenera will always have the same name as the genus. For example, some fossil tortoises belong to the genus *Hesperotestudo*, of which there are two recognized subgenera, *Hesperotestudo (Hesperotestudo)* and *Hesperotestudo (Caudochelys)*. Subgeneric names are written inside parentheses and follow the generic name.

It is optional to follow a taxonomic name with the last name or names of the author(s) who formally described it and the year in which it was proposed. An example of this is *Enhydritherium terraenovae* Berta and Morgan, 1986. This means that the species *Enhydritherium terraenovae* was described in 1986 in a scientific paper by two paleontologists named Berta and Morgan. When a species has been transferred to a different genus than the one in which it was initially proposed, the original author's name and date are placed in parentheses. For example, the name of the common Pleistocene armadillo is *Dasypus bellus* (Simpson, 1929), because Simpson placed the species in the genus *Tatu* (and not *Dasypus*) when he originally described it. This type of full citation is used sparingly in vertebrate paleontology.

Most scientific names are formed from Greek or Latin words, but this is not a requirement. However, the endings have to be "latinized," and the gender (masculine, feminine, or neuter) of the generic and trivial names must match. If a species is transferred from one genus to another of differing gender, the ending of the trivial name may have to be changed to conform with its new generic assignment. There are a variety of standard suffixes used in nomenclature, some are required, others are optional but traditionally used. If trivial names are based on the modern name of a person, then the following standard suffixes have to be used: *-i*, if the name is a man's (for example, *Pediomeryx hamiltoni*, named for Dr. Roger Hamilton); *-ae*, if the name is a woman's (for example, *Columbina prattae*, named for Dr. Ann Pratt); *-orum*, if

the name is for two or more men, or of a combination of men and women; and *-arum*, if the name is that of two or more women. If these are not used appropriately by the describing author, then they must be corrected. For example, Baskin (1981) described the new species *Barbourofelis lovei*, but stated that he was naming the species after Ron and Pat Love. Therefore the correct name for this species is *Barbourofelis loveorum*. A common ending used by paleontologist for trivial names is "*-ensis*" (or "*-ense*" if the gender of the genus is neuter). It is used when a species is named after a geographic term such as the name of a river, state, or town. For example, *Parahippus leonensis* was named after Leon County, Florida. The endings of family-group names are standardized: superfamily, -oidea; family, -idae; subfamily, -inae; tribe, -ini; subtribe, -ina. These suffixes are added to the root of the type genus to form the family-group name. Examples are *Canis,* Canidae; *Equus,* Equinae; *Trionyx,* Trionychoidea.

Scientific names can be converted to common names if so desired. Examples are bison, alligator, rhinoceros, hippopotamus, and dugong. When generic names like these are used as common names, they are neither capitalized nor italicized. There are standard ways to make common names from family-level names. The "-idae" at the end of a family name is usually shortened to "-ids" if it is to be used as a noun (Canidae becomes canids, Colubridae becomes colubrids) or "-id" for an adjective (canid features). A notable exception are families that end with "-theriidae." Their common names are made by changing the ending to "-theres," as in gomphotheres (from Gomphotheriidae) and megatheres (Megatheriidae). Subfamilies have the terminal "-inae" altered to "-ines"; for example, Dromomerycinae becomes dromomerycines.

The description of a new species includes the following information. The author(s) must state into which genus the new species is assigned. In rare cases, an author might not know for sure which genus is correct but is sure that the species is new. Then the most appropriate genus is selected and a question mark is placed in front of the generic name to indicate the uncertainty. A holotype specimen is designated for the new species and usually illustrated. Information about where the holotype was collected, who collected it and when, and where it is permanently stored (along with its unique catalog number) are among the data usually given. The holotype is the objective standard against which a species is compared to all others. If an author lists a number of type specimens but does not explicitly state which one is the holotype, then a subsequent author is obligated to designate what is called a lectotype from the series of type specimens. In all future studies, the lectotype has the same functions as a holotype. The describing author can also designate paratypes, which are specimens other than the holotype that provide additional information about the species. How many paratypes, if any at all, an author chooses to list is completely optional. In vertebrate paleontology, paratypes are often used to show features not available on the holotype. For example, if the holotype is a skull, then a lower jaw or limb element(s) might be selected as paratypes. Paratypes are usually chosen from the same locality as the holotype to help ensure that they truly belong to the same species, but this is not a requirement. However, they must be designated in the same publication when the new species name is proposed, not in a later paper. If the holotype or lectotype of a species is lost or destroyed and the availability of a type specimen is needed to solve systematic problems, then a neotype specimen can be designated to take its place. Besides listing the holotype, and optionally paratypes, the author of a new species also must provide a list of character states that distinguish the new species from others in the same genus and closely related genera. Traditionally, these are set forth in a concise, formally worded diagnosis and are more fully elaborated upon in a descriptive section. Additional specimens, if known, may be referred to the new species, either from the same locality as the holotype or from other localities. Finally, an etymology is provided to explain how the trivial name was formed. To be valid, a new scientific name must be published in a publicly available book or journal. Names proposed in newspapers, dissertations, cyberspace (for example, over the Internet), or on microfilm are not valid.

Descriptions of new taxa above the species level are simpler. For a new genus or subgenus, all that is needed is the name of a type species and a list of character states that distinguish the new taxon from others of the same family. The type species is the only objective member of the genus (or subgenus); reference of other species to the genus is subjective (and based on an analysis of evolutionary relationships). New family-group names similarly require a type genus and a list of differential character states. The family name is formed by the combination of the stem of the type genus and the appropriate suffix. In contrast, new generic names are formed from Greek and Latin roots, or latinized words from other languages, and do not necessarily have anything in common with the name of their type species. Groups above the superfamily level (order, class, phylum, etc.) do not have types, and descriptions of new names at these levels require only lists of included taxa and of distinguishing character states.

Two common problems encountered with scientific names are homonyms and synonyms. The names of two taxa are homonyms when they are spelled the same but refer to two different organisms. All generic and higher-level names of animals must be unique and differ by at least one letter from all other names in the animal kingdom. Homonyms often occur when the author of a new name unknowingly uses one previously described because they failed to search the older literature thoroughly, especially foreign journals. Checklists of proposed names are available, so one does not have to personally peruse all of the systematic literature published since Linnaeus. Of course, two workers could coin the same name independently and publish it more or less simultaneously. Homonyms like this are less egregious than the other. The following is a typical example of scientific homonyms. In 1929 two French paleontologists, Arambourg and Piveteau described a new genus of bovid (the cow and bison family) from the Pliocene of Europe and named it *Parabos*. In 1941 two Americans, Barbour and Schultz, described a new genus of bovid from Nebraska that they also named *Parabos*. These two generic names are spelled the same but refer to different species; therefore, they are homonyms. The older name, *Parabos* Arambourg and Piveteau, is called the senior homonym; the younger name, *Parabos* Barbour and Schultz, is the junior homonym. Except under special circumstances (which do not apply in this case), the rules of nomenclature require that the junior homonym be replaced with another name. In this case, Barbour and Schultz published the name *Platycerabos* in 1942 to replace *Parabos* Barbour and Schultz. This example demonstrates two common features of homonyms. First, homonyms often originate between workers from different countries, especially those that publish in different languages, or between those working with different kinds of animals (for example, insects and mammals). Second, when a case of homonymy is discovered, it is considered courteous to notify the author(s) of the junior homonym and give them the opportunity to coin the replacement name.

At the species level, homonyms are not always the fault of slipshod scholarship. The trivial name of one species can be exactly the same as that of another, providing that they belong in different genera. For example, the trivial name *floridanus* has been used as a species or subspecies name for more than eighteen different kinds of mammals as well as many other types of animals. If an author names a new species and unknowingly gives it the same trivial name as one already in use for the particular genus, it automatically becomes a junior homonym and must be replaced. A different kind of homonym at the species level occurs when two genera are synonymized (see below) into one genus and their species are combined. If in each genus there is an identically spelled trivial name, then one becomes the junior homonym to the other as the result of the synonymy. An example of this involves *Equus altidens* Reichenau, an European species of fossil horse described in 1915. In 1957 a new species of horse from Texas was described as *Onager altidens* Quinn. Because Quinn regarded *Equus* and *Onager* as distinct genera, these two species names were not homonyms to him and other workers with similar concepts regarding generic boundaries of horses. However, subsequent revisions of Pleistocene and modern horses recognize a single genus and synonymize all other genera such as *Onager* and *Asinus* with *Equus*. For workers who follow this arrangement, *Equus altidens* Reichenau and *Equus altidens* (Quinn) are homonyms, and the latter must be replaced because it does not have priority. *Equus pseudaltidens* was proposed to replace *E. altidens* (Quinn). Note that there is subjectivity in cases like this: workers who follow the synonymy must use one name for a species; workers who reject the synonymy use another.

Synonymy is when the same species or taxon has two or more different names. As was the case with homonyms, the oldest available name is usually the one selected as the correct name. An exception can be made if the oldest available name has not been used in many years and its reinstatement will upset standard usage. Synonymy can be either objective or subjective. Objective synonymy is rarer and applies to those unusual cases when the two taxa are based on the same name-bearing type. Examples of this would be if two authors used the same specimen as the holotype for two differently named species, or the same species as the type species of two different genera. Much more common is subjective synonymy, which accounts for many of the name changes in systematic nomenclature. A typical case of subjective synonymy is when two species (with different holotypes), which were previously considered distinct by their respective authors, are later considered to represent only one species. In the *opinion* of the reviser, they are synonyms, and the name applied to the species is the one with priority. The synonymy holds for everyone who agrees with this reviser's opinion, but other workers can still follow the original system. As a case example, consider the smallest horse from the Miocene Thomas Farm Site in Florida. In 1932 Simpson named the new species *Archaeohippus nanus* and selected a specimen from Thomas Farm as its holotype. Eight years previously, Hay had named a small horse from Texas *Miohippus blackbergi*. In 1942 White reviewed the Thomas Farm mammals, decided that these two represented the same species, and formally synonymized them. Hay's species name has priority over Simpson's. The fact that the holotype of *A. nanus* is more complete has no bearing on the case, nor is the fact that many more specimens are known from Thomas Farm than from Texas. White also decided that this species belonged in neither *Miohippus* nor *Archaeohippus,* but in a third horse genus, *Parahippus.* So White formally called the small Thomas Farm horse *Parahippus blackbergi* (Hay, 1924). White's synonymy has been accepted in most subsequent studies, but in 1975 Forstén transferred the species to the genus *Archaeohippus* and formed the new specific combination *Archaeohippus blackbergi* (Hay). This opinion is still accepted, but is open for change, as is the case with all other scientific names as knowledge and opinions regarding phylogenetic relationships change.

Geochronology

There are many good sources of information about geologic time scales and the history of their development. A number of books and Internet sites on this topic are listed in the references. This section will review basic concepts and present the current time scales relevant to Florida vertebrate paleontology.

Units of time can be arranged in a hierarchy on the basis of their duration. A decade is divided into years, a year into months, a month into weeks, a week into days, and so on. These are time units in everyday use. Geologists and paleontologists must deal with units of time that are far longer than those used in everyday life, with durations ranging from thousands to millions or even billions of years. Nineteenth-century geologists proposed the geologic time scale in its basic modern form (figs. 1.14, 1.15). Geologic time units, like everyday time units, are arranged in a hierarchy; each unit of geologic time is formally subdivided into two or more entities of the next lower rank in the hierarchy (fig. 1.14). For example, the Phanerozoic Eon is divided into three eras, the Paleozoic, Mesozoic, and Cenozoic. The formal names of portions of geologic time are capitalized and are unique (not repeated). Adjectives such as early, middle, late, and very late are used to indicate only a portion of a time unit (for example, the middle Pleistocene Epoch).

GEOLOGIC TIME UNITS

Eon
Era
Period
Epoch
Age

GEOLOGIC TIME-ROCK UNITS

Eonathem
Erathem
System
Series
Stage

Fig. 1.14. Standard units used in the geologic time scale. Time units refer to abstract intervals of past time, while the corresponding time-rock units refer to all rocks on Earth that formed during a time unit. For example, all rocks that formed during the Cretaceous Period belong to the Cretaceous System.

Unlike most everyday time units, geologic time units do not have predefined durations. The actual durations in years for geologic time units can vary tremendously within the same rank (compare the durations of the Neogene and Cretaceous Periods in fig. 1.15). This is because most geologic time units are not defined on the basis of time, but instead on the relative sequence of rocks and fossils that make up the global geologic record. The boundary between two consecutive time units, for example the Pliocene-Pleistocene boundary, is not defined as a certain number of years ago. Instead the boundary is defined at a certain level within a sequence of rock layers

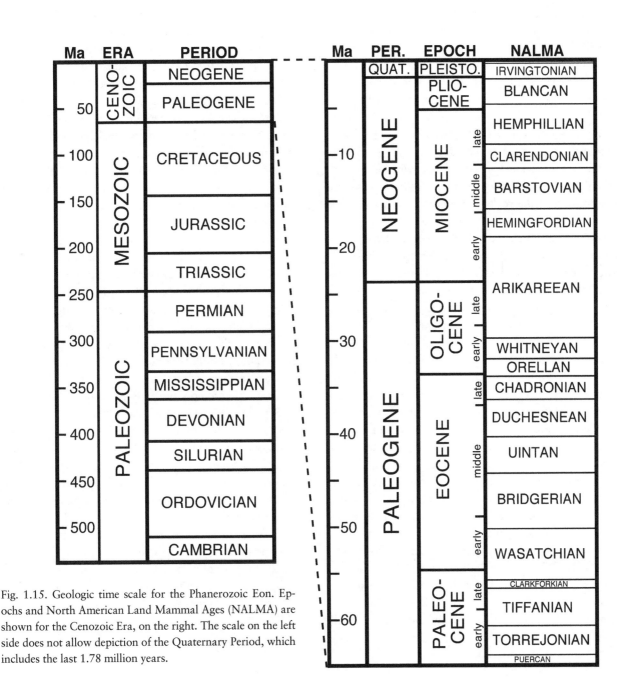

Fig. 1.15. Geologic time scale for the Phanerozoic Eon. Epochs and North American Land Mammal Ages (NALMA) are shown for the Cenozoic Era, on the right. The scale on the left side does not allow depiction of the Quaternary Period, which includes the last 1.78 million years.

at a particular location. For example, the boundary between the Pliocene and Pleistocene is defined as the base of the bed of claystone lying on top of "shale bed *e*" in the Vrica section, which is located four kilometers south of the town of Crotone in southern Italy (Van Couvering, 1997). An international committee of geologists and paleontologists determined the location and placement of this boundary after many years of work and discussion.

The current best estimate (as of 2000) for the numeric age of the Pliocene-Pleistocene boundary is 1.8 million years ago (= 1.8 Ma). However, this number does not define the boundary; the claystone in Italy does. If future research indicates that the age of this bed is instead 1.7 Ma (or some other value), then the age of the boundary will be adjusted. This is the reason that different sources will provide different numeric ages for the boundaries and durations of geologic time units. The age estimates for time unit boundaries used in this book follow the most recent major revision, that of Berggren et al. (1995).

How the age of a particular bed of rock or the fossils within the rock is determined depends on a number of factors. These include the type of rock, whether or not it contains fossils (and if so, what types of fossils), and the relationship of the rock with adjacent bodies of rock. Paleontologists have determined with a high degree of precision when many fossil species were living (their chronologic ranges). If a rock contains fossils with well-known chronologic ranges, its age must be within the intersection of the chronologic ranges of all of its species. The age of a sedimentary rock that lacks fossils is more difficult to determine. However, if the ages of the rocks lying above and below the unfossiliferous rock are known, then the latter's age must be younger than the bed of rock below it and older than the bed above it. This is what geologists refer to as the principle of superposition. The ages of most igneous and metamorphic rocks can be calculated directly by careful analysis of the proportions of radioactive isotopes and their daughter products (for example, potassium-40 and argon-40).

The time scale of figure 1.15 is global in its scope and is used by geologists worldwide. It is particularly useful for marine rocks, those deposited on the sea floor. This is because the rocks defining the boundaries of the time units are almost always marine sedimentary rocks, and the fossils used to recognize these boundaries worldwide are marine organisms. For a long time geologists and paleontologists who studied rocks deposited on continents and their fossils of land plants and animals could not directly apply the global time scale to their subjects. They devised alternate continent-wide time scales based primarily on the chronologic ranges of land-dwelling plants or animals, in particular mammals. Estimates of how these alternate time scales corresponded to the global time scale were attempted, but it was not until the 1960s to 1970s and the first widespread use of radiometric dating techniques that they could be effectively correlated.

In the late 1930s, a committee of influential vertebrate paleontologists headed by Horace Wood organized previously collected data on the geologic and chronologic ranges of North American mammals and used them to produce a continental time scale (Wood et al., 1941). They formally named eighteen intervals, which eventually came to be known as "Land Mammal Ages" (fig. 1.16). Savage (1951) later added two younger land mammal ages to those proposed by Wood et al. (1941). Subsequently proposed ages have not met with widespread acceptance. General correspondence between these North American land mammal ages and the global time scale was done using instances where beds of marine and continental rocks were found in alternating sequences, or when fossils of land mammals were found in marine rocks. However, extensive dating of volcanic ash beds using the potassium-argon radiometric method proved that some of the traditional associations of land mammal ages and the global time scale were incorrect. The corrected correlations did not come into widespread use until the middle to late 1970s (fig. 1.16). Readers of older paleontologic literature must take these changes into account. For example, many of the "Pliocene" faunas in S. J. Olsen's (1959) well-known publication *Fossil Mammals of Florida* are now regarded as Miocene.

The North American land mammal ages and the global time scale are now precisely correlated (Woodburne, 1987; Woodburne and Swisher, 1995). However, many paleontologists working with fossils of land vertebrates continue to use land mammal ages. In this book, most ages of fossil localities and geologic deposits will be reported using the name of a geologic time scale unit (usually an epoch, often modified with the appropriate adjective such as "early"), or an estimated age in millions of years before present (Ma). In certain instances, land mammal ages provide more precision and will be used, but sparingly. Anyone reading the primary scientific literature needs to be familiar with both methods for reporting ages of fossils and their general correlation (fig. 1.15).

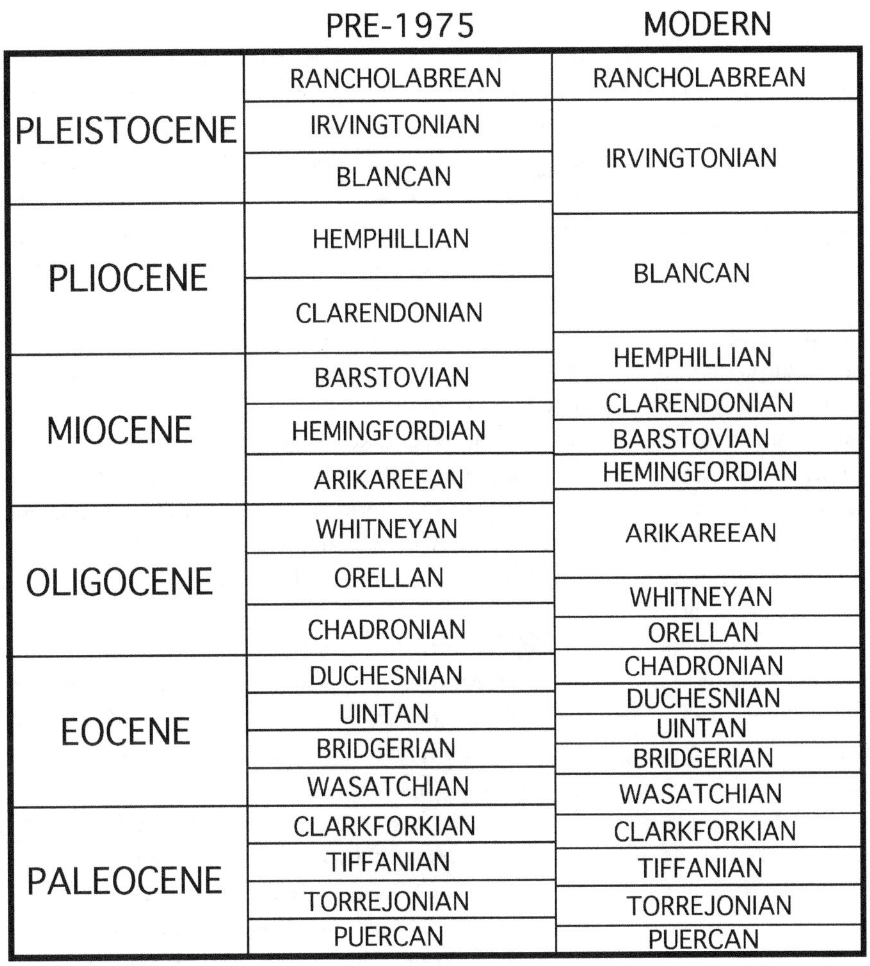

	PRE-1975	MODERN
PLEISTOCENE	RANCHOLABREAN	RANCHOLABREAN
	IRVINGTONIAN	IRVINGTONIAN
	BLANCAN	
PLIOCENE	HEMPHILLIAN	BLANCAN
	CLARENDONIAN	
MIOCENE	BARSTOVIAN	HEMPHILLIAN
		CLARENDONIAN
	HEMINGFORDIAN	BARSTOVIAN
		HEMINGFORDIAN
	ARIKAREEAN	ARIKAREEAN
OLIGOCENE	WHITNEYAN	
	ORELLAN	WHITNEYAN
	CHADRONIAN	ORELLAN
EOCENE	DUCHESNIAN	CHADRONIAN
		DUCHESNIAN
	UINTAN	UINTAN
	BRIDGERIAN	BRIDGERIAN
	WASATCHIAN	WASATCHIAN
PALEOCENE	CLARKFORKIAN	CLARKFORKIAN
	TIFFANIAN	TIFFANIAN
	TORREJONIAN	TORREJONIAN
	PUERCAN	PUERCAN

Fig. 1.16. Comparison of how the correlation of the NALMA with the global time scale changed during the 1970s. Thickness of ages is not to scale and does not indicate estimated durations (for those, see fig. 1.15). Note how the age determination of many of the land mammal ages has changed; for example, the Chadronian, Arikareean, Clarendonian, and Blancan. This means that while a fossil regarded as Clarendonian in the 1960s is today still Clarendonian, the geologic age it is assigned to has changed from early Pliocene in the 1960s to middle to late Miocene today. These changes have come about because of increased numbers of more accurate radioisotopic dates, widespread application of magnetostratigraphy (using the magnetic polarity of iron-bearing minerals in rocks to determine the ages of rocks), and better correlations between terrestrial and marine rocks.

BASICS OF FLORIDA'S GEOLOGIC HISTORY

Only a short summary of the geologic history of Florida is presented here. Readers interested in more information should consult chapters 3, 4, and 5 of the recent book edited by Randazzo and Jones (1997) or the numerous publications of the Florida Geological Survey. Although many of the latter are out of print, they can be viewed at many public and university libraries throughout the state.

The oldest rocks exposed at the surface in Florida are middle Eocene (about 42 to 44 Ma), and most are Miocene or younger. As the oldest known rocks found on the Earth are about 4 billion years old, the surficial rocks of Florida represent only the most recent 1 percent of the planet's recorded geologic history. Older rocks from Florida have been brought up by drill rigs from depths greater than a mile below the surface. This is the primary method to study rocks lying beneath the surface, and these samples reveal an eventful geologic history. Many

of the details of the early parts of this history remain poorly known.

The most common rocks in Florida are limestone, sandstone, claystone, conglomerate, and dolostone. These are all types of sedimentary rocks. Most sedimentary rocks form from deposits of loose, unconsolidated sediment that accumulate at the surface of the earth. Beaches, river and lake beds, and the piles of rock found at the base of a cliff are common examples of sedimentary deposits. After they are buried, unconsolidated sediments are turned into solid rock by compaction and cementation. In many places in Florida the sediments have never been deeply buried, so they remain unconsolidated even though they are millions of years old.

Paleozoic History of Florida (543–245 Ma)

Relatively little is known about the Paleozoic (and earlier) history of Florida, but many lines of evidence point to an association with Gondwana, not North America. Gondwana, the largest Paleozoic continent, also included what are today Africa, South America, Australia, and Antarctica, as well as southern Europe (Spain, France, and Italy) and parts of southern Asia, notably India. Florida's Gondwana heritage is revealed by comparison of its early Paleozoic rocks and fossils (Pojeta et al., 1976), and by paleomagnetic evidence that places its early Paleozoic latitude at about 50° south of the equator (Opdyke et al., 1987).

By the late Paleozoic, North America and northern and eastern Europe had combined to form a single continent known to geologists as Laurussia. Laurussia and Gondwana collided in the late Carboniferous Period (about 300 Ma), forming a long, high mountain range in the process. Rocks near the zone of collision were highly deformed (by folding, faulting, and metamorphism), and, in some cases, melted. The magma produced by the melted rocks later cooled and formed large bodies of granite. Atlanta's Stone Mountain is perhaps the best known of these. The collision zone between Gondwana and Laurussia locally runs from the western Carolinas through central Georgia into Alabama and then farther westward. The joining of Gondwana, Laurussia, and some smaller continents in the late Paleozoic meant that all of the world's landmasses were combined in a single supercontinent, which geologists call Pangaea. Florida was in the middle of this great continent, far from any oceans.

Mesozoic History of Florida (245–65 Ma)

During the Triassic Period, Pangaea began to break apart through a process geologists call rifting. In its early stages, rifting stretches continental crust until it breaks; this is accompanied by volcanic eruptions, lava flows, and earthquakes. As Pangaea separated, rifting first began locally in a northeast-southwest trending area that extends from Charleston, South Carolina, and Savannah, Georgia, to the Tallahassee-Pensacola region of the panhandle. If rifting had continued in this region, then most of Florida and southeastern Georgia would have separated from North America along with the rest of Gondwana. However, in the early Jurassic the zone of rifting shifted to what is now the eastern edge of the Bahamas, and rifting stopped in Georgia and western Florida. By the end of the early Jurassic (about 175 Ma), the newly opened Atlantic Ocean and Gulf of Mexico separated Florida from Africa and South America. Through the end of the Jurassic, much of northern Florida remained above sea level. Thick marine deposits of Jurassic salt, gypsum, and limestone indicate that southern Florida was below sea level at this time.

Rising sea levels through the Cretaceous Period eventually submerged the entire state of Florida, along with much of southern and central Alabama and Georgia. The Cretaceous shoreline was located near Macon and Columbus, Georgia. Limestone is the predominant Florida Cretaceous rock, with lesser amounts of salt, gypsum, shale, and sandstone. As parts of Florida remained above sea level during the Triassic, most of the Jurassic, and even at times in the early Cretaceous, this landscape presumably would have been inhabited by dinosaurs and other land vertebrates of the Mesozoic. No fossils of these famous animals are known from Florida, however, since the rocks that would contain them are deeply buried underneath younger rocks. The nearest finds of dinosaur fossils to Florida are late Cretaceous specimens from central Alabama and Georgia (Schwimmer and Best, 1989; Schwimmer et al., 1993).

Paleogene History of Florida (65–24 Ma)

The high sea levels of the Mesozoic continued into the Paleogene, so the basic depositional pattern observed in the Cretaceous remained in place. It is at this point in the state's geologic history that representative rocks are exposed at the surface and are available for detailed study. The most widespread bodies of Paleogene rock in Florida with surface outcrops are the Avon Park Formation (middle Eocene), Ocala Limestone (late Eocene), and Suwannee Limestone (Oligocene). All three were deposited under marine conditions, and their vertebrate fossils are oceanic fish, sharks, reptiles, whales, and dugongs. These rocks are at or near the surface in much of northern peninsular Florida and the eastern panhandle. Open-pit quarries create exposures of these Paleogene limestones, as do sinkholes, caves, and some rivers.

Sea level was higher during most of the Paleogene than at present, frequently several hundred feet higher. In the early Oligocene (about 30 Ma), sea level began a long-term decline. At this time the earth's global climate began its shift from the warm, equable, ice-free conditions that had prevailed since the middle Triassic to the cooler, temperate pattern of the modern world. Glaciers began to form in Antarctica and the temperature of the oceans dropped. Both of these caused declines in sea level. Deposition of limestone temporarily stopped in northern Florida, and dry land emerged for the first time since the Cretaceous. Sinkholes and other erosional features formed on the preexisting limestone. The oldest known fossils of land vertebrates from Florida date to this early Oligocene sea-level low stand (Patton, 1969). They were discovered near Gainesville in a pocket of sediment that filled a small sinkhole formed in Ocala Limestone. Sea level rebounded upward in the late Oligocene, and ocean waters once again covered all of Florida.

Neogene and Quaternary History of Florida (24 Ma–present)

Environmental conditions in Florida during the Neogene Period were profoundly different than during the preceding Paleogene. These differences resulted from a number of factors, including generally lower sea levels and increased input of sediment from central Alabama and Georgia. Sea level varied considerably through the Neogene. At times it was close to modern levels, and portions of the northern peninsula and panhandle were emergent. During times of higher sea level, much if not all of the state was submerged. The southern peninsula remained more or less constantly below sea level, and limestone deposition continued there with only a short halt during the Pliocene.

A strong oceanic current flowed through a deep channel or trough that ran from the central panhandle northeastward to the Savannah region of Georgia during the Cretaceous and Paleogene. This feature is variously known in the geologic literature as the Suwannee Strait, Gulf Trough, or the Georgia Channel System (Popenoe et al., 1987; Huddlestun, 1993). This current acted like the modern Gulf Stream. Sands and clays eroding from the Piedmont highlands of central Georgia and Alabama were captured by this current and could not reach peninsular Florida. The Georgia Channel System ceased near the end of the Oligocene, so Neogene rocks in Florida include more sand and clay than do those of the Paleogene.

Miocene and early Pliocene deposits through much of Florida are also relatively rich in phosphate, especially the mineral francolite. In places, the concentrations of phosphate are rich enough to make mining it an economically viable industry. The phosphate is used to manufacture fertilizer and chemicals. Phosphate is mined in two areas of Florida: the Bone Valley District east of Tampa, near the towns of Bartow and Mulberry (in southwestern Polk, northwestern Hardee, and southeastern Hillsborough Counties), and north of White Springs in Hamilton County. The phosphatic sediments in both of these regions are highly fossiliferous and produce land and marine vertebrates.

During times of relatively low sea level in the Neogene, erosion exposed Paleogene limestone to surface conditions. Limestone erodes quickly in Florida's hot, humid climate to form what is known as a karst landscape. Typical features of karst landscapes are caves, sinkholes, and lakes. These are eventually filled by boulders, sand, and clay and often contain vertebrate fossils. Especially important is that the remains of small animals—frogs, salamanders, snakes, lizards, birds, bats, rodents, and shrews—are frequently concentrated in these types of deposits. Large vertebrates can also be found, sometimes as associated or articulated skeletons. Some of Florida's richest Neogene vertebrate localities formed in this manner.

When sea levels were relatively high in the Neogene, phosphatic sand and clay were deposited in deltas, estuaries, and shallow marine coastal regions. These sediments comprise the Hawthorn Group, which is made up of mixtures of quartz sand, clay, calcite, dolomite, and francolite. Vertebrate fossils are locally abundant in Hawthorn Group sediments, and, unlike the deposits described in the preceding paragraph, tend to be mixtures of marine and land vertebrates. Associated or articulated skeletons of the terrestrial portion of the fauna are very rare but are more common among the marine component of the fauna, especially dugongs (chap. 16). The age of the Hawthorn Group ranges from the very early Miocene (possibly latest Oligocene) to early Pliocene.

The last major sea-level high stand was during the Pliocene, between 4.5 and 2.5 Ma. At this time most, if not all, of Florida was below sea level for the last time in geologic history. No fossils of land animals are known from this interval, only marine vertebrates such as sharks and whales. The earth entered a new climatic phase 2.5 Ma, with widespread continental glaciers at high latitudes in both the Northern and Southern Hemispheres. The trigger that started this phase was likely the complete connection of North and South America by the Panamanian Isthmus. The two continents were isolated from each other through the early and middle Cenozoic. Their connection altered the course of major oceanic currents, thereby effecting global climates.

Over the past 2.5 million years continental glaciers have advanced and retreated many times from their polar strongholds. During glacial maxima they covered nearly all of Canada, the northern United States, and much of northern Europe, and sea levels dropped up to 400 feet below modern values. At such times the width of peninsular Florida more than doubled. The climate was somewhat cooler and drier than at present, but not severely colder. Tropical and semitropical animals such as alligators and large land tortoises lived in the state through the entire interval. Few large bodies of sediment formed during these glacial intervals (it was primarily a time of erosion), but localized sinkhole and fissure-fill deposits provide a rich record of vertebrate life during these times.

Short interglacials were times when the glaciers retreated and sea level rose to values close to or slightly higher than today. Sandy coastal marine deposits often rich in shells formed along the Atlantic and Gulf coasts of Florida. Some of the better known of these deposits are the Tamiami, Jackson Bluff, Caloosahatchee, Nashua, Bermont, Ft. Thompson, and Anastasia formations. Although principally marine, some of these formations have produced localized concentrations of land vertebrates.

In addition to coastal marine formations and sinkhole/cave deposits, a third type of locality produces late Pliocene and Pleistocene vertebrate fossils in Florida, the beds of modern rivers. Many of the state's rivers are well known for their fossils, including the Santa Fe, Ichetucknee, Aucilla, St. Johns, Waccasassa, Withlacoochee, and Peace Rivers. Unless dug directly out of riverbed sediments, fossils collected from rivers are often assemblages of different geologic ages. Holocene archaeological artifacts are also commonly found in these deposits.

The last major glacial cycle peaked about 18 ka and ended about 10 ka. Sea level reached its modern height about 7 to 8 ka. Although there had been many such transitions from glacial to interglacial conditions throughout the Pleistocene, this one was accompanied by significant faunal change. There was a mass extinction of many of the large vertebrates that had lived not just in Florida, but throughout North America. Possible causes for this extinction are discussed in chapter 2.

REFERENCES

References on Fossils and Paleontology

Benton, M. J. 1997. *Vertebrate Palaeontology.* 2d ed. Chapman and Hall, New York, 464 p. [Reprinted in 2000 by Blackwell Science.]

Brown, R. C. 1996. *Florida's Fossils: Guide to Location, Identification, and Enjoyment.* 2d ed. Pineapple Press, Sarasota, 208 p. [First edition published in 1988.]

Carroll, R. L. 1988. *Vertebrate Paleontology and Evolution.* W. H. Freeman, New York, 698 p.

Cowen, R. 2000. *History of Life.* 3d ed. Blackwell Science, Malden, Mass., 432 p.

Erickson, J. 1995. *A History of Life on Earth: Understanding our Planet's Past.* Facts on File, New York, 244 p.

Prothero, D. R. 1997. *Bringing Fossils to Life: An Introduction to Paleontology.* McGraw-Hill, New York, 457 p.

Shipman, P. 1981. *Life History of a Fossil: An Introduction to Taphonomy and Paleoecology.* Harvard University Press, Cambridge, 222 p. [Reprinted in 1993.]

Stucky, R. K. 1995. *Prehistoric Journey: A History of Life on Earth.* Roberts Rinehart, Boulder, 144 p.

Walker, C., and D. J. Ward. 1992. *The Eyewitness Handbook of Fossils.* Dorling Kindersley, New York, 320 p.

Florida Museum of Natural History, Paleontological Resources for Fossil Collectors Home Page: http://www.flmnh.ufl.edu/natsci/vertpaleo/resources/res.htm

PaleoNet Home Page: http://www.ucmp.berkeley.edu/Paleonet/

Paleontological Research Institute Home Page: http://www.englib.cornell.edu/pri/pri1.html

Society of Vertebrate Paleontology Home Page: http://www.vertpaleo.org

Talk.Origins Archive Home Page: http://www.talkorigins.org/

UCMP, Paleontology without Walls Home Page: http://www.ucmp.berkeley.edu/exhibit/exhibits.html

References on the Vertebrate Skeleton

Gilbert, B. M. 1980. *Mammalian Osteology.* Missouri Archaeological Society, Columbia, 428 p. [A source of illustrations of skeletal elements of North American mammals.]

Hildebrand, M. 1994. *Analysis of Vertebrate Structure.* 4th ed. John Wiley and Sons, New York, 657 p.

Kocsis, F. A. 1997. *Vertebrate Fossils: A Neophyte's Guide.* IBIS Graphics, Palm Harbor, Florida, 184 p. [Numerous photos of vertebrate fossils, most from Florida.]

Olsen, S. J. 1964. Mammal remains from archaeological sites. Part 1—Southeastern and southwestern United States. Papers of the Peabody Museum of Archaeology and Ethnology, Harvard University, 56(1):1–162. [Reprinted 1990; another good source of illustrations of skeletal elements of North American mammals.]

———. 1968. Fish, amphibian and reptile remains from archaeological sites. Part 1. Southeastern and southwestern United States. Appendix: the osteology of the wild turkey. Papers of the Peabody Museum of Archaeology and Ethnology, Harvard University, 56(2):1–137. [Reprinted 1996.]

———. 1979. Osteology for the archaeologist. Number 3, the American mastodon and the woolly mammoth. Number 4, North American birds: skulls and mandibles. Number 5, North American birds: postcranial skeletons. Papers of the Peabody Museum of Archaeology and Ethnology, Harvard University, 56(3–5):1–186. [Reprinted 1996.]

Rojo, A. L. 1991. *A Dictionary of Evolutionary Fish Osteology.* CRC Press, Boca Raton, Fla., 273 p.

Romer, A. S. 1956. *Osteology of the Reptiles.* University of Chicago Press, Chicago, 772 p.

Sobolik, K. D., and D. G. Steele. 1996. *A Turtle Atlas to Facilitate Archaeological Identifications.* Mammoth Site of Hot Springs, S.D., 117 p. [Drawings and photos of many turtle skeletal elements.]

References on Animal Classification and Systematics

Donoghue, M. J., J. A. Doyle, J. Gauthier, A. G. Kluge, and T. Rowe. 1989. The importance of fossils in phylogeny reconstruction. Annual Review of Ecology and Systematics, 20:431–60.

Forey, P. L., C. J. Humphries, I. J. Kitching, R. W. Scotland, D. J. Siebert, and D. M. Williams. 1992. *Cladistics: A Practical Course in Systematics.* Systematics Association Publication 10, Oxford University Press, New York, 208 p.

Harvey, P. H., and M. D. Pagel. 1991. *The Comparative Method in Evolutionary Biology.* Oxford University Press, New York, 239 p.

Hennig, W. 1966. *Phylogenetic Systematics.* University of Illinois Press, Urbana, 263 p.

International Commission on Zoological Nomenclature. 1999.

International Code of Zoological Nomenclature. 4th ed. International Trust for Zoological Nomenclature, London, 306 p.

Mayr, E. 1969. Principles of Systematic Zoology. McGraw-Hill, New York, 428 p. [Second edition, 1991.]

McKenna, M. C., and S. K. Bell. 1997. Classification of Mammals above the Species Level. Columbia University Press, New York, 631 p. [Part 1 of this book contains an excellent history and introduction to biologic classification.]

Padian, K., D. R. Lindberg, and P. D. Polly. 1994. Cladistics and the fossil record: the uses of history. Annual Review of Earth and Planetary Sciences, 22:63–91.

Panchen, A. L. 1992. Classification, Evolution, and the Nature of Biology. Cambridge University Press, New York, 350 p.

Schoch, R. M. 1986. Phylogeny Reconstruction in Paleontology. Van Nostrand Reinhold, New York, 353 p.

Smith, A. B. 1994. Systematics and the Fossil Record: Documenting Evolutionary Patterns. Blackwell Scientific, Cambridge, Mass., 223 p.

Tubbs, P. K. 1992. The International Commission on Zoological Nomenclature: what it is and how it operates. Systematic Biology, 41:135–37.

BIOSIS Biosystematics and Life Science Resources Home Page: http://www.york.biosis.org/

International Commission on Zoological Nomenclature Home Page: http://www.iczn.org/index.htm

Tree of Life (University of Arizona) Home Page: http://phylogeny.arizona.edu/tree/phylogeny.html

UCMP Phylogenetics Resources Home Page: http://www.ucmp.berkeley.edu/subway/phylogen.html

UCMP Journey into Phylogenetic Systematics: http://www.ucmp.berkeley.edu/clad/clad4.html

References on Geologic Time Scales and Geology

Anderson, B. G., and H. W. Borns. 1994. The Ice Age World. Scandinavian University Press, Oslo, 208 p.

Bambach, R. K., C. R. Scotese, and A. M. Ziegler. 1980. Before Pangea: the geographies of the Paleozoic world. American Scientist, 68:26–38.

Berggren, W. A., D. V. Kent, M.-P. Aubry, and J. Hardenbol (eds.). 1995. Geochronology, Time Scales, and Global Stratigraphic Correlation. Society for Sedimentary Geology Special Publication 54, Tulsa.

Berry, W. B. N. 1987. Growth of a Prehistoric Time Scale Based on Organic Evolution, Revised Edition. Blackwell Scientific, Palo Alto, 202 p. [The 1968 first edition is outdated but still useful.]

Hallam, A. 1989. Great Geologic Controversies. 2d ed. Oxford University Press, New York, 244 p.

Harland, W. B., R. L. Armstrong, A. V. Cox, L. E. Craig, A. G. Smith, and D. G. Smith. 1990. A Geologic Timescale 1989. Cambridge University Press, Cambridge, England. 263 p.

Imbrie, J., and K. P. Imbrie. 1986. Ice Ages: Solving the Mystery. Harvard University Press, Cambridge, Mass., 224 p.

Lemon, R. R. 1993. Vanished Worlds: An Introduction to Historical Geology. Wm. C. Brown, Dubuque, Iowa, 480 p.

MacDougall, J. D. 1996. A Short History of the Planet Earth: Mountains, Mammals, Fire, and Ice. John Wiley, New York, 256 p.

Salvador, A. (ed.). 1991. The Geology of North America. Volume J: The Gulf of Mexico Basin. The Geological Society of America, Boulder, 568 p.

Scott, T. M. 1988. The lithostratigraphy of the Hawthorn Group (Miocene) of Florida. Florida Geological Survey Bulletin, 59:1–148.

Sheridan, R. E., and J. A. Grow (eds.). 1988. The Geology of North America, Vol. I: The Atlantic Continental Margin: U.S. The Geological Society of America, Boulder, 610 p.

Stanley, S. M. 1986. Earth and Life Through Time. W. H. Freeman, New York, 690 p.

Florida Geological Survey Home Page: http://www.dep.state.fl.us/geo/

United States Geological Survey Home Page: http://www.usgs.gov/

UCMP, Geology and Geologic Time Home Page: http://www.ucmp.berkeley.edu/exhibit/geology.html

Specific References Cited in Chapter 1

Baskin, J. A. 1981. Barbourofelis (Nimravidae) and Nimravides (Felidae), with a description of two new species from the late Miocene of Florida. Journal of Mammalogy, 62:122–39.

Berggren, W. A., D. V. Kent, C. C. Swisher, and M.-P. Aubry. 1995. A revised Cenozoic geochronology and chronostratigraphy. Pp. 335–64 in W. A. Berggren, D. V. Kent, M.-P. Aubry, and J. Hardenbol (eds.), Geochronology, Time Scales, and Global Stratigraphic Correlation. Society for Sedimentary Geology Special Publication 54.

Cantino, P. D., H. N. Bryant, K De Queiroz, M. J. Donoghue, T. Eriksson, D. M. Hillis, and M.S.Y. Lee. 1999. Species names in phylogenetic nomenclature. Systematic Biology, 48:790–807.

Carroll, R. L. 1988. Vertebrate Paleontology and Evolution. W. H. Freeman, New York, 698 p.

Forey, P., and P. Janvier. 1994. Evolution of the early vertebrates. American Scientist, 82:554–65.

Giffin, E. B. 1992. Functional implications of neural canal anatomy in recent and fossil marine carnivores. Journal of Morphology, 214:357–74.

Hedges, S. B. 1994. Molecular evidence for the origin of birds. Proceedings of the National Academy of Sciences of the United States of America, 91:2621–24.

Hildebrand, M. 1974. Analysis of Vertebrate Structure. John Wiley and Sons, New York, 710 p.

Huddlestun, P. F. 1993. Revision of the lithostratigraphic units of the coastal plain of Georgia—the Oligocene. Georgia Geologic Survey Bulletin, 105:1–152.

Mayr, E. 1942. Systematics and the Origin of Species. Columbia University Press, New York, 334 p.

Neill, W. T. 1957. The rapid mineralization of organic remains in Florida, and its bearing on supposed Pleistocene records. Quarterly Journal of the Florida Academy of Sciences, 20:1–13.

Olsen, S. J. 1959. Fossil mammals of Florida. Florida Geological Survey Special Publication, 6:1–75.

Opdyke, N. D., D. S. Jones, B. J. MacFadden, D. L. Smith, P. A. Mueller, and R. D. Schuster. 1987. Florida as an exotic terrane: paleomagnetic and geochronologic investigations of lower Paleozoic rocks from the subsurface of Florida. Geology, 15:900–3.

Patton, T. H. 1969. An Oligocene land vertebrate fauna from Florida. Journal of Paleontology, 43:543–46.

Peterson, O. A. 1906. The Miocene beds of western Nebraska and eastern Wyoming and their vertebrate faunae. Annals of the Carnegie Museum, 4:21–72.

Pojeta, J., J. Kriz, and J. M. Berdan. 1976. Silurian-Devonian pelecypods and Paleozoic stratigraphy of subsurface rocks in Florida and Georgia and related Silurian pelecypods from Bolivia and Turkey. U.S. Geological Survey Professional Paper, 879:1–32.

Popenoe, P., V. J. Henry, and F. M. Idris. 1987. Gulf trough—the Atlantic connection. Geology, 15:327–32.

Randazzo, A. F., and D. S. Jones (eds.). 1997. The Geology of Florida. University Press of Florida, Gainesville, 327 p.

Romer, A. S. 1970. The Vertebrate Body. 4th ed. W. B. Saunders, Philadelphia, 601 p.

Savage, D. E. 1951. Late Cenozoic vertebrates of the San Francisco Bay region. University of California Publications in Geological Sciences, 28:215–314.

Schwimmer, D. R., and R. H. Best. 1989. First dinosaur fossils from Georgia, with notes on additional Cretaceous vertebrates from the state. Georgia Journal of Science, 47:147–57.

Schwimmer, D. R., G. D. Williams, J. L. Dobie, and W. G. Siesser. 1993. Late Cretaceous dinosaurs from the Blufftown Formation in western Georgia and eastern Alabama. Journal of Paleontology, 67:288–96.

Shipman, P., A. Walker, and D. Bichell. 1985. *The Human Skeleton.* Harvard University Press, Cambridge, 343 p.

Torrey, T. W. 1967. *Morphogenesis of the Vertebrates.* 2d ed. John Wiley and Sons, New York, 448 p.

Van Couvering, J. A. (ed.). 1997. *The Pleistocene Boundary and the Beginning of the Quaternary.* Cambridge University Press, New York, 296 p.

Vaughn, T. A. 1982. *Mammalogy.* W. B. Saunders, Philadelphia, 463 p.

Webb, S. D. 1965. The osteology of *Camelops.* Bulletin of the Los Angeles County Museum, 1:1–54.

Wood, H. E., R. W. Chaney, J. Clark, E. H. Colbert, G. L. Jepson, J. B. Reedside, and C. Stock. 1941. Nomenclature and correlation of the North American continental Tertiary. Bulletin of the Geological Society of America, 52:1–48.

Woodburne, M. O. (ed.). 1987. *Cenozoic Mammals of North America: Geochronology and Biostratigraphy.* University of California Press, Berkeley, 336 p.

Woodburne, M. O., and C. C. Swisher. 1995. Land mammal high-resolution geochronology, intercontinental overland dispersals, sea level, climate, and vicariance. Pp. 335–64 *in* W. A. Berggren, D. V. Kent, M.-P. Aubry, and J. Hardenbol (eds.), *Geochronology, Time Scales, and Global Stratigraphic Correlation.* Society for Sedimentary Geology Special Publication 54.

Zhou, X., W. J. Sanders, and P. D. Gingerich. 1992. Functional and behavioral implications of vertebral structure in *Pachyaena ossifraga (Mammalia, Mesonychia). Contributions from the Museum of Paleontology, University of Michigan,* 28:289–319.

2

Florida's Fossil Vertebrates

An Overview

Scientific study of fossil vertebrates collected in Florida began in the 1880s with Joseph Leidy's initial descriptions of Miocene and Pleistocene species. More than 1,100 different species have been recorded from the state (as listed in chap. 3), and new finds are made every year. With the exception of fragments of a Cretaceous turtle recovered far below the surface of Okeechobee County by an oil-well drill rig (Olsen, 1965), all vertebrate fossils from Florida are middle Eocene or younger (less than 45 Ma). This corresponds to the maximum age of surficial rocks and sediments in the state (chap. 1). Land vertebrates from the Oligocene, Miocene, and Pliocene of Florida are especially important because they are so rare elsewhere in eastern North America. Without the Florida record, the history of Cenozoic terrestrial life on this continent would be very incomplete.

By the middle Eocene, mammals already had dominated the world's terrestrial vertebrate faunas for about 20 million years, since the demise of the last dinosaurs (except birds, see chap. 8). The first appearances of some of today's major mammalian groups, such as bats, artiodactyls, whales, elephants, and perissodactyls, occurred during the Eocene Epoch. Beginning in the late Eocene and continuing through the Oligocene, the archaic mammals that had ruled the North American landscape during the first 20 to 30 million years of the Cenozoic were replaced by more advanced species. This coincided with climatic and floral changes, as a predominantly forested continent with warm, humid weather became more arid and open. Similar faunal turnover, although less well documented, also occurred for birds and reptiles.

Florida's Eocene fossil record, however, is derived only from marine rocks, and the vertebrate fossils of this epoch consist only of marine animals. Shark teeth, ray mouth plates, and fish bones are the most common vertebrate fossils of this time. Most of the well-known species are extinct but belong to living genera. Other types of vertebrates are less common, and any record of such specimens originating directly from Eocene limestone is of great scientific interest. These less abundant taxa include sea turtles, marine crocodiles, and a large sea snake related to the pythons. Florida Eocene mammals are limited to very primitive kinds of sea cows and whales.

There are few good natural exposures of Eocene limestone in Florida, even though great quantities of such rock underlie the entire state. Numerous rock quarries in the western half of northern peninsular Florida, including Citrus, Marion, Levy, Alachua, Suwannee, and Taylor Counties (figs. 2.1, 2.2), mine Eocene limestone and expose its otherwise hidden fossils. Other "exposures" of Eocene limestone are provided by the walls of the many caves in northern and central Florida. The only other sources in Florida for fossils of this age occur where major rivers cut into Eocene bedrock. The most productive of these are stretches of the Waccasassa River near the town of Gulf Hammock, and the Withlacoochee River near Dunnellon. Vertebrate fossils from river bottoms in Florida are notoriously mixed chronologically, with many ages represented. In these situations, the fossils must be identified before an Eocene age can be assigned to them. Their preservation or adhering bits of limestone may also support their age determination.

Oligocene rocks are much less extensive than those of Eocene age in Florida, thus fossils of that age are rarer. The early Oligocene Marianna Limestone is well exposed only in parts of Jackson County in the Florida panhandle west of Tallahassee (fig. 2.1). It produces an exclusively marine fauna, the most spectacular of which are some relatively complete bony fish (chap. 4). The second major Oligocene formation in Florida is the Suwannee Limestone, also a marine unit. The Suwannee is exposed on the Gulf Coastal margins of Jefferson, Taylor, and Dixie Counties; in a few spotty outcrops in Hernando, Pasco, and Columbia Counties; and in places along the banks of the Suwannee River in Hamilton County (hence its name). The Suwannee Formation, like

Fig. 2.1. Counties and important geographic localities in Florida.

the Marianna, has produced a number of fossil fish, especially sharks and rays, but no reported mammals or reptiles.

Oligocene land vertebrates were unknown in Florida until the mid-1960s, when the I-75 site was discovered on the southwestern outskirts of Gainesville in Alachua County (fig. 2.3). Matrix from a small sinkhole exposed during highway construction proved to contain a remarkably rich terrestrial vertebrate fauna, with an age of about 30 Ma. The discovery of the I-75 site demonstrated that terrestrial life flourished in north-central Florida at a time when it was supposed to be beneath the

sea. The fauna includes a diverse array of amphibians, reptiles, and small mammals, mostly genera and species never before recorded from eastern North America. Most of the I-75 fauna has yet to be thoroughly studied. More recently, another small Oligocene terrestrial fauna, the Cowhouse Slough site, was recovered from a small pocket of matrix, this time in northern Hillsborough County. About four million years younger than I-75 (fig. 2.3), this fauna is still under study.

It is not until the Miocene Epoch that terrestrial fossil sites became abundant in Florida (figs. 2.3, 2.4). This is related to lowered global sea levels, as discussed in chap-

Fig. 2.2. View of a limestone quarry near Newberry, western Alachua County. The white rock being quarried is marine Eocene limestone. It contains fossils of fish, sea turtles, sea snakes, crocodiles, and whales. The piles in the foreground are limestone that has been broken into small pieces in preparation for being hauled away in trucks. The vertical cliff in the background is the edge of the quarry, and the water is where quarrying has gone deeper than the water table. Sediment-filled sinkholes, caves, and crevasses in the limestone in this region (the so-called Haile quarries) also produce fossils ranging in age from Miocene to Pleistocene.

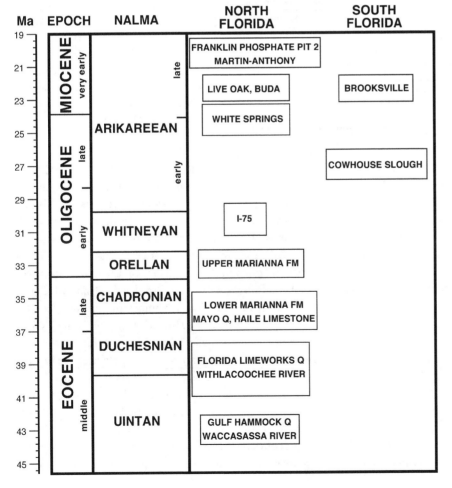

Fig. 2.3. Chronologic distribution of Eocene, Oligocene, and very early Miocene major vertebrate fossil sites and faunas in Florida. The boundary between north and south Florida is arbitrarily set at the level of Orlando, about 28° 30' N latitude. Q, quarry; FM, formation; NALMA, North American Land Mammal Ages. Fossil localities within a box are the same age; vertical height of a box indicates its possible geologic age (the taller the box, the greater the level of uncertainty).

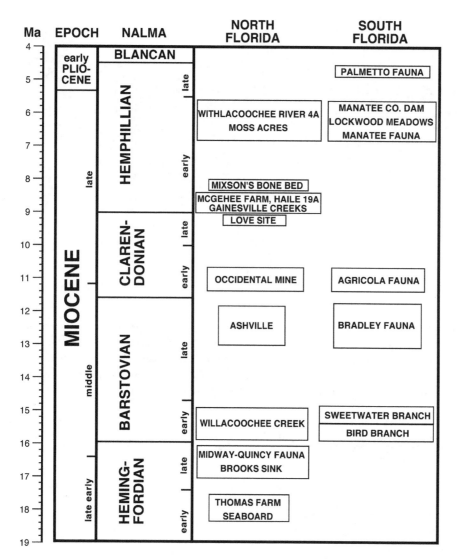

Fig. 2.4. Chronologic distribution of late early Miocene to early Pliocene major vertebrate fossil sites and faunas in Florida. The boundary between north and south Florida is arbitrarily set at the level of Orlando, about 28° 30' N latitude. *NALMA*, North American Land Mammal Ages. Fossil localities within a box are the same age; vertical height of a box indicates its possible geologic age (the taller the box, the greater the level of uncertainty).

ter 1. All parts of the Miocene are represented by several faunas or sites, although the early and late Miocene are better known than the middle Miocene. Florida in the early Miocene was covered by tropical to subtropical forest, with frost-free winters. The herbivores were mainly browsers, while the dominant carnivores were relatively short-legged and not capable of sustained chases. Florida retained the mild climate that was widespread earlier in the Cenozoic but which had deteriorated in more northern and western realms. Many of the land vertebrates known from Florida at this time are unique to the Gulf Coastal Plain and were presumably adapted to the warm, mesic climate and forested habitat of the region.

Early Miocene fossil sites in Florida formed in two distinct depositional settings: sinkhole deposits and nearshore marine environments. They are all located in the northern half of the state, with the southernmost locali-

ties in Hernando County near the town of Brooksville. In addition to the Brooksville localities, other well-known early Miocene sinkhole sites are Live Oak (or SB-1A) in Suwannee County, Buda in Alachua County, and, the most famous, Thomas Farm in Gilchrist County (fig. 2.4). Examples of nearshore marine sites are found near the town of White Springs on the Suwannee River in Hamilton County, near Brooker in Bradford County, and in the fuller's earth mines of Gadsden County northwest of Tallahassee (the Midway and Quincy sites).

Taken together, these localities provide a relatively complete picture of early Miocene vertebrate life in Florida, as all types and sizes of animals are known. While the amphibians and alligators had a very modern appearance, the mammals, birds, snakes, and (to a lesser degree) turtles were decidedly different. For example, the most common snakes were boas and pythons, song birds were very rare, and the common small rodents were heteromyids (a family now limited to southwestern deserts and Neotropical forests). The most common large herbivores were a diverse array of small- to medium-sized camels, although oreodonts, rhinos, and three-toed horses were also present. The largest carnivores were the completely extinct amphicyonids (the "bear-dogs"), while a variety of fox- to coyote-sized dogs were the most common predators. The sea cows were related to the Pacific dugong, not our familiar modern manatee.

There are a few middle Miocene localities in northern Florida (fig. 2.4), but the richest faunas of this age are found in the Bone Valley region of south-central Florida. The middle Miocene was a period of profound faunal change induced by climatic events. Savanna, a mixture of grasslands and woodlands, became the preeminent environment over much of North America. Although temperatures remained warm year-round, there were annual wet and dry seasons. Dry seasons allowed natural wildfires, which converted woodlands and thickets into grassland. Growth of woodlands at the expense of grasslands was also limited by very large herbivores, especially proboscideans, which first became common at this time. Increased grasslands allowed many more grazers than before. Horses, camels, and antilocaprids formed large herds similar to antelope and zebra of the modern African plains. The similarity with modern African savanna mammals does not end there. There were also large rhinoceroses, and a large, long-necked camel like a giraffe. The dominant carnivores were the hyena-dogs (borophagines). Unlike the early Miocene, which was a period of endemism for many Gulf Coastal Plain mammals, many middle and late Miocene large mammals had wide geographic ranges, and few were unique to Florida. Florida had numerous species in common with Texas and Nebraska, and even some with California.

Snakes, turtles, and birds took on a familiar, modern appearance during the middle and late Miocene relative to those of the early Miocene. Many were at most subtly distinct from living species. There were of course a few notably different types, such as large land tortoises, a long-snouted crocodile, and tropical birds like lily-trotters and crowned cranes. Marine mammals were decidedly different from modern forms. They included dugongs, long-snouted dolphins, and a variety of small baleen and sperm whales. The oceanic realm was dominated by the giant (up to 16 meters or 52 feet long) great white shark *Carcharodon megalodon,* which probably specialized in killing marine mammals (Gottfried et al., 1996; Purdy, 1996).

The late Miocene witnessed the culmination of the North America savanna fauna. Horses reached their maximum diversity and numerically dominated the landscape. Rhinoceroses, peccaries, camels, and proboscideans, including the bizarre shovel-tuskers with their scooplike lower tusks, were also common. Rarer were the bizarrely horned protoceratids and dromomerycines. Joining the borophagine dogs among the dominant carnivores were immigrants from Asia, most notably large cats and bears. Two kinds of ground sloths arrived from South America, heralds of a much larger influx to come in the Pliocene. The common small rodents were no longer heteromyids as in the early Miocene, but instead the evolutionary ancestors of today's native American rats and mice.

Late Miocene vertebrate fossil sites are found in two regions of Florida (fig. 2.4). Sinkhole, river, and estuarine deposits cut into Eocene limestone occur sporadically in a band running from western Alachua County south to Citrus County. These sites produce some of the richest faunas and most spectacular fossil specimens in the state. They include Mixson's Bone Bed, first worked in the 1880s; McGehee Farm; the Love site near Archer; and the Moss Acres Racetrack site near Ocala. The latter was the first Miocene site in Florida to produce articulated skeletons, including those of horses, rhinos, shovel-tuskers, and an otter (fig. 1.3A). Very late Miocene sites occur along the Gulf Coast from Hillsborough and Pinellas Counties south to Venice in Sarasota County, with Manatee County having the most sites. Their vertebrates are predominantly marine (sharks, whales, and dugongs), but isolated, waterworn, terrestrial fossils also occur, especially horse teeth.

The earliest Pliocene, about 4.5 Ma, is best represented in south-central Florida, in the Central Florida Phosphate Mining District, which encompasses southwestern Polk County, northern Hardee County, northeastern Manatee County, and southeastern Hillsborough County (figs. 2.1, 2.4). This region is better known to fossil collectors as the Bone Valley, due to its abundant vertebrate fossils, although it is not a true valley in the geographic sense. Here a complex series of earliest Pliocene fluvial and estuarine deposits produce the Palmetto Fauna (also called the Upper Bone Valley Fauna). The sediments that produce this fauna lie on top of or cut into those which contain the older Bone Valley faunas (the middle Miocene Agricola and Bradley faunas). All of these are continually exposed by very large-scale phosphate mining, mixed together on spoil piles, and then reburied as the land is reclaimed. The Palmetto Fauna contains a diverse mixture of terrestrial and marine vertebrate species. Especially noteworthy is its diverse as-

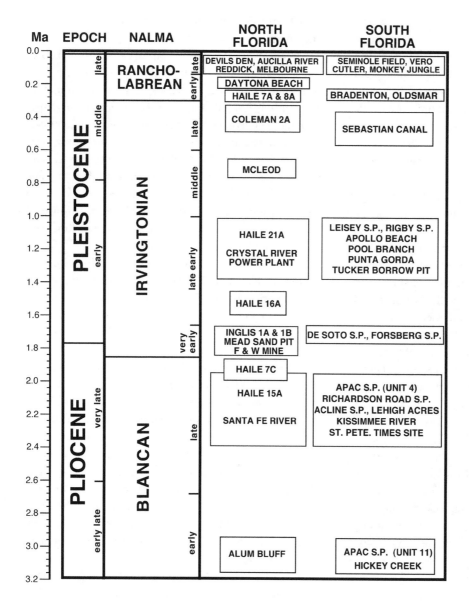

Ma	EPOCH	NALMA	NORTH FLORIDA	SOUTH FLORIDA

Fig. 2.5. Chronologic distribution of late Pliocene and Pleistocene major vertebrate fossil sites and faunas in Florida. The boundary between north and south Florida is arbitrarily set at the level of Orlando, about 28° 30' N latitude. *NALMA*, North American Land Mammal Ages; *S.P.*, shell pit. Fossil localities within a box are the same age; vertical height of a box indicates its possible geologic age (the taller the box, the greater the level of uncertainty).

semblage of marine and coastal birds. The fully marine, predominantly carbonate Tamiami Formation lies to the south and west of the Bone Valley region. The Tamiami produces shark teeth and large baleen whales, as well as numerous invertebrate fossils.

The Palmetto Fauna is one of the most significant early Pliocene faunas in North America. In the rest of the continent, the demise of the great Miocene savanna fauna was triggered by a drying climate between 7 and 4 Ma. This resulted in a change from predominantly savanna habitat to a less productive, less diverse grassland or prairie. Many groups of mammals became extinct or suffered severe declines in abundance, while others took advantage of the changes and became more common. For example, three-toed horses declined substantially, while one-toed horses more closely related to the living genus *Equus* became abundant. Shovel-tusked proboscideans and browsing artiodactyls declined while the hypsodont antilocaprids diversified. The Palmetto Fauna demonstrates that Florida was slightly out of step with the rest of North America in terms of these climatic and faunal changes. One-toed horses and antilocaprids, while present, were uncommon. In contrast to earliest Pliocene faunas of the Great Plains and West Coast, deer, peccaries, tapirs, and especially three-toed horses were common in Florida. A moister climate must have prevailed in this region, probably due to the proximity of the Gulf of Mexico, that temporarily allowed the continuation of savanna woodland habitat. This distinction was not long lived. By the late Pliocene, the terrestrial fauna of Florida again came to resemble that of Nebraska, Texas, or Arizona.

The Isthmus of Panama arose between 3 and 2.5 Ma, directly connecting the continents of North and South America for the first time since the early Cenozoic. The fauna of South America had, for the most part, evolved in isolation from the rest of the world and contained many unique elements. North America, in contrast, had engaged in extensive faunal interchange with the Old World throughout the Cenozoic, so that the two regions shared many elements of their terrestrial faunas. In the late Pliocene, South American land animals were able to spread northward to Central and North America (table 2.1). Likewise, North American taxa were free to disperse south. After 2.5 Ma, sea levels were periodically low during advances of continental glaciers, so much of what is now the Gulf of Mexico was exposed land. This formed a broad, natural highway for southern and western immigrants to enter Florida.

Late Pliocene deposits are widely scattered across Florida (fig. 2.5). In northern Florida, three major vertebrate localities are the Santa Fe River (which has produced a mixture of late Pliocene and Pleistocene fossils) and Haile 7C and 15A, near Newberry in western Alachua County. The Haile sites are sediment-filled sinkholes exposed by limestone mining. They range in age from Miocene to Pleistocene. The other primary source for late Pliocene vertebrates is from coastal deposits in southern Florida. They are principally marine sands and limestone, including thick beds rich in molluscan shells, but lenses of estuarine and lagoonal deposits containing terrestrial and freshwater fossils are not uncommon in some areas. Major sites are in Sarasota, DeSoto, and Charlotte Counties, and along the Kissimmee River.

Several Florida sites spectacularly preserve the transition from the Pliocene to the Pleistocene (fig. 2.5). Most notable are Haile 7C and the Inglis sites. Haile 7C, geographically close to Haile 15A, is a clay-filled ancient pond/sinkhole complex. Numerous complete and partial skeletons of latest Pliocene vertebrates have been collected at this site ranging from small frogs, salamanders, and snakes to giant tortoises, tapirs, and ground sloths. Aquatic turtles (snapper, soft-shell, and cooter) and alligators are the most common finds. A distinctly drier environment is sampled by the Inglis sites in Citrus County. These rich fissure-fills do have some large vertebrates (sloths, camels, deer, horse), but it is the great diversity of the smaller species that make these sites special. Most notable are the large numbers of lizards, snakes, birds, rodents, bats, and rabbits. Inglis records a relatively dry interval for the state, with many faunal elements now considered western, such as jackrabbits. Inglis also produced a wide variety of meat-eating animals, including hyenas, cheetahs, saber-toothed cats, early wolves, and the large flightless bird *Titanis*.

Pleistocene deposits and vertebrate sites are found throughout Florida (fig. 2.5). In general, the relatively dry conditions just noted for the Inglis sites continued through the early Pleistocene. Vertebrates with western affinities, such as ground squirrels, hognosed skunks, scrub jays, and prairie chicken dispersed to Florida through coastal prairie and scrub woodland habitats. Relicts of this dry period still survive in the scrub fauna and flora of central Florida (Myers, 1990; Deyrup and Eisner, 1993). The late Pleistocene saw the return of more mesic conditions and forested habitats, although

Table 2.1. South American genera that dispersed northward into the United States during the late Pliocene and Pleistocene across the Panamanian Isthmus

Late Pliocene	Early Pleistocene
Titanis, giant flightless bird	*Nothrotheriops,* ground sloth
Dasypus, armadillo	*Hydrochaeris,* capybara
Pachyarmatherium, extinct armadillo	
Holmesina, giant armadillo	
Glyptotherium, glyptodont	
Eremotherium, giant ground sloth	
Glossotherium, ground sloth	Middle Pleistocene
Desmodus, vampire bat	*Didelphis,* opossum
Neochoerus, capybara	
Erethizon, porcupine	

Note: Two other large mammals, the anteater *Myrmecophaga* and the toxodont *Mixotoxodon*, dispersed as far as Mexico but are not known from the United States. The manatee, *Trichechus*, also dispersed to Florida during this interval from South America but by an aquatic route.

lowered sea levels during glacial periods produced low water tables and active sinkhole formation. Sloths, tapirs, one-toed horses, llamas, peccaries, deer, bison, mammoth, and mastodont were the common large herbivores of the Pleistocene. Jaguar, dire wolf, and the saber-toothed cat *Smilodon* were the major carnivores, with the addition of the American lion and cougar in the late Pleistocene. Other Pleistocene faunal elements that differed greatly from those of today were storks, condors, giant tortoises, bear-sized beavers, capybaras (large aquatic rodents), porcupines, cheetah-like cats, glyptodonts, giant armadillos, and vampire bats.

With the exception of porcupine, bison, deer, and condor, all of these vertebrates became extinct in the United States at the end of the Pleistocene, although some, such as the tapir, llama, peccary, and jaguar, survive in Central or South America. The cause(s) of this mass extinction have long puzzled scientists, and no definitive explanation has emerged. Most opinion is divided among two schools of thought. According to one of these explanations, the late Pleistocene extinction is a natural, climatically caused event, much like similar extinctions in the Neogene. The distribution of plants and small vertebrates in the late Pleistocene suggest that climate was generally mild south of the main continental glacier, lacking rapid fluctuations in temperature and prolonged periods of subfreezing temperatures. After the retreat of the glacier, climate became less predictable, with major Arctic cold fronts bringing freezing temperatures and spring snowstorms to much of North America. According to this hypothesis, most large herbivores were unable to adapt to the abrupt changes in climate (and ensuing vegetational changes), and died out. Evidence from Greenland ice cores demonstrates that most of the change in climate occurred over a very short interval of time, less than 40 years, about 11,650 years ago (Taylor, 1999).

The other school of thought, while agreeing that such climatic changes were occurring, believes that this was not enough to drive most species to extinction. Instead, a new factor, the appearance of proficient, big-game hunting humans (the Paleoindians) is the proximate cause of the extinctions. Kill sites, butchered bone, and artifacts certainly attest to the presence of humans in North America in the latest Pleistocene and that they were hunting many of the species known to have subsequently become extinct. In Florida, direct evidence has been found of Paleoindian predation on mammoth, mastodont, horse, bison, and giant tortoise. As humans at this technological level are not known to have caused widespread extinction of large mammals on continents, a special scenario has been designed to explain how it happened. According to this hypothesis, the efficient big-game hunters of the Pleistocene spread out from Alaska in a wave, killing such a high proportion of most species that extinction became inevitable. As these large mammals had never been hunted by humans, they were relatively easy prey, unlike those of the Old World, which had evolved along with man. A wave of devastation rapidly swept across North and South America, perhaps taking only a few centuries. In the wake of the wave, human population density was low due to lack of abundant prey but was sufficient to wipe out stragglers missed by the main force.

Most scientists who have investigated this extinction fall into one of these two schools or try and blend them together (the double-whammy approach). The issue is not conclusively settled. For those interested in the topic, the most recent summary of the debate is provided in Martin and Klein (1984). Whatever the cause, the Pleistocene extinctions left North and South America highly depauperate in large mammals compared to any other time in the Cenozoic, or relative to other continents. No doubt immigration and evolution would solve this imbalance if the fauna was left undisturbed for a few million years. Of course, the chances of this occurring are remote.

REFERENCES

Deyrup, M., and T. Eisner. 1993. Last stand in the sand. Natural History, 102(12):42–47.

Frailey, D. 1978. An early Miocene (Arikareean) fauna from northcentral Florida (the SB-1A local fauna). Occasional Papers of the Museum of Natural History, the University of Kansas, 75:1–20.

Gottfried, M. D., L. J. V. Compagno, and S. C. Bowman. 1996. Size and skeletal anatomy of the giant "megatooth" shark *Carcharodon megalodon*. Pp. 55–66 *in* A. P. Klimley and D. G. Ainley (eds.), *Great White Sharks: The Biology of* Carcharodon carcharias. Academic Press, San Diego.

Hulbert, R. C. 1993. Taxonomic evolution in North American Neogene horses (subfamily Equinae): the rise and fall of an adaptive radiation. Paleobiology, 19:216–34.

Janis, C. M., K. M. Scott, and L. L. Jacobs. 1998. *Evolution of Tertiary Mammals of North America. Volume 1: Terrestrial Carnivores, Ungulates, and Ungulatelike Mammals.* Cambridge University Press, Cambridge, 691 p.

MacFadden, B. J. 1997. Fossil mammals of Florida. Pp. 119–37 *in* A. F. Randazzo and D. S. Jones (eds.), *The Geology of Florida.* University Press of Florida, Gainesville.

MacFadden, B. J., and S. D. Webb. 1982. The succession of Miocene (Arikareean through Hemphillian) mammal localities and faunas in Florida. Pp. 186–99 *in* T. M. Scott and S. B. Upchurch (eds.), *Miocene of the Southeastern United States.* Florida Geological Survey Special Publication 25.

Marshall, L. G., S. D. Webb, J. J. Sepkoski, and D. M. Raup. 1982. Mammalian evolution and the Great American Interchange. Science, 215:1351–57.

Martin, P. S., and R. G. Klein. 1984. *Quaternary Extinctions: A Prehistoric Revolution.* University of Arizona Press, Tucson, 892 p.

Meyers, R. L. 1990. Scrub and high pine. Pp. 150–93 *in* R. L. Myers and J. J. Ewel (eds.), *Ecosystems of Florida.* University of Central Florida Press, Orlando.

Morgan, G. S. 1993. Mammalian biochronology and marine-non-marine correlations in the Neogene of Florida. Pp. 55–66 *in* V. A. Zullo, W. B. Harris, T. M. Scott, and R. W. Portell (eds.), *The Neogene of Florida and Adjacent Regions.* Florida Geological Survey Special Publication 37.

———. 1994. Miocene and Pliocene marine mammal faunas from the Bone Valley Formation of Central Florida. Pp. 239–68 *in* A. Berta and T. A. Deméré (eds.), *Contributions in Marine Mammal Paleontology Honoring Frank C. Whitmore, Jr.* San Diego Natural History Society, San Diego.

Morgan, G. S., and R. C. Hulbert. 1995. Overview of the geology and vertebrate biochronology of the Leisey Shell Pit local fauna, Hillsborough County, Florida. Bulletin of the Florida Museum of Natural History, 37:1–92.

Olsen, S. J. 1965. *Vertebrate Fossil Localities in Florida.* Florida Geological Survey Special Publication 12, 28 p.

Purdy, R. W. 1996. Paleoecology of fossil white sharks. Pp. 67–78 *in* A. P. Klimley and D. G. Ainley (eds.), *Great White Sharks: The Biology of* Carcharodon carcharias. Academic Press, San Diego.

Scott, T. M., and W. D. Allmon. 1992. *The Plio-Pleistocene Stratigraphy and Paleontology of Southern Florida.* Florida Geological Survey Special Publication 36, 194 p.

Simpson, G. G. 1929. The extinct land mammals of Florida. Florida Geological Survey Annual Report, 20:229–79.

———. 1930. Tertiary land mammals of Florida. Bulletin of the American Museum of Natural History, 59:149–211.

Stehli, F. G., and S. D. Webb. 1985. *The Great American Biotic Interchange.* Plenum Press, New York, 532 p.

Taylor, K. 1999. Rapid climate change. American Scientist, 87:320–27.

Webb, S. D. 1974. *Pleistocene Mammals of Florida.* University Presses of Florida, Gainesville, 270 p.

———. 1977. A history of savanna vertebrates in the New World. Part 1. North America. Annual Review of Ecology and Systematics, 8:355–80.

Webb, S. D., R. C. Hulbert, and W. D. Lambert. 1995. Climatic implications of large-herbivore distributions in the Miocene of North America. Pp. 91–108 *in* E. S. Vrba, G. H. Denton, T. C. Partridge, and L. H. Burckle (eds.), *Paleoclimate and Evolution, with Emphasis on Human Origins.* Yale University Press, New Haven.

3

Checklist of Florida's Fossil Vertebrates

The following list presents the names and chronologic ranges for fossil vertebrates from Florida. All records were taken from or checked against the primary literature; changes in systematic nomenclature (synonymies, etc.) were taken from subsequent revisions; and the geologic ages were adjusted to modern standards. Some species are known to exist or have been identified in some collections but have not yet made their way to the published record. They are omitted from the checklist because their presence cannot be verified and acknowledged. An earlier version of this checklist was published in *Papers in Florida Paleontology, Number 6* (Hulbert, 1992). This version corrects some errors and omissions in the earlier one and includes new information published through the end of April 2000. An updated version of the checklist can be found on the Internet at http://www. flmnh.edu/natsci/vertpaleo/list.htm.

Excluding problematic taxa, the present checklist contains 1,159 total species and comprises 42 species of chondrichthyans, 113 actinopterygians, 49 amphibians, 486 reptiles (149 "traditional" reptiles and 337 birds), and 469 mammals. In terms of total species, this represents a 22 percent gain from the 1992 checklist, which had 953 species. All taxonomic groups showed net increases compared to the 1992 checklist, with the greatest percentage gain in the actinopterygians (an increase of 135 percent). Hulbert (1992) noted that bony fish were the poorest represented group in the Florida fossil record, and they undoubtedly still are. The exceptionally large increase in this group primarily resulted from identification of the fossil fish from three Gulf Coast faunas (Scudder et al., 1995; Emslie et al., 1996; Karrow et al., 1996). The two best-known groups, mammals and birds, still had about 12 percent and 26 percent growth, respectively, compared to the 1992 checklist, reflecting continuing strong research programs. In contrast, amphibians and nonavian reptiles only had about 7 percent growth, which is due to the current low number of active

workers in Florida on these groups compared to previous decades.

The checklist uses a number of stylistic conventions. If an author identified a particular genus from a certain time period (for example, *Canis* sp.), while in another study the same genus and a species was listed (for example, *Canis latrans*) from the same time period, then these records were combined into one listing if the former material could represent the species listed in the latter study (in this case, *C. latrans*). If the original author specifically excluded the possibility that the material was *C. latrans,* then there will be a separate listing for that record (in this case, *Canis* sp.). If a taxon from a certain time period was provisionally identified in the original and all subsequent publications (usually using the abbreviation "cf."), then a question mark ("?") is placed on the left-hand side of the time period in question. For example, if the range of a taxon is given as "?lMIO, lPLIO-ePLEIST," then it has been positively identified from the late Pliocene and early Pleistocene, but only provisionally identified from the late Miocene. A provisionally identified taxon is usually, but not always, securely identified to the taxon of next higher rank. Thus a questionably identified species is correctly identified to genus, a questionably identified genus to family or subfamily, etc. A question mark to the right of the time period signifies uncertainty regarding the age of the record. For example, a listing of "lOLIG, MIO?" means that the taxon is definitely known from the late Oligocene and possibly from the Miocene. When two time periods are separated by a hyphen, then the range of the particular taxon is known to be continuous between them. So a listing of "mMIO-ePLEIST" means the taxon is also known from the Pliocene. If there is good reason to suspect that the taxon in question was absent during that epoch, then its range would instead be given as "mMIO-lMIO, ePLEIST." No indication is made whether or not a species is still living in the state. For example, if a modern species is known as a fossil from the late Pliocene but not the Pleistocene, its chronologic range will only be given as "lPLIO." Chronologic ranges as given apply only to the fossil record in Florida, not to any other regions.

Ages of taxa were listed using units of the geologic time scale rather than land mammal ages so that marine faunas could be included. Figure 1.15 shows the correlation between land mammal ages and Cenozoic epochs (see also figs. 2.3–2.5). Abbreviations used in the checklist are as follows: EOC = Eocene; OLIG = Oligocene; MIO = Miocene; PLIO = Pliocene; PLEIST = Pleistocene; e = early; ve = very early; m = middle; l = late; vl = very late; sp. = species (singular); spp. = species (plural); n. sp. = new species; and indet. = indeterminate. The checklist is annotated with footnotes (located at the end of the list) that provide information and citations of the more recent changes in nomenclature.

The higher-level classification is primarily based upon revisions recently published in the scientific literature. Since the publication of Hulbert (1992), comprehensive revisions of higher-level classification have been published for fish (Nelson, 1994) and mammals (McKenna and Bell, 1997). The classification in this version of the checklist generally follows these two studies. No similar work covering those tetrapod groups which are not

mammals (that is, amphibians, reptiles, and birds) has appeared, although numerous studies of their evolutionary relationships have been published in this decade (for example, Carroll, 1995; De Braga and Rieppel, 1997; Gauthier, 1994; Lauren and Reisz, 1995, 1997; Lee, 1995, 1997; Novas, 1996; Parrish, 1993; and a number of papers in the volume edited by Schultze and Trueb, 1991). The classification in the checklist for these groups at the ranks between family and class generally follows Benton (1997).

To reconcile differences between the classification rankings of Nelson (1994) and McKenna and Bell (1997), the ranks of the names at the class, legion, and cohort levels had to be shifted from those used by McKenna and Bell (1997). Names of mammals at the levels of order, family, and tribe generally follow McKenna and Bell (1997). However, subdivisions within the Equidae follow Hulbert (1989); within the Sirenia follow Domning (1994); and within Cetacea follow Fordyce and Barnes (1994). Nelson (1994) used the ranks grade, division, subdivision, and series in his classification; these are replaced in the checklist by grandclass, infraclass, magnorder, and grandorder, respectively, following the system of prefixes used by McKenna and Bell (1997).

Checklist

Phylum CHORDATA
 Subphylum VERTEBRATA [vertebrates]
  Superclass GNATHOSTOMATA [jawed vertebrates]
   Grandclass CHONDRICHTHIOMORPHI
    Class CHONDRICHTHYES [sharks and rays]
     Subclass ELASMOBRANCHII
      Superorder EUSELACHII
       Order HEXANCHIFORMES
        Family HEXANCHIDAE [cow sharks, sixgill sharks]
         Notorynchus primigenius .... ?MIO?
       Order ORECTOLOBIFORMES [carpet sharks]
        Family GINGLYMOSTOMATIDAE [nurse sharks]
         Ginglymostoma delfortriei[1] ....vlOLIG-PLIO, ePLEIST?
         Ginglymostoma cirratum ....PLIO-lPLEIST
       Order LAMNIFORMES
        Family ODONTASPIDIDAE [sand tiger sharks]
         Carcharias macrota[2] .... lmEOC-eOLIG
         Carcharias cuspidata .... leOLIG-mMIO
         Carcharias taurus .... mMIO-lPLEIST

Family LAMNIDAE [mackerel, mako, and white sharks]

Otodus obliquus...lmEOC-lEOC

Isurus praecursor... lEOC

Isurus desori.. lEOC-mMIO

Isurus hastalis[3] ....eMIO-PLIO, ePLEIST?

Isurus oxyrinchus... ePLEIST-lPLEIST

Carcharodon sokolowi[4] ....lmEOC-leOLIG

Carcharodon subauriculatus...vlOLIG-veMIO

Carcharodon megalodon .... mMIO-ePLIO

Carcharodon carcharias ....?MIO, ePLIO-lPLEIST

Order CARCHARHINIFORMES

Family CARCHARHINIDAE [requiem sharks]

Galeocerdo alabamensis .... leOLIG

Galeocerdo contortus .... vlOLIG-lMIO

Galeocerdo aduncus .... vlOLIG-mMIO

Galeocerdo cuvier .... mMIO-lPLEIST

Carcharhinus n. sp. .... leOLIG-vlOLIG

Carcharhinus leucas ....eMIO-lPLEIST

Carcharhinus egertoni ....eMIO-vePLIO

Carcharhinus plumbeus.. MIO, ePLEIST

Carcharhinus limbatus ....eMIO-PLEIST

Carcharhinus acronotus .... mMIO-PLEIST

Carcharhinus obscurus ....mMIO, PLEIST

Carcharhinus brevipinna ....lPLIO

Rhizoprionodon terraenovae..leOLIG-lPLEIST

Negaprion brevirostris..leOLIG-lPLEIST

Sphyrna mokarrane.. PLEIST

Family HEMIGALEIDAE [snaggletoothed or weasel sharks]

Hemipristis serra ....leOLIG-ePLIO, ePLEIST?

Order RAJIFORMES

Suborder PRISTOIDEI

Family PRISTIDAE [sawfishes]

Pristis sp.. mEOC-lPLEIST

Suborder RAJOIDEI

Family RHINIDAE [guitarfish]

Rhynchobatus sp.. OLIG-ePLEIST

Family RAJIDAE [skates]

Raja sp. ....?lOLIG, MIO-lPLEIST

Suborder MYLIOBATOIDEI

Family DASYATIDAE [stingrays]

Dasyatis sp... vlOLIG-lPLEIST

Family MYLIOBATIDAE

Subfamily MYLIOBATINAE [eagle rays]

Myliobatis sp...lOLIG-lPLEIST

Aetomylaeus sp. .... MIO

Aetobatus sp..eMIO-lPLEIST

Aetobatus narinarie... PLEIST

Subfamily RHINOPTERINAE [cownose rays]

Rhinoptera sp... MIO-PLEIST

Rhinoptera bonasus...lPLIO-ePLEIST

Subfamily MOBULINAE [manta and devil rays]
Plinthicus stenodon .... vlOLIG-mMIO
Grandclass TELEOSTOMI
 Class ACTINOPTERYGII [ray-finned bony fish]
  Subclass CHONDROSTEI
    Order ACIPENSERIFORMES
     Family ACIPENSERIDAE [sturgeons]
      Acipenser sp. .... lPLEIST
  Subclass NEOPTERYGII
   Infraclass GINGLYMODI
    Order SEMIONOTIFORMES
     Family LEPISOSTEIDAE [garfish]
      Lepisosteus sp. .... vlOLIG-lPLEIST
      Lepisosteus osseuse .... PLEIST
      Lepisosteus oculatus .... ?ePLEIST
      Atractosteus spatula .... lMIO-lPLEIST
   Infraclass HALECOMORPHI
    Order AMIIFORMES
     Family AMIIDAE [bowfin or mudfish]
      Amia calva .... MIO-lPLEIST
   Infraclass TELEOSTEI
    Magnorder ELOPOMORPHA
     Order ELOPIFORMES
      Family ELOPIDAE [tenpounders]
       Elops saurus ....ePLEIST
      Family MEGALOPIDAE [tarpons]
       Megalops atlanticus .... mMIO-lPLEIST
     Order ANGUILLIFORMES
      Family ANGUILLIDAE [eels]
       Anguilla rostratae .... PLEIST
      Family MURAENIDAE [moray eels]
      genus and sp. indet. ....PLIO
    Magnorder CLUPEOMORPHA
     Order CLUPEIFORMES
      Family CLUPEIDAE [herrings and shad]
      genus and sp. indet.................................lPLIO-mPLEIST
       Alosa sp...lPLIO
       Harengula sp. ....lPLIO
       Jenkinsia sp. ....?lPLIO
       Opisthomena oglinum......................................lPLIO
       Sardinella sp. ....lPLIO
       Dorosoma petenense .... lPLEIST
    Magnorder EUTELEOSTEI
     Superorder OSTARIOPHYSI
      Order CYPRINIFORMES
       Family CYPRINIDAE [minnows]
        Notemigonus chrysoleucas ....ePLEIST
       Family CATOSTOMIDAE [suckers]
        Erimyzon sucetta..?ePLEIST
      Order SILURIFORMES

Family ICTALURIDAE [freshwater and bullhead catfishes]

Ameiurus spp.[5] .... MIO-lPLEIST

Ameiurus natalis...ePLEIST

Ameiurus nebulosus ....ePLEIST

Pylodictus sp. .... lPLEIST

Family ARIIDAE [sea catfish]

Arius sp.. MIO

Arius felis ....lPLIO-lPLEIST

Bagre sp. ....PLIO-lPLEIST

Bagre marinus ....ePLEIST

Superorder PROTACANTHOPTERYGII

Order ESOCIFORMES

Family ESOCIDAE [pikes]

Esox sp... ePLEIST-lPLEIST

Superorder CYCLOSQUAMATA

Order AULOPIFORMES

Family SYNODONTIDAE [lizardfishes]

Synodus foetens.. ?lPLIO

Synodus sp. ....lPLIO

Superorder PARACANTHOPTERYGII

Order OPHIDIIFORMES

Family OPHIDIIDAE [cusk-eels]

Ophidion sp. ....lPLIO

Order BATRACHOIDIFORMES

Family BATRACHOIDIDAE [toadfishes]

Opsanus sp..lPLIO-mPLEIST

Superorder ACANTHOPTERYGII

Grandorder MUGILOMORPHA

Order MUGILIFORMES

Family MUGILIDAE [mullets]

Mugil sp. ....lPLIO-lPLEIST

Grandorder ATHERINOMORPHA

Order ATHERINIFORMES

Family ATHERINIDAE [silversides]

genus and sp. indet. ....lPLIO

Menidia sp.. ?ePLEIST

Order BELONIFORMES

Family BELONIDAE [needlefishes]

genus and sp. indet. ....lPLIO

Strongylura marina.. PLEIST

Family HEMIRAMPHIDAE [halfbeaks]

Hyporhamphus sp...ePLEIST

Order CYPRINODONTIFORMES

Family FUNDULIDAE [killifish, topminnows]

Fundulus grandis ....?ePLEIST

Fundulus seminolis..ePLEIST

Fundulus majalis ....ePLEIST

Fundulus sp.. mPLEIST

Family POECILLIIDAE [mollies, guppies]

genus and sp. indet..ePLEIST

       Gambusia affinis .... mPLEIST

       Poecilia latipinna .... mPLEIST

    Family CYPRINODONTIDAE [pupfish]

       Cyprinodon variegatus ....ePLEIST

       Floridichthys sp. .... ?ePLEIST

       Jordanella floridae .... mPLEIST

Grandorder PERCOMORPHA

  Order BERYCIFORMES

    Family HOLOCENTRIDAE [squirrelfishes]

       Holocentrites ovalis .... lEOC-eOLIG

  Order GASTEROSTEIFORMES

    Family SYNGNATHIDAE [pipefishes, seahorses]

       genus and sp. indet. ....lPLIO

  Order SCORPAENIFORMES

    Family TRIGLIDAE [searobins]

       Prionotus sp. .... lPLIO-lPLEIST

  Order PERCIFORMES

   Suborder PERCOIDEI

     Superfamily PERCOIDEA

      Family CENTROPOMIDAE [snooks]

       Centropomus sp. .... MIO-lPLEIST

      Family MORONIDAE [temperate basses]

       Morone sp. .... ?ePLEIST

       Morone saxatilis .... lPLEIST

      Family SERRANIDAE [sea basses, groupers]

       genus and sp. indet. ....lEOC, lPLIO

       Diplectrum formosum ....lPLIO

       Epinephalus sp. ....lPLIO

      Family CENTRARCHIDAE [sunfish, bream, bass]

       Lepomis sp. .... MIO-lPLEIST

       Lepomis gulosus ....ePLEIST-lPLEIST

       Lepomis auritus .... ?ePLEIST

       Lepomis microlophus ....ePLEIST-mPLEIST

       Pomoxis nigromaculatus ....ePLEIST-lPLEIST

       Micropterus sp. .... lPLEIST

       Micropterus salmoides ....ePLEIST-mPLEIST

      Family CARANGIDAE [jacks, pompanos]

       Caranx sp. .... ?MIO, PLIO-lPLEIST

       Caranx hippos ....ePLEIST

       Trachinotus sp. .... ?ePLEIST

      Family LUTJANIDAE [snappers]

       Hypsocephalus atlanticus .... lEOC

       Lutjanus avus .... eOLIG

       Lutjanus sp. .... MIO-lPLEIST

      Family GERREIDAE [mojarras]

       Eucinostomus sp. ....lPLIO

       Eucinostomus gula ....lPLIO

      Family HAEMULIDAE [grunts]

       genus and sp. indet. ....lPLIO

       Haemulon sp. ....lPLIO

Orthopristis chrysopterus..lPLIO

Family SPARIDAE [porgies, pinfishes]

    genus and sp. indet...vlOLIG, lPLIO

    Diplodus sp. .... lMIO

    Lagodon rhomboides .... ?lMIO, PLIO-lPLEIST

    Archosargus sp.. MIO-lPLEIST

    Archosargus probatocephalus.................................... ?mMIO, lPLIO-lPLEIST

    Calamus sp..ePLEIST

Family SCIAENIDAE [drums]

    genus and sp. indet.. OLIG-lPLIO

    Pogonias sp. .... vlOLIG

    Pogonias cromis .... mMIO-lPLEIST

    Bairdiella chrysoura ....lPLIO, ?ePLEIST

    Cynoscion sp..lPLIO

    Cynoscion nebulosus .... ?ePLEIST

    Equetus sp...lPLIO

    Leiostomus xanthurus ....lPLIO

    Menticirrhus sp. ....lPLIO

    Micropogonias sp. .... lPLIO-ePLEIST

    Stellifer lanceolatus..lPLIO

    Sciaenops ocellatus ....lPLIO-lPLEIST

Suborder LABROIDEI

Family LABRIDAE [wrasses, hogfishes]

    genus and sp. indet.. MIO-lPLEIST

    Lachnolaimus maximus..lPLIO-ePLEIST

    Halichoeres sp. ....lPLIO

Family SCARIDAE [parrotfishes]

    Sparisoma sp. ....? vlOLIG

Suborder ACANTHUROIDEI

Family EPHIPPIDAE [spadefishes]

    Chaetodipterus faber .... ePLEIST-lPLEIST

Suborder SCOMBROIDEI

Family SPHYRAENIDAE [barracudas]

    Sphyraena sp. .... lEOC-lPLEIST

    Sphyraena barracuda... mMIO-lPLEIST

Family SCOMBRIDAE [tunas, mackerels]

    genus and sp. indet...PLIO-lPLEIST

    Thunnus sp.. lPLEIST

Family XIPHIIDAE [swordfish, marlin]

    Cylindracanthus sp.[6] .... lEOC

    Makaira sp. ....vePLIO, PLEIST

Order PLEURONECTIFORMES

Family BOTHIDAE [lefteye flounders]

    genus and sp. indet..LPLIO-ePLEIST

Order TETRAODONTIFORMES

Superfamily BALISTOIDEA

Family BALISTIDAE [triggerfishes, filefishes]

    Balistes sp..?vlOLIG, ePLEIST-lPLEIST

Family OSTRACIIDAE [boxfishes]

    Lactophrys sp..ePLEIST

Superfamily TETRAODONTOIDEA
    Family TETRAODONTIDAE [puffers]
        Sphoeroides sp..?vlOLIG
        genus and sp. indet. ....lPLIO
    Family DIODONTIDAE [porcupinefishes, burrfishes]
        Diodon sp. ....mEOC-lPLEIST
        Diodon circumflexus .... MIO
        Chilomycterus sp.. MIO
        Chilomycterus schoepfi..ePLEIST

Class SARCOPTERYGII [lobe-finned fishes and tetrapods]
  Subclass TETRAPODA [tetrapods]
   Infraclass AMPHIBIA [amphibians]
    Legion LISSAMPHIBIA
     Cohort BATRACHIA
      Order CAUDATA (=Urodela) [salamanders]
       Suborder SIRENOIDEA
        Family SIRENIDAE [sirens]
            Siren sp...?lOLIG, lMIO-mPLEIST
            Siren hesterna ....leMIO
            Siren simpsoni ....elMIO
            Siren lacertina...lPLIO-lPLEIST
            Pseudobranchus vetustus..elMIO
            Pseudobranchus robustus .... lPLEIST
       Suborder SALAMANDROIDEA
        Family BATRACHOSAUROIDIDAE [extinct large mudpuppies]
            Batrachosauroides dissimulans ....leMIO-emMIO
        Family PROTEIDAE [mudpuppies, waterdogs]
            Necturus sp. .... lPLEIST
        Family AMPHIUMIDAE [amphiumas]
            Amphiuma sp.. lMIO
            Amphiuma means.. lPLEIST
        Family SALAMANDRIDAE [newts]
            Notophthalmus robustus..leMIO
            Notophthalmus sp.. lMIO-lPLEIST
            Notophthalmus viridescens .... lPLEIST
        Family AMBYSTOMATIDAE [common salamanders]
            Ambystoma sp... ePLEIST-lPLEIST
            Ambystoma tigrinum.. lPLEIST
        Family PLETHODONTIDAE [lungless salamanders]
            genus and sp. indet.. lMIO
            Plethodon glutinosus .... lPLEIST
      Order ANURA (=SALIENTIA) [frogs and toads]
       Family PELOBATIDAE [spadefoot toads]
            Scaphiopus sp.. eOLIG
            Scaphiopus holbrooki................................... ?eMIO, ePLEIST-lPLEIST
       Family LEPTODACTYLIDAE
            Eleutherodactylus sp.[7] ....?lPLEIST
       Family BUFONIDAE [toads]
            Bufo sp...eOLIG-ePLEIST
            Bufo praevius ....leMIO

Bufo tiheni ....elMIO

Bufo n. sp. ....vlPLIO

Bufo terrestris ....ePLEIST-lPLEIST

Bufo woodhousei ....lPLEIST

Bufo quercicus ....lPLEIST

Family HYLIDAE [treefrogs]

Proacris mintoni ....leMIO

Acris barbouri ....leMIO

Acris gryllus ....lPLEIST

Pseudacris ornata ....lPLEIST

Pseudacris nigrita ....lPLEIST

Hyla sp. ....leMIO-lMIO

Hyla goini ....leMIO

Hyla miofloridana ....leMIO

Hyla cinerea ....ePLEIST-lPLEIST

Hyla baderi ....lPLEIST

Hyla femoralis ....lPLEIST

Hyla gratiosa ....lPLEIST

Hyla squirella ....lPLEIST

Family MICROHYLIDAE [narrow-mouthed toads]

Gastrophryne carolinensis ....?leMIO, ePLEIST-lPLEIST

Family RANIDAE [common frogs]

genus and sp. indet. ....eOLIG

Rana abava ....leMIO

Rana miocenica ....leMIO

Rana bucella ....leMIO

Rana utricularia[8] ....?leMIO-lMIO, lPLEIST

Rana catesbeiana ....?lMIO-lPLIO, ePLEIST-lPLEIST

Rana areolata ....ePLEIST-lPLEIST

Rana grylio ....lPLEIST

Infraclass AMNIOTA [reptiles, birds, and mammals]

Superlegion REPTILIA [reptiles and birds]

Legion ANAPSIDA

Order TESTUDINES (=CHELONIA)[9] [turtles, tortoises]

Superfamily, genus and sp. indet. ....CRETACEOUS

Superfamily CHELYDROIDEA

Family CHELYDRIDAE [snapping turtles]

Macroclemys auffenbergi ....elMIO

Macroclemys temmincki ....lPLIO-lPLEIST

Chelydra sp. ....lPLIO

Chelydra serpentina ....lPLIO-lPLEIST

Superfamily CHELONIOIDEA

Family DERMOCHELYIDAE [leatherback sea turtles]

Psephopherus sp. ....vePLIO

Family CHELONIIDAE [sea turtles]

genus and sp. indet. ....mEOC, mMIO-lPLEIST

Chelonia sp. ....vePLIO

Chelonia mydas ....ePLEIST-lPLEIST

Caretta sp. ....vePLIO

Caretta caretta ....ePLEIST-lPLEIST

Eretmochelys sp. .... vePLIO

Lepidochelys sp. .... vePLIO

Superfamily TRIONYCHOIDEA

Family TRIONYCHIDAE [softshelled turtles]

Apalone sp.[10] .... eMIO-lPLIO

Apalone ferox...?lMIO-vePLIO, ePLEIST-lPLEIST

Family KINOSTERNIDAE [mud and musk turtles]

genus and sp. indet. ....emMIO

Kinosternon sp. ....lMIO-ePLEIST

Kinosternon bauri ....lPLIO-lPLEIST

Kinosternon subrubrum .... lPLEIST

Sternotherus minor...lPLIO-lPLEIST

Sternotherus odoratus .... lPLEIST

Superfamily TESTUDINOIDEA

Family EMYDIDAE [cooters, sliders, box turtles]

Subfamily EMYDINAE

Clemmys sp...?lMIO, lPLEIST

Terrapene sp..eMIO-lPLIO

Terrapene carolina...lPLIO-lPLEIST

Subfamily DEIROCHELYINAE

Graptemys barbouri..lPLIO-lPLEIST

Malaclemys sp..?lPLIO

Deirochelys sp..leMIO

Deirochelys carri ....elMIO

Deirochelys reticularia...lPLIO-lPLEIST

Pseudemys sp.[11] ....eOLIG-eMIO

Pseudemys caelata ....elMIO

Pseudemys williamsi...elMIO

Pseudemys concinna..lPLIO-lPLEIST

Pseudemys nelsoni...lPLIO-lPLEIST

Pseudemys floridana .... lPLEIST

Trachemys sp..vlMIO

Trachemys inflata ....vlMIO-vePLIO

Trachemys platymarginata ....lPLIO

Trachemys scripta... ?lPLIO, ePLEIST-lPLEIST

Family TESTUDINIDAE [tortoises]

Subfamily XEROBATINAE

Floridemys nanus .... OLIG?, MIO?

Gopherus sp...eMIO-lPLIO

Gopherus polyphemus..ePLEIST-lPLEIST

genus and sp. indet. ....eMIO

Hesperotestudo (*Hesperotestudo*) sp.[12] .... mMIO, lPLIO-ePLEIST

Hesperotestudo (*Hesperotestudo*) *alleni* ....lMIO-?vePLIO

Hesperotestudo (*Hesperotestudo*) *mylnarskii* ....ePLEIST-mPLEIST

Hesperotestudo (*Hesperotestudo*) *incisa*.. lPLEIST

Hesperotestudo (*Caudochelys*) *tedwhitei*..leMIO

Hesperotestudo (*Caudochelys*) *hayi*............................?mMIO, lMIO-vePLIO

Hesperotestudo (*Caudochelys*) *crassiscutata*............................. ?lPLIO, PLEIST

Legion DIAPSIDA

Sublegion LEPIDOSAURIA

Order SQUAMATA [snakes, lizards, worm "lizards"]
  Suborder IGUANIA[13]
    Family IGUANIDAE [iguanas]
      genus and sp. indet..........................eMIO
      Aciprion sp..........................eOLIG
    Family POLYCHRIDAE [anoles]
      Anolis sp. ....?eMIO-lMIO
      Anolis carolinensis..........................mPLEIST-lPLEIST
    Family TROPIDURIDAE
      Leiocephalus sp..........................eMIO, ?emMIO
    Family PHRYNOSOMATIDAE
      Sceloporus undulatus..........................vlPLIO-lPLEIST
  Suborder GEKKOTA
    Family GEKKONIDAE [gekkos]
      genus and sp. indet..........................eMIO
  Suborder SCINCOMORPHA
    Family SCINCIDAE [skinks]
      genus and sp. indet..........................eMIO-lMIO
      Eumeces sp..........................eMIO-lMIO
      Eumeces carri..........................vlPLIO
      Eumeces inexpectatus..........................mPLEIST-lPLEIST
      Eumeces fasciatus..........................?lPLEIST
    Family XANTUSIDAE [night lizards]
      Paleoxantusia sp..........................eOLIG
    Family TEIIDAE [whiptail and racerunner lizards]
      genus and sp. indet..........................?eOLIG, eMIO
      Cnemidophorus sp. ....eMIO, ?emMIO, ePLEIST
      Cnemidophorus sexlineatus..........................lPLEIST
  Suborder ANGUIMORPHA
    Family ANGUIDAE [glass and alligator lizards]
      Peltosaurus sp. ....eOLIG, eMIO
      genus and sp. indet..........................eMIO
      Ophisaurus sp..........................eMIO-lMIO
      Ophisaurus ventralis ....lMIO, vlPLIO-lPLEIST
      Ophisaurus compressus..........................ePLEIST-lPLEIST
      Gerrhonotus sp. ....?vlPLIO
    Family HELODERMATIDAE [beaded lizards, gila monster]
      Heloderma sp..........................eMIO, ?emMIO
  Suborder AMPHISBAENIA
    Family AMPHISBAENIDAE [worm "lizards"]
      genus and sp. indet..........................eOLIG, eMIO
      Rhineura floridana ....?vlPLIO, lPLEIST
  Suborder SERPENTES [snakes]
    Superfamily SCOLECOPHILIA
    Family TYPHLOPIDAE [blind snakes]
      Typhlops sp. ....?eOLIG, eMIO-lMIO
    Superfamily ANILIOIDEA
    Family ANILIIDAE
      genus and sp. indet..........................?eMIO
    Superfamily BOOIDEA

Family PALAEOPHIDAE [extinct sea boas]
  Pterosphenus schucherti .... lEOC
Family BOIDAE [boas, pythons]
  Subfamily ERYCINAE
    genera and spp. indet. (3–4 spp.) .... eOLIG
    Anilioides minuatus .... leMIO
    Calamagras floridanus .... leMIO
    Ogmophis pauperrimus .... leMIO
  Subfamily TROPIDOPHIINAE
    Tropidophis sp. .... ?ePLEIST
  Subfamily BOINAE
    Pseudoepicrates stanolseni .... leMIO
    Boa barbouri[14] .... leMIO
Superfamily COLUBROIDEA
  Family COLUBRIDAE [nonpoisonous snakes]
  Subfamily INDETERMINATE
    Tantilla sp. .... vlPLIO
    Tantilla coronata .... lPLEIST
  Subfamily COLUBRINAE
    Floridaophis aufferbergi .... eOLIG
    Nebraskophis oligocenicus .... eOLIG
    Paraoxybelis floridanus .... leMIO
    genus and sp. indet. .... emMIO
    Coluber sp. .... lMIO-lPLIO
    Coluber constrictor .... vlPLIO-lPLEIST
    Drymarchon sp. .... lPLIO
    Drymarchon corais .... vlPLIO-lPLEIST
    Masticophis sp. .... lPLIO
    Masticophis flagellum .... vlPLIO-lPLEIST
    Opheodrys vernalis .... vlPLIO
    Opheodrys aestivus .... lPLEIST
  Subfamily LAMPROPELTINAE
    Pseudocemophora antiqua .... leMIO
    Cemophora coccinea .... vlPLIO-lPLEIST
    Elaphe sp. .... lMIO-ePLEIST
    Elaphe guttata .... lPLIO-lPLEIST
    Elaphe obsoleta .... vlPLIO-lPLEIST
    Lampropeltis getulus .... ?lMIO, lPLIO-lPLEIST
    Lampropeltis triangulum .... vlPLIO-lPLEIST
    Pituophis melanoleucas .... vlPLIO-lPLEIST
  Subfamily NATRICINAE
    genus and sp. indet. .... emMIO
    Nerodia sp[15] .... lMIO, ePLEIST
    Nerodia fasciata .... lPLIO
    Nerodia cyclopion .... lPLIO-lPLEIST
    Nerodia erythrogaster .... lPLIO-lPLEIST
    Nerodia taxispilota .... lPLIO-lPLEIST
    Regina sp. .... lMIO-lPLIO
    Regina intermedia .... vlPLIO
    Regina alleni .... lPLEIST

Thamnophis sp. .... lMIO
Thamnophis sirtalis .... ?vlPLIO, lPLEIST
Virginia sp. .... vlPLIO
Storeria dekayi .... lPLEIST
Subfamily XENODONTINAE
Diadophis elinorae .... lMIO-vlPLIO
Diadophis punctatus .... lPLEIST
Dryinoides sp. .... ?lMIO, ?vlPLIO
Heterodon sp. .... lMIO-lPLIO
Heterodon brevis .... elMIO
Heterodon nasicus .... vlPLIO
Heterodon platyrhinos .... vlPLIO-lPLEIST
Heterodon simus .... mPLEIST-lPLEIST
Stilosoma vetustum .... lMIO
Stilosoma extenatum .... vlPLIO-lPLEIST
Paleofarancia brevispinosus .... vePLIO
Farancia sp. .... lPLIO
Farancia abacura .... vlPLIO-lPLEIST
Rhadinaea flavilata .... ?vlPLIO, lPLEIST
Carphophis amoenus .... mPLEIST-lPLEIST
Family ELAPIDAE [cobras, coral snakes]
Micrurus sp. .... lMIO, ePLEIST
Micrurus fulvius .... ?vlPLIO, lPLEIST
Family VIPERIDAE [pit vipers, rattlesnakes]
Agkistrodon sp. .... lPLIO
Agkistrodon piscivorus .... ePLEIST-lPLEIST
Crotalus sp. .... lMIO-lPLIO
Crotalus adamanteus .... vlPLIO-lPLEIST
Sistrurus sp. .... ?lMIO
Sistrurus miliarius .... vlPLIO-lPLEIST
Sublegion ARCHOSAURIA
Supercohort CRUROTARSI
Cohort CROCODYLOTARSI
Superorder CROCODYLOMORPHA
Order CROCODYLIA
Suborder BREVIROSTRES
Family CROCODYLIDAE [crocodiles][16]
genus and sp. indet. .... mEOC
Gavialosuchus americanus .... ?eMIO, mMIO-lMIO
Family ALLIGATORIDAE [alligators, caimans]
Alligator olseni .... leMIO
Alligator mississippiensis .... ?mMIO, elMIO-lPLEIST
Supercohort DINOSAURIA
Cohort AVES[17] [birds]
Superorder NEORNITHES
Grandorder NEOGNATHAE
Order GAVIIFORMES
Family GAVIIDAE [loons]
Gavia palaeodytes .... vePLIO
Gavia concinna .... vePLIO-ePLEIST

Gavia pacifica .... v|PLIO
Gavia immer .... v|PLIO-l|PLEIST
Gavia sp. .... ePLEIST
Gavia artica .... ePLEIST

Order PODICIPEDIFORMES
  Family PODICIPEDIDAE [grebes]
    Rollandia sp. .... el|MIO
    Tachybaptus sp. .... el|MIO
    Tachybaptus dominicus .... v|PLIO-ePLEIST
    Podiceps sp. .... el|MIO-vePLIO
    Podiceps dixi .... ?l|PLIO,ePLEIST-l|PLEIST
    Podiceps auritus .... l|PLEIST
    Pliodytes lanquisti .... vePLIO
    Podilymbus podiceps .... ?vePLIO, l|PLIO-l|PLEIST
    Podilymbus wetmorei .... l|PLEIST

Order PROCELLARIIFORMES
  Family DIOMEDEIDAE [albatroses]
    Diomedea anglica .... vePLIO
  Family PROCELLARIIDAE [shearwaters, petrels]
    Puffinus micraulax .... eMIO?
    Puffinus sp. .... vePLIO
    Puffinus puffinus .... l|PLEIST

Order PELICANIFORMES
  Family SULIDAE [boobies, gannets]
    genus and sp. indet. .... eMIO-mMIO
    Sula universitatis .... eMIO?
    Sula guano .... vePLIO
    Sula phosphata .... vePLIO
    Morus peninsularis .... vePLIO
  Family PELICANIDAE [pelicans]
    Pelicanus sp. .... vePLIO
    Pelicanus erythrorhynchos .... ?ePLEIST
  Family PHALACROCORACIDAE [cormorants]
    Phalacrocorax sp. .... el|MIO, vePLIO-ePLEIST
    Phalacrocorax wetmorei .... ?v|MIO, vePLIO
    Phalacrocorax filyawi .... l|PLIO
    Phalacrocorax idahensis[18] .... v|PLIO
    Phalacrocorax auritus .... ?l|PLIO, ePLEIST-l|PLEIST
  Family ANHINGIDAE [anhingas]
    Anhinga subvolans[19] .... le|MIO
    Anhinga grandis .... el|MIO, ?l|PLEIST
    Anhinga sp. .... vePLIO
    Anhinga beckeri .... v|PLIO-l|PLEIST
    Anhinga anhinga .... v|PLIO-l|PLEIST

Order ARDEIFORMES
  Family ARDEIDAE [herons, egrets]
    Ardea sp. .... l|MIO, l|PLIO-ePLEIST
    Ardea polkensis .... vePLIO
    Ardea alba .... l|PLIO-l|PLEIST
    Ardea herodias .... ePLEIST-l|PLEIST

Ardeola sp. ....elMIO

Egretta sp. ....elMIO-lPLIO

Egretta subfluvia ....vlMIO

Egretta caerulea.. lPLEIST

Egretta thula .... lPLEIST

Egretta tricolor ....?ePLEIST, lPLEIST

Nycticorax fidens ....elMIO

Nycticorax violaceus ....lPLIO

Nycticorax nycticorax .... lPLEIST

Butorides validipes ....lPLIO

Butorides striatus (=*B. virescens*)... vlPLIO, lPLEIST

Botaurus hibbardi ....?lPLIO

Botaurus lentiginosus ....lPLIO-lPLEIST

Ixobrychus sp. ....lPLIO-mPLEIST

Ixobrychus exilis ....lPLIO-lPLEIST

Nyctanassa violacea .... lPLEIST

Order CICONIIFORMES

Family CICONIIDAE [storks, wood storks]

Propelargus olseni ....leMIO

Ciconia sp. ....elMIO-vePLIO

Ciconia maltha ....vlPLIO-lPLEIST

Mycteria sp...elMIO

Mycteria americana .... lPLEIST

Family TERATORNITHIDAE [teratorns]

Teratornis merriami... ePLEIST-lPLEIST

Teratornis incredibilis ....?ePLEIST

Family VULTURIDAE [condors, American "vultures"]

Pliogyps charon..elMIO

Gymnogyps sp.[20] .... ?lPLIO-?vlPLIO

Gymnogyps kofordi ....ePLEIST

Gymnogyps californianus .... lPLEIST

Aizenogyps toomeyae .... vlPLIO

Coragyps atratus .... vlPLIO-lPLEIST

Coragyps occidentalis...mPLEIST-lPLEIST

Cathartes aura... lPLEIST

Order ANSERIFORMES

Family ANATIDAE [ducks, geese, swans]

Subfamily, genus and sp. indet.. lMIO-vePLIO

Subfamily DENDROCYGNINAE

Dendrocygna sp. .... elMIO, lPLIO

Subfamily ANSERINAE

Branta sp. .... lMIO

Branta dickeyi...ePLEIST

Branta canadensis... ePLEIST-lPLEIST

Cygnus buccinator... ePLEIST-lPLEIST

Cygnus sp...ePLEIST

Olor sp..ePLEIST

Olor columbianus .... lPLEIST

Subfamily TADORNINAE

Anabernicula sp..ePLEIST

Anabernicula minuscula .... ?lPLIO

Anabernicula gracilenta .... lPLIO-ePLEIST

Subfamily ANATINAE

    Anas sp. .... elMIO

    Anas crecca .... lPLIO-lPLEIST

    Anas discors .... lPLIO-lPLEIST

    Anas clypeata .... ?lPLIO, vlPLIO

    Anas cyanoptera .... ?lPLIO

    Anas americana .... ?lPLIO, ePLEIST-lPLEIST

    Anas platyrhyncos .... ePLEIST-lPLEIST

    Anas strepera .... ePLEIST-lPLEIST

    Anas itchtucknee[21] .... lPLEIST

    Anas fulvigula .... lPLEIST

    Anas acuta .... lPLEIST

    Aythya sp. .... vePLIO-lPLIO

    Aythya marila .... ePLEIST

    Aythya americana .... ePLEIST-lPLEIST

    Aythya affinis .... ePLEIST-lPLEIST

    Aythya collaris .... ePLEIST-lPLEIST

    Aythya valisineria .... lPLEIST

    Bucephala ossivallis .... vePLIO, ?ePLEIST

    Bucephala albeola .... ?lPLIO, vlPLIO-lPLEIST

    Bucephala clangula .... lPLEIST

    Oxyura dominica .... ?vePLIO

    Oxyura hulberti .... lPLIO

    Oxyura jamaicensis .... lPLEIST

    Helonetta brodkorbi .... lPLIO

    Mergus merganser .... lPLIO?, lPLEIST

    Mergus serrator .... lPLEIST

    Aix sponsa .... vlPLIO-lPLEIST

    Lophodytes cucullatus .... ePLEIST-lPLEIST

    Somateria spectabilis .... ?ePLEIST

    Clangula hyemalis .... lPLEIST

    Spatula clypeata .... lPLEIST

Order ACCIPITRIFORMES (=FALCONIFORMES)

  Family ACCIPITRIDAE [hawks, eagles, true vultures]

    Promilio brodkorbi .... leMIO

    Promilio epileus .... leMIO

    Promilio floridanus .... leMIO

    Buteo sp. .... ?leMIO, lMIO-ePLEIST

    Buteo lineatus .... vlPLIO-lPLEIST

    Buteo lagopus .... mPLEIST, ?lPLEIST

    Buteo jamaicensis .... mPLEIST-lPLEIST

    Buteo platypterus .... mPLEIST-lPLEIST

    genus and sp. indet. .... elMIO-vePLIO

    Aquila n. sp. .... vePLIO-ePLEIST

    Aquila chrysaetus .... mPLEIST-lPLEIST

    Haliaeetus sp. .... ?vePLIO

    Haliaeetus leucocephalus .... lPLEIST

    Amplibuteo n. sp.A .... vlPLIO

Amplibuteo n. sp.B..........ePLEIST
Amplibuteo woodwardi..........mPLEIST-lPLEIST
Accipiter cooperi..........vlPLIO-lPLEIST
Accipiter striatus..........lPLEIST
Buteogallus fragillus..........vlPLIO
Buteogallus urubitinga..........vlPLIO
Circus cyaneus..........?ePLEIST
Neophrontops slaughteri..........vlPLIO
Spizaetus sp...........lPLIO
Spizaetus grinnelli..........lPLEIST
Family PANDIONIDAE [ospreys]
   Pandion lovensis..........elMIO
   Pandion sp...........vePLIO
   Pandion haliaetus..........lPLEIST
Family FALCONIDAE [falcons, caracaras]
   Falco columbarius..........vlPLIO-lPLEIST
   Falco sparverius..........vlPLIO-lPLEIST
   Falco peregrinus..........lPLEIST
   Milvago chimachima[22]..........lPLEIST
   Polyborus plancus (=*Caracara plancus*)..........lPLEIST
Order GALLIFORMES
  Family CRACIDAE [chachalacas]
   Boreortalis laesslei..........leMIO
   genus and sp. indet...........ePLEIST
  Family PHASIANIDAE [pheasants, quail, turkeys]
   Subfamily TETRAONINAE [grouse]
    Bonasa umbellus..........lPLEIST
    Tympanuchus cupido..........lPLEIST
   Subfamily MELEAGRIDINAE [turkeys]
    Rhegminornis calobatus[23]..........leMIO
    genus and sp. indet...........elMIO
    Meleagris sp...........?vePLIO
    Meleagris leopoldi or *M. anza*..........lPLIO-ePLEIST
    Meleagris gallopavo..........mPLEIST-lPLEIST
   Subfamily ODONTOPHORINAE [quail]
    Colinus virginianus[24]..........lPLIO-lPLEIST
    Neortyx peninsularis..........lPLEIST
Order GRUIFORMES
  Family RALLIDAE [rails, coots]
   new genus and sp...........lMIO
   Rallus sp...........elMIO
   Rallus sp. A..........lPLIO-ePLEIST
   Rallus sp. B..........lPLIO
   Rallus longirostris..........?lPLIO, ePLEIST-mPLEIST
   Rallus elegans..........?lPLIO, ePLEIST-lPLEIST
   Rallus limicola[25]..........lPLIO-lPLEIST
   Coturnicops noveboracensis..........mPLEIST-lPLEIST
   Gallinula sp...........lPLIO
   Gallinula chloropus[26]..........mPLEIST-lPLEIST
   Fulica americana..........?lPLIO, vlPLIO-lPLEIST

Laterallus sp. ....lPLIO

Laterallus exilis[27] ....?ePLEIST, lPLEIST

Porphyrula sp. ....lPLIO

Porphyrula martinica .... lPLEIST

Porzana carolina ....vlPLIO, lPLEIST

Family ARAMIDAE [limpkins]

Aramus guarauna .... lPLEIST

Family GRUIDAE [cranes]

Probalearica crataegensis...leMIO

Aramornis sp...elMIO

Grus sp...elMIO, ePLEIST

Grus canadensis ....lPLIO-lPLEIST

Grus americanus ....vlPLIO-lPLEIST

genus and sp. indet. .... vePLIO

Family PHORUSRHACIDAE [extinct flightless birds]

Titanis walleri ....lPLIO

Order CHARADRIIFORMES

Family PLATALEIDAE [ibises, spoonbills]

genus and sp. indet...lMIO, ePLEIST

Plegadis pharangites ....?lMIO

Ajaia chione ....ePLEIST

Ajaia ajaia .... lPLEIST

Eudocimus leiseyi .... lPLIO-ePLEIST

Eudocimus albus ....ePLEIST-lPLEIST

Family JACANIDAE [lily-trotters]

Jacana farrandi..elMIO

Jacana spinosa.. lPLEIST

Family HAEMATOPODIDAE [oystercatchers]

Haematopus sulcatus.. vePLIO

Family CHARADRIIDAE [plovers]

Charadrius vociferus .... lPLEIST

Vanellus chilensis[28] .... lPLEIST

Family SCOLOPACIDAE [sandpipers]

genus and sp. indet...elMIO

Actitis sp. ....?elMIO

Actitis macularia ....lPLIO, lPLEIST

Arenaria sp...?elMIO

Calidris 2 spp. ....elMIO

Calidris rayi...elMIO

Calidris pacis... vePLIO

Calidris penepusilla .... vePLIO

Calidris alba ....?lPLIO

Calidris canutus...?lPLIO

Calidris pusilla ....?lPLEIST

Limosa ossivallus .... vePLIO

Limosa fedora ....?ePLEIST

Philomachus sp. .... vePLIO

Limnodromus sp..lPLIO, lPLEIST

Limnodromus scolopaceus ....vlPLIO-lPLEIST

Gallinago sp..lPLIO

Gallinago gallinago .... vlPLIO-lPLEIST
Scolopax hutchensi .... lPLIO-ePLEIST
Scolopax minor ....mPLEIST-lPLEIST
Tringa sp. .... lPLIO
Tringa flacipes .... lPLEIST
Tringa melanoleuca .... lPLEIST
Tringa solitaria .... lPLEIST
Numenius americanus .... lPLEIST
Totanus melanoleucas .... lPLEIST
Totanus flavipes .... lPLEIST
Family RECURVIROSTRIDAE [avocets, stilts]
Himantopus n. sp. ....lPLIO
Recurvirostra sp. .... lPLIO-ePLEIST
Recurvirostra americana .... lPLEIST
Family PHOENICOPTERIDAE [flamingos]
Phoenicopterus sp. .... elMIO, lPLIO
Phoenicopterus floridanus .... vePLIO
Phoenicopterus copei .... ePLEIST
Phoenicopterus ruber .... ePLEIST
Family LARIDAE [gulls]
Larus elmorei .... vePLIO
Larus perpetuus .... lPLIO
Larus lacus .... lPLIO
Larus sp. .... lPLEIST
Stercorarius sp. .... lPLIO
Family ALCIDAE [auks, mures, puffins]
3 genera and spp. indet. .... vePLIO
genus and sp. indet. .... ePLEIST
Australca grandis .... vePLIO
Pinguinus sp. .... vePLIO
Order COLUMBIFORMES
Family COLUMBIDAE [pigeons, doves]
Columbina prattae ....leMIO
Columba sp. ....vlPLIO
Columba fasciata .... lPLEIST
Zenaida macroura ....vlPLIO-lPLEIST
Zenaida sp. ....?lPLEIST
Ectopistes migratorius .... lPLEIST
Order CUCULIFORMES
Family CUCULIDAE [cuckoos]
Coccyzus americanus ....vlPLIO-lPLEIST
Order STRIGIFORMES
Family TYTONIDAE [barn owls]
new genus and sp. ....elMIO
Tyto alba ....vlPLIO-lPLEIST
Family STRIGIDAE [owls]
Bubo sp. .... vePLIO
Bubo virginianus ....vlPLIO-lPLEIST
Asio sp. ....vlPLIO
Asio priscus ....ePLEIST

Asio flammeus .... lPLEIST
Glaucidium explorator .... vlPLIO
Glaucidium sp. .... vlPLIO
Speotyto cunicularia .... vlPLIO, lPLEIST
Speotyto megalopeza ....ePLEIST
Otus asio .... vlPLIO-lPLEIST
Strix varia .... vlPLIO, lPLEIST
Aegolius acadicus .... lPLEIST

Order CORACIIFORMES
    Family, genus and sp. indet. ....leMIO
    Family ALCEDINIDAE [kingfishers]
      Ceryle torquata .... vlPLIO
      Ceryle alcyon .... lPLEIST
    Family MOMOTIDAE [motmots]
      genus and sp. indet. ....elMIO

Order PICIFORMES
    Family PICIDAE [woodpeckers]
      Colaptes auratus... vlPLIO-lPLEIST
      Melanerpes erythrocephalus .... ?vlPLIO, ePLEIST-lPLEIST
      Melanerpes carolinus.. lPLEIST
      Picoides villosus .... vlPLIO
      Picoides borealis .... lPLEIST
      Dryocopus sp. .... vlPLIO, lPLEIST
      Campephilus dalquesti ....ePLEIST
      Sphyrapicus sp..ePLEIST
    Family CAPITONIDAE [barbets]
      genus and sp. indet. ....leMIO

Order PASSERIFORMES [perching and song birds]
  Suborder, family, genus and sp. indet..elMIO
  Suborder TYRANNI [suboscines]
    Parvorder TYRANNIDA
      Family TYRANNIDAE [tyrant flycatchers]
        Tyrannus tyrannus... lPLEIST
  Suborder PASSERI [oscines]
    Parvorder CORVIDA
      Superfamily CORVOIDEA
        Family CORVIDAE [crows, jays]
          Corvus sp. ....vlPLIO
          Corvus brachyrhyncos...............................?ePLEIST, lPLEIST
          Corvus ossifragus .... vlPLIO, lPLEIST
          Protocitta ajax... mPLEIST
          Pica pica[29] ....?mPLEIST, lPLEIST
          Aphelocoma coerulescens.......................... vlPLIO, lPLEIST
          Cyanocitta cristata .... vlPLIO, lPLEIST
          Henocitta brodkorbi.................................... lPLEIST
      Family LANIIDAE [shrikes]
        Lanius ludovicianus.. lPLEIST
      Family VIREONIDAE [vireos]
        Vireo griseus.. lPLEIST
    Parvorder PASSERIDA

Superfamily MUSCICAPOIDEA
 Family MUSCICAPIDAE [thrushes, robins]
  genus and sp. indet. .... vlPLIO
  Catharus sp. .... vlPLIO
  Catharus minimus .... lPLEIST
  Turdus sp.A .... vlPLIO
  Turdus sp.B .... vlPLIO
  Hylocichla mustelina .... ?vlPLIO
  Sialis sialis .... lPLEIST
 Family MIMIDAE [mockingbirds, thrashers]
  Dumetella carolinensis .... vlPLIO
  Toxostoma rufum .... vlPLIO, lPLEIST
  Mimus polygottos .... lPLEIST
Superfamily SYLVIOIDEA
 Family HIRUNDINIDAE [swallows]
  Progne subis .... mPLEIST-lPLEIST
  Tachycineta speleodytes .... lPLEIST
 Family SITTIDAE [nuthatches]
  Sitta pusilla .... lPLEIST
 Family TROGLODYTIDAE [wrens]
  Cistothorus brevis .... lPLEIST
  Cistothorus platensis .... lPLEIST
  Troglodytes aedon .... lPLEIST
Superfamily PASSEROIDEA
 Family EMBERIZIDAE
 Subfamily PARULINAE [wood warblers]
  genus and sp. indet. .... leMIO
  Vermivora celata .... ?vlPLIO
  Dendroica sp. .... mPLEIST
  Geothlypis trichas .... lPLEIST
 Subfamily CARDINALINAE [cardinals, grosbeaks]
  genus and sp. indet. .... vlPLIO-ePLEIST
  Cardinalis cardinalis .... vlPLIO-lPLEIST
  Passerina sp. .... ?ePLEIST
  Pheucticus ludovicianus .... mPLEIST
 Subfamily EMBERIZINAE [towhees, American sparrows]
  "*Palaeostruthus*" *eurius*[30] .... elMIO
  Ammodramus maritimus .... vlPLIO
  Ammodramus savannarum .... lPLEIST
  Ammodramus henslowi .... lPLEIST
  Chondestes grammacus .... ?vlPLIO
  Junco hyemalis .... vlPLIO-lPLEIST
  Melospiza melodia .... vlPLIO-ePLEIST
  Melospiza georgiana .... lPLEIST
  Passerculus sandwichensis .... vlPLIO-lPLEIST
  Spizella pusilla .... ?vlPLIO, mPLEIST-lPLEIST
  Spizella passerina .... lPLEIST
  Zonotrichia albicollis .... ?vlPLIO
  Zonotrichia leucophrys .... ?vlPLIO-ePLEIST
  Passerella sp. .... ePLEIST

Aimophila aestivalus .... mPLEIST

Pooecetes gramineus .... mPLEIST

Pipilo erythrophthalmus ....mPLEIST-lPLEIST

Subfamily ICTERINAE [grackles, blackbirds, orioles]

genus and sp. indet. .... lPLIO-ePLEIST

Agelaius sp. .... ?vlPLIO

Agelaius phoeniceus ....?lPLIO, vlPLIO-lPLEIST

Euphagus sp. ....vlPLIO

Euphagus cyanocephalus ....vlPLIO

Sturnella magna .... vlPLIO, lPLEIST

Molothrus ater .... vlPLIO-lPLEIST

Pandanaris floridana ....mPLEIST-lPLEIST

Cremaster tytthus .... lPLEIST

Quiscalus quiscula .... lPLEIST

Quiscalus mexicanus .... lPLEIST

Dolichonyx oryzivorus .... lPLEIST

Superlegion SYNAPSIDA

Legion MAMMALIA [mammals]

Sublegion THERIIFORMES

Supercohort THERIA

Cohort MARSUPIALIA (=METATHERIA) [marsupials]

Magnorder AMERIDELPHIA

Order DIDELPHIMORPHA

Family DIDELPHIDAE [opossums]

genus and sp. indet. .... eOLIG

Peratherium sp. .... eMIO-emMIO

Didelphis virginiana ....mPLEIST-lPLEIST

Cohort PLACENTALIA (=EUTHERIA) [placentals or eutherians]

Magnorder XENARTHRA

Order CINGULATA

Superfamily DASYPODOIDEA

Family DASYPODIDAE [armadillos]

Dasypus bellus ....lPLIO-lPLEIST

Family indeterminate

Pachyarmatherium leiseyi[31] .... lPLIO-ePLEIST

Superfamily GLYPTODONTOIDEA

Family PAMPATHERIIDAE [pampatheres, giant armadillos]

Holmesina floridanus .... lPLIO-ePLEIST

Holmesina septentrionalis ....mPLEIST-lPLEIST

Family GLYPTODONTIDAE [glyptodonts]

Glyptotherium arizonae ....lPLIO-ePLEIST

Glyptotherium floridanum ....mPLEIST-lPLEIST

Order PILOSA

Suborder PHYLLOPHAGA [sloths]

Infraorder MYLODONTA

Family MYLODONTIDAE

Subfamily LESTODONTIDAE

Tribe THINOBADISTINI

Thinobadistes segnis ....elMIO

Thinobadistes wetzeli ....vlMIO

Tribe GLOSSOTHERINI
    Glossotherium chapadmalense .... elPLIO
    Paramylodon harlani[32] .... vlPLIO-lPLEIST
Infraorder MYLODONTA
  Family MEGATHERIIDAE
    Subfamily MEGATHERIINAE
    Tribe MEGATHERIINI
      Eremotherium eomigrans .... lPLIO-ePLEIST
      Eremotherium laurillardi[33] .... lPLEIST
    Tribe NOTHROTHERIINI
      Nothrotheriops texanus..ePLEIST
  Family MEGALONYCHIDAE
    Pliometanastes protistus .... elMIO-vlMIO
    Megalonyx curvidens... vePLIO
    Megalonyx leptostomus... lPLIO
    Megalonyx wheatleyi..ePLEIST-emPLEIST
    Megalonyx jeffersoni..lmPLEIST-lPLEIST
Magnorder EPITHERIA
Superorder PREPTOTHERIA
Grandorder ANAGALIDA[34]
  Order LAGOMORPHA [rabbits, pikas]
    Family LEPORIDAE [rabbits, hares]
    Subfamily PALAEOLAGINAE[35]
      genus and sp. indet.. eOLIG
      Palaeolagus sp...veMIO
    Subfamily ARCHAEOLAGINAE
      Hypolagus tedfordi.. lmMIO
      Hypolagus sp...elMIO
      Hypolagus ringoldensis .... vePLIO
    Subfamily LEPORINAE
      Nekrolagus sp. .... vePLIO
      Sylvilagus webbi .... lPLIO-ePLEIST
      Sylvilagus floridanus..ePLEIST-lPLEIST
      Sylvilagus palustris ....ePLEIST-lPLEIST
      Sylvilagus palustrellus...vlPLEIST
      Lepus townsendii ....?ePLEIST
      Lepus sp..ePLEIST-mPLEIST
  Order RODENTIA [rodents]
  Suborder SCIUROMORPHA
    Superfamily APLODONTOIDEA
    Family MYLAGAULIDAE [extinct burrowing rodents]
      Mesogaulus sp..leMIO
      Mylagaulus sp. ....emMIO
      Mylagaulus elassos ....elMIO
      Mylagaulus kinseyi..elMIO
  Infraorder SCIURIDA
    Family SCIURIDAE [squirrels]
    Subfamily SCIURINAE [tree and ground squirrels]
    tribe, genus and sp. indet. .... veMIO
    Tribe SCIURINI

Protosciurus sp. ....leMIO
Sciurus sp. ....ePLEIST
Sciurus carolinensis ....mPLEIST-lPLEIST
Sciurus niger .... lPLEIST
Tribe MARMOTINI
    Miospermophilus sp. ....?leMIO
    Protospermophilus sp. ....?emMIO
    Spermophilus sp. .... lPLEIST
Tribe TAMIINI
    Nototamias hulberti ....leMIO
    Tamias sp. ....?eMIO
    Tamias aristus .... lPLEIST
Subfamily PTEROMYINAE [flying squirrels]
    Petauristodon pattersoni ....leMIO
    Miopetaurista webbi[36] ....vePLIO-lPLIO
    Glaucomys volans ....?vlPLIO, mPLEIST-lPLEIST
Infraorder CASTORIMORPHA
  Family EUTYPOMYIDAE
    genus and sp. indet. ....?eOLIG, lOLIG
    Anchitheriomys sp. .... lmMIO
  Family CASTORIDAE [beavers]
    Subfamily CASTOROIDINAE
    genus and sp. indet. .... veMIO
    Eucastor sp. .... lmMIO
    Eucastor planus .... ?elMIO
    Castoroides leiseyorum ....ePLEIST
    Castoroides ohioensis ....mPLEIST-lPLEIST
    Subfamily CASTORINAE
    Castor canadensis ....lPLIO, lPLEIST
Suborder MYOMORPHA
Infraorder MYODONTA
  Superfamily DIPODOIDEA
    Family DIPODIDAE [jumping mice]
    Zapus sp. ....ePLEIST
  Superfamily MUROIDEA
    Family MURIDAE [mice, voles]
    Subfamily PARACRICETODONTINAE
    genus and sp. indet. ....eOLIG-eMIO
    Leidymys sp. .... ?veMIO
    Subfamily CRICETODONTINAE
    Copemys sp. ....emMIO-elMIO
    new genus and sp. ....elMIO
    Abelmoschomys simpsoni..................................elMIO
    Subfamily SIGMODONTINAE [New World mice]
    Tribe NEOTOMINI
    Neotoma sp..vlPLIO
    Neotoma floridana ....mPLEIST-lPLEIST
    Tribe PEROMYSCINI
    Reithrodontomys sp. ....vlPLIO
    Reithrodontomys humulis ....vlPLIO-lPLEIST

Reithrodontomys fulvecens .... lPLEIST

Podomys n. sp. .... ePLEIST

Podomys floridanus[37] .... mPLEIST-lPLEIST

Ochrotomys nuttalli .... mPLEIST-lPLEIST

Peromyscus gossypinus .... lPLEIST

Peromyscus polionotus .... lPLEIST

Tribe ORYZOMYINI

 Oryzomys palustris .... mPLEIST-lPLEIST

Tribe SIGMODONTINI

 Sigmodon medius .... lPLIO

 Sigmodon minor .... vlPLIO

 Sigmodon curtisi .... vlPLIO

 Sigmodon libitinus .... ePLEIST

 Sigmodon bakeri .... mPLEIST

 Sigmodon hispidus .... lPLEIST

Subfamily ARVICOLINAE[38] [voles, lemmings]

Tribe indet.

 Atopomys salvelinus .... vlPLIO-ePLEIST

Tribe ARVICOLINI

 Microtus (Pedomys) australis .... ePLEIST

 Microtus aratai .... mPLEIST

 Microtus (Pitymys) hibbardi .... lmPLEIST

 Microtus (Pitymys) pinetorum .... lPLEIST

 Microtus (Microtus) pennsylvanicus .... lPLEIST

Tribe ONDATRINI

 Ondatra idahoensis .... vlPLIO

 Ondatra annectens .... ePLEIST

 Ondatra zibethicus .... lPLEIST

Tribe LEMMINI

 Synaptomys n. sp. .... ePLEIST

 Synaptomys australis .... mPLEIST-lPLEIST

Tribe NEOFIBRINI

 Neofiber leonardi .... mPLEIST

 Neofiber alleni .... mPLEIST-lPLEIST

Infraorder GEOMORPHA

Superfamily EOMYOIDEA

Family EOMYIDAE

 genus and sp. indet. .... eOLIG

 new genus and sp. A .... vlOLIG-veMIO

 new genus and sp. B .... eMIO

Superfamily GEOMYOIDEA

Family indet.

 Jimomys sp. .... ?eMIO

Family GEOMYIDAE

Subfamily GEOMYINAE [pocket gophers]

 Geomys propinetis .... lPLIO-vePLEIST

 Geomys pinetis .... lePLEIST-lPLEIST

 Thomomys orientalis .... mPLEIST-lPLEIST

Subfamily HETEROMYINAE [pocket mice, kangaroo rats]

 genus and sp. indet. .... eOLIG

Proheteromys several new spp.. eMIO-emMIO

Proheteromys floridanus ....leMIO

Proheteromys magnus ....leMIO

Perognathus minutus.. ?emMIO

Suborder HYSTRICOGNATHA

Infraorder HYSTRICOGNATHI

Family ERETHIZONTIDAE [porcupines]

Erethizon poyeri...vlPLIO

Erethizon kleini..vlPLIO

Erethizon dorsatum..ePLEIST-lPLEIST

Parvorder CAVIIDA

Superfamily CAVIOIDEA

Family HYDROCHOERIDAE [capybaras]

Neochoerus dichroplax ....lPLIO

Neochoerus sp...ePLEIST

Neochoerus pinckneyi ....mPLEIST-lPLEIST

Hydrochaeris holmesi...lPLIO-lPLEIST

Grandorder FERAE

Order CARNIVORA [carnivorans]

Suborder CANIFORMIA

Infraorder CYNOIDEA

Family CANIDAE [dogs, wolves, foxes]

Subfamily HESPEROCYONINAE

Mesocyon sp... ?veMIO

Osbornodon iamonensis[39] ....eMIO

Subfamily BOROPHAGINAE[40]

Phlaocyon achoros ....veMIO

Phlaocyon leucosteus..veMIO

Cynarctoides lemur ....veMIO

Cormocyon haydeni .... ?veMIO

Desmocyon matthewi...leMIO

Euoplocyon spissidens..leMIO

Metatomarctus canavus[41] ....leMIO

Aelurodon taxoides .... lmMIO

Epicyon haydeni[42] ....elMIO-vePLIO

Epicyon saevus...elMIO

Carpocyon limosus...vlMIO

Borophagus pugnator...elMIO-vePLIO

Borophagus orc...vlMIO

Borophagus hilli... vePLIO

Borophagus dudleyi..vePLIO

Borophagus diversidens..lPLIO

Subfamily CANINAE

Tribe indet.

Leptocyon sp... ?lMIO

Tribe VULPINI [foxes]

Vulpes sp...lMIO-vePLIO, lPLEIST

new genus and sp. ....vlMIO

Urocyon sp...ePLEIST

Urocyon minicephalus.. mPLEIST

Urocyon cinereoargenteus ....mPLEIST-lPLEIST
Tribe CANINI [dogs, wolves]
    Eucyon davisi ....?vePLIO
    Canis lepophagus ....lPLIO
    Canis edwardii .... vlPLIO-ePLEIST
    Canis armbrusteri... ePLEIST-mPLEIST
    Canis dirus ....lmPLEIST-lPLEIST
    Canis rufus .... lPLEIST
    Canis latrans[43] .... lPLEIST
    Canis familiaris .... lPLEIST
Infraorder ARCTOIDEA
  Parvorder URSIDA
   Superfamily AMPHICYONOIDEA
    Family AMPHICYONIDAE [bear-dogs]
     Subfamily DAPHOENINAE
      Daphoenus sp...?eOLIG
      Daphoenodon notionastes...veMIO
     Subfamily TEMNOCYONINAE
      Mammacyon obtusidens.. ?veMIO
     Subfamily AMPHICYONINAE
      Amphicyon longiramus[44] ....leMIO
      Amphicyon pontoni...leMIO
      Cynelos caroniavorus[45] ....leMIO
      Pliocyon robustus.. lmMIO
      Ischyrocyon gidleyi...elMIO
   Superfamily URSOIDEA
    Family HEMICYONIDAE
     Phoberocyon johnhenryi[46] ....leMIO
    Family URSIDAE [bears]
     Subfamily indet.
      Indarctos oregonensis..vlMIO
      Agriotherium schneideri .... vePLIO
     Subfamily TREMARCTINAE
      Plionarctos sp.. vePLIO
      Arctodus pristinus .... lPLIO-emPLEIST
      Tremarctos floridanus...lmPLEIST-lPLEIST
     Subfamily URSINAE
      Ursus americanus .... lPLEIST
      Ursus arctos ....?vlPLEIST
   Superfamily PHOCOIDEA
    Family PHOCIDAE [true or earless seals]
     Subfamily PHOCINAE
      genus and sp. indet... lmMIO
      indet. small phocine.. vePLIO
      Phocanella pumila .... vePLIO
     Subfamily MONACHINAE
      Callophoca obscura..ePLIO-lPLIO
      Monachus tropicalis ....ePLEIST-lPLEIST
    Family ODOBENINAE [walruses]
     Trichecodon huxleyi..?ePLIO

genus and sp. indet. ....ePLEIST
Parvorder MUSTELIDA
    Family MUSTELIDAE [weasels, skunks, otters, etc.]
      Subfamily OLIGOBUNINAE
        Oligobunis floridanus ....veMIO
        Paroligobunis frazieri ....veMIO
      Subfamily LUTRINAE [otters]
        Tribe ENHYDRINI
          Enhydritherium terraenovae ....vlMIO-vePLIO
        Tribe LUTRINI
          Satherium piscinarium ....lPLIO
          Lutra canadensis .... ePLEIST-lPLEIST
      Subfamily MEPHITINAE [skunks]
        genus and sp. indet. .... vePLIO
        Spilogale putorius .... ePLEIST-lPLEIST
        Conepatus sp. .... mPLEIST
        Conepatus leuconotus ....mPLEIST-lPLEIST
        Conepatus robustus .... lPLEIST
        Mephitis mephitis .... lPLEIST
      Subfamily MUSTELINAE [weasels, minks]
        Miomustela sp.[47] ....leMIO
        Sthenictis lacota .... ?elMIO
        Plionictis sp. ....elMIO
        Trigonictis macrodon ....lPLIO
        Trigonictis cookii ....ePLEIST
        Mustela frenata .... ePLEIST-lPLEIST
        Mustela vison .... lPLEIST
      Subfamily LEPTARCTINAE
        Leptarctus ancipidens ....leMIO
        Leptarctus progressus ....mMIO?
        Leptarctus n. sp. ....elMIO
      Subfamily MELLIVORINAE
        Beckia sp.[48] ....elMIO
      Subfamily GULONINAE [wolverines]
        Plesiogulo marshalli .... vePLIO
    Family PROCYONIDAE [raccoons, coatis]
        Arctonasua floridana ....elMIO
        Arctonasua eurybates .... vePLIO
        Paranasua biradica ....elMIO
        Nasua sp. ....?vePLIO
        Procyon n. sp. ....lPLIO-ePLEIST
        Procyon lotor ....mPLEIST-lPLEIST
Suborder FELIFORMIA
    Family indet.
        Palaeogale sp. ....?eOLIG
        Stenogale sp.[49] ....leMIO
    Family NIMRAVIDAE [false saber-tooths or paleofelids]
        genus and sp. indet. ....veMIO
        Barbourofelis whitfordi ....lmMIO
        Barbourofelis loveorum[50] ....elMIO

Family FELIDAE [true cats]
 Subfamily FELINAE [cats]
  Nimravides galiani[51] .... elMIO
  Lynx rexroadensis .... vePLIO-lPLIO
  Lynx rufus.. ePLEIST-lPLEIST
  Puma concolor .... vlPLEIST
  Leopardus pardalis... lPLEIST
  Leopardus amnicola[52] .... lPLEIST
 Subfamily PANTHERINAE [roaring cats]
  Panthera onca .... mPLEIST-lPLEIST
  Panthera atrox .... vlPLEIST
 Subfamily ACINONYCHINAE [cheetahs]
  Miracinonyx inexpectatus .... lPLIO-ePLEIST, ?mPLEIST
  Miracinonyx trumani .... ?lPLEIST
 Subfamily MACHAIRODONTINAE [saber-toothed cats]
  Tribe MACHAIRODONTINI
   Machairodus sp. .... vlMIO-vePLIO
   Xenosmilus hodsonae[53] .... lPLIO-ePLEIST
   Dinobastis serus .... lPLEIST
  Tribe SMILODONTINI
   Megantereon hesperus .... vePLIO
   Smilodon gracilis .... lPLIO-emPLEIST
   Smilodon fatalis.. lmPLEIST-lPLEIST
 Family HYAENIDAE [hyenas]
  Chasmaporthetes ossifragus .... ?lPLIO
Grandorder LIPOTYPHLA [insectivores]
 Order ERINACEOMORPHA
  Superfamily ERINACEOIDEA
   Family ERINACEIDAE [hedgehogs]
    Amphechinus sp. .... veMIO
    Lanthanotherium sp. .... emMIO
  Superfamily TALPOIDEA
   Family TALPIDAE [moles]
    genus and sp. indet.. leMIO-lMIO
    Scalopoides sp. .... ?leMIO
    Scalopus aquaticus .... lPLIO-lPLEIST
 Order SORICOMORPHA
  Superfamily, family, genus and sp. indet.. eOLIG
  Superfamily SORICOIDEA
   Family SORICIDAE [shrews]
    Subfamily, genus and sp. indet..eMIO-lMIO
   Subfamily LIMNOECINAE
    Limnoecus n. sp. .... leMIO
   Subfamily SORICINAE
    Cryptotis parva .... lPLIO-lPLEIST
    Blarina carolinensis[54] .... ePLEIST-lPLEIST
    Sorex longirostris .... lPLEIST
Grandorder ARCHONTA
 Order CHIROPTERA [bats]
  Suborder MICROCHIROPTERA

Family EMBALLONURIDAE [sac-winged bats]
    2 new genera and spp. ....lOLIG
    new genus and sp. ....eMIO
Infraorder YANGOCHIROPTERA
  Superfamily NOCTILIONOIDEA
    Family MORMOOPIDAE [ghost-faced bats]
      new genus and sp. ....lOLIG
      Mormoops megalophylla... lPLEIST
      Pteronotus pristinus ....?lPLEIST
    Family PHYLLOSTOMIDAE [American leaf-nosed bats]
      Subfamily DESMODONTINAE [vampire bats]
        Desmodus archaeodaptes ....ePLEIST
        Desmodus stocki[55] .... lPLEIST
  Superfamily VESPERTILIONOIDEA
    Family MOLOSSIDAE [free-tailed and mastiff bats]
      Tadarida sp. ....lPLIO
      Tadarida brasiliensis.. lPLEIST
      Eumops glaucinus[56] .... ?lPLIO, lPLEIST
      Eumops underwoodi .... lPLEIST
    Family NATALIDAE [funnel-eared bats]
      genus and sp. indet. ....lOLIG
      new genus and sp. ....eMIO
    Family VESPERTILIONIDAE [common bats]
      new genus and sp. ....lOLIG
      genus and sp. indet. ....eMIO-lMIO
      Tribe MYOTINI
        Miomyotis floridanus ....leMIO
        Suaptenos whitei ....leMIO
        Myotis sp..ePLEIST
        Myotis austroriparius ....ePLEIST-lPLEIST
        Myotis grisescens.. lPLEIST
      Tribe PLECOTINI
        Plecotus rafinesquii.. mPLEIST
      Tribe VESPERTILIONINI
        Pipistrellus subflavus ....ePLEIST-mPLEIST
        Eptesicus fuscus.. lPLEIST
      Tribe NYCTICEINI
        Nycticeius humeralis .... lPLEIST
      Tribe LASIURINI
        Lasiurus intermedius .... lPLEIST
        Lasiurus borealis .... lPLEIST
        Lasiurus seminolus...?lPLEIST
      Tribe ANTROZOINI
        Antrozous sp. ....vlPLIO
Order PRIMATES [lemurs, monkeys, apes]
  Parvorder ANTHROPOIDEA
    Family HOMINIDAE [humans]
      Homo sapiens..vlPLEIST
Grandorder UNGULATA
  Mirorder EPARCTOCYONA

Order ARTIODACTYLA [artiodactyls]
  Suborder SUIFORMES (=BUNODONTIA)
    Superfamily SUOIDEA
      Family TAYASSUIDAE [peccaries]
        genus and sp. indet...lOLIG, veMIO, emMIO, vlMIO
        Floridachoerus olseni[57] ....leMIO
        new genus and sp. ....elMIO
        Mylohyus longirostris..elMIO
        Mylohyus elmorei.. vePLIO
        Mylohyus floridanus..lPLIO
        Mylohyus fossilis[58] ....ePLEIST-lPLEIST
        Catagonus brachydontus .... vePLIO
        Platygonus bicalcaratus ....lPLIO
        Platygonus vetus..ePLEIST
        Platygonus cumberlandensis .... mPLEIST
        Platygonus compressus .... lPLEIST
        Tayassu sp.[59] ....?lPLEIST
    Superfamily OREODONTOIDEA[60]
      Family OREODONTIDAE (=MERYCOIDODONTIDAE) [oreodonts]
        Subfamily MERYCOCHERINAE
          Eporeodon occidentalis .... eOLIG
          Merycochoerus chelydra...veMIO
        Subfamily PHENACOCOELINAE
          Oreodontoides oregonensis .... eOLIG
          Merycoides harrisonensis..veMIO
        Subfamily TICHOLEPTINAE
          Merychyus elegans..leMIO
          Ticholeptus zygomaticus ....?emMIO
    Superfamily ENTELODONTOIDEA [extinct giant "hogs"]
      Family ENTELODONTIDAE
        Daeodon sp.[61] ....eMIO
  Suborder SELENODONTIA
    Infraorder TYLOPODA
      Superfamily CAMELOIDEA
        Family CAMELIDAE [camels, llamas]
          Subfamily AEPYCAMELINAE [giraffe camels]
            genus and sp. indet...veMIO
            very small new genus and sp...veMIO
            Oxydactylus sp...veMIO
            Nothokemas waldropi...veMIO
            Nothokemas floridanus[62] ....leMIO
            Aepycamelus major ....elMIO
          Subfamily FLORIDATRAGULINAE
            Floridatragulus dolichanthereus[63] ....leMIO
            Floridatragulus barbouri ....leMIO
          Subfamily CAMELINAE
            Tribe LAMINI [llamas]
              Hemiauchenia minima...elMIO
              Hemiauchenia n. sp.. vePLIO
              Hemiauchenia blancoensis...lPLIO

Hemiauchenia n. sp. (dwarf form) .... vlPLIO
Hemiauchenia macrocephala[64] .... vlPLIO-lPLEIST
Palaeolama mirifica .... ePLEIST-lPLEIST
Tribe CAMELINI [camels]
   Procamelus sp. .... lmMIO
   Procamelus grandis .... elMIO
   Megatylopus sp. .... vePLIO
Superfamily PROTOCERATOIDEA
  Family PROTOCERATIDAE [protoceratids]
    genus and sp. indet. .... eOLIG
    Prosynthetoceras texanus[65] ....eMIO
    Synthetoceras tricornatus ....elMIO
    Kyptoceras amatorum .... vePLIO
Infraorder RUMINANTIA
  Superfamily TRAGULOIDEA [chevrotains, mouse deer]
   Family HYPERTRAGULIDAE
    Nanotragulus sp. .... eOLIG
    Nanotragulus loomisi ....veMIO
   Family LEPTOMERYCIDAE
    genus and sp. indet. .... lOLIG-eMIO
  Superfamily GELOCOIDEA
   Family GELOCIDAE [extinct hornless ruminants]
    genus and sp. indet. .... mMIO
    Pseudoceras sp. ....elMIO
    new genus and sp. ....vlMIO
  Superfamily CERVOIDEA
   Family MOSCHIDAE [musk deer]
    Subfamily BLASTOMERYCINAE
     Parablastomeryx floridanus ....leMIO
     Machaeromeryx gilchristensis ....leMIO
   Family ANTILOCAPRIDAE [pronghorns]
    Subfamily COSORYCINAE
     genus and sp. indet. ....?mMIO
    Subfamily ANTILOCAPRINAE
     genus and sp. indet. ....elMIO
     Hexobelomeryx simpsoni[66] .... vePLIO
     Subantilocapra garciae .... vePLIO
     Capromeryx arizonensis ....lPLIO
   Family PALAEOMERYCIDAE
    Subfamily DROMOMERYCINAE [dromomerycines]
     genus and sp. indet. ....leMIO
     Rakomeryx sp. ....emMIO
     Bouromeryx americanus ....?emMIO
     Cranioceras sp. .... lmMIO
     Pediomeryx hamiltoni ....elMIO
     Pediomeryx hemphillenisis ....vlMIO
   Family CERVIDAE [deer, elk, moose]
    Subfamily ODOCOILINAE
     Eocoileus gentryorum .... vePLIO
     Odocoileus virginianus ....lPLIO-lPLEIST

Blastocerus extraneus[67] .... lPLEIST
    Superfamily BOVOIDEA
      Family BOVIDAE [cattle, goats, antelope]
        Subfamily BOVINAE [cattle, bison]
          Bison sp. .... lPLIO
          Bison latifrons .... mPLEIST
          Bison antiquus .... lPLEIST
Order CETE
  Suborder CETACEA [whales, dolphins, porpoises]
    Infraorder ARCHAEOCETI [archaic toothed whales]
      Family BASILOSAURIDAE
        Zygorhiza kochii .... lEOC
        Pontogeneus brachyspondylus[68] .... lEOC
        Basilosaurus cetoides .... lmEOC-lEOC
    Infraorder AUTOCETI
      Parvorder MYSTICETI [baleen whales]
        Family CETOTHERIIDAE [extinct small baleen whales]
          genus and sp. indet. .... leMIO
          3 genera and 4 spp. .... mMIO
          genus and sp. indet. .... vlMIO
        Family ESCHRICHTIIDAE [gray whales]
          Eschrichtius robustus .... PLEIST
        Family BALAENIDAE [right whales]
          genus and sp. indet. .... vlMIO
          Eubalaena sp. .... vlPLIO
          Eubalaena glacialis .... PLEIST
        Family BALAENOPTERIDAE [rorquals, humpbacks]
          genus and sp. indet. .... vlMIO
          Megaptera novaeangliae .... PLIO-PLEIST
          Balaenoptera physalus .... PLIO?
          Balaenoptera floridana .... vePLIO-lPLIO
          Balaenoptera sp. .... vePLIO, PLEIST
      Parvorder ODONTOCETI [toothed whales, dolphins]
        Superfamily indet.
        Family INIIDAE [river dolphins]
          Goniodelphis hudsoni .... vePLIO
         Family PONTOPORIIDAE [river dolphins]
           genus and sp. indet. .... vePLIO
        Superfamily PLATANISTOIDEA
        Family PLATANISTIDAE [long-beaked dolphins]
          large genus and sp. indet.[69] .... mMIO
          Pomatodelphis sp. .... leMIO
          Pomatodelphis inaequalis .... mMIO-elMIO
          Pomatodelphis bobengi[70] .... mMIO-elMIO
        Superfamily DELPHINOIDEA
        Family KENTRIODONTIDAE [extinct dolphins]
          Lophocetus sp. .... mMIO
          Hadrodelphis sp. .... ?mMIO-lMIO
          Delphinodon mento .... ?mMIO
         Family DELPHINIDAE [dolphins]

genus and sp. indet. 1 .... vePLIO

Globicephala baereckeii ....PLEIST?

genus and sp. indet. 2 ....ePLEIST

Stenella sp. ....ePLEIST

Tursiops sp. ....ePLEIST

Pseudorca crassidens ....ePLEIST

Superfamily PHYSETEROIDEA

  Family PHYSETERIDAE [sperm whales]

    Scaldicetus sp. .... mMIO

    Physeterula sp. ....vlMIO-lPLIO

    Kogiopsis floridana... vePLIO

    Physeter catodon .... PLEIST

Superfamily ZIPHIOIDEA

  Family ZIPHIIDAE [beaked whales]

    genus and sp. indet. .... mMIO

    Mesoplodon longirostris..emMIO

    Mesoplodon sp.. vePLIO

    Ninoziphius platyrostris .... vePLIO

    Ziphius cavirostris.. PLEIST

Mirorder ALTUNGULATA

  Order PERISSODACTYLA [perissodactyls]

    Suborder HIPPOMORPHA

      Superfamily EQUOIDEA

        Family EQUIDAE [horses]

        Unnamed subfamily

          Miohippus sp.. eOLIG

        Subfamily ANCHITHERIINAE

          Anchitherium clarencei ....leMIO-emMIO

          Hypohippus chico ....emMIO

          Hypohippus affinis ....?mMIO

        Subfamily EQUINAE

        Unnamed tribe or tribes

          Archaeohippus n. sp. ....vlOLIG

          Archaeohippus blackbergi...leMIO-emMIO

          Parahippus sp...eMIO-mMIO

          "*Parahippus*" *leonensis* ....leMIO

          "*Merychippus*" *gunteri*...leMIO-emMIO

          "*Merychippus*" *primus* ....?emMIO

        Tribe HIPPARIONINI

          Acritohippus isonesus..?emMIO

          Merychippus brevidontus ....?emMIO

          Merychippus californicus..?lmMIO

          "*Merychippus*" *goorisi* ....emMIO

          Pseudhipparion n. sp. ....emMIO

          Pseudhipparion curtivallum... lmMIO

          Pseudhipparion skinneri ....elMIO-vlMIO

          Pseudhipparion simpsoni.. vePLIO

          Neohipparion trampasense...lmMIO-elMIO

          Neohipparion eurystyle[71] ....vlMIO-vePLIO

          Hipparion tehonense ....?lmMIO-vlMIO

Aphelops sp...leMIO-mMIO

Aphelops malacorhinus ....elMIO

Aphelops mutilus...vlMIO-vePLIO

Peraceras hessei .... mMIO

Subtribe RHINOCEROTINA

Teleoceras medicornutum.. mMIO

Teleoceras proterum ....elMIO

Teleoceras hicksi.. vePLIO

Superfamily TAPIROIDEA

Family TAPIRIDAE [tapirs]

Miotapirus sp. .... ?eMIO

Tapirus polkensis[76] .... lMIO-vePLIO

Tapirus simpsoni .... lMIO

Tapirus sp. or spp. .... lMIO-vePLIO

Tapirus n. sp...lPLIO

Tapirus haysii[77] .... lPLIO-lePLEIST

Tapirus veroensis...mPLEIST-lPLEIST

Order URANOTHERIA

Suborder TETHYTHERIA

Infraorder BEHEMOTA

Parvorder PROBOSCIDEA [proboscideans]

Superfamily MAMMUTOIDEA

Family MAMMUTIDAE [mastodonts]

Zygolophodon tapiroides .... ?emMIO, lmMIO

Mammut sellardsi[78] .... vePLIO

Mammut americanum ....lPLIO-lPLEIST

Superfamily ELEPHANTOIDEA

Family GOMPHOTHERIIDAE [gomphotheres]

Subfamily GOMPHOTHERIINAE

Tribe GOMPHOTHERIINI

Gomphotherium calvertense...?mMIO

Gomphotherium sp. ....elMIO

Gomphotherium simplicidens................................ vePLIO, ?lPLIO

Tribe AMEBELODONTINI [shovel-tuskers]

Amebelodon floridanus ....elMIO

Amebelodon britti .... vlMIO, vePLIO?

Platybelodon sp...vlMIO

Subfamily RHYNCHOTHERIINAE

Rhynchotherium simpsoni[79] .... ?vlMIO, vePLIO

Rhynchotherium praecursor ....lPLIO

Cuvieronius tropicus ....lPLIO-lmPLEIST

Family ELEPHANTIDAE [elephants, mammoths]

Mammuthus haroldcooki[80] .... ePLEIST

Mammuthus columbi[81] ....mPLEIST-lPLEIST

Infraorder SIRENIA [sea cows, manatees, dugongs]

Family PRORASTOMIDAE

genus and sp. indet.. mEOC-lEOC

Family PROTOSIRENIDAE

Protosiren sp. .... mEOC

Family DUGONIDAE [dugongs]

Subfamily HALITHERIINAE

Metaxytherium sp. .... vlOLIG

Metaxytherium crataegense .... leMIO-mMIO

Metaxytherium floridanum .... mMIO-lMIO

Subfamily DUGONGINAE

Undescribed small dugongid .... mMIO

Crenatosiren olseni .... vlOLIG-veMIO

Dioplotherium manigaulti .... vlOLIG-mMIO

Dioplotherium allisoni .... mMIO

Corystosiren varguezi .... ePLIO

Undescribed genus and sp.[82] .... lPLIO

Family TRICHECHIDAE [manatees]

Trichechus sp.. lPLIO?, ePLEIST-lPLEIST

Trichechus manatus .... lPLEIST

NOTES

1. Includes records of *Ginglymostoma serra*, which is an Eocene species. Some Florida fossils may represent the genus *Nebrius*, but definitive study is needed to determine this. Scudder et al. (1995) recorded this extinct species from Leisey Shell Pit (along with the extant *G. cirratum*) but noted that the specimens were possibly reworked from Miocene strata.

2. Placed in the genus *Striatolamna* by some workers on fossil sharks.

3. The questionable early Pleistocene record of this species, from the Leisey Shell Pit, probably represents reworking from older, Miocene strata (Scudder et al., 1995).

4. The generic allocation of this species of shark, and the following two taxa, is controversial. Traditionally, they are regarded as extinct relatives of the modern great white shark (*Carcharodon carcharias*) and included in the genus *Carcharodon*. Many workers on fossil sharks, however, place them in a separate evolutionary lineage, together with *Otodus*, and do not regard them as especially closely related to *C. carcharias*. Under this evolutionary scenario, these species are assigned to an extinct genus, *Carcharocles* (see chap. 4 for detailed discussion and references). *Carcharodon sokolowi* includes records formerly listed as *Carcharodon auriculatus*.

5. The bullhead catfishes are now placed in a separate genus, *Ameiurus*, rather than in *Ictalurus*.

6. Sometimes classified in the extinct family Blochiidae.

7. This record is based on the preliminary faunal list of the Cutler Hammock local fauna, Dade County, published by Emslie and Morgan (1995). This genus is not native to Florida (it ranges from Argentina to Mexico and the West Indies), but two species have been introduced by humans into Florida. Further work needs to be done to confirm the presence of this taxon in the Florida fossil record.

8. The name for the leopard frog in Florida has been changed several times over the years. Older published reports included it in *Rana pipiens*, a name now reserved for northern populations. Another name applied to the same species is *Rana sphenocephala*. A. E. Pratt (pers. comm.) considers the early Miocene records of leopard frogs from Thomas Farm unjustifiable.

9. Recently DeBraga and Rieppel (1997) suggested that the Testudines should be removed from the Anapsida and placed within the Diapsida close to the Lepidosauria. The traditional allocation of the group to the Anapsida is retained until this new hypothesis can be further tested. See also Lee (1997).

10. Meylan (1987) concluded that North American softshelled turtles warranted distinct generic status from the European *Trionyx* and revived the old name *Apalone* for them.

11. At various times herpetologists have recognized 1, 2, or 3 genera for the *Pseudemys*-group of emydid turtles. The opinion of Seidel and Smith (1986) that all three, *Chrysemys*, *Pseudemys*, and *Trachemys* are distinct genera is followed here. In the original Plaster Jacket on turtles, species of all three were placed in *Chrysemys*.

12. Gaffney and Meylan (1988) and Meylan (1995) accorded *Hesperotestudo* full generic rank. Previously it was usually considered a subgenus of *Geochelone*. Under this new classification, there are two recognized tortoise lineages in Florida, both recognized as subgenera: *Hesperotestudo* (*Hesperotestudo*) and *Hesperotestudo* (*Caudochelys*).

13. The allocation of iguanian lizards to families follows the revision of Frost and Etheridge (1989).

14. Kluge (1988) revived this species from its synonymy with *Pseudoepicrates stanolseni*, and referred it to the extant genus *Boa*. He furthermore stated that it was very close, if not conspecific, with the living *B. constrictor*.

15. In some older publications, the species here referred to

Nerodia and *Regina* were placed in *Natrix*, which is now limited to Old World water snakes.

16. The presence of the South American crocodylid *Charactosuchus* in Florida and South Carolina, always a tentative identification, was refuted by Langston and Gasparini (1997). See chapter 7.

17. The sequence of avian "orders" used here is traditional and not phylogenetically accurate. See Olsen (1985) and Cracraft (1988) for two highly divergent opinions on the higher relationships of bird families and orders.

18. According to Emslie (1998), the early records of *Phalacrocorax idahensis* from the Bone Valley are incorrectly assigned to this species (they belong to some other extinct, large cormorant), but it is present at Inglis 1C.

19. This species was transferred from *Phalacrocorax* to *Anhinga* by Becker (1986). It is the oldest known record of the family.

20. These Pliocene records of fossil condors may represent *Aizenogyps toomeyae* instead of *Gymnogyps* but are too incomplete to be certain (Emslie, 1998).

21. This fossil species of duck is known only from a single specimen, and Campbell (1980) was equivocal regarding its validity.

22. The fossil species *Falco readei* Brodkorb 1959 was transferred from *Falco* to *Milvago* by Campbell (1980). Emslie (1998) later synonymized it with the living South American species *Milvago chimachima*, but recognized the Florida form as a valid subspecies.

23. This extinct genus and species was originally placed in its own family and then in the Jacanidae. It was later placed in the Phasianidae and possibly in the subfamily Meleagridinae (Steadman, 1980).

24. Includes records of *Colinus suilium*, based on Steadman (1984) and Emslie (1998).

25. Includes records of *Porzana auffenbergi*, based on Olson (1974).

26. Includes records of *Gallinula brodkorbi*, based on Olson (1974), although Campbell (1980) suggested that both might be valid.

27. Includes records of *Laterallus guti*, based on Olson (1974).

28. Includes records of *Dorypaltus prosphatus* Brodkorb, according to Emslie (1998).

29. Includes records of *Protocitta dixi* Brodkorb, according to Emslie (1998).

30. Steadman (1981) invalidated this species because its holotype (and only known specimen) is not diagnostic. He also regarded the genus *Palaeostruthus* as a junior synonym of the extant *Ammodramus*, but did not believe that the fossil was complete enough to be confidently placed in that genus either. For now it seems best to leave this record standing only to indicate the presence of some type of sparrow in the late Miocene of Florida.

31. McKenna and Bell (1997) placed this genus and species in the family Glyptodontidae, subfamily Glyptatelinae. Downing and White (1995) originally described it as an aberrant armadillo, without familial allocation, and this designation is retained here.

32. McDonald (1995) discussed the confused nomenclature of mylodont sloths and concluded that *Paramylodon*, rather than *Glossotherium*, was the most appropriate genus for this species.

33. The more familiar species names *Eremotherium mirabile* and *Eremotherium rusconii* were synonymized with *E. laurillardi* by Cartele and De Iuliis (1995). Also probably includes records of *Megatherium hudsoni*, which is a dubious name that is based on an isolated claw from the Bone Valley of Florida. The holotype of *M. hudsoni* is nondiagnostic because it could belong to either of the known species of *Eremotherium* from Florida.

34. See McKenna and Bell (1997, 105) for preference of this name over Glires.

35. McKenna and Bell (1997) did not subdivide the Leporidae

into subfamilies, but the traditional allocation of the leporid genera into three subfamilies is used in the checklist.

36. This species was originally assigned to the genus *Cryptopterus,* which is now regarded as the junior synonym of *Miopetaurista* (e.g., McKenna and Bell, 1997).

37. The Florida mouse has traditionally been allocated to the genus *Peromyscus* in the subgenus *Podomys.* Mammalogists now recognize *Podomys* as a distinct genus.

38. The voles, lemmings, and relatives were often placed in the subfamily Microtinae (or family Microtidae) by many North American systematists. However, the family-group name Arvicolinae has priority and is now in general use.

39. Includes records of *Cynodesmus nobilis, Paradaphoenus tropicalis,* and *Parictis bathygenus. Cynodesmus iamonensis* was referred to *Osbornodon* and the Hesperocyoninae by Wang (1994).

40. Classification of borophagine canids follows the recent revision of Wang et al (1999).

41. Includes records of *Tomarctus thomasi, Nothocyon insularis,* and *Tomarctus canavus.*

42. Includes records of *Epicyon validus,* following Baskin (1998b).

43. Includes records of *Canis riviveronis.*

44. Includes records of *Amphicyon intermedius.*

45. Includes records of *Parictis* or *Absonodaphoenus bathygenus.*

46. Formerly *Hemicyon (Phoberocyon) johnhenryi;* Hunt (1998) restored *Phoberocyon* to full generic rank.

47. According to Baskin (1998a), this genus *is* present at Thomas Farm, but the specimens figured and questionably referred to *Miomustela* by Olsen (1956) do *not* belong in this genus and instead are close to the Old World aeluroid *Stenogale.*

48. Baskin (1998a) refers this record to *Hoplictis* sp., but I am here following McKenna and Bell (1997), who listed both *Beckia* and *Hoplictis* as valid and restricted the range of *Hoplictis* to Europe.

49. Includes records of ?*Miomustela* by Olsen (1956), following Baskin (1998a).

50. Spelling of this species name was amended by Hulbert (1992) to follow the code of zoological nomenclature.

51. Placed in the subfamily Felinae by McKenna and Bell (1997); it was formerly classified in the Machairodontinae by Baskin (1981).

52. Regarded as an extinct subspecies of *"Felis" wiedii* by Werdelin (1985).

53. *Xenosmilus hodsonae* was described from the Haile 21A locality and is known only from Florida. Other early Pleistocene Florida records of *Homotherium* may also represent *Xenosmilus,* but this remains to be investigated.

54. Includes Florida records of *Blarina brevicauda.*

55. Includes records of *Desmodus magnus* Gut, now considered a junior synonym.

56. Includes published records of *Molossides floridanus,* now considered a subspecies of *Eumops glaucinus.*

57. *Floridachoerus* was listed as a synonym of *Hesperhyus* by McKenna and Bell (1997, 397) but regarded as valid by Wright (1998).

58. Other species names for Pleistocene *Mylohyus* have been used for Florida specimens, including *M. nasutus, M. gidleyi, M. browni, M. pennsylvanicus, M. lenis, M. tetragonus,* and *M. exortivus.*

59. Most early references of Florida fossil material to the living genus *Tayassu* are clearly in error and instead apply to *Mylohyus. Tayassu* is retained in this list based on the record from Melbourne (Gazin, 1950), which has not been specifically refuted in the literature. However, it is unlikely that this genus really occurred in Florida.

60. The Oreodontoidea are shown classified in the Suiformes (as in McKenna and Bell, 1997), but many paleontologists would classify them within the Tylopoda instead. The names for oreodont subfamilies, genera, and species used here follow Lander (1998) and differ in most cases from those listed in older literature.

61. This genus is now regarded as the senior synonym of the more familiar generic names *Dinohyus* and *Ammodon.*

62. Includes records of *Paratylopus grandis.*

63. Includes records of *Hypermekops olseni.*

64. Includes the Leisey Shell Pit sample referred to *Hemiauchenia seymourensis* by Webb and Stehli (1995).

65. Includes records of *Syndyoceras australis* and *Synthetoceras douglasi* from Thomas Farm and *Miolabis* cf. *tenuis* from Midway.

66. *Hexobelomeryx* is regarded as the senior synonym of *Hexameryx* following Simpson (1945), although McKenna and Bell (1997) listed both genera. The point is moot until a thorough phylogenetic analysis of the family is published.

67. *Blastocerus* is a genus of deer otherwise known only from South America. This Florida species is based on a single jaw. Some workers prefer to regard this specimen as an aberrant individual of *Odocoileus* that happens to look like *Blastocerus,* because no additional specimens have turned up in the last half century.

68. Many records of Eocene archaeocete whales from Florida have been uncritically referred to the genus *Basilosaurus.* Examination of diagnostic elements shows that most instead belong to the smaller *Zygorhiza,* which lacks the very elongated vertebrae of its larger relative. Other, larger specimens but without elongated vertebrae properly belong in the genus *Pontogeneus. Basilosaurus* appears to have been relatively rare in Florida.

69. Morgan (1986) demonstrated that the holotype of *Megalodelphis magnidens* was actually part of a crocodile, so this genus and species can no longer be applied to the giant long-beaked dolphin of the middle Bone Valley. Not enough material is known for it to be described or accurately assigned to a family.

70. Morgan (1994) synonymized *Schizodelphis depressus* with *Schizodelphis bobengi* and transferred it to the genus *Pomatodelphis.*

71. Includes records of *Neohipparion phosphorum.*

72. Includes records of the invalid name *Nannippus minor.*

73. Includes records of *Nannippus phlegon.*

74. Includes records formerly allocated to *Cormohipparion sphenodus* based on Woodburne (1996).

75. Late Pliocene (Blancan) *Equus* from Florida has often been referred to the Western species *E. simplicidens,* but the specimens are significantly smaller than typical material of that species and more critical comparisons are needed.

76. In the original description of this species, and in other general papers, S. J. Olsen implied that this species was a member of one of the older (i.e., middle Miocene) Bone Valley faunas and not the typical Upper Bone Valley or Palmetto Fauna. The type specimens lack any stratigraphic information. Referred specimens are not known from any of the recently collected in situ Miocene faunas. However, specimens of this species have been collected from early Pliocene deposits in the Bone Valley and from late Miocene sites in the Withlacoochee River. Ongoing research by Hulbert indicates that it is a dwarf *Tapirus* rather than a member of the Miocene genus *Tapiravus.*

77. Includes records of *Tapirus copei.* All published middle and late Pleistocene records of *T. haysii* or *T. copei* from Florida are now regarded as invalid and considered to represent instead large individuals of *Tapirus veroensis.*

78. Following McKenna and Bell (1997), *Pliomastodon* is regarded as a junior synonym of *Mammut,* and this species is referred to *Mammut.*

79. This species name is retained because of its familiarity and better holotype specimen, but Osborn's *"Serridentinus"*

brewsterensis is probably a senior synonym of *R. simpsoni*. It is the common gomphothere of the Palmetto Fauna.

80. Most specialists agree that the proper generic designation for all American mammoths is *Mammuthus*. Webb and Dudley (1995) referred early Pleistocene *Mammuthus* from North America to the species *M. hayi*, but Webb (pers. comm.) has since decided that this species name is not valid.

81. Includes records of *Mammuthus floridanus* of Osborn.

82. The presence of an undescribed late Pliocene dugongid from the Caloosahatchee Formation in Charlotte County was noted by Morgan (1994). When described, it would be the youngest member of its family from the Western Atlantic.

References

Baskin, J. A. 1981. *Barbourofelis* (Nimravidae) and *Nimravides* (Felidae), with a description of two new species from the late Miocene of Florida. Journal of Mammalogy, 62:122–39.

———. 1998a. Mustelidae. Pp. 152–73 *in* C. Janis, K. M. Scott, and L. L. Jacobs (eds.), *Evolution of Tertiary Mammals of North America*. Volume 1, *Terrestrial Carnivores, Ungulates, and Ungulatelike Mammals*. Cambridge University Press, Cambridge.

———. 1998b. Evolutionary trends in the late Miocene hyena-like dog *Epicyon* (Carnivora, Canidae). Pp. 191–214 *in* Y. Tomida, L. J. Flynn, and L. L. Jacobs (eds.), *Advances in Vertebrate Paleontology and Geochronology*. National Science Museum Monographs, No. 14, Tokyo.

Becker, J. J. 1986. Reidentification of "*Phalacrocorax*" *subvolans* Brodkorb as the earliest record of the Anhingidae. Auk, 103:804–8.

Benton, M. J. 1997. *Vertebrate Palaeontology*. 2d ed. Chapman and Hall, New York, 464 p.

Campbell, K. E. 1980. A review of the Rancholabrean avifauna of the Itchtucknee River, Florida. Natural History Museum of Los Angeles County Contributions in Science, 330:119–29.

Carroll, R. C. 1995. Phylogenetic analysis of Paleozoic choanates. Bulletin du Muséum National d'Historie naturelle de Paris, 4ème série, 17:389–445.

Cartelle, C., and G. De Iuliis. 1995. *Eremotherium laurillardi*: the panamerican late Pleistocene megatheriid sloth. Journal of Vertebrate Paleontology, 15:830–41.

Cracraft, J. 1988. The major clades of birds. Pp. 339–61 *in* M. S. Benton (ed.), *The Phylogeny and Classification of the Tetrapods*, Volume 1, *Amphibians, Reptiles, Birds*. Clarendon Press, Oxford, England.

DeBraga, M., and O. Rieppel. 1997. Reptile phylogeny and the interrelationships of turtles. Zoological Journal of the Linnean Society, 120:281–354.

Domning, D. P. 1994. A phylogenetic analysis of the Sirenia. Pp. 177–89 *in* A. Berta and T. Deméré (eds.), *Contributions in Marine Mammal Paleontology Honoring Frank C. Whitmore, Jr.* San Diego Natural History Society, San Diego.

Downing, K. F., and R. White. 1995. The cingulates (Xenarthra) of Leisey Shell Pit 1A (Irvingtonian), Hillsborough County, Florida. Bulletin of the Florida Museum of Natural History, 37:375–96.

Emslie, S. D. 1998. Avian community, climate, and sea-level changes in the Plio-Pleistocene of the Florida Peninsula. Ornithological Monographs, No. 50, 113 p.

Emslie, S. D., and G. S. Morgan. 1995. Taphonomy of a late Pleistocene carnivore den, Dade County, Florida. Pp. 65–83 *in* D. W. Steadman and J. I. Mead (eds.), *Late Quaternary Environments and Deep History: A Tribute to Paul S. Martin*. The Mammoth Site of Hot Springs, S. Dak., Scientific Papers, vol. 3.

Emslie, S. D., W. D. Allmon, F. J. Rich, J. H. Wrenn, and S. D. de France. 1996. Integrated taphonomy of an avian death assemblage in marine sediments from the late Pliocene of Florida. Palaeogeography, Palaeoclimatology, Palaeoecology, 124:107–36.

Fordyce, R. E., and L. G. Barnes. 1994. The evolutionary history of whales and dolphins. Annual Review of Ecology and Systematics, 22:419–55.

Frost, D. R., and R. Etheridge. 1989. A phylogenetic analysis and taxonomy of iguanian lizards (Reptilia: Squamata). University of Kansas Museum of Natural History Miscellaneous Publication, 81:1–65.

Gaffney, E. S., and P. A. Meylan. 1988. A phylogeny of the turtles. Pp. 157–219 *in* M. S. Benton (ed.), *The Phylogeny and Classification of the Tetrapods*. Volume 1, Amphibians, Reptiles, Birds. Clarendon Press, Oxford, England.

Gauthier, J. A. 1994. The diversification of the amniotes. Pp. 129–59 *in* D. R. Prothero and R. M. Schoch (eds.), *Major Features of Vertebrate Evolution*. The Paleontological Society, Short Courses in Paleontology 7.

Gazin, C. L. 1950. Annotated list of fossil mammals associated with human remains at Melbourne, Fla. Journal of the Washington Academy of Science, 40:397–404.

Hulbert, R. C. 1989. Phylogenetic interrelationships and evolution of North American late Neogene Equinae. Pp. 176–96 *in* D. R. Prothero and R. M. Schoch (eds.), *The Evolution of Perissodactyls*. Oxford University Press, New York.

———. 1992. A checklist of the fossil vertebrates of Florida. Papers in Florida Paleontology, 6:1–35.

Hunt, R. M. 1998. Ursidae and Amphicyonidae. Pp. 174–227 *in* C. Janis, K. M. Scott, and L. L. Jacobs (eds.), *Evolution of Tertiary Mammals of North America*. Volume 1, *Terrestrial Carnivores, Ungulates, and Ungulatelike Mammals*. Cambridge University Press, Cambridge.

Karrow, P. F., G. S. Morgan, R. W. Portell, E. Simons, and K. Auffenberg. 1996. Middle Pleistocene (early Rancholabrean) vertebrates and associated marine and non-marine invertebrates from Oldsmar, Pinellas County, Florida. Pp. 97–133 *in* K. M. Stewart and K. L. Seymour (eds.), *Palaeoecology and Palaeoenvironments of Late Cenozoic Mammals*. University of Toronto Press, Toronto.

Kluge, A. G. 1988. Relationships of the Cenozoic boine snakes *Paraepicrates* and *Pseudoepicrates*. Journal of Vertebrate Paleontology, 8:229–30.

Lander, B. 1998. Oreodontoidea. Pp. 402–25 *in* C. Janis, K. M. Scott, and L. L. Jacobs (eds.), *Evolution of Tertiary Mammals of North America*. Volume 1, *Terrestrial Carnivores, Ungulates, and Ungulatelike Mammals*. Cambridge University Press, Cambridge.

Langston, W., and Z. Gasparini. 1997. Crocodilians, *Gryposuchus*, and the South American gavials. Pp. 113–54 *in* R. F. Kay, R. H. Madden, R. L. Cifelli, and J. J. Flynn (eds.), *Vertebrate Paleontology in the Neotropics: The Miocene Fauna of La Venta, Colombia*. Smithsonian Institution Press, Washington.

Laurin, M., and R. R. Reisz. 1995. A reevaluation of early amniote phylogeny. Zoological Journal of the Linnaean Society, 113:165–223.

———. 1997. A new perspective of tetrapod phylogeny. Pp. 9–59 *in* S. S. Sumida and K. L. M. Martin (eds.), *Amniote Origins: Completing the Transition to Land*. Academic Press, San Diego.

Lee, M. S. Y. 1995. Historical burden in systematics and the interrelationships of "parareptiles." Biological Reviews, 70:459–547.

———. 1997. Pareiasaur phylogeny and the origin of turtles. Zoological Journal of the Linnean Society, 120:197–280.

McDonald, H. G. 1995. Gravigrade xenarthrans from the early

Pleistocene Leisey Shell Pit 1A, Hillsborough County, Florida. Bulletin of the Florida Museum of Natural History, 37:345–73.

McKenna, M. C., and S. K. Bell. 1997. *Classification of Mammals above the Species Level*. Columbia University Press, New York, 631 p.

Meylan, P. A. 1987. The phylogenetic relationships of soft-shelled turtles (Family Trionychidae). Bulletin of the American Museum of Natural History, 186:1–101.

———. 1995. Pleistocene amphibians and reptiles from the Leisey Shell Pit, Hillsborough County, Florida. Bulletin of the Florida Museum of Natural History, 37:273–97.

Morgan, G. S. 1986. The so-called giant Miocene dolphin *Megalodelphis magnidens* Kellogg (Mammalia: Cetacea) is actually a crocodile (Reptilia: Crocodilia). Journal of Paleontology, 60:411–17.

———. 1994. Miocene and Pliocene marine mammal faunas from the Bone Valley Formation of Central Florida. Pp. 239–68 *in* A. Berta and T. A. Deméré (eds.), *Contributions in Marine Mammal Paleontology Honoring Frank C. Whitmore, Jr.* San Diego Natural History Society, San Diego.

Nelson, J. S. 1994. *Fishes of the World*. 3d ed. John Wiley and Sons, New York, 600 p.

Novas, F. E. 1996. Dinosaur monophyly. Journal of Vertebrate Paleontology, 16:723–41.

Olsen, S. J. 1956. A small mustelid from the Thomas Farm Miocene. Breviora, 51:1–5.

Olson, S. L. 1974. The Pleistocene rails of North America. Condor, 76:169–75.

———. 1985. The fossil record of birds. Pp. 79–238 *in* D. S. Farner, J. R. King, and K. C. Palmer (eds.), *Avian Biology*, Volume 8. Academic Press, Orlando, Fla.

Parrish, J. M. 1993. Phylogeny of the Crocodylotarsi, with reference to archosaurian and crurotarsan monophyly. Journal of Vertebrate Paleontology, 13:287–308.

Schultze, H., and L. Trueb. 1991. *Origins of the Higher Groups of Tetrapods*. Comstock Publishing Associates, Ithaca, N.Y., 724 p.

Scudder, S. J., E. H. Simons, and G. S. Morgan. 1995. Osteichthyes and Chondrichthyes from the Leisey Shell Pit local fauna, Hillsborough County, Florida. Bulletin of the Florida Museum of Natural History, 37:251–72.

Seidel, M. E., and H. M. Smith. 1986. *Chrysemys, Pseudemys,*

Trachemys (Testudines: Emydidae): did Agassiz have it right? Herpetologica, 42:242–48.

Simpson, G. G. 1945. The principles of classification and a classification of mammals. Bulletin of the American Museum of Natural History, 85:1–350.

Steadman, D. W. 1980. A review of the osteology and paleontology of turkeys (Aves: Meleagrinae). Natural History Museum of Los Angeles County Contributions in Science, 330:131–207.

———. 1981. A re-examination of *Palaeostruthus hatcheri* (Shufeldt), a late Miocene sparrow from Kansas. Journal of Vertebrate Paleontology, 1:171–73.

———. 1984. A middle Pleistocene (late Irvingtonian) avifauna from Payne Creek, Florida. Pp. 47–52 *in* H. G. Genoways and M. R. Dawson (eds.), *Contributions in Quaternary Vertebrate Paleontology: A Volume in Memorial to John E. Guilday*. Special Publication of the Carnegie Museum of Natural History, No. 8, Pittsburgh.

Wang, X. 1994. Phylogenetic systematics of the Hesperocyoninae (Carnivora: Canidae). Bulletin of the American Museum of Natural History, 221:1–207.

Wang, X., R. H. Tedford, and B. E. Taylor. 1999. Phylogenetic systematics of the Borophaginae (Carnivora: Canidae). Bulletin of the American Museum of Natural History, 243:1–391.

Webb, S. D., and J. Dudley. 1995. Proboscidea from the Leisey Shell Pits, Hillsborough County, Florida. Bulletin of the Florida Museum of Natural History, 37:645–60.

Webb, S. D., and F. G. Stehli. 1995. Selenodont artiodactyls (Camelidae and Cervidae) from the Leisey Shell Pits, Hillsborough County, Florida. Bulletin of the Florida Museum of Natural History, 37:621–43.

Werdelin, L. 1985. Small Pleistocene felines of North America. Journal of Vertebrate Paleontology, 5:194–210.

Woodburne, M. O. 1996. Reappraisal of the *Cormohipparion* from the Valentine Formation, Nebraska. American Museum Novitates, 3163:1–56.

Wright, D. B. 1998. Tayassuidae. Pp. 389–401 *in* C. Janis, K. M. Scott, and L. L. Jacobs (eds.), *Evolution of Tertiary Mammals of North America*. Volume 1, *Terrestrial Carnivores, Ungulates, and Ungulatelike Mammals*. Cambridge University Press, Cambridge.

4

Cartilaginous and Bony Fishes

INTRODUCTION

Vertebrates evolved in aquatic environments and a majority of them still live in either fresh water or salt water. Aquatic vertebrates include both those with a continuous aquatic ancestry (fishes) and those which are secondarily aquatic because they have terrestrial ancestors. The focus of this chapter is only on those vertebrates generally called fishes. Secondarily aquatic vertebrates, such as turtles and whales, are discussed elsewhere. Fossils of fishes are extremely abundant in both freshwater and marine sediments in Florida but have received relatively little scientific inquiry. More remains to be learned about them than any other major vertebrate group.

Fossils of two basic types of fishes have been found in Florida; their scientific names are Chondrichthyes and Actinopterygii. The Chondrichthyes, more informally called the cartilaginous fishes, include the sharks, sawfishes, skates, and rays. Unlike other vertebrates, their internal skeleton lacks true bone. However, it is not purely made of soft, flexible cartilage; a relatively strong outer layer contains thousands of microscopic crystals of calcium phosphate connected by tiny threads of protein. Unless the body is rapidly buried, a dead chondrichthyan's internal skeleton will rapidly decay and disintegrate. The most commonly fossilized elements of chondrichthyans are teeth and mineralized spines, "thorns," and tiny denticles embedded in the skin. Vertebral centra of some species are heavily mineralized and preserve well. According to Nelson (1994), 359 species of sharks and 456 species of skates and rays live in the world today. Most are marine, but many inhabit shallow coastal waters and a few enter fresh water.

The Actinopterygii, informally called the ray-finned bony fishes, are a remarkably numerous and abundant group. They have a bony internal skeleton, so their fossils include a more diverse array of elements than those of chondrichthyans. Fossils of vertebrae, teeth, and spines are very common finds, but so too are bones of the skull and gill region. Few articulated or even associated fish skeletons have been found in Florida; instead, isolated specimens are the norm. This often makes their identification very difficult, as this group is exceedingly diverse. Of the 23,700 living species of actinopterygians, about 42 percent are primarily freshwater forms and 58 percent primarily marine (Nelson, 1994). Of course, most of these do not inhabit Florida waters, but many hundreds do (Briggs, 1958; Hoese and Moore, 1977). The 113 actinopterygian taxa known as fossils listed in chapter 3 represent a small fraction of a total that should exceed 1,000. On a percentage basis, that surely makes fossil bony fish the most poorly known vertebrate group in Florida. A number of factors have contributed to this large gap of knowledge. First, most paleontologists who study fossil bony fish are not especially interested in Neogene and Quaternary species (the ones most often found in Florida). Those interested in using fossils to unravel the interrelationships and phylogeny of bony fishes look to older faunas. Second, the isolated and fragmentary nature of most Florida specimens greatly deters their study. Third, partially because of the first two reasons, no paleontologist specializing in bony fish has ever been located in Florida or set up a research program in the state. Nevertheless, the Florida fossil record holds great potential for insights into the evolution of the fish fauna of the southeastern United States and the Caribbean.

The fossil records of both the Chondrichthyes and Actinopterygii extend well back into the Paleozoic Era. By the Eocene Epoch, when their fossil records begin in Florida, both groups had for the most part already "modernized." Thus all Florida fossil fishes belong to living orders and families and most to living genera. Those interested in the earlier aspects of fish evolution should consult Long (1995) and Maisey (1996).

Nelson's (1994) classification of fishes is used here and in chapter 3. As was discussed in chapter 3, some of the ranks used by Nelson (1994) were adjusted to better correspond with the classifications of other kinds of vertebrates used in this book, but this did not change the essence of his classification. Cappetta (1987) was used to locate extinct genera of Chondrichthyes within the classification of Nelson (1994).

SHARK TOOTH ANATOMY AND NOMENCLATURE

Because there is more general interest in shark teeth than other types of fish fossils, and because they are so abundant in Florida, a detailed discussion regarding their formation, structure, and variation is considered warranted. This information largely follows Cappetta (1987), and his dental terminology is used (fig. 4.1). Additional sources of information are Applegate (1965), Compagno (1988), Kent (1994), and Applegate and Espinosa-Arrubarrena (1996).

The first point that must be mentioned is the high degree of variation observed in tooth shape and size in most species of sharks; that is, most species show some degree of heterodonty (for example, see Hubbell, 1996). The amount of heterodonty varies from species to species, with some showing differences between teeth as profound as those observed in mammals. This means, obviously, that two isolated fossil shark teeth may belong to the same species even if they differ greatly in size and/or shape. Two basic types of heterodonty are observed in a single shark: differences between the upper and lower teeth at a comparable point along the jaw (this is termed dignathic heterodonty); and the differences between teeth observed along a single jaw from front to

back (called monognathic heterodonty). Most species show monognathic heterodonty to some degree, while the presence of dignathic heterodonty depends upon the type of shark (Cappetta, 1987).

Other sources of variation between the teeth of a single species of shark include differences between males and females (sexual dimorphism), differences between individuals of different ages (ontogenetic variation), and differences between individuals from different geographic regions (geographic variation). Of course these factors affect most vertebrate species, not just sharks. They are emphasized here because many fossil collectors try to identify shark teeth by matching them with too limited a comparative sample and assume that all the teeth of a species will be similar. Ontogenetic variation is a particularly vexing problem for paleontologists because some sharks change their tooth shape as they grow and mature. Characteristics that are frequently used to identify fossil shark teeth such as relative proportions of the width and height of the tooth, presence or absence of serrations, or the presence and frequency of lateral cusplets (fig. 4.1) can change dramatically over the lifetime of a shark (Compagno, 1988; Hubbell, 1996).

In those orders of sharks with a high degree of monognathic heterodonty, often several groups of adjacent teeth are of similar shape. Each zone or group of teeth has a separate function. Mammalian dentitions are divided into similar functional groups of teeth, with terms such as incisor and molar used to name them. A comparable terminology for shark dentitions was developed in the early twentieth century by M. Leriche and popular-

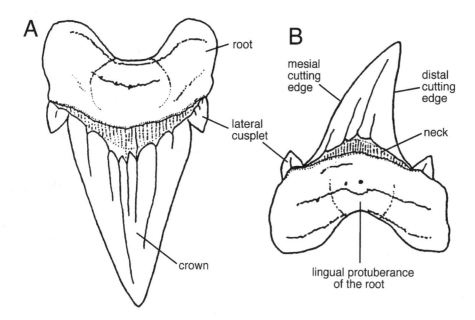

Fig. 4.1. Lingual views of shark teeth showing major morphologic features. Modified from Kent (1994), reproduced by permission of the author and publisher.

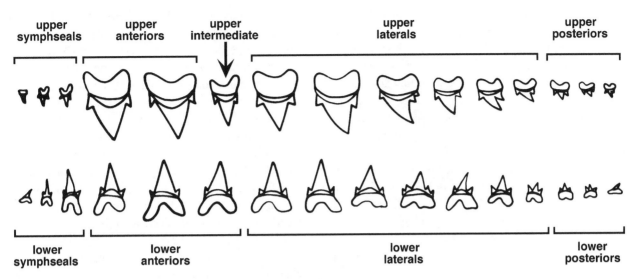

Fig. 4.2. Tooth set of a shark showing the functional tooth groups of Applegate (1965). Figure modified from Applegate and Espinosa-Arrubarrena (1996) in *Great White Sharks*, edited by Klimley and Ainley, copyright © 1996 by Academic Press; reproduced by permission of the publisher.

ized in an influential paper by Applegate (1965). This system is applicable to most members of the orders Lamniformes and Hexanchiformes, and, to a lesser degree, some genera in the Carcharhiniformes (Applegate, 1965; Compagno, 1988). In these taxa, the teeth are divided into five basic groups, which from front to back are termed symphyseal, anterior, intermediate, lateral, and posterior (fig. 4.2). As is the case with mammals, the number of teeth present in each group (if any—symphyseal and intermediate teeth are absent in some species) has systematic value. Unlike mammals, however, extinct sharks are rarely found with articulated dentitions, so the number of teeth in each functional group can be difficult to determine. The large anterior and lateral teeth are those most often collected as fossils. Isolated fossil shark teeth can sometimes be identified as to their original location in the dentition with experience and a good comparative collection of modern jaws.

Directional terms used with shark teeth differ in part from those used for mammals. The terms *labial* and *lingual* have the normal meanings, referring to the outer (external) and internal sides of the tooth, respectively. However, anterior and posterior are not used as directional terms because of the curvature of the jaw and because they are used as the names for functional tooth groups instead (see preceding paragraph). The terms *mesial* and *distal* are used to refer to the sides of the tooth, with mesial indicating the side closer to the symphysis, and distal the side farther away from the symphysis (fig. 4.1B).

A fully grown, functioning shark tooth consists of two regions, the crown and the root (fig. 4.1). Most of the crown has a thin layer of enameloid covering inner layers of dense dentine. The core of the crown may have a hollow pulp cavity or be filled with spongy dentine. The root consists entirely of spongy dentine and often bears foramina for passage of nerves and blood vessels. The boundary between the crown and root is often marked by a groove or region that lacks enameloid (called the neck); if present, the neck is larger on the lingual side of the tooth (Cappetta, 1987). In some sharks (for example, Ginglymostomatidae), a labial protuberance of enameloid, called the apron, extends over the central portion of the root. The tooth crown in modern sharks usually consists of a single, major cusp that has a pointed apex (fig. 4.1). Some species possess paired lateral cusplets that flank the main cusp; in others there are cusplets on the distal side of the cusp only. The distal and mesial sides of the cusp usually bear sharp cutting edges. The cutting edge may run the entire length of the crown (a complete cutting edge), or may be well developed only on the apex of the cusp. The surface of the cutting edge is either smooth or serrated (notched like the blade of a saw or steak knife). The orientation of the cusp may be erect (so that the tooth itself is approximately symmetrical) or pointed obliquely, usually in the distal direction (fig. 4.1B). In general, upper teeth have a greater slant than do lowers, and the amount of inclination increases going in a distal direction to the back of the jaw (fig. 4.2).

The teeth in a shark's jaw form in distinct rows. One

or a few rows may be in use at the same time. Lingual to the functional tooth row(s) are several rows of developing replacement teeth. When a tooth is lost, the next replacement tooth moves labially to take its place. Replacement teeth several rows back of the functional teeth are not fully formed, and if fossilized they consist only of a shell of enameloid without a tooth base. The series of teeth in all of the rows that occupy the same position in their respective rows are collectively referred to as a file. After a few weeks, the gum tissue holding a functional tooth in the jaw dies and the tooth is eventually shed. Given the large number of tooth files each shark has, this quickly adds up to about a thousand teeth produced and shed each year! No wonder fossil shark teeth are so abundant and that Maisey (1996) referred to sharks as "tooth factories."

FOSSIL SHARKS OF FLORIDA

Each type of shark found as a fossil in Florida and published in the scientific literature is described in this section. No doubt other genera and species are present, but they have not been recognized in published studies. Information concerning common names, ecology, feeding preferences, and geographic distributions of modern sharks is from Compagno (1984, 1988) and Bigelow and Schroeder (1948) unless specified otherwise. Chronologic ranges and classification of extinct genera and species generally follow Cappetta (1987) unless they have been revised by subsequent work. Modern sharks often have vast geographic ranges, and it is reasonable to assume the same was true in the geologic past. Therefore, references on fossil sharks from other continents frequently contain pertinent information and illustrations of species found in Florida, and many of these are listed at the end of the chapter.

Cow sharks, family Hexanchidae, are a rare component of fossil shark faunas in Florida, probably because they generally prefer deepwater habitats, depths greater than 90 meters. The broadnose sevengill shark *Notorynchus cepedianus* is an exception, as it frequents shallow coastal waters. *Notorynchus* is no longer present in the North Atlantic, but its fossils are known from both Europe and eastern North America. Because of its more inshore lifestyle, fossils of *Notorynchus* should be more common than those of deepwater cow sharks (*Hexanchus* and *Heptranchias*). Kent (1994) reported this to be the case in the Maryland-Virginia region. The mainstay of the hexanchid diet are other sharks, rays, and bony fish; despite their large size and fierce-looking teeth, there are no reported attacks on humans in open

waters (Compagno, 1984). The teeth of hexanchids have a high degree of both monognathic and dignathic heterodonty (fig. 4.3). With no distinction between anterior and lateral teeth in the lower dentition, they are collectively called anterolaterals (Ward, 1979). These teeth are the most commonly found fossils of cow sharks. They are highly distinctive, being very thin and long, with multiple sharp cusps. The largest cusp, called the acrocone, usually has some serrations on its mesial cutting edge (fig. 4.3B). Lower anterolaterals of *Hexanchus* are very long (up to more than 4 centimeters) and have numerous (usually more than six) accessory conules distal to the acrocone that decrease in height distally. The lower anterolaterals of *Heptranchias* are also very long but bear two large serrations (or cusplets) on the mesial edge of the acrocone and have four to seven accessory conules following the acrocone that first increase in height and then decrease. The anterolaterals of *Notorynchus* are not as long as the previous two (up to 3 centimeters long) and have three to six accessory conules that decrease in height distally. The symmetrical, multi-cusped lower symphyseal tooth of hexanchids (fig. 4.3A) is distinguished from superficially similar teeth in *Ginglymostoma* (see below) by having a taller central cusp or cusps and a flatter or more compressed root. The Bone Valley specimen shown in fig. 4.4 represents *Notorynchus* and probably the widespread Oligocene and Miocene species *Notorynchus primigenius* (Ward, 1979). Specimens of fossil hexanchids collected with good stratigraphic and collecting data would be of scientific importance.

Nurse sharks, family Ginglymostomatidae, are common, nearshore, primarily nocturnal sharks. Their diet consists of crustaceans, mollusks, and small bony fish. Their rarity as fossils is probably due in part to the relatively small size of their squat teeth, which are usually less than a centimeter tall. As was the case with cow sharks, nurse shark teeth are distinctively different from those of "typical" sharks (figs. 4.5, 4.6). All of the teeth of an individual are very similar, except that the central cusp is lower and more inclined distally in posterior teeth. Two genera of nurse sharks have been reported as fossils from the eastern United States, *Ginglymostoma*, which still lives here, and *Nebrius,* which is now a Indo-Pacific taxon. The characteristics that distinguish the teeth of these two genera as listed in Bigelow and Schroeder (1948), Compagno (1984), and Cappetta (1987) are somewhat at odds with each other. The crown of a nurse shark tooth has a strong apron and the central cusp is flanked laterally by a number of lateral cusplets (fig. 4.6). According to Cappetta (1987), the apron of

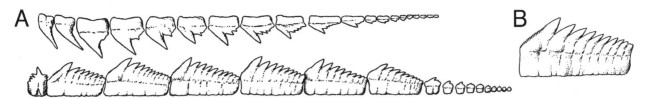

Fig. 4.3. Teeth of a modern, 11-foot-long sixgill shark, *Hexanchus griseus*, from waters off Cuba. *A*, left lateral view of upper and lower tooth rows. About 0.5×. *B*, first lower anterolateral tooth, about 1×. After Bigelow and Schroder (1948); reproduced by permission of the Peabody Museum of Natural History, New Haven, Connecticut.

Nebrius extends farther than that of *Ginglymostoma*, to the level of the base of the root or more. Cappetta (1987) also stated that the central cusp is not much larger than those of the lateral cusplets and that the crown is clearly asymmetrical in *Nebrius*. However, the tooth of the living *Nebrius ferrugineus* illustrated by Compagno (1984, 208) shows neither of these two features. Fossils of two nurse shark species have been found in Florida. The living species *Ginglymostoma cirratum* is known from the early Pleistocene (Scudder et al., 1995) and has also been found in deposits of uncertain age in the Waccasassa River and the Gainesville region (Tessman, 1969). The marine fossils from Gainesville creek beds are typically Miocene (Morgan, 1994). A second species of *Ginglymostoma* is known from the latest Oligocene through the early Pliocene and possibly to the early Pleistocene (Morgan, 1989; Tessman, 1969; Scudder et al., 1995). It was referred to *Ginglymostoma serra* by these authors, apparently following Leriche (1942) in using this as the correct species name for the Neogene North American

nurse shark. However, Cappetta (1987) stated that *G. serra* is an Eocene species, and most of the Florida fossils agree in morphology with his illustrated specimen of *Ginglymostoma delfortriei* (fig. 4.6). Teeth of this extinct species had six to twelve pairs of small lateral cusplets, while the extant *G. cirratum* has a maximum of four. One of the specimens illustrated by Tessman (1969) has the characteristics of *Nebrius* as listed by Cappetta (1987), but it is unclear whether it truly represents this genus or is from the rear of the mouth of a *Ginglymostoma* (which was the interpretation of Tessman [1969]).

Sand tiger sharks, family Odontaspididae, are large (up to about 3.5 meters in length), slow-moving, relatively common species with tall, slender teeth (fig. 4.7). They have a broad habitat but favor coastal waters. Their diet is mainly fish, but those in the genus *Carcharias* have posterior crushing teeth which allow them to also eat hard-shelled prey such as crabs. Two living genera, *Carcharias* and *Odontaspis* are currently recognized, but only one species currently lives in Florida

Fig. 4.4. Labial view of UF 130073, a left lower anterolateral tooth of *Notorynchus primigenius* from Brewster Mine, Polk County, Miocene or Pliocene. Note the smaller number of accessory conules than in *Hexanchus* (fig. 4.3B). About 2×.

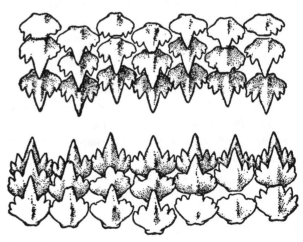

Fig. 4.5. Partial upper and lower tooth rows of a modern 2-foot-long nurse shark, *Ginglymostoma cirratum*, from waters off Cuba. About 5×. After Bigelow and Schroder (1948); reproduced by permission of the Peabody Museum of Natural History, New Haven, Connecticut.

Fig. 4.6. Labial view of UF 17925, tooth of *Ginglymostoma delfortriei* from Palmetto Mine, Polk County, probably Miocene. About 2×.

waters, *Carcharias taurus.* The generic name *Carcharias* has had a tortuous history and for about twenty-five years, from the mid-1960s to late 1980s, was formally rejected by the International Commission on Zoological Nomenclature (ICZN). During the "banned" years, the names *Eugomphodus* and *Synodontaspis* were used in place of *Carcharias* (Compagno, 1984; Cappetta, 1987). The name *Carcharias* has been reinstated by the ICZN and is currently in use (for example, Long, 1992; Kent, 1994). Cappetta (1987) also recognized a number of extinct sand tiger shark genera, including *Striatolamna.* The teeth of *Striatolamna* are like those of *Carcharias* except with strongly striated (grooved) enamel on the lingual side of the cusp and weaker lateral cusplets. Long (1992) felt that these distinctions were insufficient to justify their generic separation, so he synonymized *Striatolamna* with *Carcharias,* and that synonymy is followed here (see Siverson [1995] for a very different opinion). Thus only a single sand tiger shark genus is recognized from Florida as a fossil, *Carcharias,* with about four or five species. *Carcharias macrotus* and *Carcharias hopei* are Eocene species. *C. macrotus* has striated enamel on the lingual side of the crown, weak lateral cusplets, and relatively broad teeth for a member of this genus, especially the laterals. *C. hopei* has smooth lingual enamel and anterior teeth with a single pair of strong, lingually curved lateral cusplets (two pairs on lateral teeth). *Carcharias* teeth are common, sometimes abundant, at Miocene and Pliocene localities in peninsular Florida including the Bone Valley region and Gainesville creeks (fig. 4.8B, C). Cappetta (1987) recognized two extinct species from this time interval, *Carcharias acutissima* and *Carcharias cuspidata,* assigning teeth with lingual striations to *C. acutissima* and those with smooth lingual enamel to *C. cuspidata.* Because *C. acutissima* graded into *C. taurus* during the Pliocene (Cappetta, 1987), I follow Tessman (1969) in recognizing *C. taurus* in the Neogene of Florida instead of *C. acutissima. C. taurus*

and *C. cuspidata* both have a small pair of lateral cusplets, a nearly complete cutting edge, a tall (up to 4 centimeters), sigmoidally curved cusp (when seen in distal or mesial view), and a prominent lingual bulge on the central portion of the root (fig. 4.7C). Lateral teeth have shorter, broader crowns than the anterior teeth and either one or two pairs of lateral cusplets. The two species differ in the presence (*C. taurus*) or absence (*C. cuspidata*) of lingual striations, although over time they became weaker and less prominent in the *C. taurus* lineage, sometimes making it difficult to distinguish the two.

The mackerel sharks, family Lamnidae, are large to very large (length of adults greater than 3 meters, rarely to 7 meters in the great white), fast-swimming, wide-ranging, partially warm-blooded sharks. There are only three living genera (*Lamna, Isurus,* and *Carcharodon*) and five living species (*Isurus* and *Lamna* have two each; *Carcharodon* is monotypic). Their diet is mostly large bony fish, squid, rays, and smaller sharks, while larger *Carcharodon* individuals feed extensively on marine mammals, primarily seals and sea lions. The adult teeth of the three extant lamnid genera are easily distinguished (fig. 4.9): those of *Carcharodon* have broadly triangular cusps, coarsely serrated cutting edges, and lack lateral cusplets; those of *Lamna* are also triangular and rather

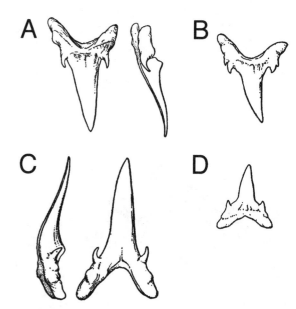

Fig. 4.7. Upper and lower teeth of a modern sand tiger shark, *Carcharias taurus. A,* anterior and labial views of the third upper anterior tooth; *B,* labial view of the third upper lateral tooth; *C,* anterior and labial views of the second lower anterior tooth; *D,* labial view of the third lower lateral tooth. About 2×. After Bigelow and Schroder (1948); reproduced by permission of the Peabody Museum of Natural History, New Haven, Connecticut.

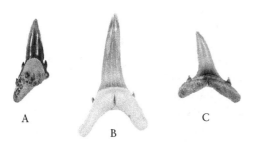

Fig. 4.8. Fossil sand tiger shark teeth (Odontaspididae) from Florida, about 1×. *A*, UF 93080, *Carcharias hopei* from D. H. Bell site, Lafayette County, late Eocene. *B, C,* UF 55976, *Carcharias taurus* from Fort Green Mine, Polk County, early Pliocene.

broad but have smooth cutting edges and a pair of lateral cusplets; and those of *Isurus* lack lateral cusplets and serrations and are narrower and more spikelike than those of the other two. Juveniles of both *Isurus* and *Carcharodon* have teeth with lateral cusplets, and juvenile *Carcharodon* teeth can lack serrations in whole or in part and are much narrower than adult teeth (Hubbell, 1996). The lateral cusplets in lamnids differ from those of odontaspidids in being broadly triangular rather than spikelike. Two of the three lamnid species now residing in Florida waters are known from fossils, the great white, *Carcharodon carcharias,* and the shortfin mako, *Isurus oxyrinchus.*

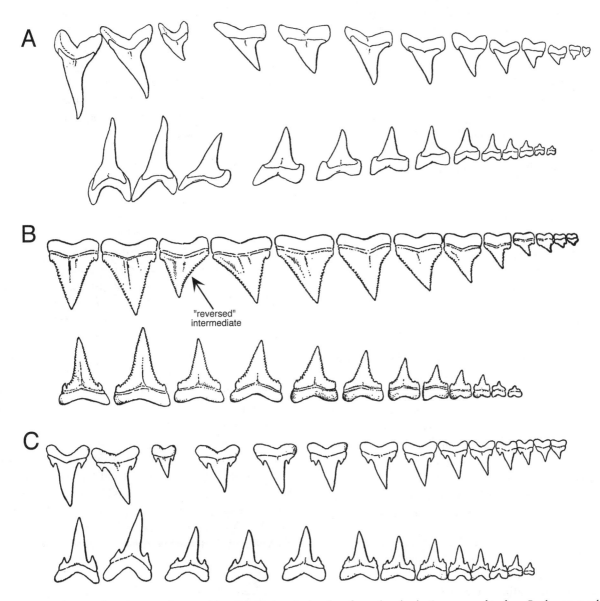

Fig. 4.9. Tooth sets of modern specimens of lamnid sharks. *A*, the shortfin mako shark, *Isurus oxyrhinchus; B*, the great white shark, *Carcharodon carcharias; C*, the porbeagle shark, *Lamna nasus.* About 1×. After Bigelow and Schroder (1948); reproduced by permission of the Peabody Museum of Natural History, New Haven, Connecticut.

In contrast to the rather straightforward systematics of modern lamnid sharks, the situation regarding extinct lamnids and closely related lamniforms is more complicated and contentious. Many extinct genera and species have been named, often based on isolated teeth. Paleontologists studying these sharks disagree about which species are valid, how they are related, and the generic names to be applied to various species. This is an unfortunate (but hardly unique) situation because their large teeth are some of the most frequently collected fossils. No specific studies have been published on specimens from Florida, but these large sharks had such huge geographic ranges that analyses conducted on specimens from anywhere in the world are applicable to fossil lamnids from our state. The evidence used by workers on both sides of the major arguments is described in the following paragraphs.

Large (8 to 17 centimeters tall), broadly triangular, serrated shark teeth are known practically worldwide from the Eocene through the Pliocene (figs. 4.10–4.12). Following Gottfried et al. (1996), the sharks producing these fossils are informally called megatooth sharks, a systematically neutral term. While arguments exist about the number of valid megatooth shark species, a common problem in numerous fossil lineages, these are relatively trivial compared to the more fundamental disagreement regarding the evolutionary relationships of megatooth sharks and modern lamnids. According to one hypothesis, megatooth sharks are more closely related to *Carcharodon carcharias* than they are to any other Cenozoic shark and are therefore classified in the genus *Carcharodon*. This will be referred to as the "*Carcharodon*-hypothesis." In the second widely held hypothesis, megatooth sharks are thought to be more closely related to the extinct shark genera *Otodus* and *Parotodus* than they are to any living genus, and that *Carcharodon carcharias* is more closely related to *Isurus* (in particular to the extinct species *Isurus hastalis*) than to the megatooths. Those workers who favor this second hypothesis currently place the megatooths in the genus *Carcharocles*, so it will be called the "*Carcharocles*-hypothesis."

Arguments and evidence used to support the *Carcharocles*-hypothesis include (but are not limited to) the following:

1. The extremely large size of megatooth shark teeth is most similar among Cenozoic sharks to *Otodus*.

2. The shape of the roots and the development of foramina in the roots in megatooth sharks is more similar to those of *Otodus* than those of *Carcharodon*

carcharias, while the same characters in the latter favor those of *Isurus*.

3. The serrations in megatooth sharks are fine, those of *C. carcharias* are large or coarse (fig. 4.10), suggesting independent acquisition.

4. Shared presence of serrations in megatooths and *C. carcharias* is not important because serrations have evolved independently in many broad-toothed shark lineages (*Carcharhinus, Galeocerdo, Hemipristis,* etc.); "missing-link" specimens with weak or partial serrations of both *Otodus* and *Isurus* are known.

5. The broadly triangular teeth of *C. carcharias* are more similar in overall shape and size to *Isurus hastalis* than those of the megatooths.

6. Juvenile teeth of *C. carcharias* resemble those of *Isurus* more than those of megatooth shark juveniles.

7. *Otodus* and megatooth shark teeth both have very broad necks on their lingual sides; teeth of *C. carcharias* and *Isurus* have relatively narrow necks.

8. The reduction or loss of lateral cusplets on adult teeth occurred independently a number of times in lamniform evolution (including *Isurus* and *Parotodus*), so this similarity between *C. carcharias* and advanced megatooth sharks has little weight.

The *Carcharocles*-hypothesis in its modern form originated with Casier (1960), who proposed the generic name *Procarcharodon* for megatooth sharks and suggested their evolutionary closeness with *Otodus* and the derivation of *C. carcharias* from *I. hastalis*. This concept received relatively little support from paleontologists until the 1970s, when a number of workers began to consistently use *Procarcharodon* for megatooth sharks (for example, Moody, 1972; Steurbaut and Herman, 1978; Longbottom, 1979; and Case, 1981). DeMuizon and DeVries (1985) supported the evolution of *C. carcharias* from *I. hastalis* on the basis of specimens from Peru. Cappetta (1987) resurrected *Carcharocles* (a name originally proposed in 1923 but little used subsequently) for megatooth sharks, replacing Casier's *Procarcharodon,* but otherwise he supported Casier's (1960) ideas regarding the evolutionary relationships of megatooth sharks. Subsequent studies using *Carcharocles* for megatooth sharks include Case and Cappetta (1990), Long (1992, 1993), and Kent (1994).

Arguments put forth to support the *Carcharodon*-hypothesis include the following:

1. Teeth of extremely large individuals of *C. carcharias* begin to resemble those of megatooths in such characters as finer serrations, crown proportions, shape of the

roots, and a broader neck. Thus the resemblance between *Otodus* and megatooth sharks may only be size related and not an indication of a close relationship.

2. The first anterior tooth is symmetrical in *C. carcharias* and megatooth sharks; it is asymmetrical in *Isurus*.

3. The intermediate tooth in *C. carcharias* and megatooth sharks has a unique orientation in that it is inclined mesially, not distally as in *Isurus, Otodus,* and other lamnids.

4. Vertebrae of *C. carcharias* resemble those of megatooth sharks more than they do those of *Isurus*.

5. The first upper anterior tooth in *C. carcharias* and megatooth sharks is the largest tooth in the dentition, while in *Isurus* the second lower anterior tooth is the largest tooth.

6. Resemblance in general tooth shape between *C. carcharias* and *I. hastalis* resulted from convergent evolution due to a similar feeding ecology (diet of marine mammals).

7. Significant differences in tooth shape between *C. megalodon,* the last of the megatooth sharks, and *C. carcharias* do exist and indicate that the two are not an ancestor-descendant pair (as was once claimed). Instead the two represent end members of species lineages whose closest common ancestor lived in the Paleogene.

Papers that have favored the *Carcharodon*-hypothesis over the *Carcharocles*-hypothesis include Keyes (1972), Welton and Zinsmeister (1980), Bendix-Almgreen (1983), Dockery and Manning (1986), Uyeno et al. (1989), Applegate and Espinosa-Arrubarrena (1996), Gottfried et al. (1996), and Purdy (1996). Both hypotheses are supported by paleontologists who have studied shark teeth for many years. Some of the claims presented by each side appear to have merit, while others are weak or shared primitive character states. My opinion is that neither side has presented conclusive evidence to substantially falsify the opposition's hypothesis. The rules of scientific nomenclature require use of a generic name with that of a species. *Carcharodon* will be used in this book for megatooth sharks because it is the more conservative approach (this genus was consistently used with these species for 130 years until the 1970s and by many workers since the 1970s) and because the arguments related to the condition of the intermediate tooth and vertebrae appear to have some merit. It should be emphasized that many of the arguments in favor of the *Carcharodon*-hypothesis rely upon correctly identifying the original tooth position for isolated teeth or associated dentitions, because articulated dentitions of megatooths are very rare.

Megatooth sharks range in age from middle Eocene to

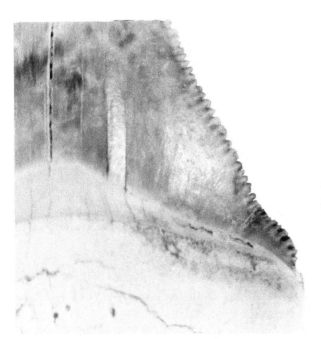 

Fig. 4.10. Comparison of magnified serrated margins of the teeth of *Carcharodon megalodon (left)* and *Carcharodon carcharias (right)*. Specimens are those shown in fig. 4.11C and 4.11D. The serrations of the extant, smaller-bodied *C. carcharias* are deeper and larger. About 5×.

early Pliocene in Florida, with two commonly recognized species, one in the Paleogene, *Carcharodon auriculatus,* and one from the Neogene, *Carcharodon megalodon.* However several recent studies have suggested that the holotype specimen of *C. auriculatus* actually belongs to a smaller, morphologically distinct species instead of the common Paleogene megatooth (Case and Cappetta, 1990; Applegate and Espinosa-Arrubarrena, 1996; Purdy, 1996). Replacement names for the Paleogene megatooth include *Carcharodon angustidens* and *Carcharodon sokolowi,* with Case and Cappetta (1990) favoring the latter.

Teeth of *Carcharodon sokolowi* (or whatever name is eventually applied to this species) have been collected from most of the Eocene and lower Oligocene limestone units of Florida. Specimens are either collected directly from limestone bedrock (often in quarries or caves) or from the alluvium of rivers that have eroded through these rocks. The anterior teeth of *C. sokolowi* have narrow, thick, triangular cusps with fine serrations and broad, triangular lateral cusplets that have slightly stronger serrations than the main cusp (fig. 4.11B). Upper lateral teeth are inclined distally (fig. 4.11A), but otherwise similar. The roots have large lobes and deep notches.

The megatooth shark lineage is not well represented in Florida from the late Oligocene to the early Miocene. Morgan (1989) listed *Carcharodon angustidens* from the very late Oligocene White Springs fauna collected along the Suwannee River in northern peninsular Florida. However, the specimens have broadly triangular cusps and reduced lateral cusplets, character states found in the megatooth species of this time variously called *Carcharodon subauriculatus* or *Carcharodon chubutensis* (Applegate and Espinosa-Arrubarrena, 1996; Kent, 1994). This species forms a good morphological intermediary between *C. sokolowi* and *C. megalodon.*

Carcharodon megalodon was the youngest and largest species of megatooth shark (fig. 4.12). The tallest known anterior upper teeth are about 17 centimeters (almost 7 inches) tall, although most specimens range

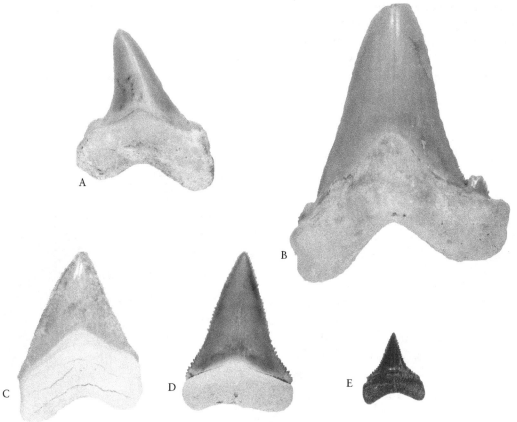

Fig. 4.11. Lingual views of teeth of fossil great white sharks, genus *Carcharodon,* from Florida. *A,* UF 49047, upper lateral tooth of *C. sokolowi* from Limestone Products Quarry, Alachua County, middle Eocene; *B,* UF 13237, lower anterior tooth of *C. sokolowi* from Buda Pit, Alachua County, late Eocene; *C,* UF 107675, upper lateral tooth of subadult *C. megalodon* from Hookers Prairie Mine, Polk County, middle Miocene; *D,* UF 131981, upper anterior tooth of *C. carcharias* from Deans Trucking Pit, Sarasota County, early Pleistocene; *E,* UF 82193, lower lateral tooth *C. carcharias* from Leisey Shell Pit, Hillsborough County, early Pleistocene. All about 1×.

Fig. 4.12. Lingual views of two teeth of the extinct giant great white shark, *Carcharodon megalodon*. *A*, UF 131982, lower anterior tooth from Deans Trucking Pit, Sarasota County, late Miocene; *B*, uncataloged UF specimen, a posterior tooth from Payne Creek Mine, Polk County, early Pliocene. Despite the great size difference in these two teeth, it is possible that the posterior tooth came from a just as large, if not larger, individual than the anterior tooth. Size varies greatly with tooth position in *Carcharodon*. Both about 1×.

between 5 and 15 centimeters. Some past estimates of the total length of *C. megalodon* have topped 30 meters, but the most recent estimates suggest a length of 13 to 17 meters for adult females and 10.5 to 14 meters for males (Gottfried et al., 1996). The largest individuals probably weighed in excess of sixty tons. An active, predaceous shark of this size would prey mostly on marine mammals, especially whales. Indeed, fossil whale bones with scratches and gouges made by *C. megalodon* teeth are not uncommon. The teeth of *C. megalodon* have fine serrations, wide necks, and the cutting edges are usually straight to slightly convex in lingual view (figs. 4.10A; 4.12). Lower teeth have more deeply notched roots than the uppers. Most teeth lack lateral cusplets, although vestigial ones are rarely present. Juvenile teeth have the same broadly triangular shaped cusps as the adults. The youngest well-dated records of *C. megalodon* in Florida are early Pliocene, from the Palmetto Fauna of the Bone Valley region. The species is also found in the Miocene strata of the same region. According to Purdy (1996), most Bone Valley specimens are juveniles and subadults, with full adult-sized teeth uncommon. He speculated that the area was a nursery region for the species because of its abundance of relatively small whales. *Carcharodon carcharias*, the extant great white, first appeared in the Miocene, but its fossils do not become abundant until the early Pleistocene (fig. 4.11D, E), following the demise of *C. megalodon*.

Accompanying *C. sokolowi* teeth in Eocene limestone are two other lamnids, the large *Otodus obliquus* (fig. 4.13) and a relatively small, narrow-toothed mako, *Isurus praecursor* (fig. 14.14A). The anterior and lateral teeth of *O. obliquus* can exceed 10 centimeters (four inches) in height. The anterior and lower lateral teeth are symmetrical and sharply pointed; upper laterals were relatively low crowned and curved slightly distally. A large bulge or protuberance is present on the lingual side of the root. Most specimens have smooth cutting edges and bear a single pair of lateral cusplets, but rare individuals have two or even three pairs. If the artificial tooth set of *O. obliquus* figured by Applegate and Espinosa-

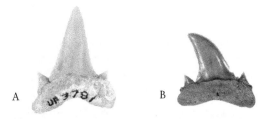

Fig. 4.13. Lingual views of two teeth of the extinct mackerel shark *Otodus obliquus* from Eocene of Florida. Note that these two teeth are from immature individuals and that teeth of this species get much larger. *A*, UF 3781, from walls of the Haile 7 quarry, Alachua County.; *B*, UF 95770, from Ichetucknee River, Columbia County. About 1×.

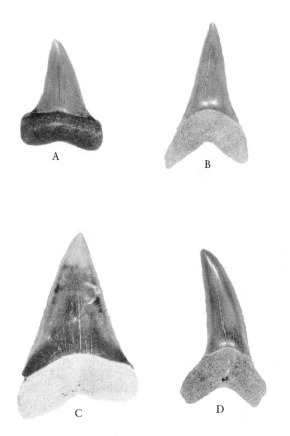

Fig. 4.14. Lingual views of fossil mako sharks (genus *Isurus*) from Florida. *A,* UF 116001, upper anterior tooth of *I. praecursor* from D. H. Bell site, Lafayette County, late Eocene; *B,* UF 3773, lower anterior tooth of *I. hastalis* from UF campus, Alachua County, probably Miocene; *C,* UF 17978, upper anterior tooth *I. hastalis* from Manatee County Dam site, Manatee County, late Miocene; *D,* UF 344, *I. oxyrinchus* from Peace River, Hardee County, Pliocene or Pleistocene. About 1×.

Arrubarrena (1996) is correct, then this species differed from *Carcharodon* and *Isurus* by having three upper anterior teeth and a lower symphyseal tooth (both primitive character states for lamniform sharks). Cappetta (1987) and Kent (1994) also illustrated teeth of this extinct large shark.

Isurus hastalis was one of the more abundant of the larger sharks in the Miocene and Pliocene. Its upper teeth are broadly triangular with smooth, unserrated cutting edges (fig. 4.14C). The lower teeth are narrower and have deeper root notches (fig. 4.14B). According to Applegate and Espinosa-Arrubarrena (1996), *I. hastalis* differed from *Carcharodon* not only in the condition of the intermediate tooth (discussed above), but also in that the first upper anterior tooth of *I. hastalis* was curved distally and not symmetrical. A few broken and water-worn specimens of *I. hastalis* were found at the Leisey

Shell Pit (early Pleistocene), but they were likely eroded out of older rocks and redeposited (Scudder et al., 1995). Otherwise, *I. hastalis* ranged between the early Miocene and early Pliocene, with the youngest records derived from the Palmetto Fauna.

The remaining sharks to be discussed all belong to the order Carcharhiniformes, the largest group of living sharks with seven families and more than 200 species (Nelson, 1994). Most carcharhiniform sharks inhabit coastal, tropical to warm temperate waters, so it is not surprising that this group is well represented in the Florida fossil record. Indeed, the carcharhiniform genera *Negaprion, Galeocerdo,* and *Carcharhinus* alone account for 60 to 80 percent of the fossil shark teeth at most Neogene localities in Florida (Webb and Tessman, 1968; Tessman, 1969). Carcharhiniform sharks are also responsible for the majority of unprovoked attacks on humans, not the generally larger and more feared lamnids (Compagno, 1988).

Carcharhiniform sharks are as yet unreported from Eocene rocks in Florida. This is surely an artifact, because the group is well represented at Eocene localities in Texas, Louisiana, and Georgia (Case, 1981; Manning and Standhardt, 1986; Westgate, 1989). The oldest fauna in Florida with carcharhiniform sharks is the I-75 site (late early Oligocene) of Alachua County, which produced five genera: the hemigaleid *Hemipristis;* and the carcharhinids *Negaprion, Galeocerdo, Rhizoprionodon,* and *Carcharhinus* (Tessman, 1969). These five genera are known from all successively younger marine rocks in the state and still inhabit Florida's offshore waters with the exception of *Hemipristis. Hemipristis* no longer lives in the Atlantic, surviving only in the Indian and Pacific Oceans (Compagno, 1988). The only additional genus of carcharhiniform known from Florida as a fossil is the hammerhead *Sphyrna,* which first appears in the Pleistocene.

The tiger shark, *Galeocerdo cuvier,* is a large (up to 7 meters long), wide-ranging, warm-water species. Its diet is unselective and commonly includes other sharks, rays, bony fish, sea turtles, crabs, gastropods, seals, and carrion. The stomach contents of tiger sharks often include a wide variety of garbage and human junk (Bigelow and Schroeder, 1948; Tessman, 1969). In *G. cuvier,* the upper and lower teeth are similar (very low degree of dignathic heterodonty). The broad, low cusps point obliquely in the distal direction, more so in posterior teeth (fig. 4.15G). Only the symphyseal teeth have erect crowns. Both mesial and distal cutting edges are serrated, most coarsely on the distal heel. The distal cutting edge is also sharply notched. Some of the large serrations may themselves bear finer serrations, a pattern known as complex or secondary serration. Adult teeth are from 2 to 3 centimeters tall. Juvenile teeth are similar to adults except that the cusp is narrower, longer, and more oblique and there are fewer coarse serrations. Fossil teeth resembling those of the living *G. cuvier* first appear in the late Miocene (Tessman, 1969). A large sample of *G. cuvier* teeth was collected from Pliocene sediments on the bed of Hickey Creek in Lee County in association with the skeleton of a baleen whale (fig. 4.15E–G; Morgan and Pratt, 1983). It is likely that tiger sharks were scavenging the whale's carcass.

At least four extinct tiger shark species have been recorded from Eocene and early Oligocene localities in eastern North America, *Galeocerdo alabamensis*, *Galeocerdo latidens*, *Galeocerdo eaglesomi*, and *Galeocerdo clarkensis* (Case, 1981; Westgate, 1984, 1989; Manning and Standhardt, 1986; Kent, 1994). It is unclear at present how many of these species are valid. Only *G. alabamensis* has been reported from Florida (Tessman, 1969). *G. alabamensis* teeth are smaller than those of *G. cuvier*, with slender crown tips that are more

vertical. The pattern of serrations is the same as in the modern tiger shark. *G. alabamensis* was replaced in the late Oligocene by two new kinds of tiger sharks, *Galeocerdo aduncus* and *Galeocerdo contortus*. The teeth of *G. aduncus* are intermediate between those of *G. alabamensis* and *G. cuvier* in size (maximum height about 2 centimeters) and have narrower cusps than *G. cuvier* (fig. 4.15C). The teeth of *G. contortus* have a much more narrow and slender cusp, weaker serrations that are usually not complex, and the mesial cutting edge is twisted or sigmoidally curved in mesial view (fig. 4.15A, B). Specific separation between *G. aduncus* and *G. contortus* has varied. Some workers, such as Tessman (1969), have regarded *G. contortus* as the lower teeth and *G. aduncus* the upper teeth of a single species. Others have suggested that they could represent males and females of one species. Evidence that these two hypotheses are unlikely and both species are valid includes the following:

1. *G. aduncus* is known from many regions of the world, but *G. contortus* is only known from North America (Cappetta, 1987).
2. Modern *Galeocerdo* lacks significant dignathic heterodonty and sexual dimorphism (Compagno, 1988), making it unlikely that a close fossil relative would show either pattern.
3. Both upper and lower teeth can be recognized by the shapes of the roots in large samples of either *G. aduncus* or *G. contortus* teeth (Kent, 1994).

G. contortus persisted to the end of the Miocene in Florida, while *G. aduncus* graded into *G. cuvier* during this epoch.

The lemon shark, *Negaprion*, is a moderately large (up to 3.5 meters long), heavily bodied, unaggressive, shallow water shark typically found in tropical waters. Fish form the basis of its diet. The teeth of *Negaprion* have narrow, moderately tall, unserrated cusps. Both the long mesial and distal shoulders on the upper teeth have notches or serrations in adults (fig. 4.16B), and juveniles have lateral cusplets (not present in adults). Lower teeth typically lack all serrations and have more slender and erect cusps than the uppers (fig. 4.16A, C). Some paleontologists distinguish a Miocene species, *Negaprion eurybathrodon*, from the modern species (Longbottom, 1979; Cappetta, 1987), but Tessman (1969) identified all Florida fossil lemon sharks as the extant *Negaprion brevirostris*. Its oldest fossils in Florida are late early Oligocene, and it is quite common in most Neogene shark-bearing localities (fig. 4.17). Typically 25 to 40 percent of the shark teeth greater than 0.5 centimeters tall belong to *Negaprion* (Tessman, 1969). However, this

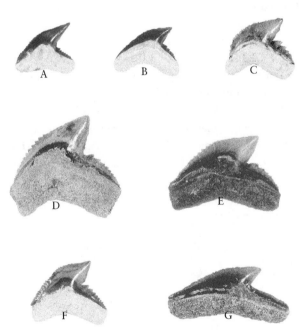

Fig. 4.15. Lingual views of fossil tiger sharks, genus *Galeocerdo*, from Florida. A–B, UF specimens of *G. contortus* from Phosphoria Mine, Polk County, middle Miocene; C, UF specimen of *G. aduncus* from Phosphoria Mine, middle Miocene; D–G, *G. cuvier*; D, UF 82191, from Leisey Shell Pit, Hillsborough County, early Pleistocene; E–G, UF 50780, from Hickey Creek Whale site, Lee County, early Pliocene. All about 1×.

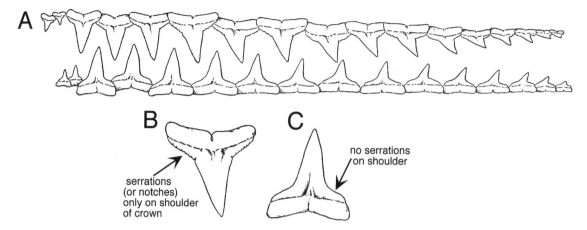

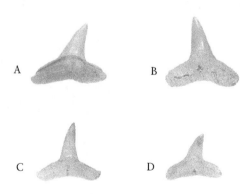

Fig. 4.16. Teeth of a modern, 10-foot-long lemon shark, *Negaprion brevirostris,* from Florida in labial view. *A,* upper and lower tooth rows; *B,* upper lateral tooth; *C,* lower anterior tooth. *A* about 1×; *B–C* about 2×. After Bigelow and Schroder (1948); reproduced by permission of the Peabody Museum of Natural History, New Haven, Connecticut.

apparent abundance may be inflated by misidentification of lower teeth of *Carcharhinus* as *Negaprion.*

Gray sharks, genus *Carcharhinus,* comprise a group of about thirty living species that are primarily coastal and tropical to warm temperate in their distribution (Compagno, 1988). Of these, eleven or twelve live in Florida waters, and several of these are among the state's most abundant medium- to large-sized sharks. These include the bull shark (*Carcharhinus leucas*), sandbar shark (*Carcharhinus plumbeus*), blacknose shark (*Carcharhinus acronotus*), blacktip shark (*Carcharhinus limbatus*), dusky shark (*Carcharhinus obscurus*), and silky shark (*Carcharhinus falciformis*). Adult body length ranges between 1 and 4 meters in this genus. *Carcharhinus* is common in shallow waters, and *C. leucas* in particular readily enters bays, estuaries, and rivers. The diet of all species is primarily fish. *Carcharhinus* teeth are moderate sized, usually shorter than 2.5 centimeters (fig. 4.18). The upper teeth of all living *Carcharhinus* are broader than their lowers, usually bear more oblique cusps, and often have better developed serrations. Serrations are usually present on both the distal and mesial cutting edges of upper teeth but are absent, irregular, or only partially developed on lowers, depending on the species (Compagno, 1988). Isolated teeth of *Carcharhinus* are often difficult to identify to species, even if complex measuring systems are used (Naylor and Marcus, 1994). Upper teeth are more distinctive than lowers, which if lacking serrations can easily be confused with those of *Negaprion.* Several species groups are recognized whose upper teeth show relatively definitive intergroup differences. Teeth of *C. leucas, C. plumbeus, C. longimanus,* and *C. obscurus* have broadly triangular

cusps, low to moderate distal notch depth, and no notch on the mesial cutting edge (fig. 4.18A–F). Teeth of *C. limbatus* and *C. brevipinna* have relatively narrow, erect to semi-erect cusps and notches on both the mesial and distal cutting edges producing long shoulders like those found in *Negaprion* (fig. 4.18H). Teeth of *C. acronotus* and *C. falciformis* also have narrow cusps, but with more oblique orientations so the mesial notch is weak, the distal notch is deep, and there may be distal cusplets (fig. 4.18G). Secure identification to a particular species within any of these groups requires a large series of modern comparative specimens.

Fossil teeth identifiable as *Carcharhinus* are known from the early Oligocene through the Pleistocene in Florida; most specimens are between 0.5 and 3 centimeters tall (fig. 4.19). A number of species, both living and ex-

Fig. 4.17. Lingual views of four teeth of the lemon shark, *Negaprion brevirostris,* from Manatee County Dam site, Manatee County, late Miocene. *A–B,* UF 17982, two upper teeth; *C–D,* UF 17979, two lower teeth. All about 1×.

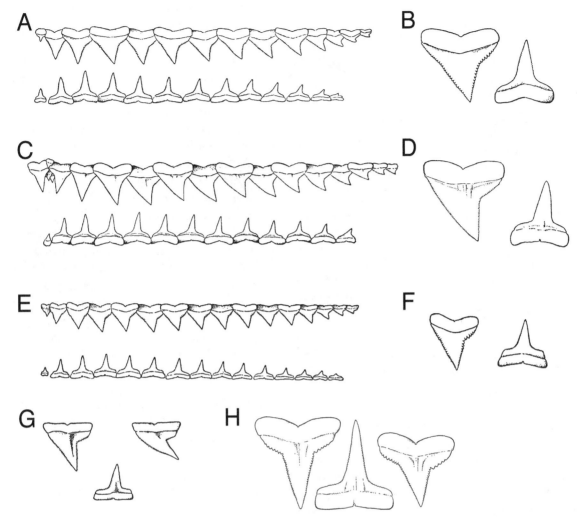

Fig. 4.18. Teeth of modern specimens of the genus *Carcharhinus* in labial view. *A*, upper and lower tooth rows of a 2.5-foot-long bull shark, *C. leucas*, from Florida; *B*, third upper and second lower teeth of the same individual as in *A*; *C*, upper and lower tooth rows of a 5-foot-long sandbar shark, *C. plumbeus*, from Massachusetts; *D*, fourth upper and third lower teeth of the same individual as *C*; *E*, upper and lower tooth rows of a 3-foot-long dusky shark, *C. obscurus*, from Massachusetts; *F*, third upper and fourth lower teeth of the same individual as *E*; *G*, three left teeth of a 2-foot-long blacknose shark, *C. acronotus*, from Florida. The two uppers are the third and eighth teeth, the lower is the third. Note the distinctly notched distal cutting edge. *H*, fourth and tenth upper and fourth lower teeth of the small blacktip shark, *C. limbatus*. *A*, *C*, and *E* about 1×; *B*, *F* about 2.5×; *D* about 2×; *G*, *H* about 3×. After Bigelow and Schroder (1948); reproduced by permission of the Peabody Museum of Natural History, New Haven, Connecticut.

tinct, have been recognized (chapter 3), but further study is probably necessary to confirm many of the identifications. The most commonly identified species has been *C. leucas,* however considering that *C. plumbeus, C. acronotus,* and *C. limbatus* are almost equally common in modern waters as *C. leucas,* their rarity as fossils might stem partially from misidentifications.

The sharpnose shark, *Rhizoprionodon terraenovae,* is the most common small shark (body length less than 1.5 meters) recovered as a fossil in Florida. Modern *R. terraenovae* is locally abundant in the Gulf of Mexico and the Caribbean, where it inhabits very shallow wa-

ters, never more than a few kilometers from land. Its prey are chiefly small fish and crustaceans. *Rhizoprionodon* teeth (figs. 4.20, 4.21) have sharp cusps that are sharply oblique with a strongly notched distal cutting edge. Only the symphyseal teeth are erect and not notched. The distal heel bears fine serrations or indentations in adults, but serrations are absent on the main cusp. The teeth are rarely greater than 4 millimeters tall but can be common if screen-washing methods are used. Several other shark genera have teeth that are very similar to *Rhizoprionodon,* including *Scoliodon,* small *Sphyrna,* and even the posterior teeth of *Negaprion.* Small teeth with distal

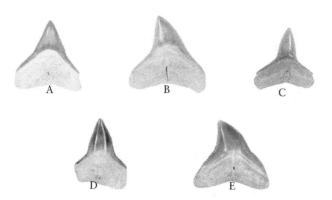

Fig. 4.19. Lingual views of fossil shark teeth of the genus *Carcharhinus* from Bone Valley District, Polk County, middle Miocene or early Pliocene. This genus is one of the most commonly found fossil sharks in Florida. *A–C*, UF 17855, *C. leucas* from Palmetto Mine. *A* and *B* are upper teeth, *C* is a lower; *D*, UF 17909, *C. egertoni* from Kingsford Mine; *E*, UF 18000, *C. obscurus* from Kingsford Mine. All about 1×.

cusplets but otherwise like *Rhizoprionodon* could belong to *Galeorhinus, Paragaleus,* or the extinct genus *Physogaleus* (Cappetta, 1987).

The snaggletooth shark, *Hemipristis elongatus,* is the only living species of the genus and inhabits the Indian Ocean and the Red Sea. Adults range from 1.5 to 2.5 meters in length. Fossil snaggletooth sharks had a much wider distribution and reached significantly larger sizes, perhaps up to 5 meters in total length (Compagno, 1988). *Hemipristis serra* had a circumglobal distribution

during the Miocene and Pliocene and is most abundant in shallow marine, tropical deposits (Cappetta, 1987). This species is known in Florida from the Oligocene to the early Pleistocene (Scudder et al., 1995). The teeth of *Hemipristis* display very strong dignathic heterodonty. The upper teeth of *H. serra* are serrated, with the serrations ending just below the tip of the crown and the distal serrations are coarser than the mesial ones. Upper anterior teeth are erect and narrowly triangular (fig. 4.22B); while upper lateral teeth are broadly triangular and strongly curved distally (fig. 4.22A). The mesial cutting edge is convex, while the distal edge is concave. Anterior lower teeth are tall, narrow, and some have small lateral cusplets (fig. 4.22C–E). The cutting edge is present only near the apex of the crown. Lower lateral teeth are inclined distally, have narrow cusps and mesial cusplets and serrations. Both uppers and lowers have a strong lingual central protuberance on the root. Lower anterior teeth somewhat resemble those of *Carcharius* but are bulkier and have shorter roots and a more limited cutting edge. *Hemipristis* upper teeth are easily distinguished from other large, serrated shark teeth. Teeth of *Carcharodon* are more erect and do not have dramatically different serrations on the mesial and distal edges; teeth of *Galeocerdo* have a strong distal notch.

The Florida fossil record has produced three distinct shark faunas. In the Eocene, four genera and five species are known, and all the species are extinct. *Carcharodon sokolowi* is the best represented of the five in collections, but this may reflect only its large size, which makes it

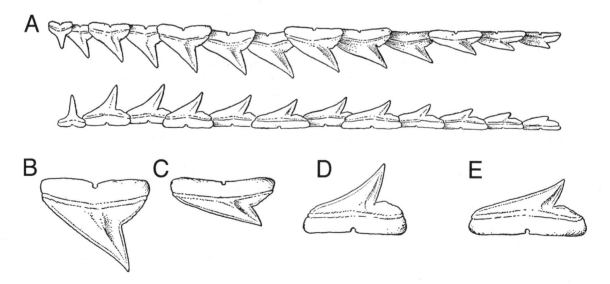

Fig. 4.20. Teeth of a modern sharp-nosed shark, *Rhizoprionodon terraenovae* in labial view. *A*, upper and lower tooth rows; *B*, fourth upper tooth; *C*, tenth upper tooth; *D*, fourth lower tooth; *E*, eighth lower tooth. *A* about 3×; *B–E* about 6×. After Bigelow and Schroder (1948); reproduced by permission of the Peabody Museum of Natural History, New Haven, Connecticut.

easier to find. A transitional Oligocene fauna contains a mixture of survivors from the Eocene and first occurrences of taxa more typical of the Neogene. The second major shark fauna is characteristic of the Miocene and early Pliocene. Its dominant forms include *Carcharodon megalodon, Isurus hastalis, I. desori, Hemipristis serra, Negaprion brevirostris,* and numerous species of *Carcharhinus.* The period of transition between the second and third shark faunas occurred during the late Pliocene and early Pleistocene and is coincident with the onset of global cooling and increased glaciation. *C. megalodon, I. hastalis,* and *H. serra* became extinct and were replaced by *Carcharodon carcharius* and *Isurus oxyrhinchus,* while the various species of *Carcharhinus* and *N. brevirostris* continue to be most abundant.

FOSSIL SAWFISHES, GUITARFISHES, SKATES, AND RAYS OF FLORIDA

The order Rajiformes includes the sawfishes, guitarfishes, skates, stingrays, eagle rays, and devil rays (Nelson, 1994). This group of chondrichthyans are characterized by flattened bodies, expanded pectoral fins, ventrally located gill slits, and dorsally located eyes. Common fossils of this group include isolated teeth, vertebral centra, and enlarged dermal denticles that form spines, thorns, and the "teeth" of the sawfish rostrum. More rarely, a partial or even a whole tooth plate is recovered. Identification of living species in these groups is based mostly on soft anatomy and coloration, so generic or even familial identification is often the best one can do with most rajiform fossils without detailed study.

The Rajiformes originated in the Jurassic, and all the major living subgroups are known by the Cretaceous (Cappetta, 1987). They are common in shallow coastal

Fig. 4.21. Lingual views of four teeth of *Rhizoprionodon terraenovae* from Northwest Palmetto Mine, Polk County, probably early Pliocene. UF specimens. About 2.5×

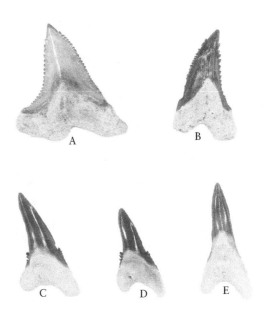

Fig. 4.22. Lingual views of five teeth of the snaggle-toothed shark, *Hemipristis serra. A–B,* UF 17981, two upper teeth from Manatee County Dam site, Manatee County, late Miocene. *C–E,* UF specimens of lower teeth from Palmetto Mine, Polk County, early Pliocene. Uppers are strongly serrated in this genus, but serrations in lowers are either limited or absent altogether. All about 1×.

waters, and many species enter bays and estuaries. This explains their abundance in the Florida fossil record. Five families are presently known from Florida as fossils (chapter 3). Because this group has been critically studied even less than the sharks, their past diversity no doubt was much greater at all taxonomic levels than our present state of knowledge would indicate. Excluding some members of the Myliobatoidea, the teeth of this group are all very small, usually less than a centimeter, and are typically recovered only by screen-washing matrix. As was the case with sharks, the tooth consists of a crown and root. The crown may bear a sharp, pointed cusp or a have a flat occlusal surface. Sexual dimorphism in tooth form is common.

The sawfish family Pristidae contains two living genera, *Anoxypristis* and *Pristis* (figs. 4.23, 4.24). All fossils of sawfishes from Florida have been referred rather uncritically to *Pristis.* Ironically, one of the oldest records, a partial rostrum encased in Eocene limestone, is one of the most complete. Otherwise sawfish fossils are typically either the isolated rostral teeth (which are enlarged, specialized dermal denticles) or the spool-like vertebrae (fig. 4.24). Oral teeth are less than 3 millimeters wide and very poorly known. The primary difference between the rostral teeth of the two living genera is that in *Pristis*

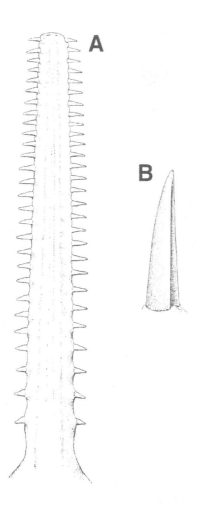

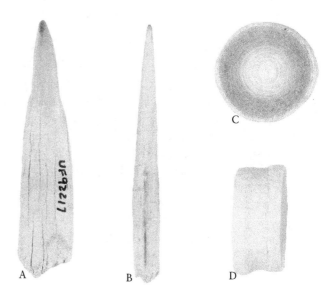

Fig. 4.24. Fossil sawfish (*Pristis* sp.) elements from Florida. *A*, dorsal, and *B*, anterior views of UF 92217, a "saw tooth" from Gainesville Creek, Alachua County, Miocene; *C*, anterior or posterior, and *D*, lateral views of an uncataloged UF specimen, a vertebra from Hookers Prairie Mine, Polk County, middle Miocene. Both about 1×.

Fig. 4.23. *A*, rostral "saw" of a modern sawfish, *Pristis pectinatus*, from southern Florida. *B*, "saw tooth" from the same individual as *A*. Note concave posterior edge. Usually only isolated "saw teeth" preserve as fossils, as they contain mineralized tissues. They are not true teeth but specialized dermal spines. A about 0.25×, B about 1×. After Bigelow and Schroder (1953); reproduced by permission of the Peabody Museum of Natural History, New Haven, Connecticut.

the posterior edge is concave, while in *Anoxypristis* both the anterior and posterior edges are sharp and neither is concave (Cappetta, 1987). Sawfishes use their long, "tooth"-bearing rostrum to stun or injure fish by thrashing it back and forth quickly in a dense school. They have a more elongate, sharklike body than skates or rays, but the head region is flattened. The saw sharks (Pristiophoridae) have a similar elongate rostrum with enlarged toothlike denticles but a more sharklike body. The two families differ in that the rostral teeth of pristids are embedded in deep sockets, while those of saw sharks are weakly embedded, and those of pristids are all about the

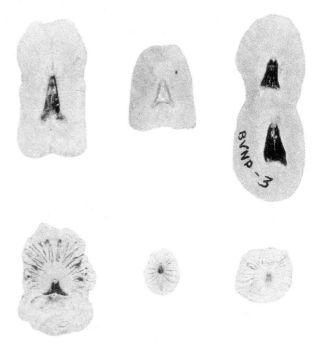

Fig. 4.25. Dorsal views of dermal spines or thorns from either the Palmetto or North Palmetto Mines, Polk County, early Pliocene. UF collection. They belong to either the skates (Rajidae) or the stingrays (Dasyatidae). All about 1×.

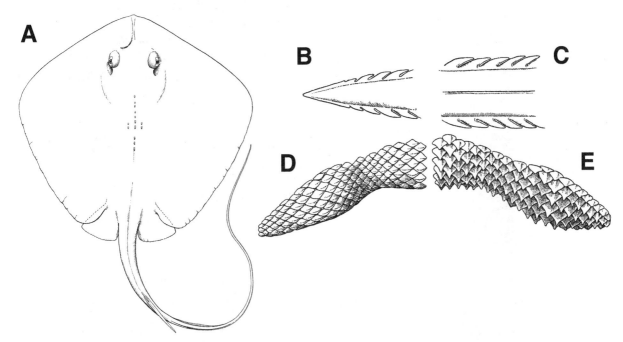

Fig. 4.26. *A*, Dorsal view of a modern stingray, *Dasyatis say,* from Florida showing the long whiplike tail with its tail spine; *B*, tip; *C*, middle section of a tail spine from *Dasyatis say; D,* upper left tooth band of a male *D. say; E,* upper right tooth band of a female *D. say.* Note the sexual dimorphism in tooth morphology. *A* about 0.25×, *B–E* about 1.5×. After Bigelow and Schroder (1953); reproduced by permission of the Peabody Museum of Natural History, New Haven, Connecticut.

same size, while in the saw sharks small and large rostral teeth alternate along the rostrum (Nelson, 1994).

Two families of guitarfishes are known, the more cosmopolitan Rhinobatidae and the Rhinidae, which is currently restricted to the Indian and Pacific Oceans (Nelson, 1994). However, it is a member of the latter family, the genus *Rhynchobatus,* whose small teeth (less than 5 millimeters wide) are relatively common in Florida deposits ranging in age from late Oligocene to early Pleistocene (Scudder et al., 1995). The overall body form of a guitarfish, as is the case with the sawfish, is less heavily modified than those of typical skates and rays.

Most living skates (Rajidae) are placed in the genus *Raja.* Skates are bottom-dwellers that move by undulating their enlarged pectoral fins. They lack the large tail spines found in myliobatoids, but have variously arranged rows of small spines or thorns on the dorsal surface of the body and on the tail. These spines are enlarged dermal denticles and so are structurally like teeth. They consist of a base of dentine and usually have a raised point or stripe of enameloid (fig. 4.25). The dentine and enameloid usually preserve as different colors due to their differing chemical composition. Skate dermal spines are difficult to distinguish from those of stingrays. Skate teeth are small and buttonlike.

Stingrays (Dasyatidae) differ from skates by having a narrow, whiplike tail that usually contains one or more large, barbed tail spines (fig. 4.26A–C). These are used for protection, as many an unfortunate swimmer who has stepped on a buried stingray has learned. The tail spine, like the dermal spines, is an enlarged enameloid-covered dermal denticle. Fossil spines often reach lengths of 5 to 10 centimeters, although long specimens are frequently broken (fig. 4.27). Stingrays resemble skates in having numerous small teeth (fig. 4.26D, E); dermal thorns are also present but in smaller numbers. Florida fossil material of stingrays is often uncritically assigned to the extant genus *Dasyatis.* However, two other related ray families live on the Florida coast: the round stingrays, Urolophidae, and the butterfly rays, Gymnuridae (Nelson, 1994). The three families cannot be distinguished based on isolated spines or thorns. Critical analysis of the tiny teeth will be necessary to prove which families and genera were actually present.

Eagle and cownose rays (Myliobatidae) have tail spines like stingrays but differ by lacking the dermal thorns of that group and skates (Bigelow and Schroeder, 1953; Nelson, 1994). Isolated tail spines of these groups usually cannot be distinguished. The common genera of eagle rays are *Myliobatis, Aetomylaeus,* and *Rhinoptera.*

Fig. 4.27. Dorsal view of UF 55968, an unusually large and complete tail spine of a ray, either a stingray (Dasyatidae) or an eagle ray (Myliobatidae). From Fort Green Mine, Polk or Hardee County, early Pliocene. About 1×.

Aetobatus is the only recognized genus of cownose rays. All four have enlarged teeth for crushing and grinding hard-shelled prey. They are very abundant as fossils. In the living ray these teeth are combined into a single functional structure called a dental plate (figs. 4.28, 4.29). The pattern of the teeth making up the dental plate differs among the genera, so articulated fossil teeth can be identified to genus relatively easily. Normally, however, the dental plate falls apart after death, since the teeth are not fused together, and isolated teeth are the most common fossils. *Myliobatis* and *Aetomylaeus* both have tooth plates composed of seven files of teeth, three pairs of lateral files with relatively small teeth on either side of the much larger teeth of the median file. Teeth from the median file are much broader than long. The two genera differ in that the occlusal surface of the median file is notably convex in *Aetomylaeus* but flat or slightly

concave in *Myliobatis*. The lateral file teeth of *Myliobatis* are symmetrical, while those of *Aetomylaeus* are not (fig. 4.30A). The dental plate of *Rhinoptera* is similar to those of *Myliobatis* and *Aetomylaeus*, but the number of files of teeth varies, from 7 up to 19, teeth of the median file

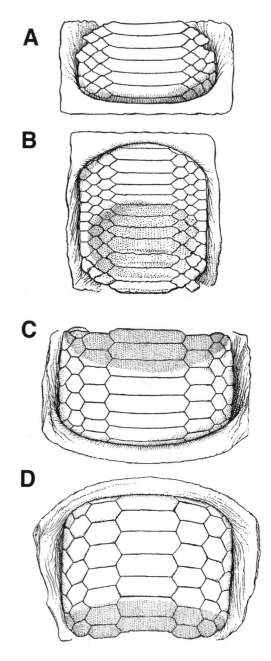

Fig. 4.28. Dental plates (made of articulated teeth) of modern eagle rays (Myliobatidae). *A*, upper, and *B*, lower dental plates of *Myliobatis freminvillii*; *C*, upper, and *D*, lower dental plates of *Rhinoptera bonasus*. *A–B*, about 2×; *C–D* about 1×. After Bigelow and Schroder (1953); reproduced by permission of the Peabody Museum of Natural History, New Haven, Connecticut.

One other type of ray has been recognized as a fossil from Florida: the extinct genus *Plinthicus*. Teeth of *Plinthicus* are most abundant in Miocene deposits of the eastern United States and are also known from Europe (Cappetta, 1987; Bor, 1990). The crowns of *Plinthicus* teeth are notably compressed, and the occlusal surface may bear a narrow groove. Articulated dental plates indicate that the overall tooth pattern was similar to *Rhinoptera*, although its evolutionary affinities are thought to lie with the manta and devil rays (Cappetta, 1987). Morgan (1989) published the only record of *Plinthicus* from Florida, but it is present at many Miocene fossil sites that contain marine fish.

FOSSIL BONY FISHES OF FLORIDA

Actinopterygians inhabit almost every ecological role in the aquatic environment possible for animals. Many species are restricted to either fresh, salt, or brackish water, while others are more tolerant to different levels of salinity. Some species live in different types of water during

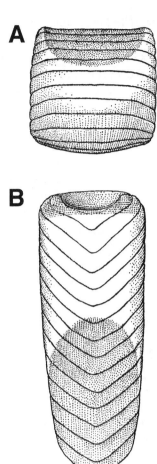

Fig. 4.29. *A*, upper, and *B*, lower dental plates of the cownose ray, *Aetobatus narinari*. About 1×. After Bigelow and Schroder (1953); reproduced by permission of the Peabody Museum of Natural History, New Haven, Connecticut.

are proportionally not as broad but are longer, and the teeth of the first lateral file (next to the median file) are broader than they are long, but not as broad as those of the median file (fig. 4.28C, D). *Aetobatus* has the most specialized dental plate of any ray, as all of the lateral files are absent and only the median file is present (fig. 4.29). The individual teeth of the lower dental plate are strongly curved or V-shaped (and because of this unique shape can easily be recognized; fig. 4.30D). The individual teeth making up the upper dental plate of *Aetobatus* are much straighter, with only a slight backward curve and narrowing of the teeth at their lateral ends. Isolated upper teeth of *Aetobatus* are more difficult to distinguish from the median teeth of other rays, especially broken specimens. The root is very high, higher than the crown in the middle part of the tooth, and the lateral end of the tooth is rounded, not angular, because it did not articulate with lateral teeth.

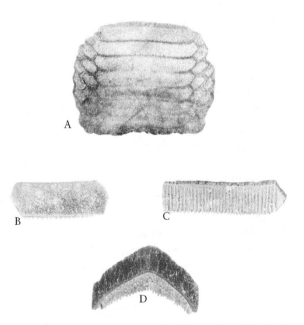

Fig. 4.30. Fossil dental plate and isolated teeth of the family Myliobatidae from Florida. Isolated ray teeth are extremely abundant at many fossil sites in the state. *A*, occlusal view of UF 92301, a dental plate of *Aetomylaeus* sp. from Gainesville Creek, Alachua County, Miocene; *B*, occlusal view of UF 28751, an isolated tooth of *Rhinoptera* sp. from Phosphoria Mine, Polk County, middle Miocene; *C*, UF 28751, an isolated tooth of a myliobatid from Phosphoria Mine, Polk County, middle Miocene showing the root side of the tooth; *D*, occlusal view of UF 64173, an isolated lower tooth of *Aetobatus* sp. from Polk County. *A* about 2×; *B–D* about 1×.

various developmental stages of their life cycle. Other ecological factors, such as temperature, current velocity, availability of oxygen, water depth, and substrate type, also play important roles in the distribution of fish species. This means that fossil fish, provided they can be identified, are useful paleoecological indicators for these same factors. Not surprisingly, the most abundant bony fish found as fossils in Florida live in those environments that produce the most common fossil sharks and rays—shallow, nearshore marine, and estuarine.

Sturgeons and paddlefishes (order Acipenseriformes) are the only living North American representatives of an ancient group of bony fish known as chondrosteans. Chondrosteans are characterized by a long fleshy lobe in the dorsal half of the tail fin; broad-based, stiff, paired fins; thick scales made of successive layers of bone, dentine, and a special enamel-like form of dentine called ganoine (ganoid scales); a notochord that is unrestricted by vertebral centra in the adult stage; and a maxilla that is firmly attached to the skull and is the main tooth-bearing element of the upper jaw. Chondrostean fishes were preeminent during the Paleozoic, but for the most part became extinct in the early Mesozoic (Long, 1995). The Acipenseriformes is specialized in its reduction of dermal cranial elements (including loss of the premaxilla), and many of the skull bones are cartilaginous even in adults. They also reduce the bony scales, and in *Acipenser* these are limited to five rows of large scutes, which are modified scales. It is these scutes which are rarely known as fossils (fig. 4.31). Sturgeons are large fish that live in rivers and coastal environments.

Two living genera of gar are recognized: *Lepisosteus* and *Atractosteus*. Both first appeared in the Cretaceous and are considered "living fossils" because they have changed so little for the past 60 million years (Wiley, 1976). They both have elongated bodies covered with shiny, diamond-shaped ganoid-covered scales. The dorsal tail fin bears a fleshy lobe, but it is smaller than in

Fig. 4.31. UF 12867, ganoid scale of the sturgeon *Acipenser* sp. from Ponte Verde Inlet, St. Johns County, Pleistocene. About 1×.

chondrostean fish. The long snout with its numerous teeth is the most prominent feature of the skull. The rostrum bears a series of tooth-bearing infraorbital bones; the maxilla is reduced and does not bear teeth. All of the dermal skull bones are thick and have sculptured ridges; many have an outer layer of enameloid in a pattern that varies among the species. Vertebrae are opisthocoelus, making them easy to distinguish from the amphicoelus vertebrae of other fish. Gars are now found in North America and Cuba. They had a wider distribution in the Cretaceous and Paleogene, ranging to India (Wiley, 1976). *Lepisosteus* is mostly limited to slow-moving fresh water, but *Atractosteus spatula* (the alligator gar) also ranges into estuaries and even open waters of the Gulf of Mexico. They are highly predacious, mostly eating other fish, but large alligator gars are also known to eat birds.

Gars are abundantly represented in Neogene deposits in Florida and elsewhere in North America. Their bones are dense, preserve well, and are easily recognized. Gar scales are often very abundant, for example, at the Love and Leisey 1A sites (fig. 4.32A). Isolated skeletal elements cannot always be confidently identified to either *Atractosteus* or *Lepisosteus*. An extinct Pleistocene species of *Atractosteus*, *A. lapidosteus*, was named by O. P. Hay from the Vero site. In his review of the gars, Wiley (1976) regarded it as specifically indeterminate but probably the same as the living species *A. spatula*. The current range of *A. spatula* extends eastward in the Gulf of Mexico only as far as Choctawhatchee Bay in the panhandle. However, in the Pleistocene it was much more widely distributed across Florida, extending south of Tampa Bay and on the Atlantic Coast (Scudder et al., 1995). Fossils of longnose gar, *Lepisosteus osseus*, can be recognized by their elongate skull bones if the appropriate elements are recovered. Otherwise, the various species of *Lepisosteus* have usually not been recognized as fossils except at the generic level because of the difficulty in distinguishing them.

Amia calva, the bowfin, is the only living genus and species of the order Amiiformes (Nelson, 1994). It is a predaceous fish that lives in lakes and sluggish streams of eastern North America (Grande and Bemis, 1998). It is moderately common in Pleistocene freshwater deposits in Florida. The texture of the head bones and the short vertebrae permit ready identification (fig. 4.33B). The large gular plate that covers the throat region is also a distinctive element of the bowfin (fig. 1.4A). Every skeletal element of the bowfin was illustrated in detail by Grande and Bemis (1998). Like gars, *Amia* is a living fossil that has remained virtually unchanged for many

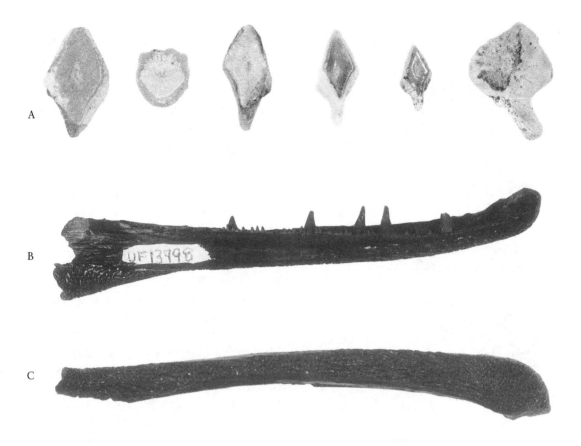

Fig. 4.32. Fossil gars (Lepisosteidae) from Florida. *A*, ganoid scales, showing the variety of shapes, from Love site, Alachua County, late Miocene; *B*, labial, and *C*, ventral views of UF 13998, right dentary of *Atracosteus spatula* from Ichetucknee River, Columbia County, late Pleistocene. *A* about 1×; *B*, *C*, about 0.67×.

millions of years (Maisey, 1996). It is less common as a fossil than gars, probably due to its more limited habitat.

The vast majority of actinopterygians belong to the infraclass Teleostei and are commonly referred to as teleosts. They include about 20,000 living species divided among some 400-plus families (Nelson, 1994). In Florida alone there are about 1,200 living species of teleosts. Much of their evolutionary success relates to modifications of the jaws involving increased mobility of the premaxillae (fig. 4.34) and of the caudal fin for more efficient swimming. Teleosts include both freshwater and saltwater species, herbivores and carnivores. The group first appeared in the middle Triassic, about 235 Ma (Maisey, 1996).

The jaws of teleosts tend to be shorter and more flexibly hinged than in more primitive ray-finned fishes. A supraoccipital bone, not present in earlier types, forms a prominence at the back of the skull, and two supramaxillary bones are located dorsal to the maxilla, another novel teleost feature (fig. 1.4B). The maxilla lacks teeth in advanced teleosts and is used as a lever to move the premaxilla. The tail fin has no fleshy lobe and specializations of the last few vertebrae increase the fin's efficiency (Maisey, 1996). The paired fins are narrow at their bases and quite flexible; in many teleosts the pelvic fins have moved forward, to lie ventral to the pectoral fins. The scales are thin, lack the layers of dentine and ganoine seen in the gar and other primitive actinopterygians, and overlap extensively. All of the following bony fish are teleosts.

Fig. 4.33. Fossil bowfin, *Amia calva*, from Love site, Alachua County, late Miocene. UF specimens. *A*, dorsal view of partial right dentary; *B*, posterior view of a vertebra. About 1×.

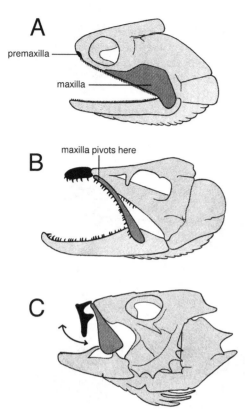

Fig. 4.34. Types of jaws found in ray-finned bony fish. *A*, primitive morphology with large, fixed premaxilla and maxilla. This is the basic jaw structure found in sturgeon, gars, and bowfin. *B*, basic teleost jaw morphology with a more movable maxilla and premaxilla, but both elements still bear teeth. This kind of jaw is found in tarpon, salmon, and pikes, among others. *C*, advanced teleost jaw morphology with reduced, highly movable, edentulous maxilla that functions as a lever to move the premaxilla. This type of jaw is found in acanthopterygian fish such as snapper, sunfish, and drum. Figures from *Analysis of Vertebrate Structure* by Milton Hildebrand, copyright © 1974 by John Wiley & Sons, Inc. Reprinted by permission of John Wiley & Sons, Inc.

At present a relatively small number of teleosts are known from the Eocene through the Pleistocene of Florida. Most of these have just been identified but not thoroughly studied. Of the many subgroups of teleosts, few are known from Florida as fossils (see chap. 3 for a complete listing). These include the Elopamorpha (tarpons, eels), Clupeomorpha (herring, sardines, anchovies), Esociformes (pikes), Ostariophysi (carp, minnows, suckers, catfish), and the Acanthopterygii, a very diverse group of advanced fishes characterized by prominent fin spines.

Elopamorph fishes are grouped together by their unique larval form, the leptocephalus. It is very elongate, thin, translucent, and bears large teeth. Tarpon

(*Megalops atlanticus*) is the most common fossil form of this group. Young tarpon inhabit fresh water, while adults (up to 2.4 m in length) are most common in shallow coastal waters such as bays and estuaries. The thin, large scales of tarpon have been recognized in many late Neogene nearshore marine sites in Florida (fig. 4.35A). Another common elopamorph is the eel (*Anguilla*), which has to date only been identified as a fossil in Florida from the early Pleistocene Leisey Shell Pit (Scudder et al, 1995).

Clupeomorph fishes are characterized by a peculiar jaw joint and an enlarged swim bladder that extends anteriorly into the skull. They are primarily small marine fishes that form large schools and feed on plankton. Their poor fossil record in Florida is probably related to their small size. The late Pliocene Richardson Road Shell Pit, Sarasota County, produced more kinds of clupeomorph fishes (five) than all of the rest of Florida's fossil sites combined (Emslie et al., 1996). All were members of the herring and shad family, the Clupeidae.

The Ostariophysi had their origins in the Cretaceous and include a vast number (about 6,500 species) of primarily freshwater fishes such as minnows, shiners, piranhas, tetras, and catfish (Nelson, 1994). They are characterized by having a complex series of bones (the Weberian apparatus) that transmits high-frequency sounds from the swim bladder to the inner ear. Catfishes (order Siluriformes) are the most common members of this group to be recognized in Florida as fossils. Four genera of catfishes occur as fossils in the Pliocene and Pleistocene. *Bagre* and *Arius* are marine, while *Ameiurus* (formerly included in *Ictalurus*) is found in fresh or brackish waters. Another genus of freshwater catfish, *Pylodictus olivaris,* the flathead catfish, is known from the Pleistocene. *Arius* and *Ameiurus* are also known from the Miocene. *Arius felis,* the sea catfish, is sometimes placed in the genus *Galeichthyes*. Large pectoral spines are the most common, easily recognized fossil remains of catfishes (Lundberg, 1975). Pectoral spines of *Ameiurus* have toothlike projections only on one edge of the spine, whereas those of *Bagre* and *Arius* are serrated on both edges. It is sometimes possible to distinguish species of these genera by details of the pectoral spine (fig. 4.36). The basioccipital (a bone from the base of the skull) of the sea catfish forms a distinctive cross shape, so that it is also called the crucifix fish.

Esociform fishes are long-bodied, predaceous, freshwater fish with no fin spines, posteriorly located pelvic fins, and a toothless maxilla. The pikes and pickerels are placed in the genus *Esox*. The large-toothed jaw elements (dentary and premaxillae) of *Esox* are not un-

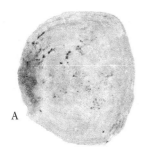

Fig. 4.35. *A*, scale of the tarpon, *Megalops atlanticus*, from Love site, Alachua County, late Miocene. *B*, UF/FGS 4062, lateral view of partial left dentary of *Esox*, from Ichetucknee River, Columbia County, late Pleistocene. Note the large, sharp teeth. Both about 1×.

common in freshwater Pleistocene deposits (fig. 4.35B; Cavender et al., 1970).

The majority of teleosts represented by Florida fossils belong in the Acanthopterygii. This group originated near the end of the Mesozoic and includes a vast number of marine and freshwater fishes (Maisey, 1996). They are characterized by stiff spines that project from the dorsal fins, edentulous maxillae, and usually have well-developed pharyngeal teeth. Locomotion is provided by powerful strokes of the caudal fin, with less of the sinuous body movement seen in other fish. Nearly one hundred families of these spiny-finned fishes live in Florida waters today, but only about a quarter of these are presently known as fossils. Some of the better-known representatives of these families are discussed below.

The living squirrelfishes (Holocentridae) occur in marine waters, especially around coral reefs. *Holocentrites ovalis* is an extinct squirrelfish known from Eocene and Oligocene limestone near Marianna in Jackson County (Conrad, 1941; Dunkle and Olsen, 1959). The unusual ear cavity, the mucus channels in the head, and the serrated and striated bones distinguish *Holocentrites* from other kinds of fish (fig. 4.37). The excellent preservation of the fish from the Marianna Limestone must be attributed to the quiet water situation in which they were deposited.

Two fossil snappers (Lutjanidae), *Hypsocephalus atlanticus* and *Lutjanus avus,* also derive from the limestone beds near Marianna (Gregory, 1930; Swift and Ellwood, 1972). The holotype of *L. avus* consists of most of the skull and the scaly surface of the body (fig. 4.38). This fish was about 2 feet long. It is readily recognized as a snapper by its very strong caniniform teeth. It does not differ greatly from the living red snapper, *Lutjanus campechanus,* that lives in marine waters off Florida today. More incomplete remains of *Lutjanus* are present in Miocene and younger deposits. *Hypsocephalus atlanti-*

cus, an extinct genus and species, is known mostly from a partial skull (figs. 4.39, 4.40). It belongs to a living group of Pacific snappers that is no longer present in the Atlantic (Swift and Ellwood, 1972).

Many species of sunfishes (Centrarchidae) live in Florida's lakes and streams. The shellcracker or redear sunfish, *Lepomis microlophus*, is commonly found as a fossil. It is readily recognized in freshwater deposits of Miocene, Pliocene, or Pleistocene age by its diagnostic pharyngeal grinding mill (figs. 4.41, 4.42). Less common

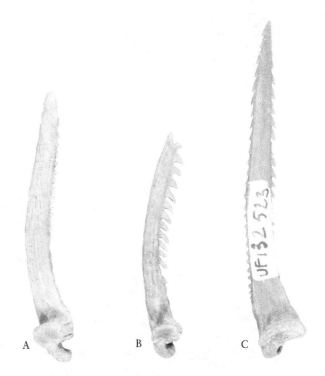

Fig. 4.36. Pectoral spines of fossil catfish from Florida. *A, Ameiurus* sp., cf. *A. brunneus* from Haile 15A, Alachua County, late Pliocene; *B, Ameiurus serricanthus* from Haile 15A; *C, Arius felis* from Rock Springs, Orange County, late Pleistocene. All about 2×.

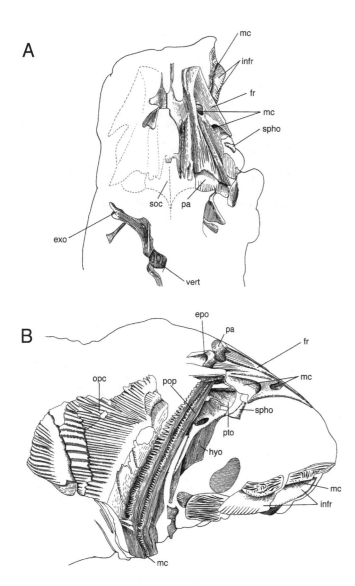

Fig. 4.37. Dorsal and right lateral views of UF/FGS 5776, a partial skull of *Holocentrites ovalis* from Marianna Limestone, Jackson County, early Oligocene. Abbreviations: *epo*, epiotic; *exo*, exoccipital; *fr*, frontal; *hyo*, hyomandibular; *infr*, infraorbital; *mc*, mucus channel; *opc*, opercular; *pa*, parietal; *pop*, preopercular; *pto*, pterotic; *soc*, supraoccipital; *spho*, sphenotic; and *vert*, vertebra. About 2×. After Dunkle and Olsen (1959); reproduced by permission of the Florida Geological Survey, Tallahassee.

are the remains of other species of *Lepomis*, bass (*Micropterus*), and crappie (*Pomoxis*).

Fossil jacks (Carangidae, genus *Caranx*) have been found in many Miocene through Pleistocene sites in Florida. These records are often based on the inflated hyperostotic or "tilly" bones, which are found on the dorsal fin supports, neural arches of the vertebrae, and shoulder girdle of some species of jacks (fig. 4.43). It is not known at what size these growths appear, whether or not they grow with the fish, and how widely among the jacks they are distributed. Similar swollen bony elements are also found in certain species of the Ephippidae, Sciaenidae, Sparidae, and Trichiuridae (Fierstine, 1968; Hewitt, 1983). Since many of these families are known to occur in fossil sites on the basis of diagnostic cranial elements or vertebrae, caution should be exercised in uncritically identifying all inflated fish fossils to the Carangidae (Tiffany et al., 1980). Indeed, the wide range of morphologies exhibited by tilly bones suggests multiple sources from several different types of fish. Further information on tilly bones can be found in Konnerth (1966), Olsen (1971), Weiler (1973), and Hewitt (1983).

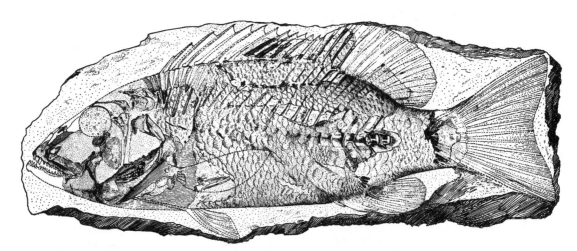

Fig. 4.38. Partially restored specimen of the fossil snapper *Lutjanus avus* from Marianna Limestone, Jackson County, early Oligocene. About 0.33×. After Gregory (1930); reproduced by permission of the Florida Geological Survey, Tallahassee.

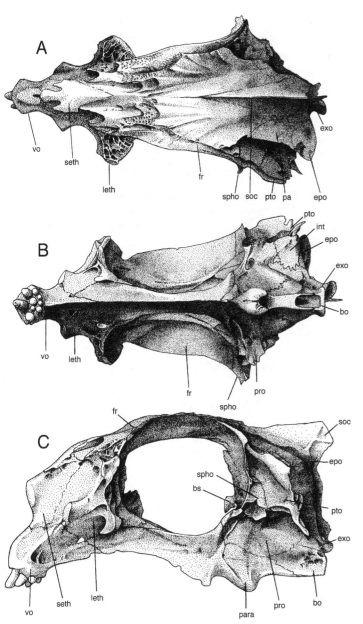

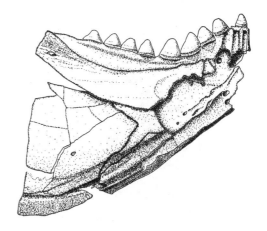

Fig. 4.40. Lateral view of right dentary and partial articular of the fossil snapper *Hypsocephalus atlanticus*. Same specimen as in fig. 4.39. About 3×. After Swift and Ellwood (1972); reproduced by permission of Camm Swift.

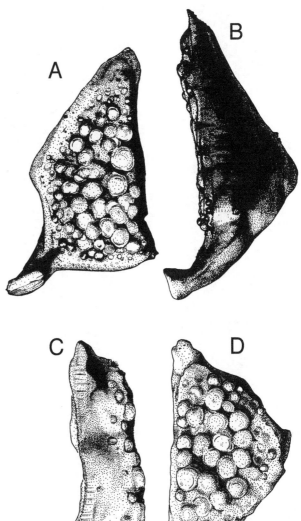

Above: Fig. 4.39. Partial skull of the fossil snapper *Hypsocephalus atlanticus* from Miltons Cave, Jackson County, late Eocene. *A,* dorsal, *B,* ventral, and *C,* lateral views of LACM 27859. Abbreviations: *bo,* basioccipital; *bs,* basisphenoid; *epo,* epiotic; *exo,* exoccipital; *fr,* frontal; *int,* intercalar; *leth,* lateral ethmoid; *pa,* parietal; *para,* parasphenoid; *pro,* prootic; *pto,* pteriotic; *seth,* supraethmoid; *soc,* supraoccipital; *spho,* sphenotic; *vo,* vomer. All about 2×. After Swift and Ellwood (1972); reproduced by permission of Camm Swift.

Right: Fig. 4.41. Pharyngeal grinding mills of the redear sunfish, *Lepomis microlophus. A,* occlusal (dorsal), and *B,* medial views of the left lower pharyngeal; *C,* medial, and *D,* occlusal (ventral) view of the left upper pharyngeal. About 5×. After Colburn et al. (1991); reproduced by permission of the Illinois State Museum, Springfield.

Fig. 4.42. Occlusal view of right lower pharyngeal grinding mill of *Lepomis microlophus* from Haile 15A, Alachua County, late Pliocene. Compare with functionally similar structure in drums (fig. 4.46), which differ by their larger teeth. About 2×.

The porgies (Sparidae) are moderate-sized, omnivorous coastal fishes. Their front teeth are shaped like mammalian incisors or canines (fig. 4.44), while the rear teeth are flatted and molariform. These teeth are common as fossils and are often not recognized as belonging to fish. Three principal genera are known as fossils, the pinfish *Lagodon,* the sheepshead *Archosargus,* and the porgy *Diplodus.* The anterior "incisors" of these have different morphologies (Caldwell, 1958).

Twenty-two species of drum or croaker (Sciaenidae) occur along the coasts of Florida today and are among the most common nearshore fish in the Gulf of Mexico (Hoese and Moore, 1977). All are shallow-water species and most reach a foot or more in length. Many have large, distinctive calcareous ear stones (otoliths), which function in hearing. They also have "mills" of pharyngeal grinding teeth in the gill region that produce sounds used in courtship. Many of the pebblelike fish teeth in fossil deposits come from the pharyngeal grinding mills of drums (figs. 4.45, 4.46). Most have been referred to

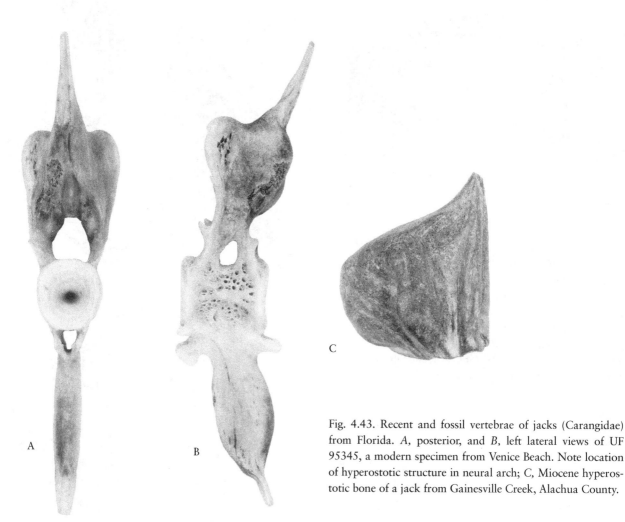

Fig. 4.43. Recent and fossil vertebrae of jacks (Carangidae) from Florida. *A,* posterior, and *B,* left lateral views of UF 95345, a modern specimen from Venice Beach. Note location of hyperostotic structure in neural arch; *C,* Miocene hyperostotic bone of a jack from Gainesville Creek, Alachua County.

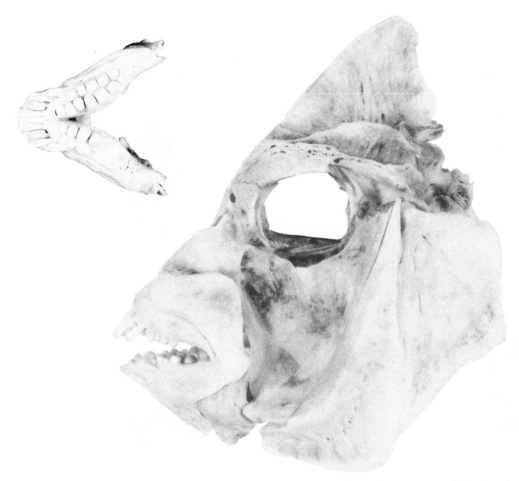

Fig. 4.44. Lateral view of the skull and occlusal view of the dentaries of a modern specimen of the sheepshead *Archosargus probatocephalus* from Florida. Note the mammal-like anterior "incisors." About 1×.

Pogonias cromis, the black drum, although other species are also represented. As the isolated teeth are similar to the molariform teeth found in porgies, they are difficult to identify when not found in place in the tooth-bearing element. As was the case with the Clupeidae, the Richardson Road Shell Pit of Sarasota County produced a diverse assemblage of the sciaenid family, with nine recognized species (Emslie et al., 1996). In addition to black drum, the silver perch, seatrout, red drum, star drum, and spot were also identified.

Fossil barracudas (Sphyraenidae) are commonly recognized by their sharp, triangular, bladelike teeth (fig. 4.47). Their teeth may be distinguished from other similar teeth by the razor-sharp cutting edge that passes the length of at least one edge of the tooth. Some are sharp on only one edge; others on both edges. Fossil barracudas are seemingly common in all marine fossiliferous deposits in Florida, from the Eocene through the Pleisto-

cene. *Sphyraena barracuda,* the modern great barracuda, lives in marine waters off Florida today and reaches a length of up to 2 meters.

Wrasses (Labridae) are abundant today in coral reefs and coastal waters. This family and the Scaridae, the parrotfishes, are closely related. They differ in that wrasses have individual teeth that are not fused to form a beak. Both families are often bright colored, possess characteristic pharyngeal teeth, move by flapping the pectoral fins, and occur around reefs and rocky areas. Wrasses first appear in the Miocene in the Florida fossil record.

Remains of porcupinefish (Diodontidae) are, next to shark and ray elements, possibly the most abundant fossilized remains in Florida. Fossil porcupinefish are usually represented by the upper or lower half of their beaklike jaws (fig. 4.48). These "beaks" in fact represent fused teeth and jawbones. A fossil species of porcupinefish, *Diodon circumflexus,* was based on an upper

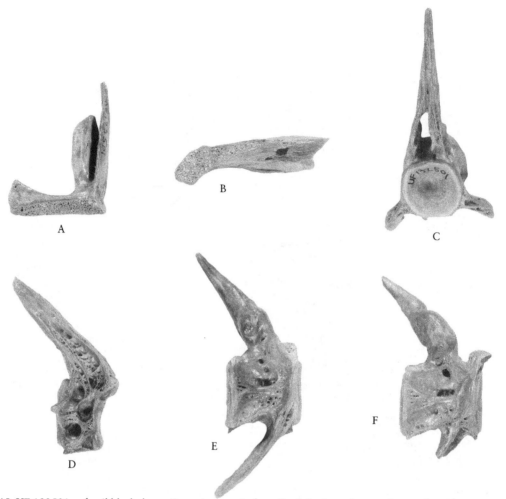

Fig. 4.45. UF 132501, a fossil black drum, *Pogonias cromis*, from Rock Springs, Orange County, late Pleistocene. *A*, medial view of left premaxilla; *B*, occlusal view of right dentary; *C*, anterior view of a vertebra; *D–F*, lateral views of three vertebrae. All about 1×.

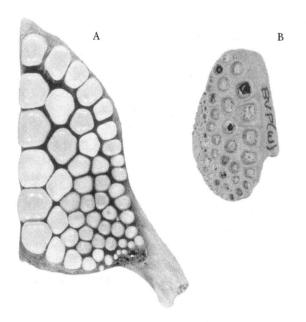

Fig. 4.46. Recent and fossil pharyngeal grinding mills of the black drum, *Pogonias cromis*. *A*, occlusal view of UF 95340, right lower pharyngeal of a modern specimen; *B*, upper pharyngeal from West Palmetto Mine, Polk County, early Pliocene. As is common in fossil specimens, the loosely attached teeth have all fallen out, but several of the unerupted, dark-colored replacement teeth are visible through breaks in the bone. Both about 1×.

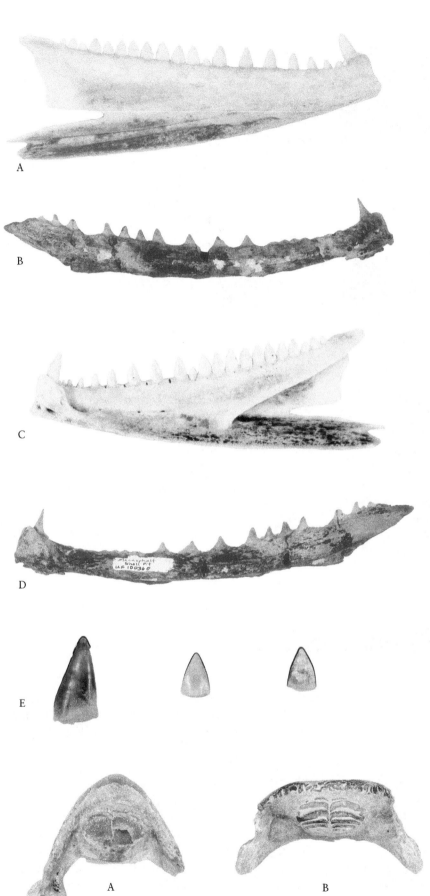

Left: Fig. 4.47. Recent and fossil barracuda, *Sphyraena barracuda*. *A,* lateral, and *C,* medial views of the right dentary of a modern specimen from the Cayman Islands. *B,* lateral, and *D,* medial views of UF 100360, a partial right dentary from APAC (=Macasphalt) Shell Pit, Sarasota County, late Pliocene; *E,* UF 58322, three barracuda teeth from Fort Green Mine, Polk County, early Pliocene. The enameloid cutting edges are more darkly preserved than the dentine and thus more clearly defined than in *A* or *C. A–D* about 0.67×; *E* about 1×.

Below: Fig. 4.48. Fossil porcupine and burr fish (Diodontidae) from Florida. *A,* occlusal view of UF 137610, fused dentaries of an indeterminate genus from Fishbone Cave, Marion County, late Eocene; *B,* occlusal view of the fused dentaries of a UF specimen of *Diodon* sp. from North Palmetto Mine, Polk County, early Pliocene; *C,* occlusal view of the fused premaxillae of a UF specimen of *Chilomycterus* sp. from Love site, Alachua County, late Miocene.

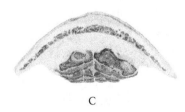

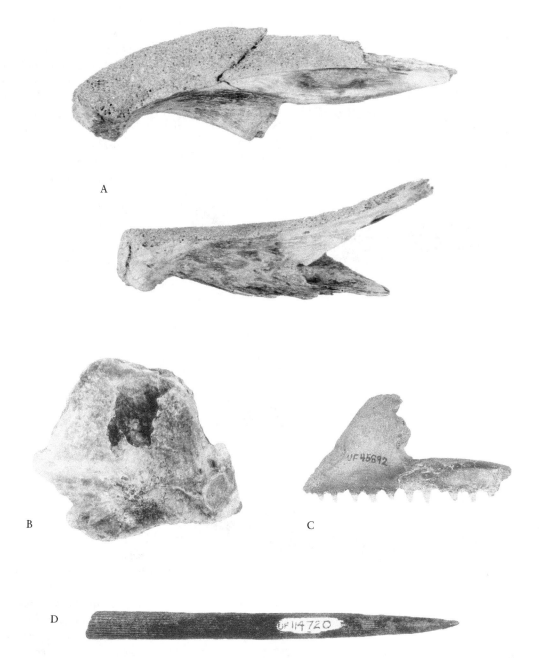

Fig. 4.49. Fossils of various large teleost fish from Florida. *A*, dentaries in occlusal and medial views of a snook (*Centropomus* sp.) from McGehee Farm site, Alachua County, late Miocene; *B*, medial view of an opercular of a mullet (*Mugil* sp.) from St. Johns Lock, Putnam County, late Pleistocene; *C*, lateral view of UF 45892, left premaxilla of a mackerel (Scombridae, genus indeterminate) from Fort Green Mine, Polk County, early Pliocene; *D*, UF 114720, distal end of the rostrum of the extinct swordfish *Cylindracanthus* sp. from Santa Fe River, Columbia County, late Eocene.

jaw from the middle Miocene near Kingston, Florida. Modern porcupinefishes live near shore, generally near reefs, where they eat shelled prey such as mollusks and hermit crabs. Triggerfish (Balistidae) are a related group of tropical, shallow water marine fishes. Their diet includes sea urchins, crabs, and octopus. They get their name from their long first dorsal spine, which is moved with a triggering mechanism involving the smaller sec-

ond spine. Fossil remains of *Balistes* include spines and teeth.

Figure 4.49 shows a few of the larger perciform marine fish known from Florida. Their modern relatives are all highly prized by today's fisherman. Marlin and mackerel have a fossil record dating back to the Pliocene, snook to the Miocene, and swordfish to the Eocene.

REFERENCES

Applegate, S. P. 1965. Tooth terminology and variation in sharks with special reference to the sand shark, *Carcharias taurus* Rafinesque. Contributions in Science, 86:1–18.

Applegate, S. P., and L. Espinosa-Arrubarrena. 1996. The fossil history of *Carcharodon* and its possible ancestor *Cretolamna*: a study in tooth identification. Pp. 19–36 *in* A. P. Klimley and D. G. Ainley, *Great White Sharks: The Biology of* Carcharodon carcharias. Academic Press, San Diego.

Bendix-Almgreen, S. E. 1983. *Carcharodon megalodon* from the upper Miocene of Denmark, with comments on elasmobranch tooth enameloid: coronoïn. Bulletin of the Geological Society of Denmark, 32:1–32.

Bigelow, H. B., and W. C. Schroeder. 1948. Fishes of the Western North Atlantic: Part 1, Sharks. Memoir of the Sears Foundation for Marine Research, Yale University, New Haven.

———. 1953. Fishes of the Western North Atlantic: Part 2, Sawfishes, guitarfishes, skates and rays. Memoir of the Sears Foundation for Marine Research, Yale University, New Haven.

Bor, T. J. 1990. A new species of mobulid ray (Elasmobranchi, Mobulidae) from the Oligocene of Belgium. Contributions to Tertiary and Quaternary Geology, 27:93–97.

Briggs, J. C. 1958. A list of Florida fishes and their distribution. Bulletin of the Florida State Museum, 2:223–318.

Caldwell, D. K. 1958. Fossil fish teeth of the family Sparidae from Florida. Quarterly Journal of the Florida Academy of Sciences, 21:113–16.

Cappetta, H. 1987. *Chondrichthyes II. Mesozoic and Cenozoic Elasmobranchii. Handbook of Paleoichthyology, Volume 3B*. Gustav Fischer Verlag, Stuttgart, 193 p.

Case, G. R. 1981. Late Eocene selachians from south-central Georgia. Palaeontographica, Abteilung A, 176:52–79.

Case, G. R., and H. Cappetta. 1990. The Eocene selachian fauna from the Fayum Depression in Egypt. Palaeontographica, Abteilung A, 212:1–30.

Casier, E. 1960. Note sur la collection des poissons Paléocènes et Éocènes de l'enclave de Cabinda (Congo). Annales du Musée Royal du Congo Belge, Serie 3, 1(2):1–48.

Cavender, T. M., J. G. Lundberg, and R. L. Wilson. 1970. Two new fossil records of the genus *Esox* (Teleostei, Salmoniformes) in North America. Northwest Science, 44:176–83.

Colburn, M. L., L. Kelly, and J. Snider. 1991. Redear sunfish in the late Holocene of Illinois. Pp. 67–79 *in* J. R. Purdue, W. E. Slippel, and B. W. Styles (eds.), *Beamers, Bobwhites, and Blue-Points: Tributes to the Career of Paul W. Parmalee*. Illinois State Museum Scientific Papers, Volume 23.

Compagno, L. J. V. 1984. *FAO Species Catalogue*. Vol. 4, *Sharks of the World*. Pt. 1, *Hexanchiformes to Lamniformes*. United Nations Development Programme, Food and Agriculture Organization of the United Nations, Rome, 249 p.

———. 1988. *Sharks of the Order Carchariniformes*. Princeton University Press, Princeton, N.J., 486 p.

Conrad, G. M. 1941. A fossil squirrelfish from the upper Eocene of Florida. Florida Geological Survey Bulletin, 22:9–25.

DeMuizon, C., and T. J. DeVries. 1985. Geology and paleontology of late Cenozoic marine deposits in the Sacaco area (Peru). Geologische Rundschau, 74:547–63.

Dockery, D. T., III, and E. M. Manning. 1986. Teeth of the giant shark *Carcharodon auriculatus* from the Eocene and Oligocene of Mississippi. Mississippi Geology, 7(1):7–19.

Dunkle, D. H., and S. J. Olsen. 1959. Description of a beryciform fish from the Oligocene of Florida. Florida Geological Survey Special Publication, 5:1–20.

Emslie, S. D., W. D. Allmon, F. J. Rich, J. H. Wrenn, and S. D. de France. 1996. Integrated taphonomy of an avian death assemblage in marine sediments from the late Pliocene of Florida. Palaeogeography, Palaeoclimatology, Palaeoecology, 124:107–36.

Fierstine, H. L. 1968. Swollen dorsal fin elements in living and fossil *Caranx* (Teleostei: Carangidae). Contributions in Science, 137:1–10.

Gottfried, M. D., L. J. V. Compagno, and S. C. Bowman. 1996. Size and skeletal anatomy of the giant "megatooth" shark *Carcharodon megalodon*. Pp. 55–66 *in* A. P. Klimley and D. G. Ainley, *Great White Sharks: The Biology of* Carcharodon carcharias. Academic Press, San Diego.

Grande, L., and W. E. Bemis. 1998. A comprehensive phylogenetic study of amiid fishes (Amiidae) based on comparative skeletal anatomy. An empirical search for interconnected patterns of natural history. Society of Vertebrate Paleontology Memoir, 4:1–690.

Gregory, W. K. 1930. A fossil teleost fish of the snapper family (Lutianidae) from the lower Oligocene of Florida. Florida Geological Survey Bulletin, 5:7–17.

Hewitt, R. A. 1983. Teleost hyperostoses: a case of Miocene problematica from Tunisia. Tertiary Research, 5:63–70.

Hildebrand, M. 1974. *Analysis of Vertebrate Structure*. John Wiley and Sons, New York, 710 p.

Hoese, H. D., and R. H. Moore. 1977. *Fishes of the Gulf of Mexico*. Texas A & M Press, College Station, 327 p.

Hubbell, G. 1996. Using tooth structure to determine the evolutionary history of the white shark. Pp. 9–18 *in* A. P. Klimley and D. G. Ainley, *Great White Sharks: The Biology of* Carcharodon carcharias. Academic Press, San Diego.

Kent, B. W. 1994. *Fossil Sharks of the Chesapeake Bay Region*. Egan Rees & Boyer, Columbia, Md., 146 p.

Keyes, I. W. 1972. New records of the elasmobranch C. *megalodon* (Agassiz) and a review of the genus *Carcharodon* in the New Zealand fossil record. New Zealand Journal of Geology and Geophysics, 15:228–42.

Konnerth, A. 1966. Tilly bones. Oceanus, 12(2):6–9.

Kozuch, L., and C. Fitzgerald. 1989. A guide to identifying shark centra from southeastern archaeological sites. Southeastern Archaeology, 8:146–57.

Leriche, M. 1942. Contribution à l'étude des faunes ichthyologiques marines des terrains tertiaires de la plaine côtière Atlantique et du centre des États-Unis; le synchronisme des formations tertiaires des deux côtes de l'Atlantique. Mémoires de la Société Géologique de France, Nouvelle Série, 45(2–4):1–112.

Long, D. J. 1992. Sharks from the La Meseta Formation (Eocene), Seymour Island, Antarctic Peninsula. Journal of Vertebrate Paleontology, 12:11–32.

———. 1993. Late Miocene and early Pliocene fish assemblages from the north central coast of Chile. Tertiary Research, 14:117–26.

Long, J. A. 1995. *The Rise of Fishes: 500 Million Years of Evolution*. Johns Hopkins University Press, Baltimore, 223 p.

Longbottom, A. E. 1979. Miocene shark's teeth from Ecuador. Bulletin of the British Museum (Natural History), Geology, 32:57–70.

Lundberg, J. G. 1975. The fossil catfishes of North America. University of Michigan Papers on Paleontology, No. 11, 51 p.

Maisey, J. G. 1996. *Discovering Fossil Fishes*. Henry Holt and Co., New York, 223 p.

Manning, E., and B. R. Standhardt. 1986. Late Eocene sharks and rays of Montgomery Landing, Louisiana. Pp. 133–61 *in* J. A. Schiebout and W. van den Bold (eds.), *Montgomery Landing Site, Marine Eocene (Jackson) of Central Louisiana*. Gulf Coast Association of Geological Societies, Baton Rouge.

Moody, R. T. J. 1972. The Turtle Fauna of the Eocene Phosphates of Metlaoui, Tunisia. Proceedings of the Geologists' Association (London), 83:327–36.

Morgan, G. S. 1989. Miocene vertebrate faunas from the Suwannee River Basin of north Florida and south Georgia. Pp. 26–53 in G. S. Morgan (ed.), *Miocene Paleontology and Stratigraphy of the Suwannee River Basin of North Florida and South Georgia*. Southeastern Geological Society Guidebook No. 30.

———. 1994. Miocene and Pliocene marine mammal faunas from the Bone Valley Formation of Central Florida. Pp. 239–68 in A. Berta and T. A. Deméré (eds.), *Contributions in Marine Mammal Paleontology Honoring Frank C. Whitmore, Jr.* San Diego Natural History Society, San Diego.

Morgan, G. S., and A. E. Pratt. 1983. Recent discoveries of late Tertiary marine mammals in Florida. Plaster Jacket, 43:1–30.

Naylor, G. J. P., and L. F. Marcus. 1994. Identifying isolated shark teeth of the genus *Carcharhinus* to species: relevance for tracking phyletic change through the fossil record. American Museum Novitates, 3109:1–53.

Nelson, J. S. 1994. *Fishes of the World, 3rd Edition*. John Wiley and Sons, New York, 600 p.

Olsen, S. J. 1971. Swollen bones in the Atlantic cutlassfish, *Trichiurus lepturus*. Copeia 1971:174–75.

Purdy, R. W. 1996. Paleoecology of fossil white sharks. Pp. 67–78 in A. P. Klimley and D. G. Ainley, *Great White Sharks: The Biology of* Carcharodon carcharias. Academic Press, San Diego.

Scudder, S. J., E. H. Simons, and G. S. Morgan. 1995. Osteichthyes and Chondrichthyes from the Leisey Shell Pit local fauna, Hillsborough County, Florida. Bulletin of the Florida Museum of Natural History, 37:251–72.

Siverson, M. 1995. Revision of the Danian cow sharks, sand tiger sharks, and goblin sharks (Hexanchidae, Odontaspididae, and Mitsukurinidae) from southern Sweden. Journal of Vertebrate Paleontology, 15:1–12.

Steurbaut, E., and J. Herman. 1978. Biostratigraphie et poissons fossiles de la formation de l'argile de Boom (Oligocène moyen du Bassin belge). Geobios, 11:297–325.

Swift, C., and B. Ellwood. 1972. *Hypsocephalus atlanticus*, a new genus and species of lutjanid fish from marine Eocene limestones of northern Florida. Contributions in Science, 230:1–29.

Tessman, N. 1969. The fossil sharks of Florida. Master's thesis, University of Florida, Gainesville, 132 p.

Tiffany, W. J., R. E. Pelham, and F. W. Howell. 1980. Hyperostosis in Florida fossil fishes. Florida Scientist, 43:44–49.

Uyeno, T., O. Sakamoto, and H. Sekine. 1989. Description of an almost complete tooth set of *Carcharodon megalodon from a middle Miocene bed in Saitama Prefecture, Japan. Bulletin of the Saitama Museum of Natural History, 7:73–85.*

Ward, D. J. 1979. *Additions to the fish fauna of the English Palaeogene. 3. A review of the hexanchid sharks with a description of four new species. Tertiary Research, 2:111–29.*

Webb, S. D., and N. Tessman. 1968. *A Pliocene vertebrate fauna from low elevation in Manatee County, Florida. American Journal of Science, 266:777–811.*

Weiler, W. 1973. *Durch Hyperostose verdickte Fischknochen aus dem oberen Sarmat von Nord-Carolina, USA. Senckenbergiana Lethaea, 53:469–77.*

Welton, B. J., and W. J. Zinsmeister. 1980. *Eocene neoselachians from the La Meseta Formation, Seymour Island, Antarctic Peninsula. Contributions in Science, 329:1–10.*

Westgate, J. W. 1984. *Lower vertebrates from the late Eocene Cow Creek local fauna, St. Francis County, Arkansas. Journal of Vertebrate Paleontology, 4:536–46.*

———. 1989. *Lower vertebrates from an estuarine facies of the middle Eocene Laredo Formation (Claiborne Group), Webb County, Texas. Journal of Vertebrate Paleontology, 9:282–94.*

Wiley, E. O. 1976. *The phylogeny and biogeography of fossil and recent gars (Actinopterygii: Lepisosteidae). Miscellaneous Publication, University of Kansas Museum of Natural History, 64:1–111.*

5

Amphibia

Frogs, Toads, and Salamanders

Introduction to Tetrapods and Amphibians

Colonization of the land by plants and animals, one of the major events in the history of life on earth, occurred in the Paleozoic Era. The oldest fossils of land plants are Ordovician (see fig. 1.15 for time scale; Gray and Shear, 1992). Land plants remained small and simple until the Devonian Period, when they diversified into numerous shrub- and tree-sized forms. Many terrestrial arthropods (including scorpions, centipedes, spiders, and insects) also first appeared during the middle Paleozoic. Vertebrates were relative latecomers to this process, not appearing on land until the late Devonian, and fully terrestrial vertebrates did not evolve until the Carboniferous. The evolutionary transition from aquatic lobe-finned fishes such as *Eusthenopteron* and *Panderichthyes* to early tetrapods such as *Ichthyostega* and *Acanthostega* is reasonably well documented by fossils and capably described in many accessible books (for example, Long, 1995; Benton, 1997; Cowen, 2000).

Tetrapods are the vertebrate group comprising modern and fossil amphibians, reptiles, birds, and mammals, as well as a few fossil taxa that do not easily fit into any of these groups. To contrast them from aquatic fishes, tetrapods are often described as being the "land vertebrates," but this is an overgeneralization. The earliest tetrapods were primarily aquatic (Coates and Clack, 1991; Coates, 1996), and throughout tetrapod history there have been many partial to fully aquatic taxa. Familiar living examples include the crocodiles and alligators, most turtles, penguins, seals, and whales.

What then characterizes tetrapods as a group? The etymological derivation of the word *tetrapod* from its Greek roots is "four feet." This is most appropriate because the paired limbs of tetrapods terminate in a jointed series of either wrist (carpal) or ankle (tarsal) skeletal elements and then individual digits composed of a metapodial and phalanges (figs. 1.8, 5.1, 5.2). The paired lateral fins of fish lack these structures. Of course some

tetrapods, such as snakes, have secondarily reduced or even completely lost their limbs, but this does not exclude them from membership in the Tetrapoda. Other characteristics of tetrapods also relate to the limbs. The shoulder (pectoral) and pelvic skeletal elements are strengthened and their connections with the axial skeleton are improved. Notably the ischium of the pelvis is directly connected to one or more sacral vertebrae. A number of flaplike opercular bones are present in the rear portion of the fish skull. These protect the gills, but also directly connect to the pectoral elements. Although the earliest tetrapods retained some of these opercular bones (they were later lost in the course of tetrapod evolution), they were reduced in size and no longer in articulation with the shoulder elements. This allowed independent movement of the forelimb and head.

Following their first appearance in the late Devonian, tetrapod diversity increased dramatically in the Mississippian Period, with more than twenty recognized families. They varied from small salamander-, lizard-, and snakelike forms to large, bulky plodders (some were more than 2 meters long). Which of these early tetrapods can truly be called amphibians depends on interpretation of their evolutionary relationships with the modern amphibians, the Lissamphibia. A number of studies of the evolutionary relationships of Paleozoic amphibians concluded that the extinct Temnospondyli are the closest relatives of the Lissamphibia (for example, Panchen and Smithson, 1988; Milner, 1990; Bolt, 1991), and this idea has diffused to most current textbooks. This hypothesis has been challenged by the results of Laurin and Reisz (1997), which offers a different group of Paleozoic amphibians, the Lepospondyli, as the closest relatives of lissamphibians. These results, if substantiated by further work, imply a vastly different pattern for Paleozoic tetrapod evolution than was previously theorized. The Lissamphibia comprises three living orders: the Gymnophonia, Caudata, and Anura. The elongate, worm- or snakelike Gymnophonia (also called caecilians or apodans) are the poorest known of the living amphibians. Most are burrowers, with one aquatic family. More than 160 species of caecilians live in the tropical regions of Asia, Africa, and South and Central America (Duellman and Trueb, 1986). They have a very poor fossil record that begins in the Jurassic (Jenkins and Walsh, 1993) and are not known from Florida.

The two other lissamphibian orders, the Caudata and Anura, are common in Florida today, and all amphibian fossils found in the state belong to either one or the other. According to Ashton and Ashton (1988), twenty-eight living species of anurans (frogs and toads) and twenty-four species of caudatans (salamanders, newts, and sirens) are native to the state. A full listing of species represented as fossils is found in chapter 3. The Caudata and Anura have been separate groups since the Triassic Period and may have diverged even earlier than that. With one exception, Florida's fossil amphibians generally resemble living forms, and most are classified into extant genera.

ANURAN ANATOMY AND NATURAL HISTORY

Most of this section is based on information presented in Duellman and Trueb (1986) or Ashton and Ashton (1988). Anurans have a number of specialized skeletal features related to their peculiar hopping mode of locomotion (figs. 5.1, 5.2). The hindlimb is much longer than the forelimb. The radius and ulna are fused into a single element, as are the tibia and fibula. The astragalus and calcaneum are elongated and fused proximally and distally. Four digits are present on the forelimb and five on the hindlimb. The three elements of the pelvis—the ilium, ischium, and pubis—do not fuse together (so they are recovered separately as fossils), and the pubis is usually cartilaginous. The ilium has a long shaft that articulates with the transverse process of the sacral vertebra (fig. 5.2A). The shape of the ilium, especially the morphology of the anterior shaft, is most often used to identify fossil anurans (Holman, 1995:85–88). This structure is so useful because many leg muscles originate from it, and their development greatly depends on the individual species' ecology. Especially different are ilia of species that are primarily terrestrial and do little swimming (for example, *Bufo*, the common toad), and those of swimmers like *Rana*, the true frogs. Most other postcranial skeletal elements, although easily identified as anuran, are difficult if not impossible to identify to genus or even family.

Anurans have nine or fewer vertebrae between the skull and the sacrum and no caudal vertebrae (nor an external tail in the adult stage; *anura* means "without a tail" in Greek). Anurans from Florida have procoelus vertebrae. The lone exception is that the vertebra immediately anterior to the sacrum is amphicoelus in the Ranidae and Microhylidae. A rodlike structure, the urostyle (fig. 5.2A), ossifies posterior to the sacrum. Anteriorly it bears two (one in some primitive frogs) concav-

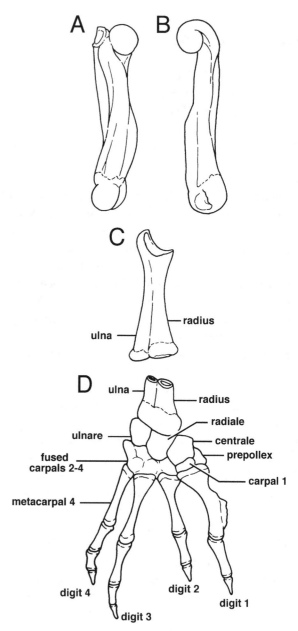

Fig. 5.1. Skeletal elements of the right forelimb of the frog *Rana*. *A*, medial, and *B*, lateral views of the humerus; *C*, dorsal view of the fused radius and ulna (radioulna); *D*, dorsal view of the manus. After Duellman and Trueb (1986); reproduced by permission of the McGraw-Hill Companies, New York.

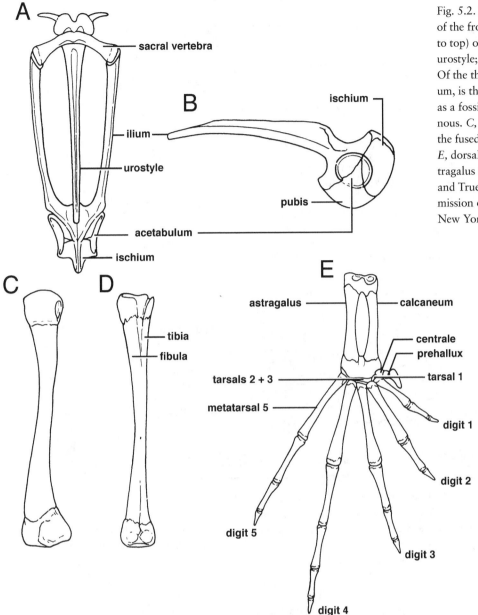

Fig. 5.2. Pelvic girdle and right hind limb of the frog *Rana*. *A*, dorsal view (anterior to top) of pelvis, sacral vertebra, and the urostyle; *B*, left lateral view of the pelvis. Of the three elements, the largest, the ilium, is the one most commonly recovered as a fossil. The pubis is often cartilaginous. *C*, ventral, and *D*, dorsal views of the fused tibia and fibula (the tibiofibula). *E*, dorsal view of the pes. Note fused astragalus and calcaneum. After Duellman and Trueb (1986); reproduced by permission of the McGraw-Hill Companies, New York.

ities that articulate with one or two convex projections on the sacral vertebra.

The skull of anurans is broad, flattened, and approximately D-shaped (fig. 5.3A–C). Much of the skull is open. Small, spatulate, bicuspid teeth are found on the premaxilla and maxilla (and possibly the vomer). These are completely absent in some families (for example, Bufonidae), and enlarged in a few of the more carnivorous species to form fangs. The lower jaw contains three pairs of bony elements (fig. 5.3D, E). The mandibular symphysis is formed by a piece of ossified cartilage called the mentomeckelian bone. The other two mandibular elements are the dentary and angulosplenial. The latter is formed by fusion of the angular, splenial, and articular elements and forms the jaw joint with the skull. Almost all anurans lack lower teeth. In Florida, anuran fossil cranial elements are almost always found separated from one another and are difficult to identify to family. Parasphenoids, frontoparietals, and angulosplenials are among the cranial elements most commonly recovered as fossils.

Most anurans simply lay their numerous eggs in water and lack any type of parental care, although some exceptions do exist. After hatching from the egg, the anuran begins life as a specialized aquatic form, the tadpole. Tadpoles of anurans differ from larval salamanders in

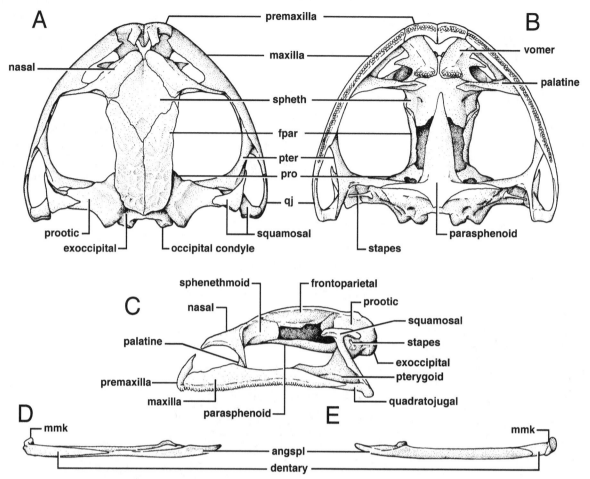

Fig. 5.3. Skull and mandibular bones of the frog *Gastrotheca*. *A*, dorsal, *B*, ventral, and *C*, left lateral views of the skull; *D*, lateral, and *E*, medial views of the left mandible. Abbreviations: *anglspl*, angulosplenial (the combined angular, splenial, surangular, and articular bones); *fpar*, frontoparietial (combined frontal and parietal); *max*, maxilla; *mmk*, mentomeckelian bone; *pmax*, premaxilla; *pro*, prootic; *prsph*, parasphenoid; *pter*, pterygoid; *qj*, quadratojugal; *spheth*, sphenethmoid. After Duellman and Trueb (1986); reproduced by permission of the McGraw-Hill Companies, New York.

lacking limbs (except just prior to transformation) and true teeth. Most have rows of rasping structures in their mouths that they use to scrape or filter algae. Tadpoles of some species prey on other tadpoles, even those of the same species. Fossil tadpoles are rare and not known from Florida. They require extraordinary preservation conditions, as the skeleton is entirely cartilaginous. During transformation to the adult stage, limbs are grown and the gills and tail are lost. While one typically thinks of anurans as aquatic creatures, many of Florida's anurans spend little time in water as adults except to breed. The true, narrowmouth, and spadefoot toad families are primarily terrestrial, while the treefrogs (Hylidae), as the name implies, live in trees and shrubs. Only the true frogs (Ranidae) spend most of their lives in aquatic habitats.

FOSSIL FROGS AND TOADS OF FLORIDA

Anurans are well represented as fossils in Florida and are relatively well studied. Their fossils are found at almost every productive site, except those with a strong marine influence or those that only produce large specimens. Anuran fossils are sometimes found in nearshore marine or brackish water sites (environments where frogs do not live), presumably as a result of transport after death by water currents or predators. The oldest known anurans from Florida come from the Oligocene I-75 site near Gainesville. Three families are present, the Pelobatidae, Bufonidae, and Ranidae. These and anurans from some latest Oligocene to earliest Miocene localities have as yet not been thoroughly studied.

Early Miocene frogs and toads are well known thanks

to Gilchrist County's Thomas Farm site (fig. 5.4). The frog fauna from Thomas Farm is quite remarkable. Nine genera and thirteen species have been recognized at Thomas Farm, although some of these records are probably invalid. Of the total, one genus and eight species are now extinct. All five anuran families native to modern Florida are present. The Leptodactylidae, a tropical family common in Caribbean islands has also been reported from Thomas Farm, but this identification is now discounted. The most diverse family is the Hylidae (fig. 5.4A–D). Thomas Farm hylids include the extinct genus *Proacris,* which is thought to be related to *Acris,* the extant cricket frog. An extinct species of cricket frog, *Acris barbouri,* is also present at the site, as are three extinct species of *Hyla,* the living genus of treefrog. A small toad (*Bufo praevius;* fig. 5.4E) is also extinct, as are two of three true frogs (*Rana;* fig. 5.4F). The toad is the most abundant anuran at Thomas Farm. Also, three living species, the southern leopard frog (*Rana utricularia*), the eastern spadefoot toad (*Scaphiopus holbrooki*), and the eastern narrowmouth toad (*Gastrophryne carolinensis*) have all been tentatively identified from Thomas Farm. If correctly identified, these species have lived in the Southeast for the last 18 million years.

The middle Miocene represents another gap in our knowledge of Florida anurans, but frogs and toads are known from the late Miocene through the end of the Pleistocene. Most of these represent either modern species or are closely related to living forms. Two extinct species have been named from this interval, a late Miocene toad, *Bufo tiheni* (fig. 5.5) and a small, late Pleistocene treefrog, *Hyla baderi.* The latter was named from Arredondo, which is a sinkhole/fissure-fill complex southwest of Gainesville in Alachua County. Ten addi-

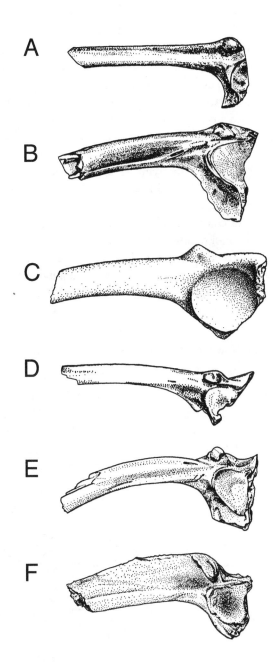

Fig. 5.4. Lateral views of anuran ilia from the Thomas Farm site, Gilchrist County, early Miocene. *A,* right ilium of *Acris barbouri,* about 16× (reversed to appear as from the left side); *B,* left ilium of *Hyla miofloridana,* about 5×; *C,* right ilium of *Microhyla* sp., about 12× (reversed). *D,* right ilium of *Hyla goini,* about 6× (reversed); *E,* left ilium of *Bufo praevius,* about 4×. This small toad is the most common anuran at Thomas Farm. *F,* left ilium of *Rana miocenica,* about 3×. *A–B, D–F* after Holman (1967); reproduced by permission of the Florida Academy of Sciences, Orlando. *C* after Auffenberg (1956); reproduced by permission of the Museum of Comparative Zoology, Cambridge, Massachusetts.

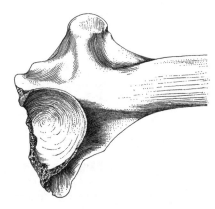

Fig. 5.5. Lateral view of UF 6363, partial right ilium of *Bufo tiheni* from Haile 6A, Alachua County, late Miocene. About 9×. After Auffenberg (1957); reproduced by permission of the Florida Academy of Sciences, Orlando.

tional species of anurans (all living species) were also found at this single site, exemplifying Florida's rich record of Pleistocene anurans. Other especially productive sites are Reddick (Marion County), Williston 3A and Devils Den (Levy County), Vero Beach (Indian River County), and Cutler Hammock (Dade County). The latter included an especially plentiful sample of *Scaphiopus holbrooki,* as well as the only fossil record of a leptodactylid frog from the state (Emslie and Morgan, 1995).

Just because they so resemble modern species does not mean that Florida's fossil frogs are without interest. In fact, because of this, paleoecologists find them useful barometers of climatic conditions. For example, the plentiful hylids and ranids at Thomas Farm indicate a fairly moist environment, while at the late Pliocene Inglis 1A site anurans especially adapted to drier conditions such as *Scaphiopus,* the spadefoot toad, and *Rana areolata,* the gopher frog, are abundant.

SALAMANDER ANATOMY AND NATURAL HISTORY

Salamanders are less specialized than anurans, retaining the primitive elongated body, long tail, and undulatory locomotive pattern of their Paleozoic ancestors. The oldest known salamander is the late Jurassic *Karaurus,* from the Republic of Kazakhstan in western Asia. Fossil salamanders are recognized in North America beginning in the Cretaceous; in Florida, from the Oligocene onward. Most salamanders are small, but aquatic forms such as *Siren* and *Amphiuma* can be up to 3 feet long.

Individuals of most families of salamanders go through an aquatic larval stage (with gills and a tail fin) after hatching from an egg laid in water, and then metamorphose into the adult form. Unlike anuran tadpoles, salamander larvae have legs and are otherwise more similar to the adult stage. A common trend in many salamander families is for incomplete metamorphosis of the aquatic larvae. They do not transform into terrestrial adults, but instead into sexually mature, fully aquatic adults that retain gills. Such a salamander is described as being neotenic. Members of the families Cryptobranchidae, Sirenidae, Amphiumidae, and Proteidae have this type of life history, as do various species of other families. Of the latter, those occurring in Florida are the newt *Notophthalmus* and some species of *Ambystoma.* In these two, neoteny is variable within a species and dependent on local environmental conditions. Species in the Plethodontidae, however, lay their eggs on land, have direct development into the adult stage, and lack an aquatic larval stage.

The vertebral column of salamanders is divided into four main regions. The single neck or cervical vertebra, the atlas, is followed by ten to sixty trunk vertebrae that typically bear transverse processes that articulate with

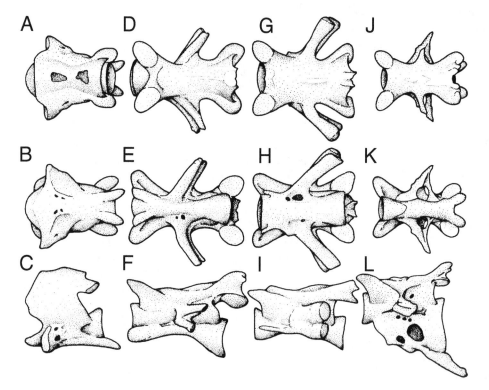

Fig. 5.6. Representative vertebrae of the salamander *Ambystoma opacum. A,* dorsal, *B,* ventral, and *C,* lateral views of the atlas; *D,* dorsal, *E,* ventral, and *F,* lateral views of the seventh dorsal vertebra; *G,* dorsal, *H,* ventral, and *I,* lateral views of the sacral vertebra; *J,* dorsal, *K,* ventral, and *L,* lateral views of the first caudal vertebra. After Duellman and Trueb (1986); reproduced by permission of the McGraw-Hill Companies, New York.

two-headed ribs. The trunk vertebrae have well-formed zygapophyses and neural arches. The trunk vertebrae of most families are amphicoelus (fig. 5.6). However, advanced families (Plethodontidae, Salamandridae) have opisthocoelus vertebrae. The single sacral vertebra is similar to the trunk vertebrae but bears enlarged transverse processes to attach with the pelvis. There can be up to one hundred caudal vertebrae, depending on the species. These do not articulate with ribs and have a ventral hemal arch as well as a dorsal neural arch. Fossil vertebrae are the most commonly collected and identified portion of the salamander skeleton.

Salamander fore- and hindlimbs are short but otherwise relatively unspecialized. Much of the pectoral and pelvic elements remains cartilaginous, as do the articular ends of the limb bones. The salamander skull is more robust than that of anurans, although it is much weaker than in Paleozoic amphibians and amniotes. In many families (for example, Amphiumidae), the premaxillae are fused into a single element. The maxilla is the main tooth-bearing unit, except in neotenic species in which it is often very reduced (for example, *Siren*) or absent altogether (*Pseudobranchus*, Proteidae).

Fossil Salamanders of Florida

With one notable exception, salamanders are typically much rarer than anurans in Florida fossil localities. The exception is the Sirenidae (sirens), which are often common in coastal freshwater deposits. Aquatic, neotenic salamanders in general have a better fossil record than more terrestrial species. This is due both to their better chances of being preserved and their larger average size. The latter makes their bones more robust and easier to find.

Two genera of sirens are known from Florida, *Pseudobranchus* and the larger *Siren*. Fossil species from the Miocene have been described for both (fig. 5.7; Goin and Affenberg, 1955). In addition to size, the two can be distinguished by the much more concave ventral margin of the centrum in *Pseudobranchus* (fig. 5.7H, 5.7L). Siren vertebrae are amphicoelus. The presence of zygapophy-

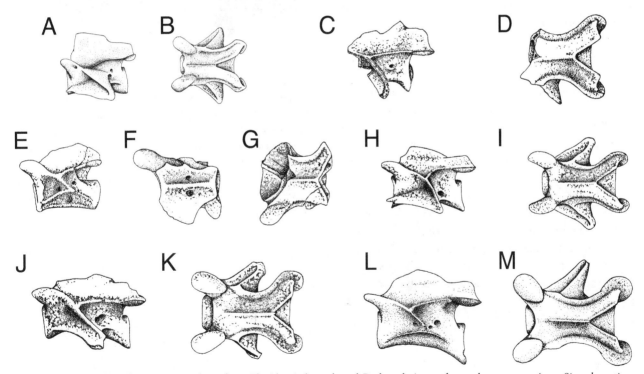

Fig. 5.7. Fossil and modern siren vertebrae from Florida. *A*, lateral, and *B*, dorsal views of a modern greater siren, *Siren lacertina*; *C*, lateral, and *D*, dorsal views of MCZ 2284, *S. simpsoni* from Haile 6, Alachua County, late Miocene; *E*, lateral, *F*, ventral, and *G*, dorsal views of MCZ 2278, *S. hesterna* from the Thomas Farm site, Gilchrist County, early Miocene; *H*, lateral, and *I*, dorsal views of the modern dwarf siren, *Pseudobranchus striatus*; *J*, lateral, and *K*, dorsal views of MCZ 2279, *P. robustus* from Haile 7A, Alachua County, late Pleistocene; *L*, lateral, and *M*, dorsal views of MCZ 2282, *P. vetustus* from Haile 6. *A–B* about 2×; *C–G* about 6×; *H–M* about 10×. After Goin and Auffenberg (1955); reproduced by permission of the Museum of Comparative Zoology, Cambridge, Massachusetts.

ses easily differentiates siren vertebrae from those of amphicoelus fish vertebrae with which they are frequently recovered. Sirens have sometimes been separated from the Caudata and placed in their own group, the Trachystomata. Modern classifications include them within the Caudata but vary in their placement relative to the other living families of salamanders. Sirens superficially resemble eels, as they have lost all trace of the hindlimbs, and the forelimbs are relatively small. They have feathery external gills and feed on aquatic invertebrates.

Batrachosauroides dissimulans is the only known Florida salamander that represents an extinct family (fig. 5.8A, B). The Batrachosauroididae ranged from the Cretaceous through the early Pliocene of North America, with additional records from Germany in the Eocene and southwestern Asia (Uzbekistan) in the late Cretaceous. They were large, elongate, probably neotenic salamanders with reduced limbs. Unlike sirens, their vertebrae are opisthocoelus (fig. 5.8A, B). *B. dissimulans* was originally described from the middle Miocene of southeastern Texas. Estes (1963) first recorded it from the early Miocene of Florida (Thomas Farm), and Bryant (1991) reported a vertebra from the middle Miocene of Gadsden County. The oldest newt from Florida, *Notophthalmus robustus,* also derives from Thomas Farm. According to Estes, this 18 Ma species differs from modern forms by its more robust vertebrae (hence the species name) with stubbier transverse processes (fig. 5.8C–F).

As was the case with the anurans, late Miocene and younger fossil salamanders closely resemble modern species. With one exception, all late Pleistocene records are referred to living species. *Pseudobranchus robustus,* an extinct dwarf siren, is recorded from several late Pleistocene sites in northern peninsular Florida (fig. 5.7J, K), although its distinctness was questioned by Lynch (1965). *Ambystoma tigrinum,* the tiger salamander, and *Amphiuma means,* the congo "eel," are the most commonly recorded salamanders from the Pleistocene of Florida after *Siren.* As is the case with most small vertebrates, recovery of salamanders usually depends on screen-washing matrix.

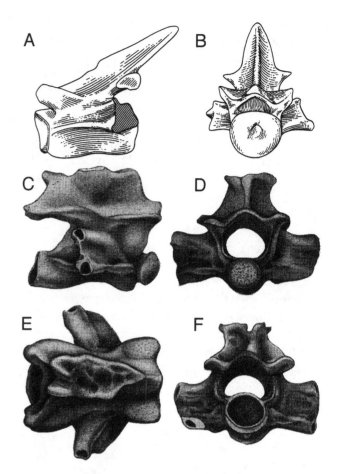

Fig. 5.8. Miocene salamander vertebrae from Florida. *A,* left lateral, and *B,* anterior views of UF 111741, *Batrachosauroides dissimulans* from Gunn Farm Mine, Gadsden County, middle Miocene; *C,* right lateral, *D,* anterior, *E,* dorsal, and *F,* posterior views of MCZ 3384, *Notophthalmus robustus* from the Thomas Farm site, Gilchrist County, early Miocene. *A–B* about 3×; *C–F* about 20×. *A–B* after Bryant (1991); reproduced by permission of the Society of Vertebrate Paleontology. *C–F* after Estes (1963); reproduced by permission of the Florida Academy of Sciences, Orlando.

References

Ashton, R. E., and P. S. Ashton. 1988. *The Handbook of Reptiles and Amphibians of Florida*. Part Three, *The Amphibians*. Windward Publishing Co., Miami, 191 p.

Auffenberg, W. Remarks on some Miocene anurans from Florida, with a description of a new species of *Hyla*. Breviora, 52:1–11.

———. 1957. A new species of *Bufo* from the Pliocene of Florida. Quarterly Journal of the Florida Academy of Sciences, 20:14–20.

Benton, M. J. 1997. *Vertebrate Palaeontology*, 2d ed. Chapman and Hall, New York, 464 p.

Bolt, J. R. 1991. Lissamphibian origins. Pp. 194–222 *in* H.-P. Schultze and L. Trueb (eds.), *Origins of the Higher Groups of Tetrapods: Controversy and Consensus*. Cornell University Press, Ithaca, N.Y.

Bryant, J. D. 1991. New early Barstovian (middle Miocene) vertebrates from the upper Torreya Formation, eastern Florida panhandle. Journal of Vertebrate Paleontology, 11:472–89.

Coates, M. I. 1996. The Devonian tetrapod *Acanthostega gunnari* Jarvik; postcranial anatomy, basal tetrapod interrelationships and patterns of skeletal evolution. Transactions of the Royal Society of Edinburgh: Earth Sciences, 87:363–421.

Coates, M. I., and J. A. Clack. 1991. Fish-like gills and breathing in the earliest known tetrapod. Nature, 352:234–36.

Cowen, R. 2000. *History of Life*, 3d ed. Blackwell Science, Malden, Mass., 432 p.

Duellman, W. E., and L. Trueb. 1986. *Biology of the Amphibia*. McGraw-Hill, New York, 670 p.

Emslie, S. D., and G. S. Morgan. 1995. Taphonomy of a late Pleistocene carnivore den, Dade County, Florida. Pp. 65–83 *in* D. W. Steadman and J. I. Mead (eds.), *Late Quaternary Environments and Deep History: A Tribute to Paul S. Martin*. Mammoth Site of Hot Springs, S. Dak., Scientific Papers, vol. 3.

Estes, R. 1963. Early Miocene salamanders and lizards from Florida. Quarterly Journal of the Florida Academy of Sciences, 26:234–56.

Goin, C. J., and W. Auffenberg. 1955. The fossil salamanders of the family Sirenidae. Bulletin of the Museum of Comparative Zoology, 113:495–514.

Gray, J., and W. Shear. 1992. Early life on land. American Scientist, 80:444–57.

Holman, J. A. 1967. Additional Miocene anurans from Florida. Quarterly Journal of the Florida Academy of Sciences, 30:121–40.

———. 1995. *Pleistocene Amphibians and Reptiles of North America*. Oxford University Press, New York. [Includes references to many additional studies on Pleistocene amphibians.]

Jenkins, F. A., and D. M. Walsh. 1993. An Early Jurassic caecilian with limbs. Nature, 365:246–49.

Lauren, M., and R. R. Reisz. 1997. A new perspective on tetrapod phylogeny. Pp. 9–59 *in* S. S. Martin and K. L. M. Martin (eds.), *Amniote Origins: Completing the Transition to Land*. Academic Press, San Diego.

Long, J. A. 1995. *The Rise of Fishes: 500 Million Years of Evolution*. Johns Hopkins University Press, Baltimore, 223 p. [Pages 200–209 provide a well-illustrated, general account of the fish-tetrapod transition.]

Lynch, J. D. 1965. The Pleistocene amphibians of Pit II, Arredondo, Florida. Copeia, 72–77.

Milner, A. R. 1988. The relationships and origin of living amphibians. Pp. 59–102 *in* M. S. Benton (ed.), *The Phylogeny and Classification of the Tetrapods*. Volume 1, *Amphibians, Reptiles, Birds*. Clarendon Press, Oxford, England.

———. 1990. The radiation of temnospondyl amphibians. Pp. 321–49 *in* P. D. Taylor and G. P. Larwood (eds.), *Major Evolutionary Radiations*. Clarendon Press, Oxford, England.

Panchen, A. L., and T. R. Smithson. 1988. The relationships of the earliest tetrapods. Pp. 1–32 *in* M. S. Benton (ed.), *The Phylogeny and Classification of the Tetrapods, Volume 1: Amphibians, Reptiles, Birds*. Clarendon Press, Oxford, England.

6

Reptilia I

Turtles and Tortoises

Introduction

The Amniota (reptiles, birds, and mammals) are those tetrapods that reproduce using an amniote egg. This type of egg contains specialized membranes that provide the developing embryo with a liquid environment, give oxygen in exchange for carbon dioxide, and store food (yolk) and nitrogenous waste. This innovation frees these tetrapods from depending on bodies of water in which to lay their eggs and distinguishes them from fish and amphibians (some frogs and salamanders do lay their eggs on land without an amniotic egg, but their eggs are small and the young cannot remain in the egg long before hatching). With this evolutionary breakthrough, vertebrates became completely adapted for terrestrial life. In addition to the living types of amniotes, the fossil record documents many extinct groups of late Paleozoic and Mesozoic tetrapods that are classified in the Amniota, including dinosaurs and pterosaurs. It is often impossible to tell whether or not certain extinct tetrapods had an amniote egg, because eggs do not often fossilize unless the shell is mineralized. The amniote skeleton contains a number of derived features or specializations relative to those of amphibians and other early tetrapods that are not amniotes. These allow paleontologists to classify extinct tetrapods as amniotes without knowing their method of reproduction. Among these features are having the frontal bone of the skull form the dorsal part of the orbit, one or two well-ossified occipital condyles (the structure on the back of the skull with which the first or atlas vertebra articulates), the second neck or cervical vertebra specialized as the distinctive axis, two or more sacral vertebrae (often fused together into a sacrum) that articulate with the pelvis, and a true astragalus in the ankle (Gauthier et al., 1988; Benton, 1997; Lee and Spencer, 1997).

The oldest fossils considered to be true amniotes are about 300 Ma (Pennsylvanian). Early in their history, amniotes diverged into two major lineages: the Synap-

sida (commonly called synapsids), the lineage that ultimately gave rise to mammals; and the Reptilia (=Sauropsida of Benton, 1997), which includes the other living amniote groups—turtles, crocodilians, birds, lizards, and snakes. A consensus of modern interpretations regarding the evolutionary relationships among these major amniote groups is shown in figure 6.1. The synapsid lineage was the more diverse and abundant of the two until the middle Triassic Period; then various reptile groups, most notably the dinosaurs, became the dominant land tetrapods. This chapter covers one of the major reptilian groups, the turtles and tortoises. Lizards, snakes, and crocodilians are discussed in chapter 7 and birds in chapter 8.

The reptilian or sauropsid lineage (fig. 6.1) also split into two major groups in the late Paleozoic (the Anapsida and Diapsida), plus a few genera that do not belong in either of them. The terms *anapsid, diapsid,* and *synapsid,* which are used to name the major groups of amniotes, also describe an important feature of the skull. Anapsid also describes a skull without any lateral openings on the side of the cranium posterior to the orbit. Diapsid skulls have two lateral openings, one above the other, posterior to the orbit. A synapsid skull has a single, low opening. These openings permit larger, more powerful jaw muscles for use in chewing and prey capture. They do not expose the brain, because it is covered by an internal layer of bone. The anapsid skull is the primitive condition for amniotes—synapsid and diapsid skulls evolved independently from the anapsid state.

Turtles and tortoises, the order Testudines, have traditionally been classified as members of the Anapsida. The oldest fossil turtle has the anapsid type of skull. Later turtles had modified skulls that allowed more room for jaw muscles, not by having openings like diapsids or synapsids, but by indenting the posterior margin of the back of the skull. This accomplishes the same function, but in a different way. The three other families of anapsid reptiles all became extinct by the end of the Triassic

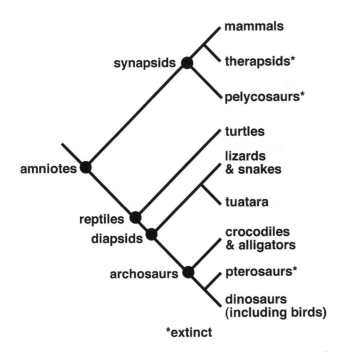

Fig. 6.1. Evolutionary relationships among major groups of amniote tetrapods. Note that some recent studies have questioned the traditional placement of the turtles (see text). Also birds are today generally regarded as a subgroup of dinosaurs.

(Benton, 1997). Recent workers on these groups collectively call them parareptiles but disagree about which parareptile group is most closely related to turtles and tortoises (Lauren and Reisz, 1995; Lee, 1997).

An alternate evolutionary history for turtles and tortoises has been proposed by Rieppel and deBraga (1996), deBraga and Rieppel (1997), and Rieppel and Reisz (1999). They placed them within the diapsid reptiles and among them, closer to the lizards and snakes than to the crocodilians and birds. They do not believe that the turtle skull is truly of the anapsid type, but that it has come to resemble it by convergent evolution. Their evidence for a diapsid relationship for turtles is based on similarities in the development and morphology of various bones in the skull (including the maxilla, jugal, and prefrontal), the wrist and ankle, and the fifth metatarsal. The traditional anapsid classification of turtles and tortoises is followed in this book because this new hypothesis has yet to be tested by other workers and even its supporters acknowledge that it has problems (Rieppel and Reisz, 1999).

The oldest fossil turtles are *Proganochelys* and *Proterochersis* from the late Triassic of Germany. Representatives of most of the modern groups of turtles appeared before the end of the Cretaceous Period. As noted in chapter 2, Florida's oldest vertebrate fossil is a small marine turtle of Cretaceous age. It was collected 9,210 feet below the surface of Okeechobee County in 1955 as a petroleum company was drilling a deep well (Olsen, 1965). Portions of the turtle's shell, shoulder region, and forelimbs were recovered. All of the fossil turtles yet found in Florida belong to a group that are called cryptodires. Cryptodire turtles retract their heads back into their shells by making a vertical bend in the neck. The other major living group of turtles, the pleurodires, bend the neck sideways to retract the head. Pleurodires today live in South America, Africa, and Australia, although they had a much wider range in the past and could eventually be found as fossils in Florida.

Four major groups (superfamilies) of cryptodires are known from Florida: Chelydroidea, the snapping turtles; Chelonioidea, the sea turtles; Trionychoidea, the softshelled and mud turtles; and the Testudinoidea, a diverse group that includes the pond turtles, box turtles, and land tortoises. The higher classification of turtles used here follows that of Gaffney and Meylan (1988); common names and ecological information for modern species follows Ashton and Ashton (1991).

TURTLE ANATOMY AND NATURAL HISTORY

Turtles have very specialized skulls (fig. 6.2) and postcranial skeletons. In the skull, the postparietal and postfrontal bones are absent, as is the lacrimal in all but the most primitive turtles. No teeth are present on the maxilla, premaxilla, and dentary; these bones are instead covered with a horny beak. The right and left dentaries are usually fused at their symphysis. The cervical and caudal vertebrae are most often opisthocoelus or procoelus; the pattern is consistent within a species but varies at higher taxonomic levels. Along the vertebral column a switch from procoelus to opisthocoelus morphology may occur (or vice versa); the intervening vertebral centrum may either be biconcave (amphicoelus) or biconvex. The trunk vertebrae are fused to the shell (fig. 6.3). The shoulder (pectoral) and pelvic elements are tripartite structures (fig. 6.4). The shoulder consists of only two elements, a "V"-shaped scapula and a rodlike coracoid. Two other shoulder bones are actually present but modified into parts of the shell (see next paragraph). The pelvis (fig. 6.4B) consists of the usual three elements (ilium, ischium, and pubis) that radiate out from the acetabulum. They are usually found separately as fossils. The corresponding elements of the fore- and hindlimb generally resemble one another; for example it can be difficult to distinguish an isolated fossil femur from a humerus (fig. 6.5). A complete comparative skeleton

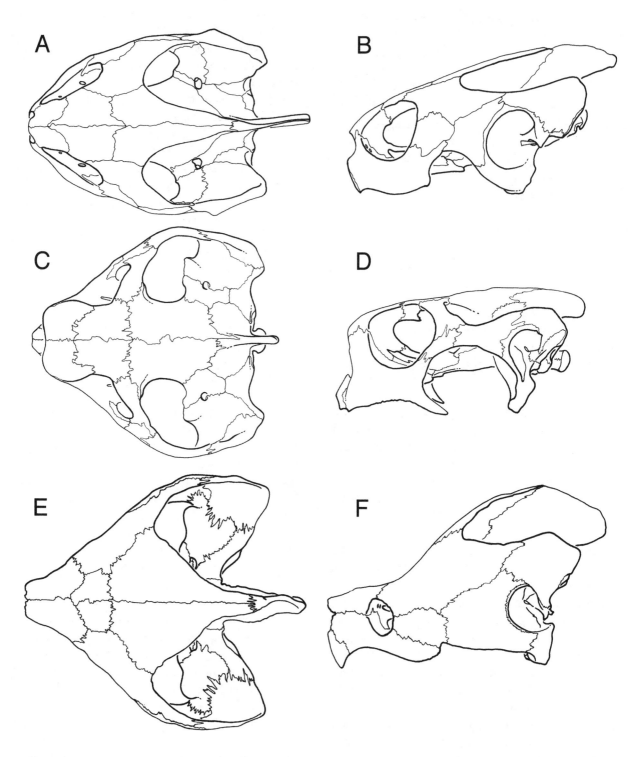

Fig. 6.2. Skulls of three important groups of turtles found in Florida. *A*, dorsal, and *B*, left lateral views of *Pseudemys concinna*, an emydid or pond turtle; *C*, dorsal, and *D*, left lateral views of *Gopherus polyphemus*, the gopher tortoise; *E*, dorsal, and *F*, left lateral views of *Macroclemys temminckii*, a chelydrid (snapping turtle). After Gaffney (1979); reproduced by permission of the American Museum of Natural History, New York.

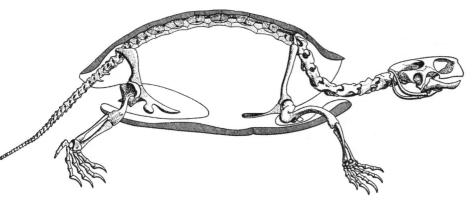

Fig. 6.3. Right lateral view of a turtle skeleton to show the relationship between the carapace and the vertebrae. The cervical and caudal series of vertebrae are free, but the dorsals and sacrals are fused to the carapace. Modified from Romer (1956); copyright © 1956 by The University of Chicago Press and reproduced by permission of the publisher.

from a modern turtle is essential to identify these bones. Bony osteoderms are found in the skin of the limbs and around the tail of some turtles, especially tortoises. These act as armor to protect these exposed areas when the limbs and tail are retracted within the shell.

The upper half of the shell of a turtle, the dome-shaped carapace, is actually formed by numerous separate bony elements joined together. A medial series of neural bones are fused with the trunk vertebrae, a paired series of costal bones are fused to the ribs, and a surrounding series of peripheral bones form the edge of the carapace (fig. 6.6A, C). The carapace elements represent new centers of bone growth within the skin (not modifications of other bones). The ventral portion of the shell, the plastron (plural is *plastra*) is formed by two transformed elements of the shoulder, the clavicle (which becomes the epiplastron) and interclavicle (entoplastron),

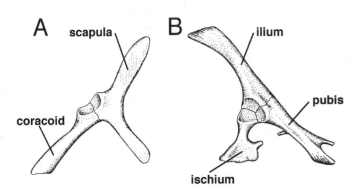

Fig. 6.4. *A*, pectoral, and *B*, pelvic girdles of a modern turtle. From *Vertebrate Paleontology and Evolution* by Carroll © 1988 by W. H. Freeman and Company. Used with permission.

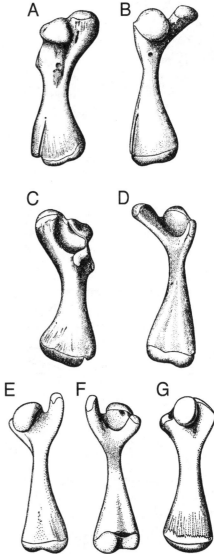

Fig. 6.5. Comparison of the humerus and femur in turtles. *A*, dorsal, and *C*, ventral views of the humerus of the Ridley sea turtle *Lepidochelys*; *B*, dorsal, and *D*, ventral views of the humerus of the softshell turtle *Apalone*; *E*, dorsal, and *F*, ventral views of the femur of *Apalone*. *G*, dorsal view of the femur of *Lepidochelys*. Modified from Romer (1956); copyright © 1956 by The University of Chicago Press and reproduced by permission of the publisher.

and by up to five additional paired elements that represent new centers of bone growth (fig. 6.6B). These are the hyoplastron, hypoplastron, xiphiplastron, and in primitive turtles several paired mesoplastra. Mesoplastra are lost in all living turtles and all of those found in Florida.

Fig. 6.7. Dorsal view of UF 63136, a nuchal carapace element of the tortoise *Hesperotestudo crassiscutata* from Haile 21A, Alachua County, early Pleistocene. The grooved lines mark the former boundaries of the epidermal scutes that once covered the shell. These scutes are made entirely of organic materials and rarely fossilize. The patterns made by the epidermal scutes is often of taxonomic importance in identifying turtle shell elements to species. About 1×.

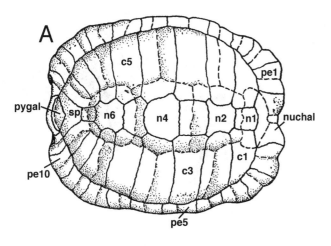

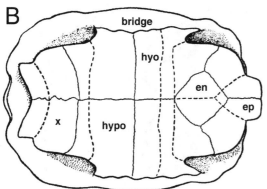

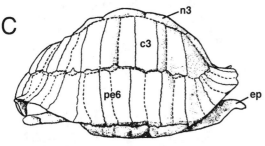

Fig. 6.6. *A*, dorsal, *B*, ventral, and *C*, right lateral views of a tortoise shell. Sutures between ossified elements shown by solid lines; outlines of epidermal scutes by dashed lines. Abbreviations: *c*, costal; *en*, entoplastron; *ep*, epiplastron; *hyo*, hyoplastron; *hypo*, hypoplastron; *nu*, nuchal; *n*, neural; *pe*, peripheral; *x*, xiphiplastron. The neurals, costals, and peripherals each consist of a numbered series. Modified from Romer (1956); copyright © 1956 by The University of Chicago Press and reproduced by permission of the publisher.

Most individual elements of the turtle shell rigidly articulate with one another in life, although some turtles have evolved movable hinges between some elements to tightly close the shell. After death, the elements of the shell easily disassociate unless the corpse is rapidly buried. The carapace and plastron are united by articulations between some of the lateral peripheral bones and dorsal projections of the hyoplastra, mesoplastra (if present), and hypoplastra (fig. 6.6B). This unifying structure is known as the bridge. In many highly aquatic turtles, the plastron is reduced and the bridge is very narrow or lost. In a living turtle, the bony shell is overlain by hardened epidermal scutes which overlap the articulations between the bones of the shell. It is the pattern made by these scutes and not those of the underlying bones of the carapace and plastron that are observed on a living turtle. The scutes are completely organic (made of the equivalent of fingernails) and do not preserve in fossils. Their shapes can still be observed on fossils because of the impressions they leave on the bones of the shell (fig. 6.7). The arrangement of both the bones and scutes of the shell are used in turtle identification and systematics.

The overall shape of a turtle's shell reveals clues as to its ecology. A high-domed shell, like that of a box turtle (*Terrapene*), is characteristic of terrestrial species. Low-domed, more streamlined shells are found in aquatic turtles. The most aquatic species, such as softshelled and sea turtles, tend to reduce and eventually lose elements of the shell. Despite the lack of teeth, most turtles are carnivorous. The major exception to this are the Testudinoidea, which are primarily herbivorous.

FOSSIL TURTLES OF FLORIDA

Snapping turtles (Chelydridae) are large, aquatic carnivores well known for their powerful bite. Both of the living Florida genera, *Chelydra* and *Macroclemys,* are known as fossils (figs. 6.8–6.10). *Chelydra serpentina* (the common snapping turtle) is found in some Pleistocene localities. Its habitat preferences are broad, so that its presence at a fossil site indicates little except the occurrence of fresh water. An exceptionally large, extinct species of *Chelydra* is a common constituent of late Pliocene faunas in central Florida (fig. 6.8B). Two species of *Macroclemys* are known from Florida: *M. auffenbergi,* from the late Miocene, and the extant alligator snapping turtle, *M. temmincki,* from the Pliocene onward. It is the largest freshwater turtle in Florida. *Macroclemys* is today restricted to waters west of the Suwannee River, but it ranged much farther in the past, as far as Sarasota. The two species of *Macroclemys* can only be separated on the basis of complete skulls or toes. *Macroclemys* differs

from *Chelydra* in having smaller, more laterally located orbits (fig. 6.8), more deeply notched posterior peripherals, larger ridges on the carapace, and in lacking osteoderms on the tail.

Sea turtles (Chelonioidea) have a long fossil history, extending back into the Mesozoic (Jurassic). Their fossils are sometimes common in shallow marine deposits in Florida, although usually represented by shell fragments. Sea turtles have very reduced plastra, and the limbs are modified into flippers (fig. 6.11). Without limb or jaw elements, or a reasonably complete piece of the shell, it is difficult to identify them much beyond "sea turtle." This is most often done by the characteristic sculpturing of the carapace and the very loose, groovelike articulation between the costal and peripheral elements of the carapace. Shell fragments of indeterminate cheloniids have been found in marine Eocene and Oligocene limestone. Five genera of sea turtles are known from the early Pliocene Palmetto Fauna (Dodd and Morgan, 1992). *Psephophorus,* an extinct genus of leatherback sea turtle, is known from a single osteoderm. In leatherback sea turtles (Dermochelyidae), the carapace lacks epidermal scutes and is reduced to a mosaic of osteoderms embed-

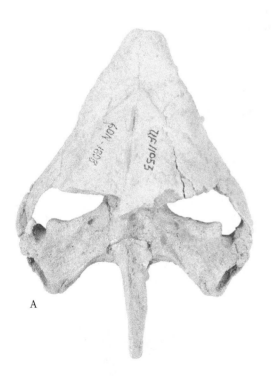
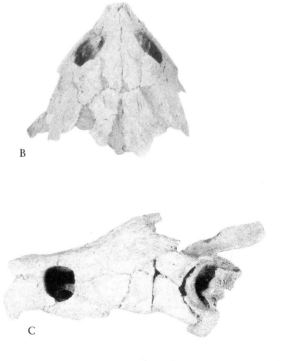

Fig. 6.8. Comparison of fossil skulls of snapping turtles. *A,* dorsal, and *C,* left lateral views of UF 11053, nearly complete skull of *Macroclemys auffenbergi* from the McGehee Farm site, Alachua County, late Miocene; *B,* dorsal view of a partial skull of *Chelydra* n. sp. from Haile 7C, Alachua County, late Pliocene. The orbits of *Chelydra* are clearly visible in dorsal view, but those of *Macroclemys* can only be seen in lateral view. All about 0.5×.

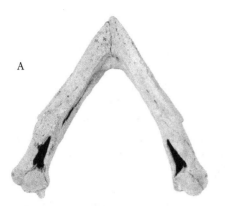

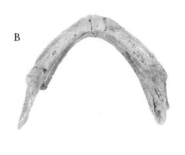

Fig. 6.9. Comparison of fossil mandibles of snapping turtles in dorsal view. Specimens are from the same two individuals shown in fig. 6.8. A, *Macroclemys auffenbergi*; B, *Chelydra* n. sp. The key difference is the more sharply angled "V" made by the jaw of *Macroclemys*. Both about 0.5×.

ded in the thick, leathery skin. An undescribed species of *Caretta* (loggerhead), as well as species of *Chelonia* (green sea turtle), *Eretmochelys* (hawksbill), and *Lepidochelys* (ridley sea turtle) make up the Bone Valley cheloniids (fig. 6.12). Of these, the herbivorous *Chelonia* is the most common, with the majority of the specimens representing subadults. The extant *Chelonia mydas* and *Caretta caretta* are also known from Pleistocene specimens. Both nest on Florida beaches today as does the Atlantic leatherback.

Softshelled turtles (Trionychidae) are freshwater carnivores which have lost the epidermal scutes of the shell. Instead a layer of soft, leathery skin covers the shell. The carapace does not have any peripheral elements, and the plastron is reduced and highly modified (fig. 6.13). The

outer surface of the shell bones is dimpled with round or oval golf ball-like depressions, allowing even small fragments to be identified as softshelled turtle. North American trionychids were usually referred to the genus *Trionyx* in the older literature (*Amyda* is another generic name that was sometimes used). Meylan (1987) reviewed this group and determined that the generic name *Apalone* was more appropriate for North American trionychids. The living species *Apalone ferox,* the Florida softshell, is recognized well back into the Neogene. It is found today and in the Pleistocene throughout most of the state. Skulls and associated shell and limb elements from the Love Site and McGehee Farm (late Miocene) are also referable to *A. ferox.* Two other species of *Apalone, A. spiniferus* and *A. muticus,* today live in the panhandle, but neither has been identified from fossils. They

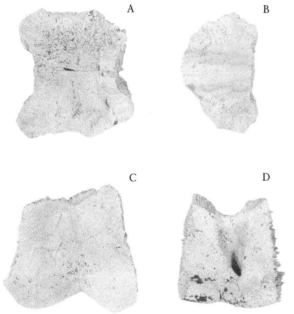

Fig. 6.10. Carapace elements of an undescribed large snapping turtle from the late Pliocene of Florida, *Chelydra* n. sp. UF specimens from Haile 7C, Alachua County. A–B, two neurals, both in dorsal view; C, dorsal-lateral view of a peripheral element; D, ventral-medial view of a peripheral element. All about 0.5×.

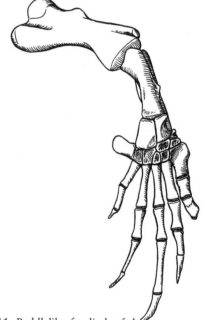

Fig. 6.11. Paddlelike forelimb of the green sea turtle, *Chelonia*. Figures from *Analysis of Vertebrate Structure* by Milton Hildebrand, copyright © 1974 by John Wiley & Sons, Inc. Reprinted by permission of John Wiley & Sons, Inc.

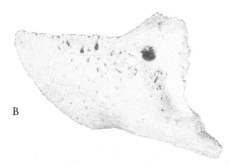

Fig. 6.12. Early Pliocene fossil sea turtles (genus *Caretta*) from the Gardinier Mine, Bone Valley region, Polk County, Florida. *A*, lateral view of UF 121947, left maxilla and premaxilla; *B*, lateral view of UF 101915, left dentary. About 0.75×. Photos courtesy of G. S. Morgan.

would most likely show up in late Pleistocene sites from northern and western Florida.

Mud and musk turtles (*Kinosternon* and *Sternotherus*) are known from several Florida Pliocene and Pleistocene deposits, but the material has not been extensively studied (fig. 6.14). These two small genera are separable on the basis of certain elements of the plastron (Sobolik and Steele, 1996). They are typically much less common in fossil deposits than emydid turtles, and only occasionally abundant (for example, Ichetucknee River).

The Emydidae, including the pond and box turtles, is the best represented family of turtles by fossils in Florida, with many deposits containing large numbers. Some of the genera are well studied, although the rarer types tend to be less so. Figures 6.15–6.21 illustrate representative recent and fossil material of this family.

Clemmys guttata (the spotted turtle) is a small aquatic emydid that lives in only a few scattered localities in northern Florida, suggesting that its past distribution was probably much more extensive. Its fossil record is very poor, with one doubtful Miocene occurrence from central Florida and a few Pleistocene records (Haile 16A and Waccasassa River).

Box turtles (*Terrapene carolina*) are very common in

Pleistocene beds and are often found complete (fig. 6.15), sometimes with the skull and whole limbs, in fissure deposits. Some of these are thought to be dens where groups of hibernating box turtles suffered mass mortality (Auffenberg, 1958, 1959). They have been intensively studied because the geographic variation of living *Terrapene* populations is easily traced back into Pleistocene samples. Gigantism is typical in Pleistocene coastal populations, which are referred to the extinct subspecies *Terrapene carolina putnami*. Generic identification aids include the plastral hinge, allowing complete closure of the shell, and the fact that in adults all of the bones of the carapace are fused together (fig. 6.15). The oldest box turtles from Florida are from the very early Miocene (late Arikareean), but they have not been identified to species. *Terrapene* is one of the most terrestrial species in this otherwise predominantly aquatic family. Its diet includes plants, fungi, and small invertebrates.

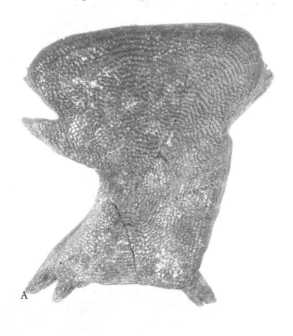

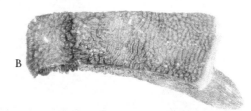

Fig. 6.13. The Florida softshelled turtle, *Apalone ferox*, from the Leisey Shell Pit, Hillsborough County, early Pleistocene. *A*, ventral view of UF 83720, fused hyo-hypoplastron; *B*, dorsal view of UF 80602, costal element. The dimpled surface texture of trionychid shell elements is easily recognizable, even in small fragments. Both about 0.5×.

A

B

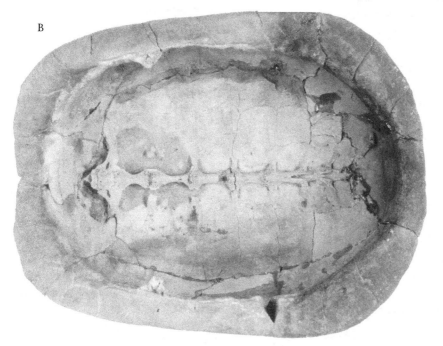

C

Fig. 6.14. The loggerhead musk turtle *Sternotherus minor,* from the Ichetucknee River, Columbia County, late Pleistocene. UF specimens. *A,* dorsal view of a nuchal element; *B,* dorsal view of a costal; *C,* ventral view of the hyo-hypoplaston. All about 1.5×.

A

B

Fig. 6.15. *A,* dorsal, and *B,* ventral views of UF 48004, carapace of the box turtle, *Terrapene carolina,* from the Aucilla River, Jefferson County, late Pleistocene. About 0.5×.

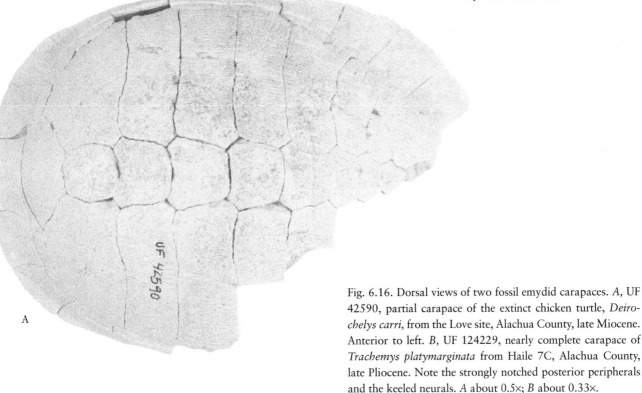

Fig. 6.16. Dorsal views of two fossil emydid carapaces. *A*, UF 42590, partial carapace of the extinct chicken turtle, *Deirochelys carri*, from the Love site, Alachua County, late Miocene. Anterior to left. *B*, UF 124229, nearly complete carapace of *Trachemys platymarginata* from Haile 7C, Alachua County, late Pliocene. Note the strongly notched posterior peripherals and the keeled neurals. *A* about 0.5×; *B* about 0.33×.

species in the same genus presently occur only in large rivers from the Apalachicola westward. However, a few Pleistocene fossils from the Suwannee and its tributaries show that it was more widespread in the past. It is never very common. It is difficult to separate from *Pseudemys concinna*. A close relative of *Graptemys* is the terrapin *Malaclemys*. This coastal turtle is very rarely recovered as a fossil, despite living in an environment favorable for preservation. No definitive fossil records from Florida exist.

Deirochelys reticulata (the living chicken turtle) is a medium-sized emydid characteristic of shallow water environments. *Deirochelys* is carnivorous, unlike most emydids, and has special adaptations of its neck vertebrae and tongue bones (hyoids) to quickly suck in small fish and tadpoles. Shell sculpturing on the carapace of

Deirochelys is similar to that found on the shells of *Pseudemys nelsoni* and *Trachemys scripta* (fig. 6.16A), but scute morphology, particularly in the nuchal bone, usually separates these three. The neurals of *Deirochelys* are proportionally much wider than in other emydid genera (fig. 6.16A). *D. reticulata* is known from several Pleistocene deposits in northern Florida, and during that time it attained a much larger size than at present

(Jackson, 1978). A late Miocene representative of the chicken turtle, the extinct species *Deirochelys carri*, is also known from northern Florida (figs. 6.16A, 6.17B). An undescribed, primitive species is present at Thomas Farm. It is intermediate between typical *Deirochelys* and *Pseudemys* in neural width (Jackson, 1978).

The closely related genera *Chrysemys*, *Pseudemys*, and *Trachemys* are represented by five living species in

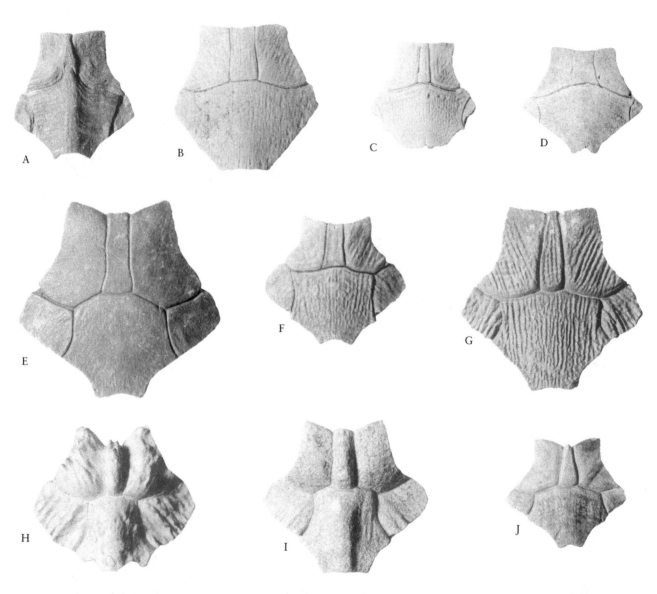

Fig. 6.17. Nuchal elements of fossil and modern emydid turtles from Florida; shown in dorsal view. Oblique lighting was used to emphasize relief. *A*, UF 21766, *Terrapene carolina* from Bradenton, Manatee County, middle Pleistocene; *B*, UF 20903, *Deirochelys carri* from the McGehee Farm site, Alachua County, late Miocene; *C*, UF 95076, *Deirochelys reticulata*, Alachua County, Recent; *D*, UF 65629, *Pseudemys williamsi* from the McGehee Farm site; *E*, UF 56427, *Pseudemys concinna* from the Santa Fe River 2A site, Gilchrist County, late Pleistocene; *F*, UF 20882, *Pseudemys caelata* from the McGehee Farm site; *G*, UF 49395, *Pseudemys nelsoni* from the Ichetucknee River, Columbia County, late Pleistocene; *H*, UF 55865, *Trachemys inflata* from Nichols Mine, Polk County, early Pliocene; *I*, UF 10048, *Trachemys platymarginata* from Haile 15A, Alachua County, late Pliocene; *J*, UF 82200, *Trachemys scripta* from the Leisey Shell Pit, Hillsborough County, early Pleistocene. All about 0.67×.

Fig. 6.18. Dorsal view of UF 11561, carapace of *Pseudemys williamsi* from the McGehee Farm site, Alachua County, late Miocene. This is the late Miocene representative of the smooth-shelled lineage of *Pseudemys* that includes the extant *P. concinna* and *P. floridana*. "*C. williamsi*" is written on the shell because the species was originally placed in the genus *Chrysemys*. About 0.5×.

Florida. The number of recognized genera in this group has varied over the years between one and three; presently all three are considered valid (Seidel and Smith, 1986). They are primarily herbivorous, eating aquatic vegetation, although a few tend to be more omnivorous. *Chrysemys picta* (the painted or eastern pond turtle) just barely enters the state in the northern Apalachicola River Valley. No fossils are known of *C. picta* from Florida, but it can be expected to have ranged farther south during cooler periods in the Pleistocene.

Three lineages of *Pseudemys* and *Trachemys* are represented in the fossil record of Florida (figs. 6.16B, 6.17–6.21). One of the three includes the living species *Pseudemys concinna* (the Suwannee cooter) and *P. floridana* (the Florida cooter). These two can be separated only by minor details of the scute pattern (and coloration in living animals). This group differs from the other two lineages (described below) in having a smooth carapace (fig. 6.18) and in details of the arrangement of the scutes on the nuchal bone (fig. 6.17D, E). *P. floridana* occurs in slow-moving bodies of water throughout the state, while streams and rivers along the Gulf Coast are inhabited by *P. concinna*. *Pseudemys williamsi* is an extinct, smooth-shelled species in this lineage that lived in rivers of northern Florida during the late Miocene (fig. 6.18). *P. concinna* is known from a number of Pleistocene river deposits in the western part of the peninsula, while *P. floridana* is recorded from many Pleistocene localities all over the state.

Pseudemys nelsoni (the Florida redbelly turtle) and the late Miocene *Pseudemys caelata* represent a lineage of emydids with rugose, sculptured shells (fig. 6.19). An early Miocene *Pseudemys* from Thomas Farm also belongs in this group. *P. nelsoni* is currently widely distributed across peninsular Florida, and this was also the case in the Pleistocene. The genus *Trachemys* represents the second emydid lineage in Florida with rugose, sculptured shells. *Trachemys scripta*, the extant yellowbelly or pond slider, has a shell generally like that of *P. nelsoni*, but the two differ in details of the scutes on the nuchal bone (fig. 6.17), and in the notched posterior peripherals found in *Trachemys* (fig. 6.16B). *T. scripta* is today restricted to the panhandle and northern peninsular Florida (south to Levy County), where it lives in ponds, sinkholes, streams, and lakes with abundant vegetation. In the Pleistocene it was both much more widespread and common than at present. The distinctly larger Pleistocene form is referred to the extinct subspecies *T. scripta petrolei* (fig. 6.20). In the late Pliocene, a similar species *(Trachemys platymarginata)* with a thicker shell and sharply keeled neural elements lived in Florida (fig. 6.16B). *Trachemys inflata* of the late Miocene and early Pliocene had an extremely thick shell (fig. 6.21) and is the common emydid of the Bone Valley Palmetto Fauna. Along with *P. floridana*, *P. nelsoni* and *T. scripta* are the most common Pleistocene turtles in Florida.

The Testudinidae, the land tortoises, are closely related to the Emydidae and are also a herbivorous group. The only living Florida tortoise is the endangered gopher tortoise (*Gopherus polyphemus*). Pleistocene deposits

A

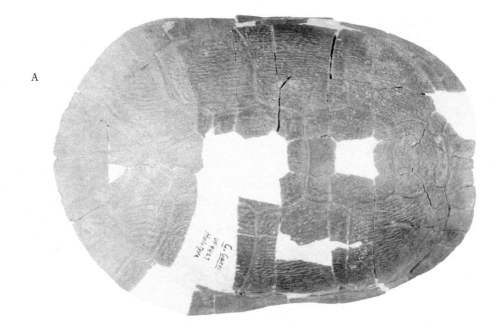

B

Fig. 6.19. Dorsal views of the carapaces of two members of the rugose-shelled lineage of *Pseudemys. A*, UF 9427, *P. caeleta* from the McGehee Farm site, Alachua County, late Miocene. *B*, UF 3500, *P. nelsoni* from the Ichetucknee River, Columbia County, late Pleistocene. Anterior to right. Contrast the relatively smooth margined peripherals with the notched ones of *Trachemys* (Figs. 6.16B and 6.21C). Both about 0.5×.

Fig. 6.20. Associated shell elements of UF 81146, a partial shell of the emydid *Trachemys scripta* from the Leisey Shell Pit, Hillsborough County, early Pleistocene. Carapace elements shown in dorsal view; plastron elements in ventral view. *A*, a costal element; *B*, a neural element; *C*, a peripheral element from the bridge region; *D*, a "free" peripheral; *E*, the left epiplastron; *F*, the left hyoplastron; *G*, the left hypoplastron. All about 1×.

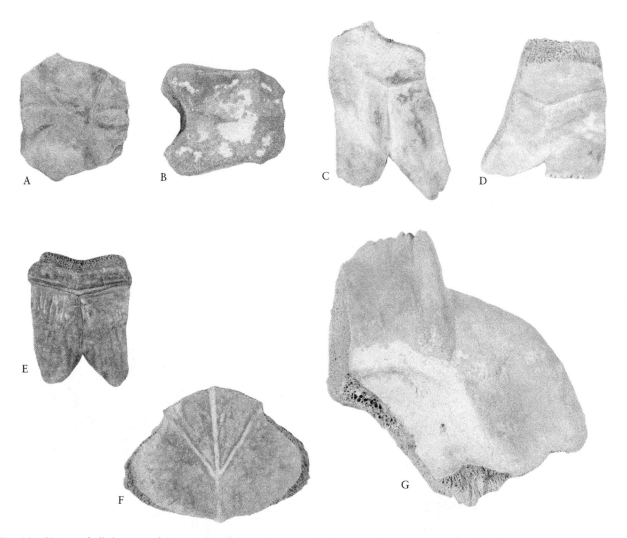

Fig. 6.21. Various shell elements of the early Pliocene emydid turtle *Trachemys inflata* from the Bone Valley District. *A–D, G,* from Fort Green Mine, either Hardee or Polk County. *E–F,* from Gardinier Mine, Polk County. *A–B,* two neural elements. *C, D,* two peripheral elements. *E,* a pygal. *F,* an entoplastron. *G,* an epiplastron in dorsal view. All others in external (dorsal for carapace, ventral for plastron) view. See figure 6.17H for the nuchal element of this species. All about 1×.

often contain remains of this thin-shelled tortoise. Entire skeletons have been obtained in a few localities. Neogene records of *Gopherus* are known but are as yet unstudied.

Floridemys nanus is an interesting and enigmatic tortoise that is probably of Miocene age. It is interesting because of its small size and possible relationships. It is enigmatic because its age is not known to any degree of certainty. It is still definitely known from only a single shell collected early this century at the Holder Phosphate Mine in Citrus County (fig. 6.22). The age of these local phosphate deposits is unknown; they have no special geological connection with the Bone Valley Formation. It is a dwarf species; the specimen is an adult only 105 millimeters long (about 4 inches). *Floridemys* may be from the early Miocene (or late Oligocene?) and appears to be

related to the extinct tortoises of the genus *Stylemys*, which, in turn, is closely related to *Gopherus. Stylemys* is very common in Oligocene and Miocene deposits of the western United States. Fragments of a tortoise at the Oligocene I-75 site may also represent *Floridemys.*

Two North American lineages of thick-shelled tortoises are recognized, each with several species. These two are now recognized as distinct subgenera, *Hesperotestudo* and *Caudochelys,* within the genus *Hesperotestudo* (Gaffney and Meylan, 1988; Meylan, 1995). In the older literature both types were usually included in the broadly defined genus *Geochelone* (or *Testudo*), but North American species are now excluded from this genus. Meylan and Sterrer (2000) showed that *Hesperotestudo* is more closely related to *Gopherus* and other

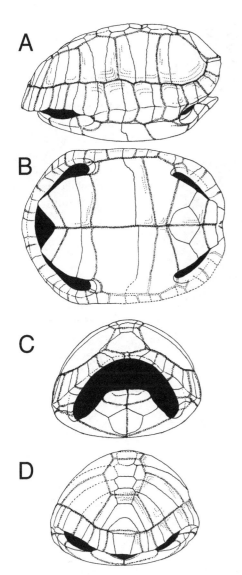

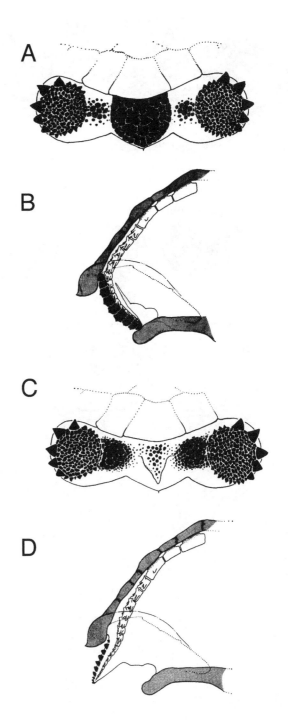

Fig. 6.22. *A,* right lateral, *B,* ventral, *C,* anterior, and *D,* posterior views of USNM 10247, carapace and plastron of the early Miocene or late Oligocene Florida dwarf tortoise, *Floridemys nanus,* from the Holder Phosphate Mine, Citrus County. About 0.5×. After Auffenberg (1963); reproduced by permission of the Florida Museum of Natural History.

North American forms than it is to either *Geochelone* or *Testudo.* Both North American subgenera can be traced back to the Oligocene.

The smaller of the two subgenera, *Hesperotestudo (Hesperotestudo),* can be recognized by its unique caudal shield of fused osteoderms that covered the dorsal surface of the tail and protected the posterior end of the animal (fig. 6.23A, B). The epiplastra and xiphiplastra are relatively thick. Adult *H. (Hesperotestudo)* reached a maximum of 60 centimeters (2 feet) in overall length. *H. incisa* is the common late Pleistocene member of this

Fig. 6.23. Comparisons of the distribution of caudal osteoderms in the tortoise subgenera *Hesperotestudo* (*Hesperotestudo*) and *H. (Caudochelys).* *A,* reconstructed posterior view of *H. (Hesperotestudo) incisa* showing caudal and limb osteoderms in solid black. Note fused caudal buckler or "hinny binder"; *B,* as in *A,* lateral view in cross section; *C,* and *D,* as in *A* and *B,* except for *H. (Caudochelys) crassiscutata.* There is no fused caudal buckler in this subgenus. After Auffenberg (1963); reproduced by permission of the Florida Museum of Natural History.

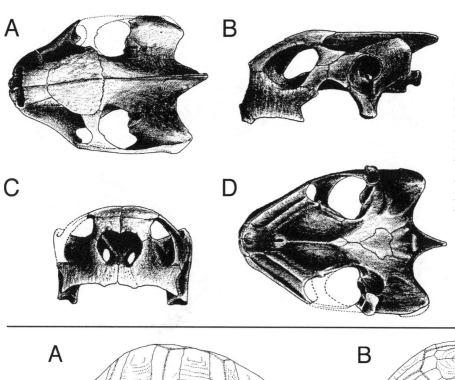

Fig. 6.24. Nearly complete skull of *Hesperotestudo incisa* from Haile 8A, Alachua County, middle Pleistocene. *A*, dorsal, *B*, left lateral, *C*, anterior, and *D*, ventral views of UF 3141. About 1×. After Auffenberg (1963); reproduced by permission of the Florida Museum of Natural History.

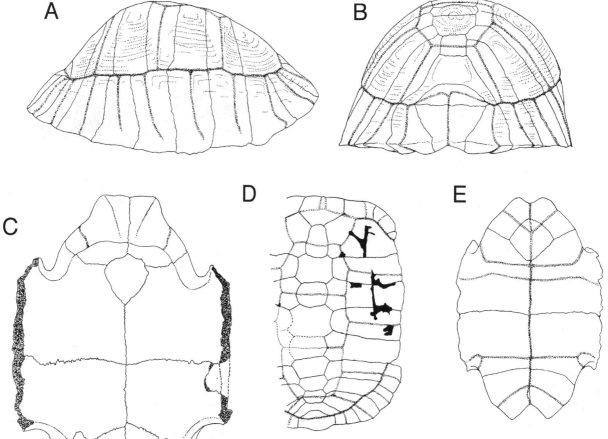

Fig. 6.25. *A*, right lateral, *B*, posterior, and *C*, ventral views of UF 3077, nearly complete carapace and plastron of *Hesperotestudo incisa* from Haile 8A, Alachua County, middle Pleistocene; *D–E*, the early Miocene tortoise, *Hesperotestudo (Caudochelys) tedwhitei*, from the Thomas Farm site, Gilchrist County, early Miocene; *D*, dorsal view of MCZ 2020, carapace; *E*, ventral view of MCZ 2020, partial plastron. Anterior toward top in both figures. *A–C* about 0.4×; *D–E* about 0.3×. After Auffenberg (1963); reproduced by permission of the Florida Museum of Natural History.

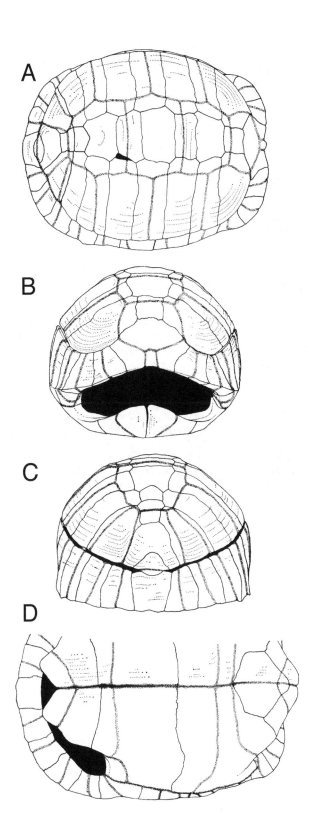

subgenus (figs. 6.24, 6.25A–C), while the closely related *H. mylnarskii* is known from the early to middle Pleistocene.

The other lineage of tortoises in Florida is the subgenus *Caudochelys* (figs. 6.25D, E, 6.26). Members of this subgenus had unmodified tail armor (fig. 6.23C, D) and grew to a much greater size than those of the subgenus *Hesperotestudo,* well over a meter in length. As juvenile *Caudochelys* individuals overlap in size with adult *Hesperotestudo,* size alone is not a good criterion for distinguishing the two (unless the fossil is from a very large individual, in which case it is *Caudochelys*). The epiplastron is proportionally thicker in the smaller *Hesperotestudo* than in *Caudochelys,* adjusting for size. *H. (Caudochelys) crassiscutata*, the common Pleistocene representative of this lineage (fig. 6.26), was as large as, or larger than, the famous Galapagos Islands tortoises. *Caudochelys* is also represented by the relatively small *H. (Caudochelys) tedwhitei* from the early Miocene (fig. 6.25D, E) and the giant *H. (Caudochelys) hayi* from the late Miocene to early Pliocene. Both subgenera of *Hesperotestudo* persisted in Florida through the very late Pleistocene. Evidence of this can be found at Devil's Den, where both *H. incisa* and *H. crassiscutata* co-occur (Holman, 1978). Both tortoise fossils and human artifacts are preserved in Little Salt Spring in Charlotte County (Clausen et al., 1979). This and other evidence suggest that the earliest people to arrive in Florida found giant tortoises living here and hunted them. Giant tortoises are important paleoecologic indicators of relatively mild winter temperatures, as they cannot withstand prolonged periods of freezing. Their presence in Florida and elsewhere throughout the southern United States during the Pleistocene Epoch is seen as evidence that winter temperatures were actually on average milder during the so-called Ice Age than at present.

Fig. 6.26. *A*, dorsal, *B*, anterior, *C*, posterior, and *D*, ventral views of UF 3151, a nearly complete carapace and plastron of a young adult individual of *Hesperotestudo (Caudochelys) crassiscutata* from Haile 8A, Alachua County, middle Pleistocene. About 0.2×. After Auffenberg (1963); reproduced by permission of the Florida Museum of Natural History.

REFERENCES

Ashton, R. E., and P. S. Ashton. 1991. *The Handbook of Reptiles and Amphibians of Florida. Part Two, Lizards, Turtles, and Crocodilians. Revised 2nd Edition.* Windward Publishing Co., Miami, 191 p.

Auffenberg, W. 1958. Fossil turtles of the genus *Terrapene* in Florida. Bulletin of the Florida State Museum, 3:53–92.

———. 1959. A Pleistocene *Terrapene* hibernaculum, with remarks on a second complete box turtle skull from Florida. Quarterly Journal of the Florida Academy of Sciences, 22:49–53.

———. 1963. Fossil testudine turtles of Florida. Genera *Geochelone* and *Floridemys.* Bulletin of the Florida State Museum, 7:53–97.

———. 1966. A new species of Pliocene tortoise, genus *Geochelone,* from Florida. Journal of Paleontology, 40:877–82.

Benton, M. J. 1997. *Vertebrate Palaeontology.* 2d ed. Chapman and Hall, New York, 464 p.

Carroll, R. L. 1988. *Vertebrate Paleontology and Evolution*. W. H. Freeman, New York, 698 p.

Clausen, C. J., A. D. Cohen, C. Emiliani, J. A. Holman, and J. J. Stipp. 1979. Little Salt Spring, Florida: a unique underwater site. Science, 203:609–14.

DeBraga, M., and O. Rieppel. 1997. Reptile phylogeny and the interrelationships of turtles. Zoological Journal of the Linnean Society, 120:281–354.

Dodd, C. K., and G. S. Morgan. 1992. Fossil sea turtles from the early Pliocene Bone Valley Formation, central Florida. Journal of Herpetology, 26:1–8.

Gaffney, E. S. 1979. Comparative cranial morphology of recent and fossil turtles. Bulletin of the American Museum of Natural History, 164:65–375.

Gaffney, E. S., and P. A. Meylan. 1988. A phylogeny of the turtles. Pp. 157–219 *in* M. S. Benton (ed.), *The Phylogeny and Classification of the Tetrapods, Volume 1: Amphibians, Reptiles, Birds*. Clarendon Press, Oxford, England.

Gauthier, J. A., A. G. Kluge, and T. Rowe. 1988. The early evolution of the Amniota. Pp. 103–55 *in* M. S. Benton (ed.), *The Phylogeny and Classification of the Tetrapods, Volume 1: Amphibians, Reptiles, Birds*. Clarendon Press, Oxford, England.

Hildebrand, M. 1974. *Analysis of Vertebrate Structure*. John Wiley and Sons, New York, 710 p.

Holman, J. A. 1978. The late Pleistocene herpetofauna of Devil's Den sinkhole, Levy County, Florida. Herpetologica, 34:228–37.

———. 1995. *Pleistocene Amphibians and Reptiles of North America*. Oxford University Press, New York. [Contains references to many other papers dealing with Pleistocene turtles from Florida.]

Jackson, D. R. 1976. The status of the Pliocene turtles *Pseudemys caelata* Hay and *Chrysemys carri* Rose and Weaver. Copeia, 1976:655–59.

———. 1978. Evolution and fossil record of the chicken turtle *Deirochelys*, with a re-evaluation of the genus. Tulane Studies in Zoology and Botany, 20:35–55.

Laurin, M., and R. R. Reisz. 1995. A reevaluation of early amniote phylogeny. Zoological Journal of the Linnean Society, 113:165–223.

Lee, M. S. Y. 1997. Pareiasaur phylogeny and the origin of turtles. Zoological Journal of the Linnean Society, 120:197–280.

Lee, M. S. Y., and P. S. Spencer. 1997. Crown-clades, key characters and taxonomic stability: when is an amniote not an amniote? Pp. 61–84 *in* S. S. Sumida and K. L. M. Martin (eds.), *Amniote Origins: Completing the Transition to Land*. Academic Press, San Diego.

Meylan, P. A. 1987. The phylogenetic relationships of soft-shelled turtles (Family Trionychidae). Bulletin of the American Museum of Natural History, 186:1–101.

———. 1995. Pleistocene amphibians and reptiles from the Leisey Shell Pit, Hillsborough County, Florida. Bulletin of the Florida Museum of Natural History, 37:273–97.

Meylan, P. A., and W. Sterrer. 2000. *Hesperotestudo* (Testudines: Testudinidae) from the Pleistocene of Bermuda, with comments on the phylogenetic position of the genus. Zoological Journal of the Linnean Society, 128:51–76.

Olsen, S. J. 1965. Vertebrate fossil localities in Florida. Florida Geological Survey Special Publication 12, 28 p.

Rieppel, O., and M. DeBraga. 1996. Turtles as diapsid reptiles. Nature, 384:453–55.

Rieppel, O., and R. R. Reisz. 1999. The origin and early evolution of turtles. Annual Review of Ecology and Systematics, 30:1–22.

Romer, A. S. 1956. *The Osteology of the Reptiles*. University of Chicago Press, Chicago, 772 p.

Rose, F. L., and Weaver, W. G. 1966. Two new species of *Chrysemys* (= *Pseudemys*) from the Florida Pliocene. Tulane Studies in Geology and Paleontology, 5:41–48.

Seidel, M. E., and H. M. Smith. 1986. *Chrysemys, Pseudemys, Trachemys* (Testudines: Emydidae): did Agassiz have it right? Herpetologica, 42:242–48.

Sobolik, K. D., and D. G. Steele. 1996. *A Turtle Atlas to Facilitate Archaeological Identifications*. Mammoth Site of Hot Springs, S. Dak., 117 p. [Although the drawings are crude and the taxonomy at times outdated, a good reference for identifying isolated turtle shell elements.]

Reptilia 2

Lizards, Snakes, and Crocodilians

INTRODUCTION

As indicated in chapter 6, a fundamental division among reptiles exists between those classified as Anapsida and Diapsida (fig. 6.1). The subject of this chapter are two of the three major living groups of diapsid reptiles, the squamates (lizards and snakes) and the crocodilians (gavials, alligators, and crocodiles). The third extant diapsid group, the birds, is discussed in the following chapter. Many additional types of diapsids are known only from fossils. In the late Paleozoic, about 275 Ma, diapsids diverged into two evolutionary lineages, which are called the lepidosaurs and archosaurs (Benton, 1997). Lepidosaurs include the squamates and a few fossil groups such as the marine plesiosaurs and ichthyosaurs of the Mesozoic. The archosaur lineage is more diverse. In addition to the living crocodilians and birds, phytosaurs, pterosaurs (flying reptiles of the Mesozoic), and the traditional dinosaurs are all archosaurs.

Although squamates are the more abundant and widespread group of lepidosaur reptiles, a second living group is represented by a single species from New Zealand, the tuatara. It is the sole survivor of a once more numerous order, the Sphenodontia. No fossils of this group have ever been found in Florida. The Squamata includes the lizards and two of their evolutionary offshoots, the snakes (Caldwell and Lee, 1997) and "worm lizards" (amphisbaenids). The oldest known squamates are middle Jurassic (reports of Triassic squamates are now discredited; Benton, 1997). The six suborders of the Squamata are all known as fossils from Florida. All families of squamates native to the state have fossil representatives, and several known as fossils no longer live in Florida.

The crocodilian branch of the archosaur reptiles is presently a relatively restricted group both in terms of numbers of species and morphologic diversity. The fossil record demonstrates that this was not always so—some ancient crocodilians once lived entirely on land and oth-

ers specialized to live entirely in the ocean; the 50-foot-long *Deinosuchus* preyed upon dinosaurs and the two-foot-long *Terrestrisuchus* had the long, slender legs of a greyhound (Benton, 1997). Semiaquatic crocodilians similar to the modern species have existed since the late Jurassic, and they survived the mass extinction event at the end of the Cretaceous that devastated their dinosaur contemporaries.

ANATOMY AND NATURAL HISTORY

Representative skulls and lower jaws of squamates are illustrated in figure 7.1. Practically all the skeletal elements of squamates have been recovered as fossils in Florida, generally isolated from one another. The most commonly studied fossils of lizards from Florida are usually portions of the dentaries (the tooth-bearing portion of the lower jaw) or maxillae, less frequently vertebrae or other skull elements. The bony osteoderms found in the skin of some lizards are diagnostic at the generic level (for example, *Heloderma,* the Gila monster, and *Ophisaurus*). Due to their small size, lizard fossils are usually only found by screen-washing sediment from localities that also produce other small terrestrial vertebrates such as rodents, small birds, and toads.

The most commonly found fossils of snakes are vertebrae, ribs, and isolated parts of the jaw. Ribs are too similar to be useful in identification. Lizard and snake dentaries and maxillae are easily distinguished if teeth are present. Lizard teeth are usually erect, blunt, and sometimes multicusped (fig. 7.1C). Snake teeth are larger, pointed, and curved more posteriorly. Snake vertebrae are procoelus, robust, and each individual potentially contributes one to two hundred to the fossil record. As a practical matter, most fossil snakes are identified solely by vertebral morphology (Holman, 2000). Snakes outnumber lizards at most Florida fossil localities.

Snake vertebrae are similar enough that they are difficult to identify to genus and species without some prac-

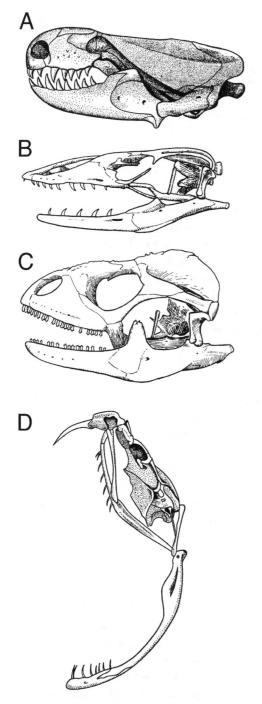

Fig. 7.1. Representative skulls and mandibles of squamate reptiles. *A*, left lateral view of *Amphisbaena*, an amphisbaenid ("worm lizard"); *B*, right lateral view of *Varanus*, a carnivorous lizard; *C*, right lateral view of *Conolophus*, a herbivorous lizard; *D*, left lateral view of *Crotalus*, a rattlesnake, in striking pose with erect fangs. *A, D* modified from Romer (1956); copyright © 1956 by The University of Chicago Press and reproduced by permission of the publisher. *B, C* from *Analysis of Vertebrate Structure* by Milton Hildebrand, copyright © 1974 by John Wiley & Sons, Inc. Reprinted by permission of John Wiley & Sons, Inc.

tice. Although no characters of the vertebrae are known that will distinguish all the Florida snake genera, complete specimens can often be identified if a comparative collection is available. Holman (1979, 1981, 2000) has provided well-illustrated guides to identify many snake vertebrae. The terminology used to describe features of snake vertebrae is shown in figure 7.2. Of all the vertebrae, those from the middle of the body (the trunk vertebrae) are the most consistent in structure and useful for identification. Cervical and caudal vertebrae are less useful. Trunk vertebrae tend to be larger, longer, and either lack or have smaller hyapophyses (hy in fig. 7.2A) than cervicals, and lack the elaborate ventral and lateral processes of caudals (Holman, 1979). Vertebrae from each of the major groups of snakes can generally be distinguished by the following characters. In the Scolecophidia, the neural spine is absent, the condyle is flattened, and the hemal keel is indistinct. In the superfamilies Anilioidea and Booidea (except Palaeophidae), neural spines are present but tend to be short and thick,

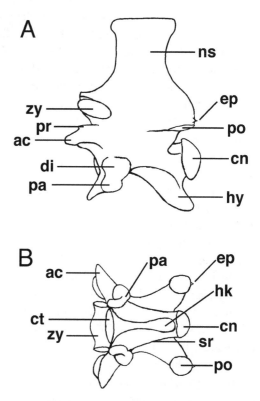

Fig. 7.2. *A*, left lateral, and *B*, ventral views of snake vertebrae, with key identifying structures labeled. Abbreviations: *ac*, accessory process; *cn*, condyle; *ct*, cotyle; *di*, diapophysis; *ep*, epizygopophyseal spine; *hk*, hemal keel (if a long spine, then it is called a hyapophysis, *hy*); *ns*, neural spine; *pa*, paradiapophysis; *po*, postzygopophysis; *pr*, prezygopophysis; *sr*, subcentral ridge; *zy*, zygosphene.

hyapophyses are absent on trunk vertebrae, and the centra are usually wider than long. In the Colubroidea, the neural spine is long and thin; a hyapophysis is present in the trunk vertebrae of the Natricinae, Elapidae, and Viperidae (but absent in other colubrid subfamilies); and the centrum is usually as long or longer than it is wide.

All snakes are carnivorous. The larger species commonly prey upon birds, amphibians, lizards, other snakes, or small mammals, while smaller snakes tend to favor invertebrates. Only a few species in Florida are poisonous and thus potentially dangerous, but the entire group is often unjustly persecuted. The native Florida lizards are insectivorous, although some fossils represent more herbivorous groups (iguanas). The books by Ashton and Ashton (1988, 1991) are great sources of information about Florida's numerous living snakes and lizards.

The crocodilian skeleton has remained without major innovations since the Mesozoic. The skull (fig. 1.5) is rather elongate, extremely so in fish-eating types like the living Indian gavial. It is also massive, well ossified, and flattened in most forms. The frontal bones are fused into a single midline element, as are the parietals (fig. 1.5B). The external nostrils are at the tip of the snout, which enables the animal to breathe with only this part of the body out of the water. The roof of the mouth is also modified for life in the water. Below the original inner opening of the nostrils, the bones of the roof of the mouth (premaxillae, maxillae, and palatines) have formed a secondary palate which extends backward (fig. 1.5A). Inhaled air passes back above the secondary palate to the posteriorly situated internal nares. Thus, air does not enter the mouth at all, but passes back separately into the throat, which can be closed off from the water in the mouth by a flap of skin. A similar secondary palate was evolved by mammals but almost never to the extent seen in crocodilians. Crocodilian teeth are basically simple cones, either pointed or blunt, and found only on the lateral margins of the jaw (fig. 7.3).

The general body shape of modern crocodilians is rather lizardlike compared to other archosaurs, with relative short, sprawling limbs. The long, flattened tail is used mainly for swimming. Crocodilians have about 9 cervical vertebrae, 15 dorsal or trunk vertebrae, 2

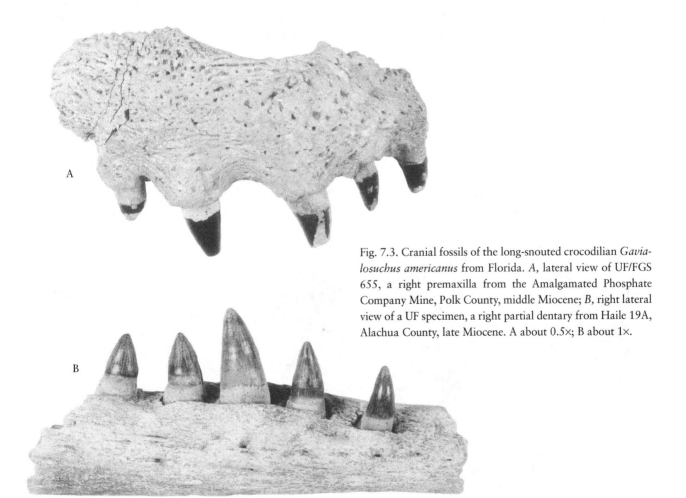

Fig. 7.3. Cranial fossils of the long-snouted crocodilian *Gavialosuchus americanus* from Florida. *A*, lateral view of UF/FGS 655, a right premaxilla from the Amalgamated Phosphate Company Mine, Polk County, middle Miocene; *B*, right lateral view of a UF specimen, a right partial dentary from Haile 19A, Alachua County, late Miocene. A about 0.5×; B about 1×.

sacrals, and 30 to 40 caudals. They are procoelus in modern crocodilians (fig. 7.4). A well-developed series of flatten osteoderms are present down the back (usually in two or more rows) and sometimes along the sides of the body (fig. 7.5). These are often pitted and may bear ridges, as in the alligator. Along with isolated teeth, the most commonly found crocodilian fossils are vertebrae and osteoderms.

The book edited by Ross (1989) is an excellent introduction to the biology and conservation of modern crocodilians, which are classified into three families, the Crocodylidae, Alligatoridae, and Gavialidae. The distinction between the first two is not sharp, and assignments, particularly of some of the fossil forms, are often uncertain. Brochu (1997, 1999) presented the results of a thorough analysis of the evolutionary relationships of modern and fossil crocodilian species. The Gavialidae is today represented by a single living species and restricted to southern Asia. However, it had a wider distribution in the Cenozoic. All gavials have very slender snouts sharply marked off from the rest of the skull. Most crocodiles also have an elongate snout, but it always blends smoothly into the back of the skull. In the alligators, the snout is always broad. In addition, the first tooth of the alligator's lower jaw (and often the fourth as

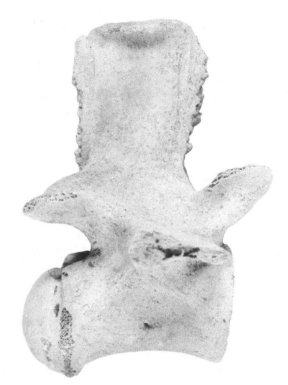

Fig. 7.4. Right lateral view of a vertebra of the late Miocene crocodilian *Gavialosuchus americanus*. Procoelus vertebrae are typical for crocodilians. UF specimen from Haile 19A, Alachua County. About 1×.

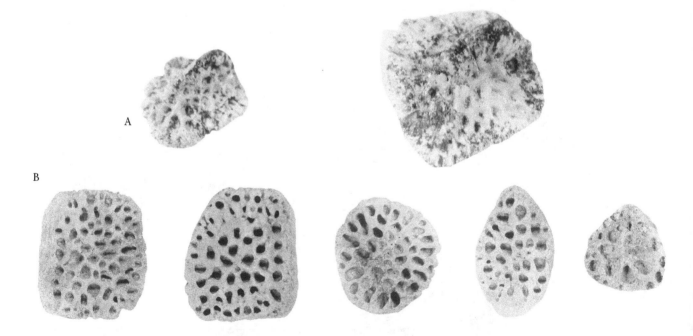

Fig. 7.5. Comparison of the two major types of dorsal osteoderms of crocodilians found as fossils in Florida. *A*, oblique view of UF 95429, two osteoderms of *Alligator mississippiensis* from the Moss Acres Racetrack site, Marion County, late Miocene. Note the prominent ridge or spine on the dorsal surface, characteristic of most *Alligator* osteoderms. *B*, dorsal view of UF 115983–115987, five osteoderms of *Gavialosuchus americanus* from Haile 19A, Alachua County, late Miocene. The osteoderms of *Gavialosuchus* lack a dorsal spine and have larger pits. *A* about 1×; *B* about 0.5×.

well) fits into a deep pit in the palate of the upper jaw; while in crocodiles these pits are absent and the upper jaw is usually notched for these long lower teeth, which are visible externally.

True gavials have not yet been reported from North America, although their fossils are known from South America and many parts of the Old World. Both crocodiles and alligators were abundant and varied in the New World throughout the Cenozoic. Species of both families still occur here, but only in tropical and subtropical regions. Of the Alligatoridae, caimans are found from Mexico to Argentina, alligators only in the southern United States and northeastern Mexico. A second species of *Alligator* lives in China. Members of the Crocodylidae are found in both freshwater and saltwater habitats in the West Indies and Central and northern South America as well as the Old World. The single living species of crocodile in the United States, *Crocodylus acutus,* is found in mangrove forests at the southern tip of Florida. However, its range extends southward throughout the Caribbean and northern South America. It has never been recorded as a fossil in Florida deposits and so has apparently entered the state only recently. The largest known specimens reach lengths of about 6 meters.

FOSSIL LIZARDS OF FLORIDA

Ten families of lizards are known as fossils from Florida (chap. 3). All of these families are extant, but several are no longer native to the state. The Phrynosomatidae is a large North American family of mainly insectivorous lizards, including earless lizards, horned lizards, and spiny lizards. The Polychridae is a diverse tropical and Caribbean group of mostly arboreal species, many with the ability to change skin color. The green anole, *Anolis carolinensis,* is the lone native species of this family in the United States, although several introduced species are now common in Florida as well. The Tropiduridae is today a South American and Caribbean group. The Iguanidae is another primarily tropical family, although a few species live in the American West. They are typically large bodied and herbivorous. These four families, and others, were often classified together in a much more broadly defined Iguanidae and still appear that way in some modern guidebooks. The classification used here follows the revision of Frost and Etheridge (1989), who split up the "classical" Iguanidae into eight families.

The remaining lizard families found in Florida include the Gekkonidae, a diverse, primarily tropical group. Most geckos are nocturnal and good climbers. The Mediterranean gecko, *Hemidactylus turcicus,* is an introduced Old World species now common in urban areas throughout much of Florida. The Scincidae (skinks) is a large, cosmopolitan group of short-legged, long-tailed lizards. The Xantusidae, the night lizards, is a small North American group no longer found in Florida. The Teiidae, the whiptails and racerunners, is principally a Central and South American group, although some fossil lizards from the Cretaceous of Asia have been referred to this family. Only one teiid is native to Florida, the common six-lined racerunner. The Anguidae, the lateral fold lizards, includes the alligator lizards of western North America and the limbless, snakelike glass lizards. The Helodermatidae is a small family that includes the poisonous Gila monster of the American Southwest.

The oldest fossil locality in Florida with lizards is the Oligocene I-75 site, which contains mostly extinct genera. Perhaps the most surprising factor about this site is the large number of lizards relative to snakes, with as many types of lizards as there are snakes; unlike today, when there are twice as many snakes as lizards. Of course this could be misleading and not representative of the entire state's fauna as it existed 30 million years ago. Not until several more faunas of similar antiquity are found and studied can such speculations be tested. The Oligocene lizards of Florida included members of at least three and possibly four families. The Anguidae are represented by a jaw and a piece of dermal armor that might belong to a limbed form called *Peltosaurus* and a single vertebra from the limbless *Ophisaurus* or a similar genus. Anguids are generally not thought of as being a major part of the lizard fauna of North America. During the early Cenozoic, however, many genera were present over much of the continent. The teeth of iguanid lizards are easily recognized by their multiple cusps. Two iguanid jaws found at the I-75 site may represent *Aciprion,* a genus that is know from other Oligocene fossil localities in the American West. The Xantusidae is represented by a jaw of the extinct genus *Paleoxantusia.* This family is now restricted to western North America, Central America, and Cuba. Fossil xantusids have also been found in Montana and South Dakota, so their current distribution is relict. One vertebra from the I-75 site may belong to a teiid lizard.

The pattern shown by early Miocene lizards in Florida is similar to that of the Oligocene, although more species are known because sites of that age are numerous and their squamate faunas are well studied (fig. 7.6A–D). The lizards of this age include the earliest skink (*Eumeces*) and gecko (an unidentified genus) from Florida, as well as a Gila monster (*Heloderma*). The iguanian lizards are very diverse, with as many as four genera present. A curly-tailed lizard similar to the modern West Indian genus *Leiocephalus* has been reported. One of the other iguanians seems to be an *Anolis,* or at least related to the anoles. A third form is a true iguanid, one of the larger herbivorous group of iguanids that includes the modern genera *Iguana, Ctenosaura,* and *Cyclura.* The Teiidae is represented by a form that has been referred to *Cnemidophorus.* A moderately large, limbed anguid lizard is relatively common. It was originally described as *Peltosaurus floridanus,* but it is now suspected that the type specimen of this species is actually from a fossil deposit in the western United States and was accidentally mixed in with some Florida fossils.

The only lizards reported from the middle Miocene of Florida belong to the Willacoochee Creek fauna of Gadsden County. Bryant (1991) noted the presence of four types: an iguanian near *Leiocephalus,* a teiid near *Cnemidophorus,* a skink, and a helodermatid (fig. 7.6E–G). Lizards are also rather rare in the late Miocene and early Pliocene, but at least one species of skink, a glass lizard, and a small iguanian (probably *Anolis*) were present. The large skink was similar to the living broad-headed skink (*Eumeces laticeps*).

The latest Pliocene Inglis 1A locality (Citrus County) produces the most diverse, best studied assemblage of lizards since those of the early Miocene (Meylan, 1982).

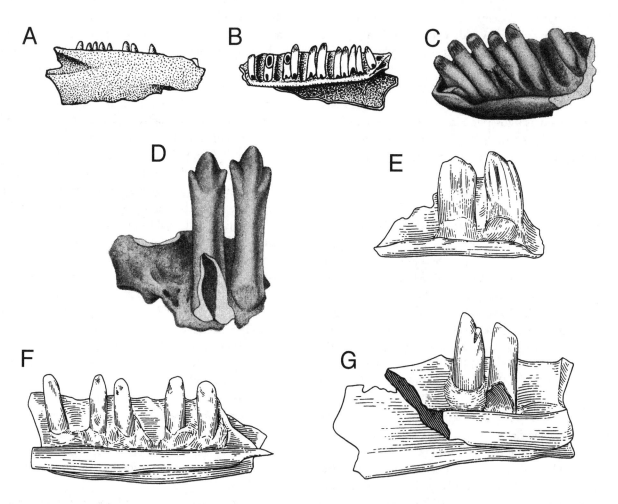

Fig. 7.6. Early and middle Miocene lizards from Florida. *A–D,* from the Thomas Farm site, Gilchrist County, early Miocene. *E–G,* from Milwhite Gunn Farm, Gadsden County, middle Miocene. *A,* lateral, and *B,* medial (lingual) views of MCZ 3382, a partial right dentary of an unidentified genus of gecko; *C,* lingual view of MCZ 3377, anterior end of a right dentary of *Eumeces* sp., a skink; *D,* lingual view of MCZ 3380, jaw fragment with two teeth of *Leiocephalus* sp., a curly tailed lizard; *E,* lingual view of UF 123632, partial maxilla of cf. *Leiocephalus* sp.; *F,* lingual view of UF 118501, a dentary of a skink; *G,* lingual view of UF 121882, partial dentary of a teiid lizard near *Cnemidophorus. A–B* about 10×; *C, D* about 20×; *E–G* about 25×. *A–D* after Estes (1963); reproduced by permission of the Florida Academy of Sciences, Orlando. *E–G* after Bryant (1991); reproduced by permission of the Society of Vertebrate Paleontology.

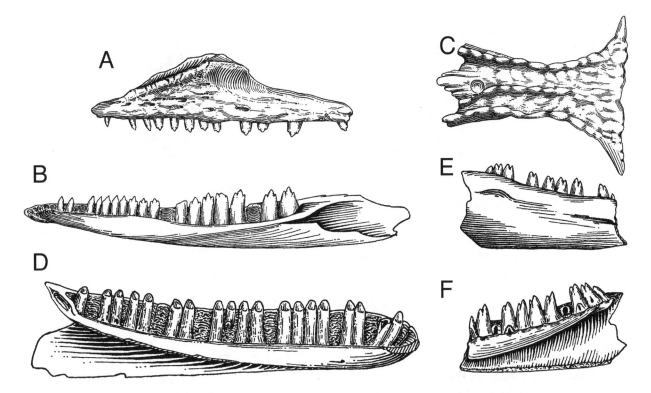

Fig. 7.7. Pleistocene lizards from Florida. *A*, lateral view of left maxilla; *B*, lingual view of right dentary, and *C*, dorsal view of frontal of UF 5829, the green anole *Anolis carolinensis* from the Winter Beach site, St. Lucie County, late Pleistocene; *D*, lingual view of UF 5086, left dentary of the skink *Eumeces*, probably *E. fasciatus*; *E*, lateral, and *F*, lingual views of UF 5089, partial right dentary of the racerunner *Cnemidophorus*, probably *C. sexlineatus*. *D–F* from Reddick 1B, Marion County, late Pleistocene. After Auffenberg (1956); reproduced by permission of the Florida Academy of Sciences, Orlando.

A total of four species are known, including an extinct species of skink (*Eumeces carri*), the most common of the modern glass lizards (*Ophisaurus ventralis*), and the fence lizard *Sceloporus undulatus*. An alligator lizard (close to the genus *Gerrhonotus*) suggests an affinity with western lizard faunas. The worm lizard or amphisbaenian *Rhineura* was also recovered from Inglis. Only identified on the basis of vertebrae, Meylan (1982) could find no differences with the modern species *R. floridana*. It has been more securely identified from several late Pleistocene localities (Holman, 1995).

Early Pleistocene fossils of lizards have only been reported from two localities in Florida, Leisey Shell Pit of Hillsborough County (Meylan, 1995) and Haile 16A of Alachua County. Only a single species, based on two vertebrae, was recorded at Leisey, the island glass lizard *Ophisaurus compressus*. It was identified on the basis of its more elongate vertebrae compared to others of its genus. The racerunner *Cnemidophorus* is present at Haile 16A, its first record since the middle Miocene. Two additional lizards appeared in Florida during the middle Pleistocene. They are two of the most common living

species, *Eumeces inexpectatus* (the southeastern five-lined skink) and *Anolis carolinensis*; this is their earliest appearance in the Florida fossil record.

The late Pleistocene lizard fauna of Florida is essentially modern (fig. 7.7). The green anole *Anolis carolinensis* is known from the most sites and ranged throughout the peninsula (as it does today). The next most commonly reported species is the eastern glass lizard *Ophisaurus ventralis*. Other lizards are only known from one or two localities, so clearly much is to be learned about the historical biogeography of this group.

FOSSIL SNAKES OF FLORIDA

Seven families of snakes are known as fossils from Florida (chap. 3). The Typhlopidae, the blind snakes, is a primitive group of small burrowers. They retain vestiges of the pelvic girdle and hindlimbs. Blind snakes no longer live in Florida but are known from Texas and the West Indies. The Aniliidae is a small, rare family, with only two remaining genera, *Anilius* in South America and *Cylindrophis* in Southeast Asia, but fossils indicate that aniliid snakes were more widespread in the past. They are only questionably identified from Florida fossils. The Palaeophidae is an extinct group of large marine snakes related to the boas. The Boidae includes the boas and pythons. This family is found in practically all tropical and subtropical parts of the world but is no longer native to Florida. The Colubridae is the largest group of living snakes, containing almost 75 percent of the described species in the world and most of the harmless, nonpoisonous varieties. The Elapidae includes poisonous snakes with erect fangs, such as the cobras and coral snakes. Their distribution is worldwide but largely subtropical and tropical; the coral snake is the only elapid to have ever inhabited Florida. The Viperidae is a group of highly specialized poisonous snakes with large, folding fangs (fig. 7.1D). This group includes the Old World vipers and the New World rattlesnakes, copperheads, and water moccasins. They are found in all continental areas except Australia.

Only one snake vertebra of Eocene age is known from Florida. It was collected in a limestone quarry in western Alachua County and belongs to the palaeophid snake *Pterosphenus schucherti*. Palaeophids are extinct marine snakes that apparently were common in Eocene seas. They are easily distinguished from all other snakes by the presence of pteropophyses, posteriodorsal projections from the neural arch (fig. 7.8). They are not related to modern sea snakes (Hydrophidae) but instead to the Boidae. Palaeophids reached lengths of 2 or more meters. They have also been found in Eocene deposits in Mississippi, Alabama, Arkansas, and Georgia, and in other Cretaceous to Eocene localities throughout eastern North America, Europe, and North Africa.

Most of the Oligocene snakes from Florida are of the erycine type of boid, with perhaps three or four species present at the I-75 site. The worm snake *Typhlops* may also be represented. Cowhouse Slough, a very late Oligocene site in Hillsborough County, produced what may be the only true aniliid snake from Florida. Most interesting among the I-75 snake fossils are two vertebrae that represent two genera of colubrids. This is the second oldest known record of colubrids in the New World (there is a slightly older record from Colorado; Holman [1999]). These represent a dispersal from Asia, apparently across the Bering Straits. It is remarkable that the now rare erycine boids once made up the majority of the Florida snake fauna and did so for many millions of years.

Primitive snakes of the boid family continue to greatly outnumber the colubrids in the early Miocene. Colubrids of this interval are represented by only two genera,

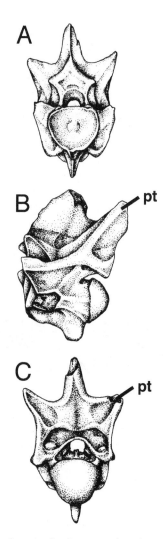

Fig. 7.8. *Pterosphenus schucherti* vertebra from the late Eocene of Twiggs County, Georgia. *A*, anterior, *B*, left lateral, and *C*, posterior views of University of Georgia (UGV) 52a. Note the pteropophyses *(pt)*, which characterize the family Palaeophidae, a marine group related to the boids. About 1×. After Holman (1977); reproduced by permission of the Society for the Study of Amphibians and Reptiles.

Paraoxybelis and *Pseudocemophora*—each with a single species (fig. 7.9G–J). *Paraoxybelis* is apparently unrelated to any snake now living in Florida, but may be close to the vine snake *(Oxybelis)* of Mexico and the Neotropics. The smaller *Pseudocemophora* seems most closely related to the living king and scarlet snakes (*Lampropeltis* and *Cemophora*). Elsewhere in North America at this time, colubrids were slightly more diverse but still outnumbered by erycine boids. Early Miocene Florida boids include two erycines, *Ogmophis* and *Calamagras* (fig. 7.9A, B). They were both about 60 centimeters long and closely related to the rubber and burrowing boas of California and Mexico. Also present were two boine boids, the extinct *Pseudoepicrates* and the extant genus *Boa. Pseudoepicrates* was a large form more than 2 meters long (fig. 7.9C, D). It is currently thought to have been related to the large constricting boas *Xenoboa* and *Boa,* rather than the tree boa *Epicrates,* for which it was named. The second large boid represented at Thomas Farm is *Boa barbouri,* an extinct species very close to the living *Boa constrictor* (Kluge, 1988). Also known from the early Miocene is a small

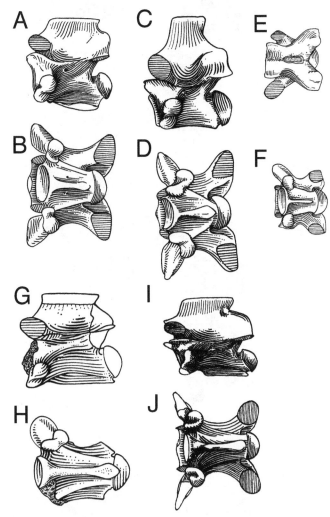

Fig. 7.9. Early Miocene snake vertebrae from the Thomas Farm site, Gilchrist County. *A,* left lateral, and *B,* ventral views of UF 5131, an erycine boid, *Ogmophis pauperrinus; C,* left lateral, and *D,* ventral views of the boine boid *Pseudoepicrates stanolseni; E,* dorsal, and *F,* ventral views of the small erycine *Anilioides minuatus; G,* left lateral view of UF 5143, the early colubrid *Paraoxybelis floridanus; H,* ventral view of UF 6007, *P. floridanus; I,* left lateral, and *J,* ventral views of UF 5744, the colubrid *Pseudocemophora antiqua. A–B, G–J* about 7×; *C–D* about 5×; *E–F* about 8×. After Auffenberg (1963); reproduced by permission of the Tulane University Museum of Natural History.

snake called *Anilioides minuatus* (fig. 7.9E, F). As the name suggests, this snake was originally thought to belong in the Aniliidae, but it is now regarded as an erycine boid.

The only middle Miocene snakes reported from Florida are three colubrids (a natricine and two colubrines; Bryant, 1991). The absence of boids is significant; as in all older faunas, they are the dominant group of snakes. Although boids persisted through the Miocene in Texas and still survive on the West Coast, they have apparently been very rare or absent in Florida since the middle Miocene.

The late Miocene marks the first occurrence of many snake genera still living in Florida (fig. 7.10). Furthermore, the distribution patterns of these genera are somewhat more similar to those of their living representatives than those of older faunas. The Colubridae is represented by about ten genera in the late Miocene (chap. 3). The various faunas (mostly from northern peninsular Florida) contain some of the most common snake genera in Florida today: the garter snake, *Thamnophis;* the racer, *Coluber;* the water snake, *Nerodia* (formerly called *Natrix*); the kingsnake, *Lampropeltis;* the hognosed snake, *Heterodon* (fig. 7.10B); the ringneck snake, *Diadophis* (fig. 7.10A); and the short-tailed snake, *Stilosoma* (fig. 7.10C). A coral snake *(Micrurus),* a pygmy rattlesnake *(Sistrurus),* and a rattlesnake *(Crotalus)* are also known. All are represented by species distinct from but similar to extant species.

Florida's Pliocene snakes include those listed for the late Miocene plus the indigo snake, *Drymarchon;* the coachwhip, *Masticophis;* the mud snake, *Farancia;* the rat snake, *Elaphe;* the crayfish snake, *Regina;* and the water moccasin, *Agkistrodon* (see figs. 7.11–7.15 for examples of some of these genera). Some of the *Lampropeltis* fossils definitely represent the kingsnake (*L. getulus*), while some of the *Elaphe* fossils are from a corn snake (*E. guttata*). Four of the modern water snake species, the green (*Nerodia cyclopion*), brown (*N. taxispilota*), red-bellied (*N. erythrogaster*), and banded (*N. fasciata*) water snakes were present. Obviously the snake fauna of Florida had taken on a very modern aspect by about 2 Ma.

The Inglis 1A fauna (latest Pliocene) has a rich and well-studied snake fauna (Meylan, 1982). With few exceptions, the specimens in this fauna can be assigned to modern species. Only *Dryinoides,* a xenodontine colubrid, belongs to an extinct genus, and its identification is tentative. Extinct species of *Diadophis* and *Regina* are also present. The species of *Regina* is one of the most interesting forms because it is intermediate between the

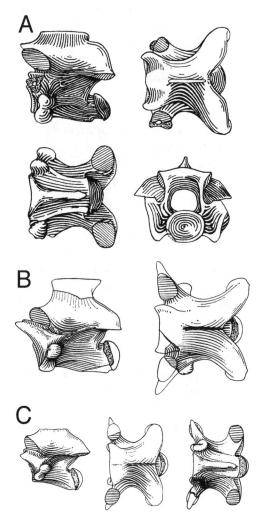

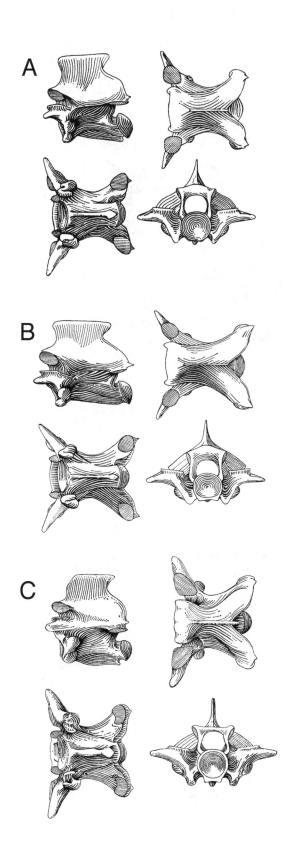

Fig. 7.10. Examples of the vertebrae of three late Miocene snakes from Haile 6A, Alachua County. *A (upper left, clockwise)*, left lateral, dorsal, anterior, and ventral views of UF 6413, the extinct ringneck snake *Diadophis elinorae; B,* left lateral and dorsal views of UF 6153, the extinct hognose snake *Heterodon brevis; C,* left lateral, dorsal, and ventral views of UF 6467, the extinct shorttailed snake *Stilosoma vetustum.* As is the case with these three species, most known late Miocene snakes from Florida belong to living genera but to extinct species. After Auffenberg (1963); reproduced by permission of the Tulane University Museum of Natural History.

Right: Fig. 7.11. Common large colubrine snakes from the Pleistocene of Florida. *A (upper left, clockwise),* left lateral, dorsal, anterior, and ventral views of UF 5200, trunk vertebra of the racer *Coluber constrictor* from Haile 7A, Alachua County, middle Pleistocene; *B,* UF 5722, trunk vertebra of the coachwhip *Masticophis flagellum* from Reddick 1B, Marion County, late Pleistocene; *C,* UF 5079, trunk vertebra of the indigo snake *Drymarchon corias* from Arredondo 1A, Alachua County, late Pleistocene. After Auffenberg (1963); reproduced by permission of the Tulane University Museum of Natural History.

advanced and primitive living members of the genus. The more primitive species (*R. grahami* and *R. septemvittata*) have slender, sharp teeth fixed firmly in the jaw. They can only eat newly molted (soft-bodied) crayfish. The advanced forms (*R. alleni* and *R. rigida*) have short, blunt teeth attached to the jaw by a hinge. The specialized teeth allow these snakes to eat hard-shelled crayfish. In the species at Inglis (*Regina intermedia*), the teeth are blunt and short but not hinged. Other interesting records include the smooth green snake (*Opheodrys vernalis*), the western hognose snake (*Heterdon nasicus*), and an indigo snake, which is more similar to the Mexican and Central American subspecies of *Drymarchon corais* than to the one present in Florida today. Meylan (1982) suggested that populations of western hognose snakes became isolated in Florida during the Pleistocene and gave rise to the southern hognose, *H. simus*. The smooth green snake is now found only as close to Florida as Houston, Texas, and northern North Carolina.

The remaining snakes from Inglis 1A (twenty-one species) are a good representation of the modern fauna. They include every species one would expect to find in a xeric setting in peninsular Florida today. Species that occur for the first time in Florida at Inglis are *Cemophora coccinea*, *Lampropeltis triangulum*, *Pituophis melanoleucus*, *Stilosoma extenuatum*, and *Rhadinaea flavilata*.

Thirteen snakes were found at the early Pleistocene Leisey Shell Pit (Meylan, 1995). All appear to represent modern species currently living in the state with one notable exception. This is a single vertebra of a small boid, the first of its family to appear since the early Miocene. It apparently does not represent a holdover from the Miocene but instead a eastern dispersal from Mexico. This is the lone record for this boid, which is most similar to the modern genera *Tropidophis* and *Exiliboa*. *Carphophis amoneus* and *Heterodon simus* appear for the first time at Haile 16A.

All of the Florida late Pleistocene snakes (figs. 7.11–7.15) belong to genera and species found in or near the peninsula today, except one. The worm snake, *Carphophis amoenus* (fig. 7.14D), lived in Florida during the Pleistocene, but currently its southern limit is just to the north of the state. In addition, the red-bellied water snake, *Nerodia erythrogaster*, also occurs in Pleistocene deposits located south of its present range limit in the Suwannee River drainage system.

Very large rattlesnake vertebrae are common in late Pleistocene deposits (fig. 7.15B). They were described as *Crotalus giganteus*, an extinct species of rattlesnake. When *C. giganteus* was first described, it was said to

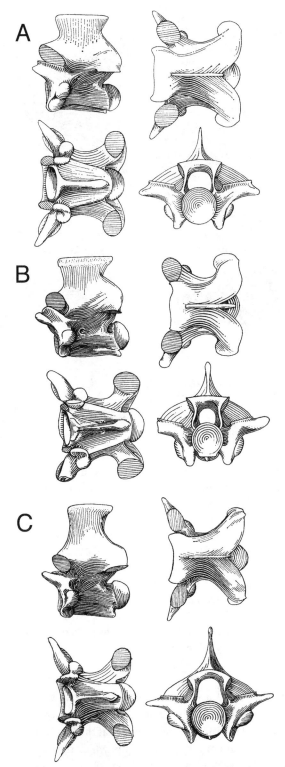

Fig. 7.12. Common lampropeltine snakes from the Pleistocene of Florida. Four views of trunk vertebrae as in figure 7.8. *A,* UF 5684, the corn snake *Elaphe guttata* from Reddick 1B, Marion County, late Pleistocene; *B,* UF 5153, the common kingsnake *Lampropeltis getulus* from Haile 7A, Alachua County, middle Pleistocene; *C,* UF 5002, the pine snake *Pituophis melanoleucus* from Reddick 1B. After Auffenberg (1963); reproduced by permission of the Tulane University Museum of Natural History.

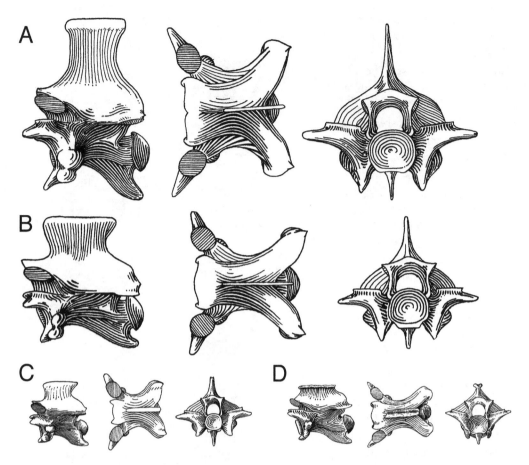

Fig. 7.13. Common natricine or water snakes from the Pleistocene of Florida. *A,* left lateral, dorsal, and anterior views of UF 4292, vertebra of the green water snake *Nerodia cyclopion* from Ichetucknee Springs, Columbia County, late Pleistocene; *B,* same three views of UF 6402, vertebra of the eastern garter snake *Thamnophis sirtalis* from Reddick 1B, Marion County, late Pleistocene; *C,* same three views of UF 5536, vertebra of the striped water snake *Regina alleni,* Winter Beach site, St. Lucie County, late Pleistocene; *D,* same three views of UF 6241, vertebra of the brown snake *Storeria dekayi* from Reddick 1B. After Auffenberg (1963); reproduced by permission of the Tulane University Museum of Natural History.

have reached a length of twelve feet, with the bulk of "a modern boa constrictor." Many subsequent authors repeated the supposed giant size that this snake attained. The centrum of the vertebra upon which the species was described is about a half inch long (12 mm). *C. giganteus* was believed to have lived in the late Pleistocene with the eastern diamondback rattler *Crotalus adamanteus.* Vertebrae of both species were described from a number of fossil sites. The body length of *C. giganteus* can be estimated by measuring vertebrae of the modern species from individuals of known length. By doing this, it can be calculated using a statistical method known as regres-

sion that the holotype of *C. giganteus* (the largest specimen ever collected) came from a snake between 5 feet, 8 inches (1.7 meters) and 8 feet, 8 inches (2.6 meters) long (Christman, 1977). Thus *C. giganteus* was not really any more giant than very large individuals of the modern species. Because the only character differentiating the extinct giant rattler from the eastern diamondback was size, Christman (1977) concluded that specimens referred to *C. giganteus* are actually just large individuals of *C. adamanteus.*

FOSSIL CROCODILIANS OF FLORIDA

The oldest crocodilians from Florida are found in Eocene limestone in the northern half of the peninsula. They are only known from a few osteoderms and fragments of jaws and cannot be identified to family. They apparently represent marine crocodilians, as they are found in the same deposits as whales, sea turtles, ocean-dwelling fish, and sharks. Discovery of specimens that are more complete is clearly desirable.

The earliest member of the Alligatoridae from Florida is *Alligator olseni* of the early Miocene. Originally described from the Thomas Farm Site (Gilchrist County), it has since been found in several other deposits of about

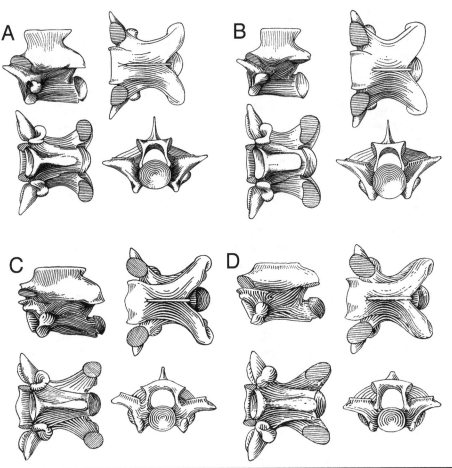

Fig. 7.14. Common xeno-
dontine snakes from the late
Pleistocene of Florida. All four
specimens are from Reddick
1B, Marion County. Same four
views as in figure 7.11. *A*, UF
6125, the eastern hognose
snake *Heterodon platyrhin-
cos*; *B*, UF 5696, the southern
hognose snake *H. simus*; *C*,
UF 6134, the yellow-lipped
snake *Rhadinaea flavilata*;
D, UF 6141, the worm snake
Carphophis amoenus. After
Auffenberg (1963); reproduced
by permission of the Tulane
University Museum of Natural
History.

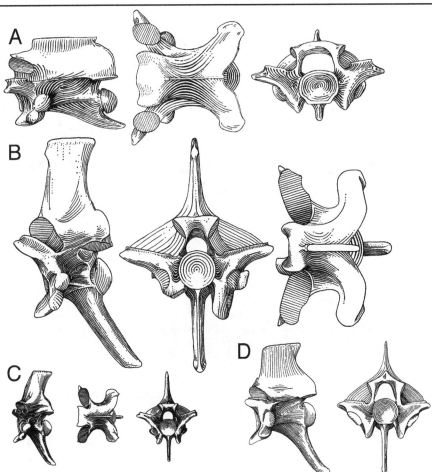

Fig. 7.15. Common poisonous
snakes from the Pleistocene of
Florida. Left lateral, dorsal, and
anterior views of each except
D, which lacks the dorsal view.
A, UF 6135, vertebra of the
eastern coral snake *Micrurus
fulvius* from Reddick 1B, Marion County, late Pleistocene; *B*,
UF 6281, vertebra of the eastern
diamondback rattlesnake *Crotalus adamanteus* from Hornsby
Springs, Alachua County, late
Pleistocene; *C*, UF 5733, vertebra of the cottonmouth *Agkistrodon piscivorus* from Paynes
Prairie 1B, Alachua County, late
Pleistocene. *D*, UF 6417, vertebra of the pigmy rattlesnake
Sistrurus miliarius from Reddick
1B. After Auffenberg (1963);
reproduced by permission of the
Tulane University Museum of
Natural History.

the same age in the panhandle. In size it seems to have been considerably smaller than the living alligator; the largest fossils obtained so far suggest a maximum length of 7 to 8 feet (2.5 meters). Other than its smaller size, *A. olseni* differs from the living *A. mississippiensis* by the fluted dorsal margin of the dentary and details of its cranial proportions. It was recognized as a valid species in the recent analysis of Brochu (1997, 1999).

Fossil remains of *Alligator* are quite plentiful in middle to late Miocene and Pliocene deposits from peninsular Florida (figs. 7.5A, 7.16A). Although these fossils have not yet been thoroughly studied, they seem to represent the living species, *A. mississippiensis.* This species is also extremely abundant in Florida Pleistocene deposits. Isolated teeth and osteoderms are the most commonly recovered elements, but all parts of the skeleton have been found including some very complete skulls. Coprolites attributed to alligators are also common (fig. 1.1D). Although modern specimens rarely attain lengths greater than 5 meters, individuals of this species may

have reached as much as 7 meters in length during the Pleistocene.

The earliest fossils of true crocodiles from Florida are also Miocene in age. Remains of this extinct species, *Gavialosuchus americanus,* have been found only in middle through late Miocene sites deposited in shallow marine bays and estuaries at the mouths of streams and rivers. Some workers prefer the name *Thecachampsa antiqua* instead of *Gavialosuchus americanus.* It is a long-snouted, fish-eating species related to *Tomistoma schlegelii,* a species now living in Thailand, Malaysia, and Indonesia and known from fossils in China. *G. americanus* attained lengths up to 12 meters and was probably most common in situations similar to those presently found at the mouths of Homosassa and Weekiwachee Springs, where fresh and salt water meet and where fish are particularly abundant. It ranged from Florida to South Carolina. A complete, articulated skeleton was collected from Haile 6 in the 1950s and is on display at the Florida Museum of Natural History. The species was

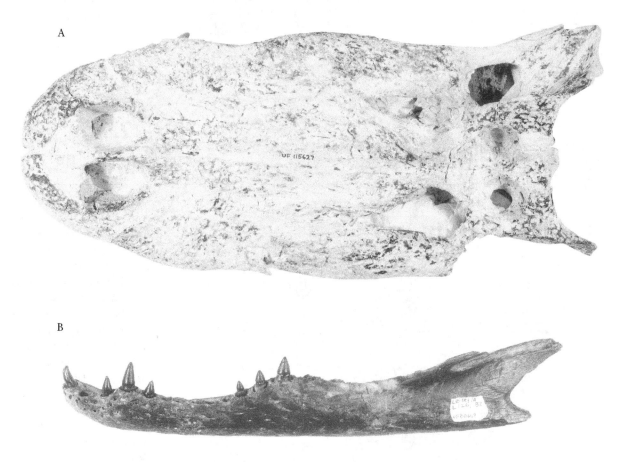

Fig. 7.16. Fossil alligators *(Alligator mississippiensis)* from Florida. *A,* dorsal view of UF 115627, nearly complete skull from the Moss Acres Racetrack site, Marion County, late Miocene; *B,* lateral view of UF 80669, a left dentary from Leisey Shell Pit, Hillsborough County, early Pleistocene. *A* about 0.4×; *B* about 0.5×.

very abundant at another Haile quarry, Haile 19A (figs. 7.3–7.6). The elongated cranial elements of *Gavialosuchus* are easily distinguished from those of *Alligator,* and its isolated teeth are usually longer and more pointed (fig. 7.3). The osteoderms of the two are also easily differentiated, as those of *Gavialosuchus* lack a dorsal ridge (which is prominent in *Alligator*) and have larger pits (fig. 7.5). Postcranial elements on the other hand are difficult to distinguish unless comparative material of both is available and even then not always possible. Alligators and crocodiles rarely live side by side in Florida today. However, *G. americanus* and *A. mississippiensis* are found together in several late Miocene deposits in Florida, although it is not known whether they occupied the same areas at the same time.

Crocodilian teeth with distinctly raised vertical ridges on the crown are occasionally found in late Miocene deposits of Florida. On the basis of a specimen from South Carolina, it seems to have been a long-snouted crocodilian similar to *Gavialosuchus. Charactosuchus* from the Miocene of South America has similar teeth, and Auffenberg (1957) suggested that the fossils from South Carolina and Florida could belong to this Neotropical genus. However, Langston and Gasparini (1997) concluded that this hypothesis was unlikely but did not completely exclude it.

REFERENCES

Ashton, R. E., and P. S. Ashton. 1988. *The Handbook of Reptiles and Amphibians of Florida.* Part One, *The Snakes.* 2d ed. Windward Publishing Co., Miami, 176 p.

———. 1991. *The Handbook of Reptiles and Amphibians of Florida.* Part Two, *Lizards, Turtles, and Crocodilians.* Rev. 2d ed. Windward Publishing Co., Miami, 191 p.

Auffenberg, W. 1954. Additional specimens of *Gavialosuchus americanus* (Sellards) from a new locality in Florida. Quarterly Journal of the Florida Academy of Sciences, 17:185–209.

———. 1956. Additional records of Pleistocene lizards from Florida. Quarterly Journal of the Florida Academy of Sciences, 19:157–67.

———. 1957. Notes on fossil crocodilians from southeastern United States. Quarterly Journal of the Florida Academy of Sciences, 20:107–13.

———. 1963. The fossil snakes of Florida. Tulane Studies in Zoology, 10:131–216.

Benton, M. J. 1997. *Vertebrate Palaeontology.* 2d ed. Chapman and Hall, New York, 464 p.

Brochu, C. A. 1997. A review of "*Leidyosuchus*" (Crocodyliformes, Eusuchia) from the Cretaceous through Eocene of North America. Journal of Vertebrate Paleontology, 17:679–97.

———. 1999. Phylogenetics, taxonomy, and historical biogeography of Alligatoroidea. Society of Vertebrate Paleontology Memoir 6:9–100.

Bryant, J. D. 1991. New early Barstovian (middle Miocene) vertebrates from the upper Torreya Formation, eastern Florida panhandle. Journal of Vertebrate Paleontology, 11:472–89.

Caldwell, M. W., and M. S. Y. Lee. 1997. A snake with legs from the marine Cretaceous of the Middle East. Nature, 386:705–9.

Christman, S. P. 1977. The status of the extinct rattlesnake, *Crotalus giganteus.* Copeia, 1977:43–47.

Estes, R. 1963. Early Miocene salamanders and lizards from Florida. Quarterly Journal of the Florida Academy of Sciences, 26:234–56.

Frost, D. R., and R. Etheridge. 1989. A phylogenetic analysis and taxonomy of iguanian lizards (Reptilia: Squamata). University of Kansas Museum of Natural History Miscellaneous Publication, 81:1–65.

Hildebrand, M. 1974. *Analysis of Vertebrate Structure.* John Wiley and Sons, New York, 710 p.

Holman, J. A. 1979. A review of North American Tertiary snakes. Michigan State University Paleontological Series, 1:200–260.

———. 1981. A review of North American Pleistocene snakes. Michigan State University Paleontological Series, 1:261–306.

———. 1995. *Pleistocene Amphibians and Reptiles of North America.* Oxford University Press, New York, 243 p.

———. 1999. Early Oligocene (Whitneyan) snakes from Florida (USA), the second oldest colubrid snakes in North America. Acta Zoologica Cracoviensia, 42:447–54.

———. 2000. *Fossil Snakes of North America.* Indiana University Press, Bloomington, 400 p.

Hutchison, J. H. 1985. *Pterosphenus* cf. *P. schucherti* Lucas (Squamata, Palaeophidae) from the late Eocene of peninsular Florida. Journal of Vertebrate Paleontology, 5:20–23.

Kluge, A. G. 1988. Relationships of the Cenozoic boine snakes *Paraepicrates* and *Pseudoepicrates.* Journal of Vertebrate Paleontology, 8:229–30.

Langston, W., and Z. Gasparini. 1997. Crocodilians, *Gryposuchus,* and the South American gavials. Pp. 113–54 *in* R. F. Kay, R. H. Madden, R. L. Cifelli, and J. J. Flynn (eds.), *Vertebrate Paleontology in the Neotropics: The Miocene Fauna of La Venta, Colombia.* Smithsonian Institution Press, Washington.

Meylan, P. A. 1982. The squamate reptiles of the Inglis 1A fauna (Irvingtonian: Citrus County, Florida). Bulletin of the Florida State Museum, 27:1–85.

———. 1995. Pleistocene amphibians and reptiles from the Leisey Shell Pit, Hillsborough County, Florida. Bulletin of the Florida Museum of Natural History, 37:273–97.

Romer, A. S. 1956. *The Osteology of the Reptiles.* University of Chicago Press, Chicago, 772 p.

Ross, C. A. 1989. *Crocodiles and Alligators.* Facts on File, New York, 240 p.

8

Reptilia 3

Birds

INTRODUCTION

Birds are archosaur reptiles (fig. 6.1) that evolved the capability of powered flight in the Mesozoic Era. Their formal scientific name is Aves. In traditional vertebrate classification, Aves was given equal rank with Mammalia and Reptilia, usually at the class level. This meant that members of the Aves were excluded from the Reptilia, even though it was realized that "reptiles" were ancestral to birds. Such classifications are no longer regarded as valid. Two basic solutions are to either include birds as a subgroup within the reptiles or classify crocodiles, dinosaurs, and pterosaurs as birds and exclude them from the Reptilia. The former is presently preferred.

Avian origins are still a matter of dispute among paleontologists. The debate centers around determining the closest relative of birds. Presently, the majority, or mainstream, opinion favors a close relationship with dinosaurs, more specifically with one of the bipedal, carnivorous theropods, not too distant from *Velociraptor* of Hollywood fame. This hypothesis in its modern incarnation was first championed by John Ostrom of Yale University (for example, Ostrom, 1974, 1975) and has been supported by many exciting new discoveries (for example, Novas and Puerta, 1997; Forster et al., 1998), including feathered Chinese dinosaurs (Chen et al., 1998; Ji et al., 1998). A somewhat unorthodox implication of this concept is that dinosaurs are actually not extinct, since by definition a taxonomic group becomes extinct only when *all* of its members (including their descendants) become extinct. Thus dinosaurs can be found in backyards across America dining at birdfeeders and are served baked or fried at most restaurants (see Dingus and Rowe [1998] for more on this theme).

A small but vocal minority opinion among paleontologists argues against the dinosaurian origin of birds. Some derive them from an early crocodilian group (Martin, 1991), others from an early, extinct archosaur lineage that is neither a crocodilian nor a dinosaur (Fe-

duccia, 1996). Whether or not a controversial fossil from Texas named *Protoavis* is considered a bird also plays a pivotal role in interpreting bird origins—many paleontologists do not regard it as a bird (Sereno, 1997).

Many well-preserved fossils collected over the past few decades now reveal that true birds (by anyone's definition) were abundant, widespread, and morphologically diverse as early as the Cretaceous (Feduccia, 1996). By the Miocene, when Florida's fossil record of birds begins, all of the major groups of birds had evolved. Fossils of birds are generally regarded as rare. Certainly they are uncommon, even totally absent, at some localities, especially those where the fossils were extensively tumbled or transported before they were buried. However, many localities in Florida have produced large numbers of avian fossils. Their recovery is greatly enhanced by careful collecting techniques and screenwashing large amounts of matrix.

Two Florida localities deserve special recognition for their avian fossils. The Reddick fissure-fills in Marion County produced more than 10,000 late Pleistocene specimens representing sixty-four species (Brodkorb, 1957). The late Pliocene Richardson Road Shell Pit in Sarasota County produced far fewer species (eleven), but more than 1,600 specimens. Most belonged to a single species of cormorant, including 137 partial to complete articulated skeletons (Emslie, 1995; Emslie et al., 1996). For more than four decades research on fossil birds from Florida was spearheaded by the late Pierce Brodkorb of the University of Florida and his graduate students. Most of the information in this chapter is a result of their activities.

AVIAN SKELETAL ANATOMY

The avian skeleton is profoundly modified for flight (fig. 8.1), which makes most of their bones instantly identifiable as that of a bird. Even those types which have secondarily abandoned flying (such as the ostrich) show

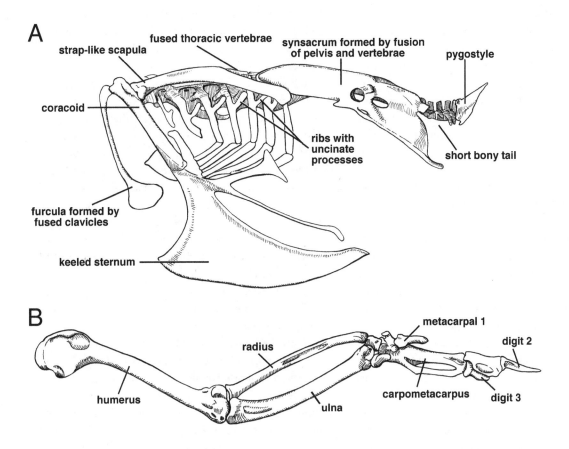

Fig. 8.1. Modifications of the avian skeleton for flight. *A,* left lateral view of postcranial skeleton without the limbs. Note the keeled sternum, synsacrum, furcula, shortened tail, and other typical avian features. *B,* the forelimb of a flying bird, showing the major skeletal elements. Figures from *Analysis of Vertebrate Structure* by Milton Hildebrand, copyright © 1974 by John Wiley & Sons, Inc. Reprinted by permission of John Wiley & Sons, Inc.

evidence for it in their anatomy. Perhaps the only skeletal elements not immediately recognizable as avian are their toe bones, which still resemble those of other reptiles. Most bird bones are hollow and thin walled to decrease weight. The internal cavity has struts of bone to provide structural integrity. Other weight-reducing features include loss of teeth (except in some Mesozoic taxa) and reduction in the number of digits and caudal vertebrae. Rigidity is also provided by fusion of many of the skull elements, so that sutures between them are not apparent in adults, and by development of a synsacrum. This single element is an amalgamation of the right and left pelves, many dorsal vertebrae, the sacral vertebrae, and the anterior caudal vertebrae (fig. 8.1A).

The pectoral (shoulder) bones of birds are highly modified to provide support for the wing. The two clavicles are fused to form the furculum (fig. 8.1A, commonly called the wishbone), which strongly articulates with the large coracoids. The coracoid structurally connects the humerus (and thus the wing) to the sternum. The coracoid has a strong notch through which passes a tendon connecting the flight muscles located on the sternum to the humerus. The sternum bears a keel for attachment of the major flight muscles (the breast meat of chicken and turkey). Birds that have reduced the capacity for flight have correspondingly small keels on their sternums. The scapula is a rather small, bladelike, and not especially diagnostic bone. The major wing bones are the humerus, ulna, radius, and carpometacarpus (fig. 8.1B). The latter element consists of the fused carpals plus the second and third metacarpals. The radius is more slender than the ulna, unlike that of most tetrapods. The phalanges of the manus are fused into one or two elements.

In the hindlimb, the femur is relatively unspecialized and resembles those of small dinosaurs. Advanced archosaurs (including birds and dinosaurs) are like mammals and differ from other tetrapods in having a distinct proximal head on the femur that is set off medially from the shaft and inserts into the acetabulum of the pelvis. The proximal tarsal elements are fused with the tibia to form the tibiotarsus, while the distal tarsals and the metatarsals are fused into a single bone called the tarsometatarsus. The latter bears the trochleae for articulation with the phalanges. The fibula is reduced to a small splint that runs along the proximal half of the tibiotarsus. The limb bones of birds do not have epiphyses, but the bones do not continue to grow, as is the case with crocodilians. In fact, bird skeletal elements generally show less variation in size within a single species than do those of a typical mammal.

Fossil birds are most commonly described from postcranial skeletal elements. Bird skulls are extremely fragile and, of course, lack the resistant teeth that make up many of the fossils of other groups. The major limb elements (for example, coracoid, humerus, carpometacarpus, tibiotarsus, tarsometatarsus) in many birds are so distinctive that they are diagnostic to the species level if they are complete and well preserved. It is not uncommon to see a valid description of a new fossil species based on a partial humerus or the distal half of a tarsometatarsus. For others, especially among the small passerines, only cranial material can differentiate among the many families. Unfortunately, elements of totally unrelated birds are sometimes superficially similar, so identification even to the family level requires a keen eye and an extensive comparative collection. Typical bird fossils from Florida are illustrated in this chapter, but a whole book would be required to show the elements for all of the major groups.

Fossil Birds of Florida

The age of the oldest bird fossils in the state is early Miocene, approximately 20 Ma. These records provide little more than a tantalizing glimpse of what must have been a much more diverse avian fauna. It is not until the late Miocene and early Pliocene that Florida's fossil birds begin to approximate their modern diversity. Even so, because these Neogene faunas come from water-laid sediments, only those types of birds associated with aquatic habitats are well represented. These include ducks, geese, herons, cranes, grebes, cormorants, anhingas, boobies, and auks. It is not until the late Pliocene and Pleistocene that a fuller picture of terrestrial birds emerges.

The most important early Miocene locality producing fossil birds in Florida is Thomas Farm. An anhinga (*Anhinga subvolans*, fig. 8.2), three kites (*Promilio floridanus, P. epileus,* and *P. brodkorbi*), a chachalaca (*Boreortalis laesslei*), a primitive turkey (*Rhegminornis calobates*), and a small ground dove (*Columbina prattae*)

have been described from this site. Additional, undescribed material shows the presence of an ibis, a hawk, a rail, a barbet, a rollerlike bird, and several small passerines including a warbler. The ground dove is the most common bird at Thomas Farm, making up 60 percent of the avian fossils. The surface texture of some of the bones indicates that they came from breeding females, and juveniles are also common. Therefore, Becker and Brodkorb (1992) concluded that *C. prattae* was nesting in the vicinity of the sinkhole that was to become the Thomas Farm deposit. The presence of passerines (the perching birds or songbirds) at Thomas Farm is important, as it is one of the oldest records of this important group in North America. They are the dominant group of terrestrial birds in modern faunas but are not known with certainty before the Oligocene (Fedducia, 1996). According to Olson (1985), most small terrestrial birds in the Paleogene were not passerines but were instead related to the modern rollers.

The other known early Miocene species all come from the Hawthorn Group. *Puffinus micraulax,* an extinct shearwater slightly smaller than Audubon's shearwater, was described from the distal half of a humerus found along Hogtown Creek in Gainesville. *Sula universitatis,* similar in size to the brown booby, is known from only a proximal carpometacarpus. It was found along a small creek near Fraternity Row on the University of Florida campus (as denoted by the species name). The presumed coastal ecology of both of these birds agrees well with the

Fig. 8.2. Anterior view of UF 4500, proximal end of a humerus of *Anhinga subvolans* from the Thomas Farm site, Gilchrist County, early Miocene. About 1×.

predominantly marine nature of the Hawthorn Group. Two additional early Miocene birds were described from a more terrestrial deposit in Tallahassee: *Propelargus olseni,* a small extinct stork, and *Probalearica crataegensis,* an extinct crane related to the crowned cranes of the Old World. It is only one of a number of birds whose modern relatives have distributions that do not include Florida, or even North America, but which are found in Florida as fossils. The ease by which birds

can cross oceanic barriers and invade new territories is evidenced by the cattle egret, *Bubulcus ibis,* which is a native of Africa and southern Asia. It has spread across the West Indies, northern South America, and southern North America in less than a century.

The published record of fossil birds in Florida next skips to the late Miocene (fig. 8.3). The Love site contains the most diverse nonmarine bird fauna known in North America older than Pleistocene (Becker, 1987).

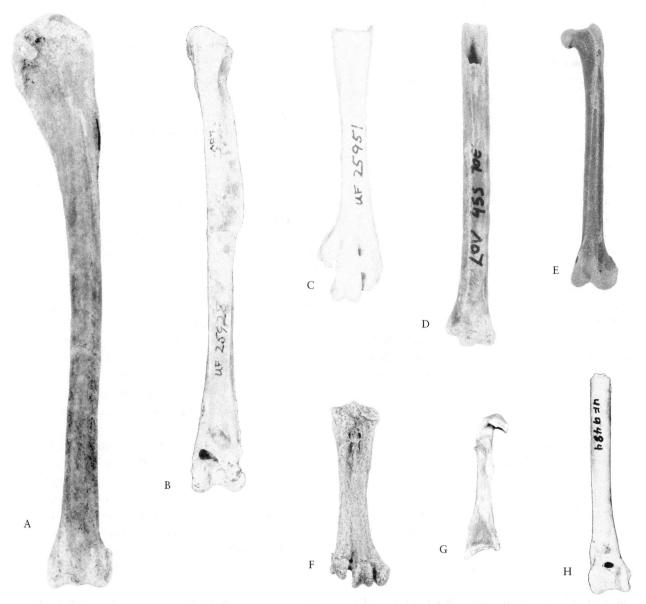

Fig. 8.3. Examples of late Miocene fossil birds from Alachua County, Florida. *A,* anterior view of UF 114600, a humerus of *Anhinga grandis* from Haile 19A; *B,* anterior view of UF 25928, a tibiotarsus of *Pandion lovensis* from the Love site; *C,* anterior view of UF 25951, distal end of a tarsometatarsus of a large anatid, possibly *Branta,* from the Love site; *D,* posterior view of UF 29677, distal end of a radius of a flamingo, family Phoenicopteridae, from the Love site; *E,* anterior view of UF 3285, a left femur of *Nycticorax fidens* from the McGehee Farm site; *F,* anterior view of UF 114588, a right tarsometatarsus of *Anhinga grandis* from Haile 19A; *G,* dorsal view of UF 11108, a left coracoid of *Jacana farrandi* from the McGehee Farm site; *H,* anterior view of UF 9484, a distal end of a tibiotarsus of an anatid from the McGehee Farm site. All are about 1× except *G,* which is about 2×.

The following are present: a grebe, cormorants, an anhinga, flamingos, an ibis, herons, a stork, ducks and geese, a vulture, an osprey, a hawk, a turkey, rails, coots, cranes, a limpkin, shorebirds, and perching birds. The heavy aquatic influence on the fauna is readily apparent. Slightly younger birds are known from a number of other localities in northern and central Florida, and these record components of essentially the same fauna. Only one passerine bird has been described from this age, *Palaeostruthus eurius,* a finchlike bird the size of a towhee. It is known from only one specimen, and the validity of this species has been questioned (Steadman, 1981).

Fish-eating forms are common in Florida's late Miocene bird fauna. *Phalacrocorax wetmorei* is a cormorant smaller than the modern double-crested cormorant. It is also known from the early Pliocene. *Anhinga grandis* is a large species (about 1.5 times the size of the living *A. anhinga*) that was less of a percher than the modern form but a better soarer (fig. 8.3A, F). The extinct osprey, *Pandion lovensis,* is the most primitive and hawklike of its family (fig. 8.3B). Its hind limb was less adapted for grasping fish than that of the modern osprey. Five herons were present, including *Nycticorax fidens* (fig. 8.3E), most similar to the modern black-crowned night heron and *Egretta subfluvia,* a very small, extinct egret (Becker, 1985).

Other interesting late Miocene birds found in Florida include *Jacana farrandi,* an extinct lily-trotter (fig. 8.3G). They are specialized members of the order Charadriiformes, which includes sandpipers, gulls, and auks. The outstanding physical feature of jacanas is very long toes and toenails (up to four inches) enabling them to walk on floating plants. Jacanas are no longer found in Florida. Another tropical bird is the motmot, which has been reported from one bone. *Pliogyps charon* is an extinct genus and species of a small condor. It is one of the older records of this group from North America (the oldest is middle Miocene from California), but ironically the so-called New World vultures and condors appear to have originated in Europe, where Eocene and Oligocene records are known. Note that this group is more closely related to storks than to either the "Old World" or true vultures and the hawks and eagles (Emslie, 1988).

The phosphate mines of the Bone Valley region have produced a very large number of bird fossils (fig. 8.4). All of these are now regarded to have originated from the early Pliocene Palmetto Fauna (fig. 2.4). Most of the species, and all of the common types, are aquatic forms, primarily from marine or coastal environments. Brodkorb (1955) once suggested that guano from large rookeries of these birds, especially the cormorants and boobies, played a role in producing the region's phosphate deposits, which make up the largest mining reserves of this important geologic resource in North America. This hypothesis is now discredited, as the phosphate was formed in the early and middle Miocene and that found in early Pliocene strata is largely reworked and enriched from older deposits.

The most abundant bird from the Bone Valley is the small cormorant *Phalacrocorax wetmorei* (fig. 8.4A–D). A second, much larger and rarer cormorant is also present in the Bone Valley. Cormorant fossils make up more than half of the bird bones found in this region. Somewhat ironically, present-day mining operations have created many lakes where one can occasionally see flocks of cormorants. The birds of course cannot know that 4.5 Ma the region was home to vast numbers of their extinct relatives.

Other birds from the Bone Valley (fig. 8.4E–M) include loons, a grebe, sulids, alcids, herons, flamingos, and *Diomedea anglica,* an extinct species of albatross originally described from Suffolk, England. An oystercatcher, *Haematopus sulcatus,* is similar to the modern America or black oystercatcher in size. Oystercatchers feed mainly on mollusks, which they pry open with chisel-shaped bills. Several extinct sandpipers are found in the Bone Valley (fig. 8.4K). *Calidris pacis* is most similar to the modern red knot, while *Erolia penepusilla* is slightly larger than the living least sandpiper. The third described species in this family, *Limosa ossivallis,* is similar to the marbled godwit. A gull, *Larus elmorei,* is known from a handful of specimens (fig. 8.4M) and is closely related to the modern ring-billed gull. Fossil records of gulls older than Pleistocene are rare. Several auks, including *Australca grandis,* which is about the size of the modern tufted puffin, are not uncommon (fig. 8.4F, G).

One of the most remarkable members of the Florida bird fauna is the extinct late Pliocene *Titanis walleri* (Brodkorb, 1963b; fig. 8.5). It was a large, predatory, flightless bird larger than an African ostrich (about 2 meters tall). It is a member of the Phorusrhacidae, a family of flightless birds that originated in South America (Chandler, 1994). *Titanis* must have dispersed to Florida in the Pliocene after the formation of the Panamanian Land Bridge. Remains of *Titanis* are only known from Florida and Texas. Most of the known specimens have come from either the Inglis 1A site or a short stretch of the Santa Fe River. Their large size, massiveness, and modifications for flightlessness often makes the fossils of *Titanis* difficult to recognize as belonging to a bird.

Late Pliocene and early Pleistocene sites in Florida have produced many other types of birds (figs. 8.6–8.8).

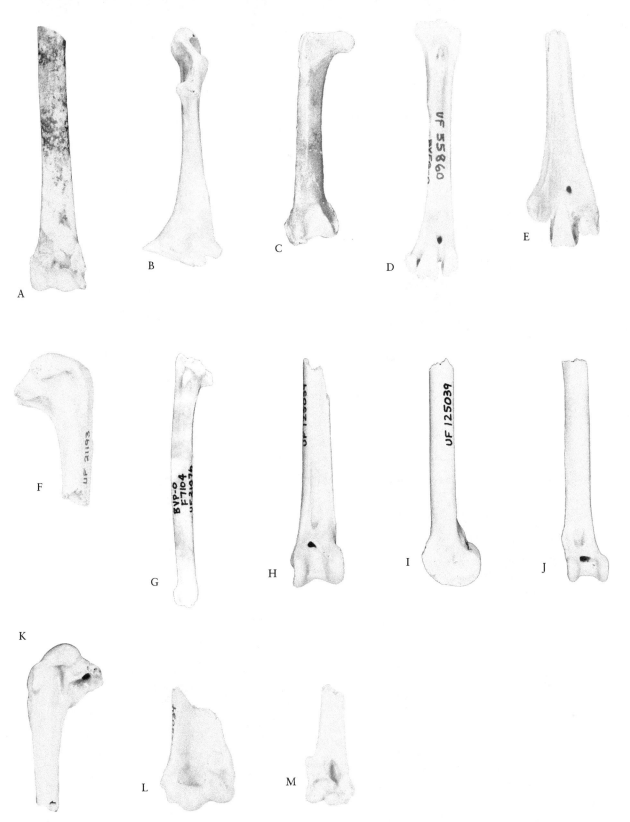

Fig. 8.4. Early Pliocene birds from the Bone Valley District. *A–D,* the cormorant *Phalacrocorax wetmorei,* the most common Bone Valley bird. *A,* posterior view of UF 53914, the distal end of a humerus; *B,* dorsal view of UF/PB 87, a right coracoid; *C,* anterior view of UF 49090, a right femur; *D,* anterior view of UF 55860, a left tarsometatarsus; *E,* posterior view of UF/PB 146, the distal end of a tarsometatarsus of the flamingo *Phoenicopterus floridanus; F,* anterior view of UF 21193, the proximal end of a right humerus of the alcid *Australca* sp.; *G,* lateral view of UF 21076, an ulna of *Australca* sp.; *H,* anterior, and *I,* medial views of UF 125039, the distal end of a tibiotarsus of the heron *Ardea polkensis; J,* anterior view of UF/PB 7747, the distal end of a tibiotarsus of a loon, *Gavia* sp.; *K,* anterior view of UF/PB 594, the proximal end of a left humerus of the sandpiper *Calidris pacis; L,* posterior view of UF 125034, the distal end of a humerus of the boobie *Sula* sp.; *M,* posterior view of UF/PB 140, the distal end of a humerus of the gull *Larus elmorei.* All are about 1× except *K,* which is about 2×.

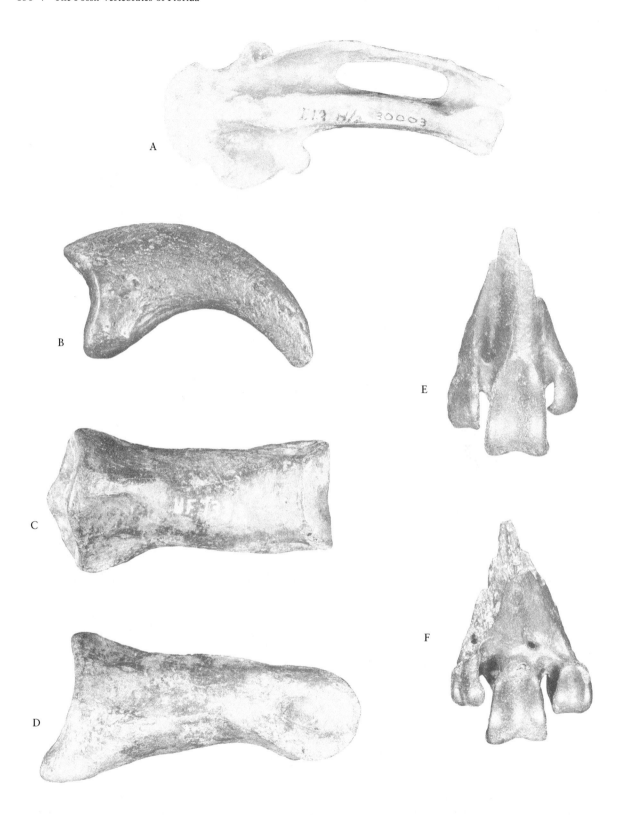

Fig. 8.5. The giant flightless bird *Titanis walleri* from the late Pliocene of Florida. *A*, lateral view of UF 30003, a left carpometacarpus from Inglis 1A, Citrus County; *B–F*, from the Santa Fe River, Gilchrist County; *B*, lateral view of UF 10417, a distal phalanx (claw) of the pes; *C*, dorsal, and *D*, lateral views of UF 7332, a proximal phalanx of pedal digit 3; *E*, anterior, and *F*, posterior views of UF 4108, the distal end of a tarsometatarsus. *A–D* about 1×; *E–F* about 0.5×.

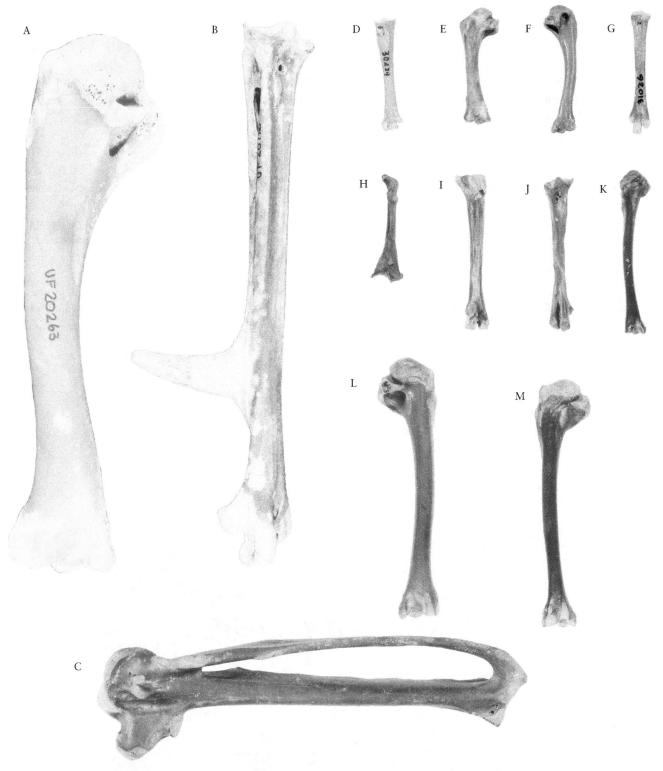

Fig. 8.6. Late Pliocene birds from Inglis 1A, Citrus County *(A–H)* and the APAC (=Macasphalt) Shell Pit, Sarasota County *(I–M). A*, anterior view of UF 20263, a left humerus of the fossil turkey *Meleagris leopoldi* (or *M. anza*); *B*, anterior view of UF 20716, a male tarsometatarsus of *M. leopoldi* or *M. anza*; its sex is indicated by the presence of the large spur on the shaft; *C*, ventral view of UF 3005, a carpometacarpus of an extinct species of golden eagle, *Aquila; D*, anterior view of UF 30224, a left tarsometatarsus of the screech owl, *Otus asio; E*, anterior view of UF 31409, a left humerus of the mourning dove, *Zenaida macroura; F–H*, the bobwhite quail, *Colinus virginianus; F*, anterior view of UF 30381, a right humerus; *G*, anterior view of UF 310226, a left tarsometatarsus; *H*, dorsal view of UF 30695, a right coracoid; *I*, posterior view of UF 94974, a left tarsometatarsus of the pied-billed grebe, *Podilymbus podiceps; J*, anterior view of UF 94970, a left tarsometatarsus of *P. podiceps; K*, anterior view of UF 94919, a left humerus of a rail, *Rallus* sp.; *L*, anterior view of UF 101415, a right humerus of a teal, possibly *Anas cyanoptera; M*, anterior view of UF 95446, a left humerus of the extinct ruddy duck, *Oxyura hulberti*. All about 1×.

Diverse, interesting bird faunas have come from both karst in-fillings (Inglis, Haile 7C) and shallow marine shell beds (APAC, Richardson Road, and Leisey Shell Pits). Unlike older faunas, living species predominate in faunas of this time. However extinct species (or even genera) are still common, including a large anhinga, *Anhinga beckeri;* pygmy goose, *Anabernicula gracilenta;* a giant goose, *Branta dickyi* (fig. 8.8A); a hawk, *Buteogallus fragilis,* related to the living great black hawk of the Neotropics; a large stork, *Ciconia maltha* (fig. 8.11); a spoonbill, *Ajaia chione;* an ibis, *Eudocimus leiseyi;* two condors, *Aizenogyps toomeyae* and *Gymnogyps kofordi* (fig. 8.8C–E); a true "Old World" vulture, *Neophrontops slaughteri;* a teratorn, *Teratornis merriami,* a member of an extinct family related to condors that includes the largest known flying bird; a turkey, *Meleagris leopoldi* (fig. 8.6A, B); a woodcock, *Scolopax hutchensi;* and a woodpecker, *Campephilus dalquesti,* slightly smaller than the recently extinct ivory-billed woodpecker. Living species present include the loon (fig. 8.10A–C), pied-billed grebe (fig. 8.6I, J), common egret, mallard (fig. 8.9B), green-winged teal, turkey and black vultures, sparrow and pigeon hawks, owls, whooping and sandhill cranes, white ibis, mourning dove, bobwhite quail (fig. 8.6F–H), yellow-billed cuckoo, flicker, red-headed woodpecker, and numerous passerines (Emslie, 1998). Among the latter are such common birds as cardinals, blue jays, crows, thrushes, and several sparrows.

One of the best studied of these faunas is the remarkable assemblage of birds from the late Pliocene APAC Shell Pit near Sarasota (Emslie, 1992). At least forty species are present at this one site, primarily taxa that inhabit coastal marshes, wetlands, and shorelines. Grebes, rails, and ducks were the most common birds (fig. 8.6I–M). Important records at APAC include the oldest

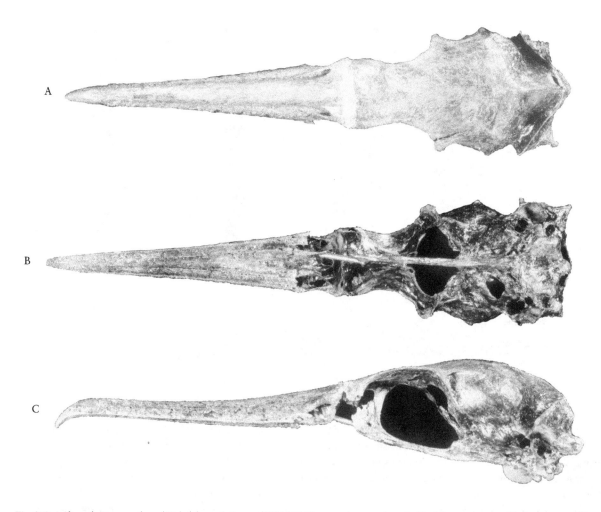

Fig. 8.7. *A,* dorsal, *B,* ventral, and *C,* left lateral views of UF 132181, a nearly complete skull of the cormorant *Phalacrocorax filyawi* from the Quality Aggregates (Richardson Road) Shell Pit, Sarasota County, late Pliocene. The discovery of numerous avian fossils, including articulated skeletons, at this quarry was one of the most important paleontological finds of the late 1980s in Florida. About 1×.

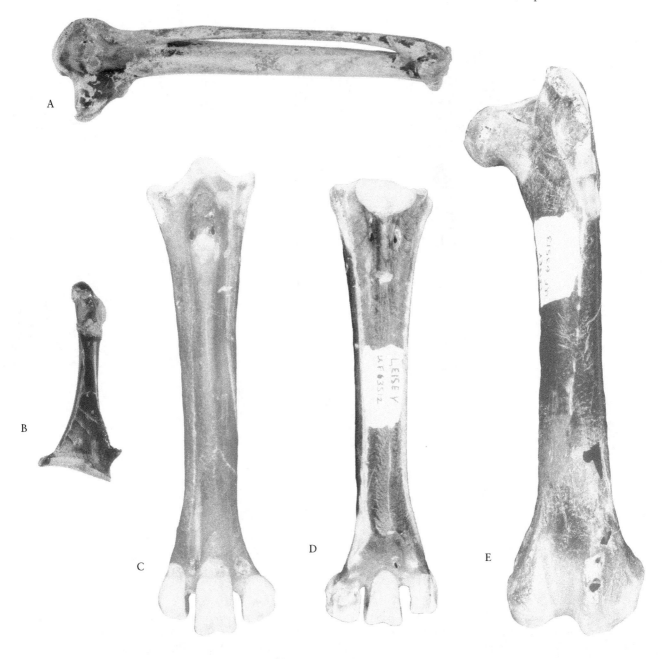

Fig. 8.8. Examples of early Pleistocene birds from the Leisey Shell Pit, Hillsborough County. *A*, ventral view of UF 87235, a right carpometacarpus of the extinct giant goose *Branta dickyi*; *B*, dorsal view of UF 65818, a right coracoid of the redhead duck, *Aythya americana*; *C*, anterior, and *D*, posterior views of UF 63512, a right tarsometatarsus of the extinct condor *Gymnogyps kofordi*; *E*, anterior view of UF 63513, a left femur of *G. kofordi*. All about 1×.

known specimens of the condor *Gymnogyps,* the pygmy goose *Anabernicula gracilenta,* and the hawk-eagle *Spizaetus.* The latter now lives in the Neotropics, Southeast Asia, and Africa, but was widely distributed in North America in the Pliocene and Pleistocene. Emslie (1992, 1998) has analyzed the fossil record of Florida's wetland birds over the past 2.5 million years and concluded that Pliocene faunas were considerably richer than modern ones in terms of numbers of species. He hypothesized

that numerous rapid Quaternary sea-level fluctuations caused drastic changes in the extent of coastal wetlands and this in turn had a deleterious effect on species richness of the region's birds.

Richardson Road Shell Pit (also known as the Quality Aggregates Shell Pit) is located on the opposite side of Interstate Highway 75 from APAC. It too has produced an important assemblage of late Pliocene bird fossils. Here, however, the fauna is dominated by one species, a

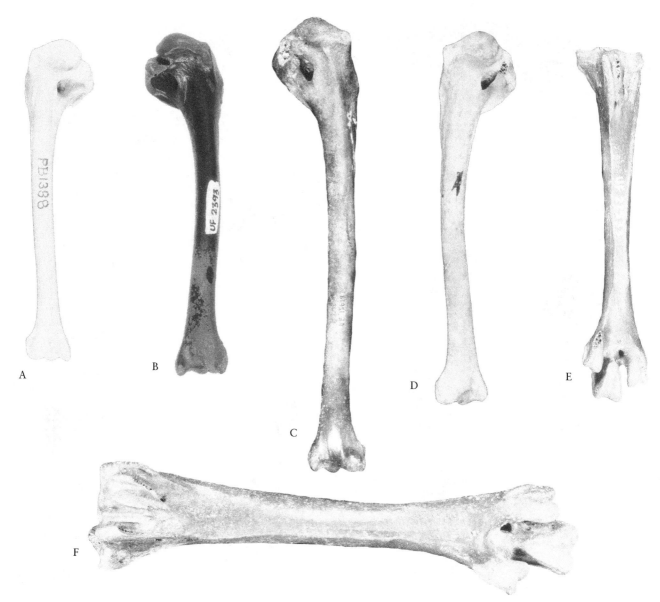

Fig. 8.9. Late Pleistocene anatids (ducks and geese) from Florida. *A,* anterior view of UF/PB 1388, a left humerus of the red-breasted merganser, *Megus serrator,* from Rock Springs, Orange County. *B–F,* from the Ichetucknee River, Columbia County; *B,* anterior view of UF 2393, a right humerus of the mallard, *Anas platyrhynchus; C,* anterior view of UF 15619, a right humerus of the whistling swan, *Olor columbianus; D,* anterior view of UF 22492, a left humerus of the Canada goose, *Branta canadensis; E,* posterior view of UF 22533, a right tarsometatarsus of *B. canadensis; F,* posterior view of UF 15656, a right tarsometatarsus of *O. columbianus.* *C–D* about 0.5×; all others about 1×.

cormorant (fig. 8.7). Described as a new species, *Phalacrocorax filyawi,* thousands of its bones, including complete skulls and about 130 articulated skeletons, were recovered. Emslie et al. (1996) presented evidence that a toxic red tide caused by a bloom of dinoflagellate algae (a type of marine plankton) was the source for the catastrophic mortality of the cormorants. *P. filyawi* is more closely related to species now living along the coasts of the northern Pacific than to any of the modern cormorants of the Atlantic Coast (Emslie, 1995). Other, much

less common birds from this site included two extinct gulls, two grebes, and a pygmy goose. As noted above, gulls have a very poor fossil record, and those found here proved to be new species in the genus *Larus,* which includes most of the typical seagulls. A third larid at this site is a large, predaceous jaegar (*Stercorarius* sp.).

Several dozen late Pleistocene localities in Florida have produced fossil birds. The major localities (with number of species of birds reported) are Arredondo (41), Haile 11B (67), Ichetucknee River (67), Reddick (64),

Rock Springs (31), and Seminole Field (47). The age of several of these localities has at times been incorrectly reported as middle Pleistocene. These and other late Pleistocene sites have produced a combined record of approximately 155 species of birds (figs. 8.9–8.11). About twenty of these are extinct and others are living species which have modern distributions far removed from Florida. These include *Olor buccinator,* the trumpeter swan, known today from Wyoming and Alaska; *Bonasa umbellus,* the ruffed grouse, from wooded North America south to the mountains of Georgia; *Tympanuchus cupido,* the greater prairie chicken, and *Numenius*

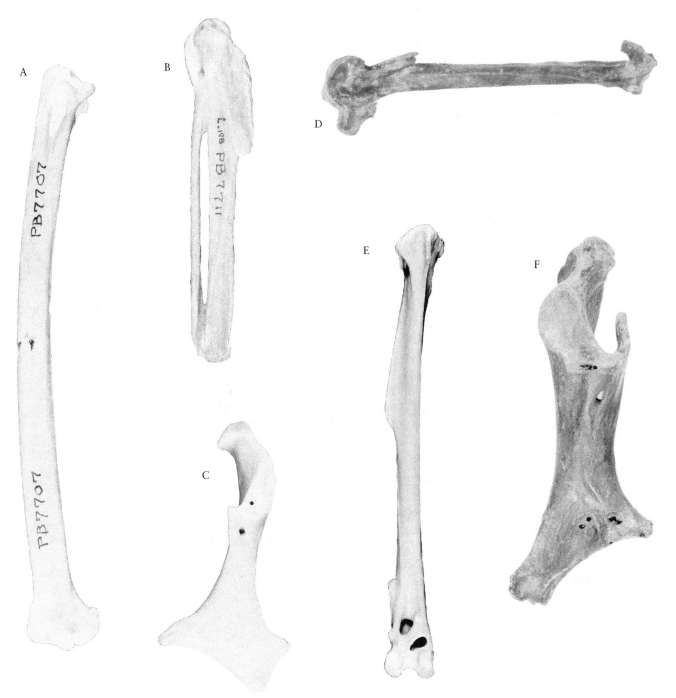

Fig. 8.10. Late Pleistocene birds from Florida. *A–C,* the common loon, *Gavia immer,* from Rock Springs, Orange County. *A,* dorsal view of UF/PB 7707, an ulna; *B,* ventral view of UF/PB 771, a right carpometacarpus; *C,* dorsal view of UF 7701, a right coracoid; *D,* dorsal view of UF 22447, a right carpometacarpus of the bald eagle, *Haliaeetus leucocephalus,* from the Ichetucknee River, Columbia County; *E,* anterior view of a UF/PB specimen, a right tibiotarsus of the osprey, *Pandion haliaetus,* from Rock Springs; *F,* dorsal view of UF 15576, a left coracoid of the sandhill crane, *Grus americanus,* from the Ichetucknee River. All about 1×.

americanus, the long-billed curlew, both from the prairie regions of North America; *Laterallus exilus,* the gray-breasted crake, *Jacana spinosa,* the lily-trotter, and *Vanellus chilensis,* the southern lapwing, all from South America; and *Pica pica,* the black-billed magpie from western North America. *Ectopistes migratorius,* the recently extinct (1914) passenger pigeon, was also present in the Pleistocene of Florida. The late Pleistocene also marks the first fossil appearance in Florida of the modern species of whistling swan (fig. 8.9C), bald eagle (fig. 8.10D), osprey (fig. 8.10E), red-tailed hawk, and a number of passerines, including such familiar birds as wrens, mockingbird, grackle, shrike, meadowlark, and numerous sparrows.

Two faunal patterns are apparent in the Pleistocene bird fauna of Florida. The first is the presence of northern species, such as *Olor buccinator* and *Bonasa umbellus.* Their presence indicates that their ranges expanded southward during glacial periods and contracted northward during the interglacials. A more important faunal pattern is related to the cyclic opening of the Gulf Coast Savanna corridor when sea levels dropped during glacial periods. Two different sets of avian species appear in the fossil record reflecting the opening of this corridor: arid western and mesic tropical species.

The western species include *Pandanaris floridanus,* a close relative of the fossil cowbird of Rancho La Brea, California; *Quiscalus mexicanus,* the great-tailed grackle; *Pica pica;* and *Gymnopgyps californicus.* Two modern species, the burrowing owl, *Speotyto cunicularia,* and the scrub jay, *Aphelocoma coerulescens,* are both essentially western forms with small relict populations in central Florida. The Neotropical species include, among others, *Caracara plancus* and *Milvago chimachima* (two species related to the Neotropical caracaras), *Protocitta ajax* and *Henocitta brodkorbi* (two jays related to the magpie jay of Mexico and South America), *Cremaster tytthus* (related to the crested oropendola), and the rail *Laterallus exilis.* Again, this pattern is also shown in the present distribution of some modern species. The snail or Everglades kite *Rostrhamus sociabilis* and the short-tailed hawk *Buteo brachyurus* are both Neotropical species with relict populations in Florida. The list of western and Neotropical birds in Florida's past continues to grow as additional localities are discovered and new species are described.

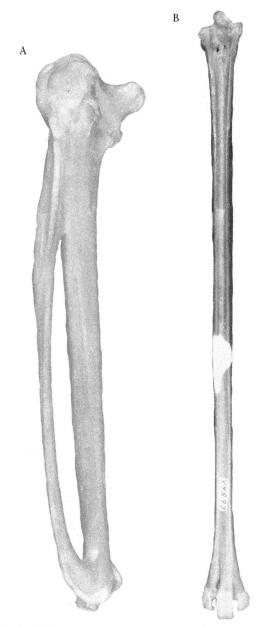

Fig. 8.11. Two specimens of the extinct stork, *Ciconia maltha,* from the Ichetucknee River, Columbia County. *A,* ventral view of UF/PB 1234, a left carpometacarpus; *B,* anterior view of UF/FGS 4897, a right tarsometatarsus. *A* about 1×; *B* about 0.5×.

REFERENCES

Becker, J. J. 1985. Fossil herons (Aves: Ardeidae) of the late Miocene and early Pliocene of Florida. Journal of Vertebrate Paleontology, 5:24–31.

———. 1986. Reidentification of *"Phalacrocorax" subvolans* Brodkorb as the earliest record of the Anhingidae. Auk, 103:804–8.

———. 1987. *Neogene Avian Localities of North America.* Smithsonian Institution Press, Washington, D.C., 171 p.

Becker, J. J., and P. Brodkorb. 1992. An early Miocene ground-dove (Aves: Columbidae) from Florida. Natural History Museum of Los Angeles County Science Series, 36:189–93.

Brodkorb, P. 1955. The avifauna of the Bone Valley Formation. Florida Geological Survey Report of Investigations, 14:1–57.

———. 1957. New passerine birds from the Pleistocene of Reddick, Florida. Journal of Paleontology, 31:129–38.

———. 1959. The Pleistocene avifauna of Arredondo, Florida. Bulletin of the Florida State Museum, 4:269–91.

———. 1963a. Fossil birds of the Alachua Clay of Florida. Florida Geological Survey Special Publication, 2:1–17.

———. 1963b. A giant flightless bird from the Pleistocene of Florida. Auk, 80:111–15.

Campbell, K. E. 1980. A review of the Rancholabrean avifauna of the Itchtucknee River, Florida. Natural History Museum of Los Angeles County Contributions in Science, 330:119–29.

Chandler, R. M. 1994. The wing of *Titanis walleri* (Aves: Phorusrhacidae) from the late Blancan of Florida. Bulletin of the Florida Museum of Natural History, 36:175–80.

Chen Peiji, Dong Zhiming, and Zhen Shuonan. 1998. An exceptionally well-preserved theropod dinosaur from the Yixian Formation of China. Nature, 391:147–52.

Dingus, L., and T. Rowe. 1997. *The Mistaken Extinction: Dinosaur Evolution and the Origin of Birds.* W. H. Freeman, New York, 332 p.

Emslie, S. D. 1988. The fossil history and phylogenetic relationships of condors. Journal of Vertebrate Paleontology, 8:212–28.

———. 1992. Two new late Blancan avifaunas from Florida and extinctions of wetland birds in the Plio-Pleistocene. Natural History Museum of Los Angeles County Science Series, 36:249–69.

———. 1995. A catastrophic death assemblage of a new species of cormorant and other seabirds from the late Pliocene of Florida. Journal of Vertebrate Paleontology, 15:313–30.

———. 1995. An early Irvingtonian avifauna from Leisey Shell Pit, Florida. Bulletin of the Florida Museum of Natural History, 37:299–344.

———. 1998. Avian community, climate, and sea-level changes in the Plio-Pleistocene of the Florida Peninsula. Ornithological Monographs, No. 50, 113 p.

Emslie, S. D., W. D. Allmon, F. J. Rich, J. H. Wrenn, and S. D. de France. 1996. Integrated taphonomy of an avian death assemblage in marine sediments from the late Pliocene of Florida. Palaeogeography, Palaeoclimatology, Palaeoecology, 124:107–36.

Feduccia, A. 1996. *The Origin and Evolution of Birds.* Yale University Press, New Haven. 420 p.

Forster, C. A., S. D. Sampson, L. M. Chiappe, and D. W. Krause. 1998. The theropod ancestry of birds: new evidence from the late Cretaceous of Madagascar. Science, 279:1915–19.

Hildebrand, M. 1974. *Analysis of Vertebrate Structure.* John Wiley and Sons, New York, 710 p.

Ji, Q., P. J. Currie, and M. A. Norell. 1998. Two feathered dinosaurs from northeastern China. Nature, 393:753–61.

Ligon, J. D. 1965. A Pleistocene avifauna from Haile, Florida. Bulletin of the Florida State Museum, 10:127–58.

Martin, L. D. 1991. Mesozoic birds and the origins of birds. Pp. 485–540 in H.-P. Schultze and L. Trueb (eds.), *Origins of the Higher Groups of Tetrapods: Controversy and Consensus.* Cornell University Press, Ithaca, N.Y.

Novas, F. E., and P. F. Puerta. 1997. New evidence concerning avian origins from the late Cretaceous of Patagonia. Nature, 387:390–92.

Olson, S. L. 1974. The Pleistocene rails of North America. The Condor, 76:169–75.

———. 1976. A jacana from the Pliocene of Florida. Proceedings of the Biological Society of Washington, 89:259–64.

———. 1985. The fossil record of birds. Pp. 79–238 in D. S. Farner, J. R. King, and K. C. Palmer (eds.), *Avian Biology, Volume 8.* Academic Press, Orlando.

Ostrom, J. H. 1974. *Archaeopteryx* and the origin of flight. Quarterly Review of Biology, 49:27–47.

———. 1975. The origin of birds. Annual Review of Earth and Planetary Sciences, 3:55–77.

Ritchie, T. L. 1980. Two mid-Pleistocene avifaunas from Coleman, Florida. Bulletin of the Florida State Museum, 26:1–36.

Sereno, P. C. 1997. The origin and evolution of dinosaurs. Annual Reviews of Earth and Planetary Sciences, 25:435–89.

Steadman, D. W. 1980. A review of the osteology and paleontology of turkeys (Aves: Meleagrinae). Natural History Museum of Los Angeles County Contributions in Science, 330:131–207.

———. 1981. A re-examination of *Palaeostruthus hatcheri* (Shufeldt), a late Miocene sparrow from Kansas. Journal of Vertebrate Paleontology, 1:171–73.

Mammalia I

Marsupials, Insectivores, Bats, and Primates

INTRODUCTION TO THE MAMMALIAN RADIATIONS

As discussed in chapter 6, the amniote vertebrates split into two major evolutionary lineages about 300 Ma (fig. 6.1). One of these, the Reptilia, has been the subject of the three previous chapters. The remaining nine chapters will describe members of the second of these two lineages, the Synapsida. All living synapsids are members of the Mammalia. Extinct groups of synapsids are known from the late Paleozoic and early Mesozoic. Perhaps the most famous of these was the fin-backed synapsid *Dimetrodon* of the Permian Period. Prior to the appearance of the dinosaurs in the late Triassic Period, synapsids were the world's dominant terrestrial vertebrates. Many of the skeletal features we think of today as characteristic of mammals actually evolved in earlier synapsids.

Mammals, or very mammal-like synapsids (depending on how one defines the group Mammalia; see Rowe, 1988), first appeared in the late Triassic, almost simultaneously with the earliest dinosaurs. These early mammals were small, nocturnal, and insectivorous. Through the remainder of the Mesozoic, mammals experienced a moderate evolutionary radiation into different groups, most of which retained a small body size. They were reviewed by Lillegraven et al. (1979). All three living groups of mammals (fig. 9.1), the monotremes, marsupials, and placentals evolved in the Mesozoic. The monotremes, which retain the primitive amniote reproductive method of laying eggs, today survive only in Australia and New Guinea. Their fossil record, which extends back to the Cretaceous Period, is very poor and limited to Australia and a single tooth from South America. They were apparently never a very successful group. Marsupials and placentals both have live birth, instead of laying eggs, and were more successful than the monotremes. Marsupials and placentals diverged from their common ancestor by the early Cretaceous.

Although marsupials have what superficially appears to be a more primitive reproductive pattern (giving birth

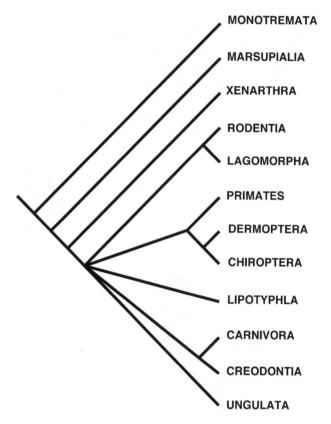

Fig. 9.1. Hypothesized evolutionary relationships among the major groups of mammals. After various studies by Novacek and Beard. Six major groups of placental mammals are recognized: the Xenarthra; the Anagalida (lagomorphs and rodents); the Archonta (primates, dermopterans, and bats); the Lipotyphla (hedgehogs, shrews, and moles); the Ferae (carnivorans and creodonts); and the Ungulata (whales, elephants, sea cows, artiodactyls, and perissodactyls). The interrelationships of these six groups are not well understood.

at a much more underdeveloped stage, followed by extensive extrauterine development, often in a "pouch"), certain features of their anatomy are derived relative to placentals. Thus marsupials are not "ancestral" to placental mammals. Marsupials experienced two great evolutionary radiations in their history. One occurred in the late Cretaceous and Paleogene of the New World, principally in South America, but also in Asia and North America to a lesser degree. One branch of this radiation dispersed to Europe in the Paleogene and spread to North Africa before going extinct. In a separate dispersal, one or more South American marsupials got to Australia via Antarctica. This launched the second major marsupial radiation, which resulted in the majority of the living Australian mammal fauna.

Placental mammals also began their own radiation in the late Cretaceous, but they truly exploded (in an evolutionary sense) in the early Paleogene following the extinction of most of the dinosaurs. Placental mammals dominated terrestrial vertebrate faunas on all the continents throughout the Cenozoic except Australia and Antarctica. Phylogenetic relationships among placental mammals remain uncertain (fig. 9.1). It is generally agreed that the Xenarthra (possibly with the pangolins, the Pholidota) are only distantly related to all other living placentals and were the first group to branch off (see chapter 10). Relationships among the remaining placentals, collectively called the epitherians, are at present unresolved. At least five separate major lineages are recognized among living epitherians:

· the Lipotyphla, which includes the hedgehogs, shrews, and moles;
· the Archonta, which includes the primates, tree-shrews, flying lemurs, and bats;
· the Ferae, which includes most of the carnivorous mammals, such as dogs, bears, seals, weasels, hyenas, and cats;
· the Anagalida, which includes the small herbivores, such as rabbits and rodents; and
· the Ungulata, which includes the large herbivores, such as elephants, manatees, horses, rhinos, whales, pigs, antelope, and deer.

Two excellent sources of general information on the ecology, distribution, anatomy, and behavior of living mammals are Macdonald (1984) and Eisenberg (1981). The most up-to-date classification of fossil and modern mammals is McKenna and Bell (1997).

MAMMALIAN SKELETAL ANATOMY

Many of the distinctive features of the mammalian skeleton and dentition were described and figured in chapter 1 and are only summarized here. The majority of them are linked, directly or indirectly, with the important physiological differences between mammals and their amniote ancestors, their much higher metabolic rate, and an internally regulated body temperature ("warm-bloodedness"). Although these features allow mammals to be active during the night and when ambient temperatures are low, they come with a demanding cost. Mammals require much more food to fuel their higher energy demands than do "cold-blooded" reptiles or amphibians. Thus many mammalian features are related to either procuring more food or utilizing energy more efficiently. These include hair, which acts as insulation to limit heat loss; multicusped teeth, which are more efficient at processing food for rapid digestion; increased sensory acuity and brain size; and limb elements adapted to an upright stance and long periods of activity.

A number of skeletal features characterize mammals. In the skull, there are two occipital condyles at the back for articulation with the atlas (first vertebra). The nasal region has ossified turbinal bones and an internal septum. The lower jaw consists predominantly (exclusively in all living mammals) of a single element, the dentary, and the joint between the skull and the lower jaw is between the squamosal and dentary. The small bones which form the posterior region of the lower jaw in other amniotes are no longer associated with the mammalian lower jaw, but are instead incorporated into the ear region of the skull and take on new functions. There are three auditory ossicles for transmission of sound vibrations between the ear drum (tympanic membrane) and the inner ear, and the auditory bulla encases the inner and middle ear regions. The teeth are generally heterodont, and the molars (and sometimes the premolars) have two or more roots. The vertebrae are generally acoelous and separated into morphologically distinct cervical, thoracic, lumbar, sacral, and caudal regions. Only the thoracic vertebrae articulate with ribs, and the sacral vertebrae fuse to form a single element, the sacrum (lost in whales and sea cows). The large limb bones, vertebral centra, pelvis, and scapula have separate ossification centers (epiphyses) for growth prior to maturity. The femur has a distinct, spherical head for insertion into the deep acetabulum of the pelvis; in adults, the pelvis is a single element representing the fused ilium, ischium, and pubis. Mammals primitively have five digits with a phalangeal formula of 2-3-3-3-3, but these numbers are modified by various groups.

FOSSIL MARSUPIALS OF FLORIDA

Representatives of the marsupial family Didelphidae are known from two time periods in Florida. From the late Oligocene to the early middle Miocene (about 29 to 16 Ma), small didelphids of the genus *Peratherium* were relatively rare members of the mammalian fauna of the state (fig. 9.2). At present they are known only by isolated teeth from several localities in northern and central Florida and have not been thoroughly studied (Wolff, 1987). The early Barstovian Florida records are among the youngest in North America prior to the family's extinction on this continent in the middle Miocene. Didelphids survived and flourished in South America through the Cenozoic, however. Upon emergence of the Panamanian Isthmus in the late Pliocene, faunal interchange of terrestrial vertebrates between North and South America recommenced on a large scale (see chap. 2 and Stehli and Webb, 1985). The genus *Didelphis* was one of the last of the South American mammals to appear in Florida, as its first occurrence is the middle Pleistocene Coleman 2A fauna. This form, apparently the living species *Didelphis virginiana,* persisted in Florida through the late Pleistocene and into the Holocene. It is not infrequent in Rancholabrean sites, but usually not very abundant.

Relatively complete cranial material of didelphids is easily distinguished from that of placentals by the greater number of incisors (five uppers, four lowers) and the presence of three premolars and four molars (fig. 1.10). The posterior ventral end of the mandible bears what is called an inflected angle; this feature is also characteristic of many other marsupials. Upper molars of *Didelphis* are triangular and distinguished by the strong labial stylar cones. Lower molars are sharply divided into trigonid and talonid regions, and the protoconid is the highest cusp. Many elements of the opossum's postcranial skeleton are also distinctive.

FOSSIL INSECTIVORES OF FLORIDA

Two groups of insectivores (= Lipotyphla of McKenna and Bell, 1997), shrews and moles, now live in Florida and both are represented by fossils. Shrews are present in the oldest known terrestrial site (early Oligocene) and many subsequent faunas. As is the case for other small mammals (marsupials, bats, rodents, and rabbits), late Paleogene and early Neogene representatives are now known but remain poorly studied. In addition to archaic shrews and moles, another type of insectivore, the hedgehog, is known from the Miocene of Florida. Hedgehogs (Erinaceidae) are now exclusively an Old World group but were present in North America during the Paleogene and early Neogene (Rich and Patton, 1975). Two genera are known from Florida, each from a single tooth. They are *Amphechinus* (very early Miocene, Alachua County) and *Lanthanotherium* (middle Miocene, Gadsden County).

Miocene shrews are known exclusively from extinct genera, several of which are undescribed. Thomas Farm has produced a species of the medium-sized *Limnoecus* (fig. 9.3), while a smaller species is known from the Brooks Sink fauna of Bradford County. Brooks Sink has also produced the oldest moles from eastern North

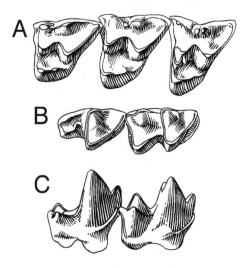

Fig. 9.2. Teeth of the Oligocene marsupial *Peratherium fugax* from South Dakota. Similar fossils are known from the early to middle Miocene of Florida. *A,* occlusal view of SDSM 31134, left M1–M3; *B,* occlusal, and *C,* labial views of SDSM 31135, right m3–m4. About 10×. Modified after Green and Martin (1976).

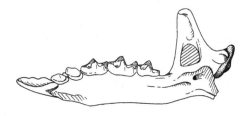

Fig. 9.3. Right dentary of the Miocene shrew *Limnoecus* in lingual view. Specimen shown is UCMP 36171 from the middle Miocene of Nebraska, but this extinct genus is known from Florida. About 5×. After Repenning (1967); reproduced by permission of the U.S. Geological Survey.

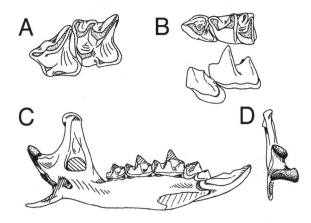

Fig. 9.4. Dentary and dentition of a modern specimen of the least shrew, *Cryptotis parva,* from Michigan. *A,* occlusal view of left P4–M1; *B,* occlusal and labial views of p4–m1; *C,* medial, and *D,* posterior views of the left dentary. *A–B* about 10×; *C–D* about 6.5×. After Repenning (1967); reproduced by permission of the U.S. Geological Survey.

America. Two types are present, one similar to *Scalopoides,* the other *Mystipterus.* These are both Miocene genera originally described from the western United States.

The modern Florida insectivore fauna comprises three species of shrews, *Blarina carolinensis, Cryptotis parva,* and *Sorex longirostris,* and one mole, *Scalopus aquaticus.* These are also the only known insectivore taxa in Florida since the Pliocene (chap. 2), when the modern fauna began to form. *B. carolinensis* (includes records formerly assigned to *B. brevicauda*) and *C. parva* are the two most common shrews found in Pleistocene sites (fig. 9.4), while *Sorex* is known from only a few localities in the northern part of the state. The common eastern mole *S. aquaticus* is also a frequent constituent of Pleistocene fissure-fill and other deposits with substantial input from owl pellets. The bones of the forelimb in moles are highly modified for digging and are instantly recognizable (fig. 9.5). Dentitions of moles differ from those of shrews by their greater size—the large anterior incisor has only one cusp (two or more in shrews)—and in the arrangement of the major cusps of the molars (fig. 9.6). The tips of shrew teeth are pigmented, which is sometimes apparent in fossils. This is not present in moles.

FOSSIL BATS OF FLORIDA

Bats (Chiroptera) are one of the most diverse and common groups of living mammals. The fossil record of bats in general is poor, but that of Florida is one of the best in the world. However, most of the older species remain

undescribed. Bats are often unjustifiably all considered to be cave dwellers that eat insects they catch on the wing. This stereotype results from a cultural bias, as many bats from northern temperate regions have this lifestyle. In tropical and semitropical regions, bats specialize in a wide variety of diets and roosting places. While some do catch flying insects, others glean them off the ground or vegetation. Some bats are carnivorous and eat small vertebrates, others eat fruit or flowers, while still other species subsist on nectar and pollen from flowers like hummingbirds. The vampire bats, which feed on the blood of vertebrates, are perhaps the most specialized.

Most elements of the bat skeleton are adapted for flight and are distinct from those of other mammals. The forelimb elements are elongated and the ulna is very reduced (fig. 9.7). Metacarpals 2 through 5 are especially long to provide much of the support for the wing membrane. As was the case with birds, the bones of the forelimb are highly specialized. The proximal end of the humerus (fig. 9.8) is especially diagnostic in distinguishing between the various kinds of bats. Although the molars

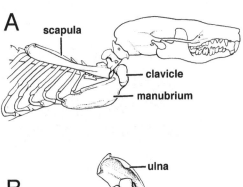

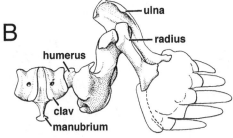

Fig. 9.5. Specializations for digging of the pectoral girdle and forelimb of the mole *Scalopus aquaticus. A,* lateral view of pectoral girdle with forelimb removed; *B,* anterior view of part of the pectoral girdle and the left forelimb. The humerus articulates with both the scapula and the clavicle. Also note the short, stout limb bones and the keeled manubrium of the sternum. Figures from *Mammalogy* by Terry A. Vaughn, copyright © 1972 by Saunders College Publishing, reproduced by permission of the publisher.

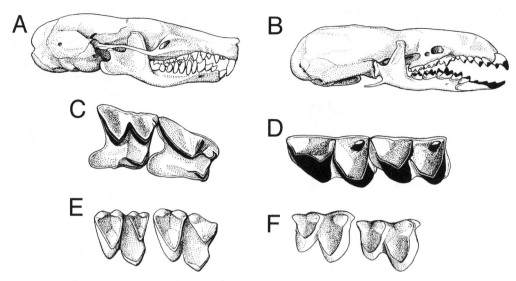

Fig. 9.6. Comparison of the skulls and teeth of moles and shrews. *A*, lateral view of the skull and dentary of the mole *Scalopus aquaticus*, about 1.5×; *B*, lateral view of the skull and dentary of the shrew *Sorex vagrans*, about 3.5×; black indicates the portions of the teeth that are covered by dark pigment in this species; *C*, occlusal view of right P4–M1 of *S. vagrans*; *D*, occlusal view of left m1–m2 of *S. vagrans*; *E*, occlusal view of the right M1–M2 of *S. aquaticus*; *F*, occlusal view of left m1–m2 of *S. aquaticus*. Figures from *Mammalogy* by Terry A. Vaughn, copyright © 1972 by Saunders College Publishing, reproduced by permission of the publisher.

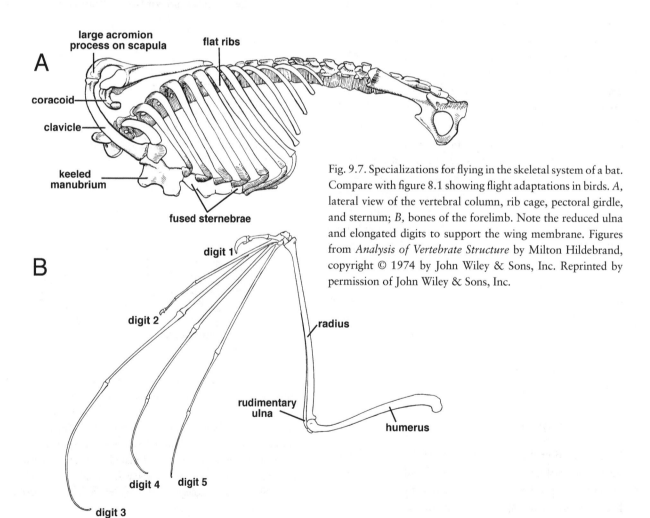

Fig. 9.7. Specializations for flying in the skeletal system of a bat. Compare with figure 8.1 showing flight adaptations in birds. *A*, lateral view of the vertebral column, rib cage, pectoral girdle, and sternum; *B*, bones of the forelimb. Note the reduced ulna and elongated digits to support the wing membrane. Figures from *Analysis of Vertebrate Structure* by Milton Hildebrand, copyright © 1974 by John Wiley & Sons, Inc. Reprinted by permission of John Wiley & Sons, Inc.

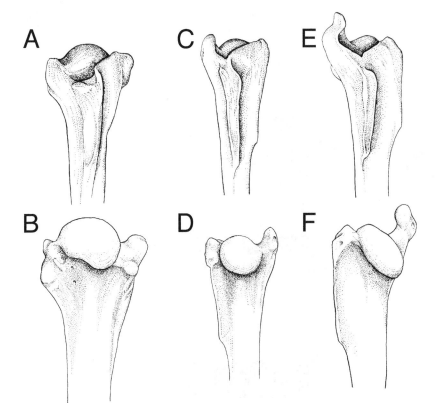

Fig. 9.8. Proximal end of the right humerus in three types of bats. The end of this bone often differs among bat genera or even species, reflecting different modes of flight. Anterior views on top row; posterior views on bottom row. *A–B, Pteropus,* an Old World "flying fox"; *C–D, Myotis lucifugus,* the little brown bat; *E–F, Molossus ater,* the red mastiff bat. Many fossil bats are described on the basis of their humeri. Figures from *Mammalogy* by Terry A. Vaughn, copyright © 1972 by Saunders College Publishing, reproduced by permission of the publisher.

of insect-eating bats superficially resemble those of insectivores, the incisors and canines are very different (fig. 9.9). Thus complete maxillae and mandibles are easily identified. Groups of bats are differentiated by the number of premolars. The cheek teeth of vampire bats are very reduced (fig. 9.9B), as they have lost all need to chew any food. They do have large, sharp upper incisors and canines to make small incisions on their prey.

The modern bat fauna of Florida consists entirely of members of the superfamily Vespertilionoidea, and most are vespertilionids. Fossil bats are most common in fissure and cave deposits, but they are also found in lesser numbers in other types of sites which produce small vertebrates. The latter are often those bats that do not roost in caves, but rather in trees.

The oldest vampire bat, *Desmodus archaeodaptes,* is known from the late Pliocene and early Pleistocene of Florida (Morgan et al., 1988). It probably migrated from South America, as *Desmodus* is a member of the primarily Neotropical family Phyllostomidae. It has been inferred that vampires evolved in South America, preying on large mammals such as sloths and notoungulates. Following the formation of the Panamanian Isthmus in

the Pliocene, these groups migrated north and vampires came with them. *D. archaeodaptes* was similar in size to the modern species *Desmodus rotundus,* which lives in Central and South America. A larger vampire, *Desmodus stocki,* lived in North America in the late Pleistocene and has been found at four sites in north-central Florida (Morgan, 1991). It was about 15 to 20 percent larger than *D. rotundus* and differed in several features of the skull. Florida specimens were originally named *Desmodus magnus,* but it was synonymized with *D. stocki. D. stocki* is one of the few small mammals to become extinct at the end of the Pleistocene, presumably as a result of losing most of its large prey species or from increasingly colder winters.

The Oligocene and Miocene bat fauna of Florida was more diverse than today, with several additional families present that are now restricted to more tropical realms. They include extinct members of the Natalidae, Emballonuridae, and Mormoopidae. As these records predate any from South America, it has been suggested that some of the groups of bats that are now restricted to Central and South America may have evolved or diversified in North America before spreading south.

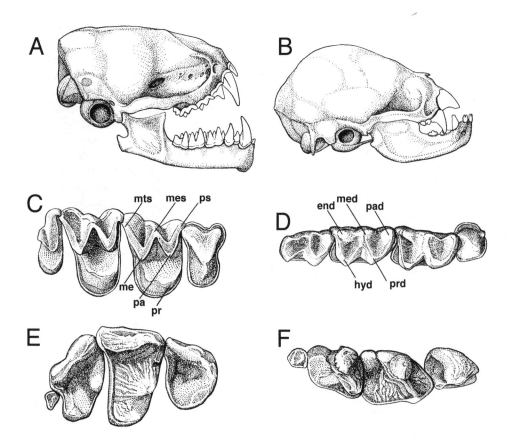

Fig. 9.9. Relationship between bat cranial and dental morphology and diet. *A,* right lateral view of the skull and dentary of the hoary bat, *Lasiurus cinereus,* an insectivorous bat; *B,* right lateral view of the skull and dentary of the common vampire, *Desmodus rotundus;* with its diet of blood, the vampire has lost nearly all of its cheek teeth but has razor-sharp anterior teeth to make incisions on its prey; *C,* occlusal view of the right P4–M3 of *L. cinereus; D,* occlusal view of the right p4–m3 of *L. cinereus; E,* occlusal view of the right P4–M3 of the Jamaican fruit-eating bat, *Artibeus jamaicensis,* a frugivore with broad, crushing teeth; *F,* occlusal view of the left p4–m3 of *A. jamaicensis.* Abbreviations (see fig. 1.11): *end,* entoconid; *hyd,* hypoconid; *med,* metaconid; *mes,* mesostyle; *me,* metacone; *mes,* metastyle; *pa,* paracone; *pad,* paraconid; *prd,* protoconid; *pr,* protocone; *ps,* parastyle. Figures from *Mammalogy* by Terry A. Vaughn, copyright © 1972 by Saunders College Publishing, reproduced by permission of the publisher.

FOSSIL PRIMATES OF FLORIDA

There is only a single known fossil primate from Florida, namely ourselves, *Homo sapiens.* Unquestionable human-made artifacts are known from the very late Pleistocene (from about 12 to 10 ka), often in association with extinct vertebrates such as mammoths, mastodons, giant sloths, and tortoises. Tools made of mastodont ivory have been found in a number of Florida's rivers. Actual skeletal material of *H. sapiens* is not known until the early Holocene, or less than 10 ka. Exactly when humans first entered North America during the late Pleistocene is a matter of considerable controversy.

During the Paleogene, several primitive families of primates are known from North America, including the Adapidae and Omomyidae. Primate diversity in North America dropped markedly during the late Eocene, and only three genera are known subsequently: *Macrotarsius* and *Rooneyia* from the early Oligocene and the nearly unpronounceable *Ekgmowechashala* from the late Oligocene to early Miocene of South Dakota and Oregon (fig. 9.10). These all belong to the family Omomyidae. (McKenna [1990] removed *Ekgmowechashala* from the omomyid primates and referred it to the family Plagiomenidae of the order Dermoptera. Living dermopterans from southeast Asia, called colugos or flying lemurs, are also members of the superorder Archonta, along with primates, bats, and tree shrews. Regardless of its familial affinities, recovery of *Ekgmowechashala* from Florida would be an important range extension.) As more early Miocene and older terrestrial faunas are discovered in

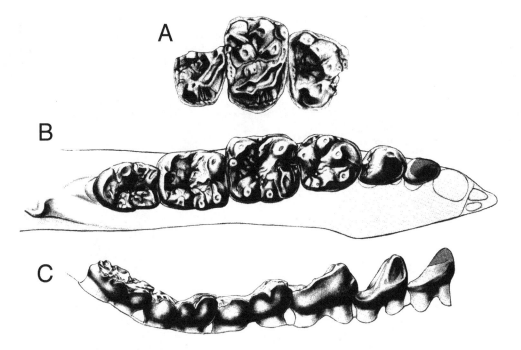

Fig. 9.10. Upper and lower dentitions of the last North American primate (or dermopteran, see text), *Ekgmowechashala*, from the late Oligocene and very early Miocene. *A*, occlusal view of right P4, M1, and partial M2 from Oregon; *B*, occlusal, and *C*, lingual views of left p2–m3 from South Dakota. This rare genus has not yet been recovered from Florida, although only a few sites of the proper age are known. About 5×. Modified from McKenna (1990) in *Dawn of the Age of Mammals in the Northern Part of the Rocky Mountain Interior, North America*, edited by T. M. Brown and K. D. Rose. Modified with permission of the publisher, the Geological Society of America, Boulder, Colorado, U.S.A. Copyright © 1990 The Geological Society of America.

Florida, the chances for recovery of one of these ancient primates increases. As paleobotanical and other environmental indicators suggest semitropical to tropical forested conditions, the presence of primates would not be unexpected.

REFERENCES

Eisenberg, J. E. 1981. *The Mammalian Radiations.* University of Chicago Press, Chicago, 610 p.

Green, M., and J. E. Martin. 1976. *Peratherium* (Marsupialia: Didelphidae) from the Oligocene and Miocene of South Dakota. Pp. 155–68 in *Athlon: Essays on Palaeontology in Honour of Loris Shano Russell.* Royal Ontario Museum, Toronto.

Hildebrand, M. 1974. *Analysis of Vertebrate Structure.* John Wiley and Sons, New York, 710 p.

Hill, J. E., and J. D. Smith. 1984. *Bats: A Natural History.* British Museum (Natural History), London, 243 p.

Jones, C. A., J. R. Choate, and H. H. Genoways. 1984. Phylogeny and paleobiology of short-tailed shrews (genus *Blarina*). Pp. 56–148 in H. H. Genoways and M. R. Dawson (eds.), *Contributions in Quaternary Vertebrate Paleontology.* Carnegie Museum of Natural History, Pittsburgh, Special Publication 8.

Lawrence, B. 1943. Miocene bat remains from Florida, with notes on the generic characters of the humerus in bats. Journal of Mammalogy, 24:356–69.

Lillegraven, J. A., Z. Kielan-Jaworowska, and W. A. Clemons.

1979. *Mesozoic Mammals: The First Two-thirds of Mammalian History.* University of California Press, Berkeley, 311 p.

Macdonald, D. 1984. *The Encyclopedia of Mammals.* Facts on File Publications, New York, 895 p.

McKenna, M. C. 1990. Plagiomenids (Mammalia: ?Dermoptera) from the Oligocene of Oregon, Montana, and South Dakota, and the middle Eocene of northwestern Wyoming. Pp. 211–34 in T. M. Bown and K. D. Rose (eds.), *Dawn of the Age of Mammals in the Northern Part of the Rocky Mountain Interior, North America.* Geological Society of America, Special Paper 243.

McKenna, M. C., and S. K. Bell. 1997. *Classification of Mammals above the Species Level.* Columbia University Press, New York, 631 p.

Morgan, G. S. 1985. Fossil bats (Mammalia: Chiroptera) from the late Pleistocene and Holocene Vero Fauna, Indian River County, Florida. Brimleyana, 11:97–117.

———. 1991. Neotropical Chiroptera from the Pliocene and Pleistocene of Florida. Bulletin of the American Museum of Natural History, 206:176–213.

Morgan, G. S., and A. E. Pratt. 1988. An early Miocene (late Hemingfordian) vertebrate fauna from Brooks Sink, Bradford County, Florida. Pp. 53–69 *in* F. L. Pirkle and J. G. Reynolds (eds.), *Heavy Mineral Mining in Northeast Florida and an Examination of the Hawthorne Formation and Post-Hawthorne Clastic Sediments.* Southeastern Geological Society Guidebook No. 29.

Morgan, G. S., O. J. Linares, and C. E. Ray. 1988. New species of fossil vampire bats (Mammalia: Chiroptera: Desmodontidae) from Florida and Venezuela. Proceeding of the Biological Society of Washington, 101:912–28.

Novacek, M. J., and A. R. Wyss. 1986. Higher-level relationships of the Recent eutherian orders. Cladistics, 2:257–87.

Novacek, M. J., A. R. Wyss, and M. C. McKenna. 1988. The major groups of eutheran mammals. Pp. 31–71 *in* M. S. Benton (ed.), *The Phylogeny and Classification of the Tetrapods*. Volume 2, *Mammals*. Clarendon Press, Oxford, England.

Repenning, C. A. 1967. Subfamilies and genera of the Soricidae. U.S. Geological Survey Professional Paper, 565:1–74.

Rich, T. H. V., and T. H. Patton. 1975. First record of a fossil hedgehog from Florida (Erinaceidae, Mammalia). Journal of Mammalogy, 56:692–96.

Rowe, T. 1988. Definition, diagnosis, and origin of Mammalia. Journal of Vertebrate Paleontology, 8:241–64.

Stehli, F. G., and S. D. Webb. *The Great American Biotic Interchange*. Plenum Press, New York, 532 p.

Vaughn, T. A. 1972. *Mammalogy*. W. B. Saunders, Philadelphia, 463 p.

Wolff, R. G. 1987. Late Oligocene-middle Miocene didelphid marsupials from Florida. Journal of Vertebrate Paleontology, 7:29A.

10

Mammalia 2

Xenarthrans

INTRODUCTION

Although xenarthrans did not arrive in Florida until relatively recently (about 9 Ma) and did not become very diverse until 2.5 Ma, they are one of the state's most characteristic orders of fossil mammals. As noted in chapter 9, xenarthrans diverged from the other placental mammals relatively early (fig. 9.1) and acquired many morphological features that seem very odd relative to more familiar placentals. South America was the home to the majority of xenarthran history. Some paleontologists have suggested that early Paleogene fossils from North America, Europe, and China are possibly xenarthrans, but most recent investigations have discounted their xenarthran status (Rose and Emry, 1993). All the xenarthrans from Florida, fossil and modern, clearly evolved from South American ancestors.

Living xenarthrans include the tree sloths (two genera), anteaters (three genera), and armadillos (eight genera). Most live in South America, although some have ranges extending into tropical Central America. Only the common long-nosed armadillo (*Dasypus novemcinctus,* also called the nine-banded armadillo) currently ranges into the United States. Extinct groups of xenarthrans include the pampatheres (or giant armadillos), the glyptodonts, and the ground sloths. The South American fossil record of xenarthrans is extensive, and many species found in Florida are believed to be closely related or even conspecific (in a few cases) to those from South America. Although fossil xenarthrans once ranged extensively in North America (a ground sloth is even known from Alaska), they were apparently most diverse and common on the coastal plains of the southeastern United States, especially in Florida.

These strange mammals have two widely used scientific names, each referring to one of the unusual characteristics of the group. One is Xenarthra, which means "strange joints." It refers to extra surfaces for articu-

lation, called xenapophyses, found on the lumbar and some thoracic vertebrae of members of this group (fig. 10.1A). Exactly how many xenarthrous vertebrae there are varies between families. Xenarthra is the currently accepted name for the group (McKenna and Bell, 1997). The second name by which the group is known is Edentata, which literally means "without teeth." This is strictly true only for the anteaters. However, the teeth are greatly reduced to simple pegs in many armadillos, and all living and most fossil xenarthrans lack the covering of enamel on their teeth. Reduced or simplified dentitions are characteristic of mammals that eat primarily ants, termites, grubs, and soft invertebrates, so this is presumably the diet of the ancestral xenarthrans. Two xenarthran lineages, the sloths and the Glyptodontoidea, later became herbivores and evolved cheek teeth with broad occlusal surfaces like other large-bodied, plant-eating mammals. To compensate for the lack of enamel, the teeth of these xenarthrans have two or more types of dentine of differing hardness (fig. 10.2). These often preserve as different colors in fossils. They function to provide the shearing edges needed on the occlusal surfaces of mammalian herbivores that result from unequal wear between enamel and dentine (and cement) in other mammals. The shearing edges function in efficiently cutting and chewing fibrous plant material. Another compensation for the lack of enamel is that the teeth of xenarthrans are hypselodont (ever-growing) and so never wear out.

Other unusual features of the xenarthran skeleton include articulation of one or more caudal vertebrae to the ischium of the pelvis (fig. 10.1B); the scapula is massive with an extra spine and long processes (fig. 10.1C); the well-ossified sternebrae and sternal ribs (a second, ventral set of ribs) have well-developed joints; and there are no teeth across the front of the jaws—instead, a long, heavy tongue protrudes through a troughlike symphysis. The teeth cannot be identified as the regular mammalian complement of incisors, canines, premolars, and molars.

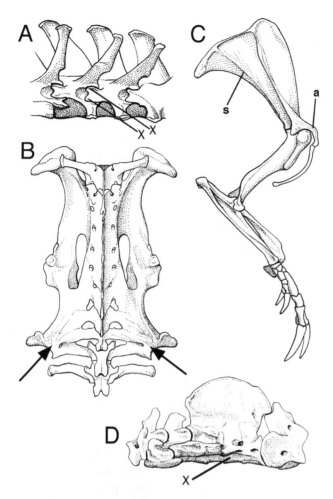

not have any deciduous teeth; the teeth of juveniles are smaller versions of the adults' that gradually increase in size with age. The walls of adult teeth are parallel sided; those of juveniles are inclined because the base is larger than the occlusal surface.

Xenarthran limb bones are typically robust, with large processes for muscle attachment. Relative to their body size, living xenarthrans are extremely strong. In addition to the sesamoid bones present in most mammals (chap. 1), other sesamoid bones are found in the feet. The feet are broad and heavy, and the distal phalanges bear large claws (glyptodont hind feet are an exception—they bear hooves instead of claws). In contrast to most mammals, the proximal phalanx is shorter than the medial phalanx of the same digit, and the medial phalanx is in turn shorter than the distal phalanx (fig. 10.1C). Caudal vertebrae tend to have very well-developed hemal arches. Bony scutes or osteoderms are found in the skin of one group of sloths (Mylodontidae) as well as in the Cingulata in which they form the articulated body armor.

The Xenarthra is made up of three major evolutionary lineages: the shelled Cingulata; the anteaters, Myrmecophagidae; and the sloths, the Phyllophaga ("leaf-eaters"). The divisions between these three groups extend well back into the Paleogene (if not further). The Myrmecophagidae and Phyllophaga are thought to be more closely related to one another based on shared de-

Fig. 10.1. Some of the skeletal specializations of xenarthrans, as seen in the armadillo *Dasypus*. *A*, xenarthrous lumbar vertebrae; the "X"s indicate the extra articulations; *B*, the synsacrum in dorsal view, anterior to top; arrows indicate articulation between ischium and caudal vertebrae; *C*, right forelimb in lateral view; the "s" indicates the secondary scapular spine, and the "a" the enlarged acromion process. Also note the large processes on the humerus and ulna (for muscle attachment), the massive claws, and the relative proportions of the phalanges on each digit. *D*, cervical vertebrae in right lateral view, "X" indicating fused axis and third through fifth cervical vertebrae. Fused cervicals are typical only in cingulate edentates. Figures from *Mammalogy* by Terry A. Vaughn, copyright © 1972 by Saunders College Publishing, reproduced by permission of the publisher.

In the cingulates (shelled xenarthrans), the teeth are simply referred to by number, N1 being the first (anteriormost) tooth, N2 the second, and so on. Sloths may have a sharp, anteriormost tooth called the caniniform. Otherwise, all their cheek teeth are referred to as molariforms, with the anteriormost being the first molariform, etc. Exactly how these teeth relate to the normal mammalian series has yet to be determined. Xenarthrans do

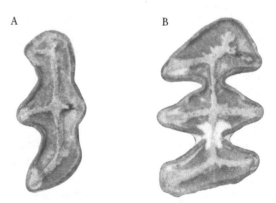

Fig. 10.2. Occlusal views of two glyptodont teeth (*Glyptotherium floridanum*, UF/FGS 6643) from Catalina Gardens, Pinellas County, late Pleistocene. *A*, left lower second tooth; *B*, left lower seventh tooth. As is characteristic of all but the earliest edentates, these teeth lack true enamel. Across the flat occlusal surface are ridges composed of osteodentine, a harder, more mineralized form of dentine. One of these ridges surrounds the tooth, and in these specimens is preserved a slightly darker color. Internal ridges of osteodentine form branching tracts in the teeth; these are preserved a lighter color than the ordinary dentine. The white material in *B* is plaster that is filling a broken section. Both are about 2×.

rived character states of the kidney, testes, and skeleton. The latter include an opening (foramen) in the scapula and a concave articular surface for the navicular on the astragalus. Together they are called the Pilosa, in reference to their having a hair-covered body rather than a shell. In contrast, the Cingulata is diagnosed by the shared presence of dermal bones modified into a shell of articulating osteoderms; the axis is fused with one or more cervical vertebrae (fig. 10.1D); the femur has large greater and third trochanters; and the tibia and fibula are fused at the proximal and distal ends (Engelmann, 1985).

FOSSIL CINGULATES OF FLORIDA

Three families of shelled xenarthrans (Cingulata) are known from Florida, the Dasypodidae, Pampatheriidae, and Glyptodontidae. Each first appeared in the late Pliocene when the first important wave of South American taxa entered North America. All are common throughout the Pleistocene, although most records consist only of isolated osteoderms. This is not unexpected, considering that each individual has hundreds more osteoderms than it does other skeletal elements. All three families became extinct in North America at the end of the Pleistocene, with the Pampatheriidae and Glyptodontidae becoming extinct in South America too. The Dasy-

podidae survived in South America (and probably Central America), and *Dasypus novemcinctus* has reclaimed most of the territory lost by its extinct relatives within the last century. Humphrey (1974) provided an interesting account of this remarkably rapid range extension.

The body armor of cingulates comes in several distinct pieces. The top of the head is covered by a shield of interlocking, thin, polygonal plates. The caudal vertebrae are surrounded by movable rings each formed by several individual osteoderms. The caudal rings are supported by the neural spine, transverse processes, and hemal arch of the enclosed vertebra. The last few rings are fused into a solid, tubelike structure in glyptodonts. Some South American glyptodonts modified this tube into an armored ball or spiked mace, but only simple tubes are known on North American species. Loose osteoderms are present in the skin of the limbs of some cingulates. These resemble analogous elements from the limbs of the tortoise *Hesperotestudo*. The largest piece of body armor is the carapace, which covers the dorsal and lateral sides of the trunk region. In dasypodids and pampatheres, the carapace is divided into three regions, the anterior shield or buckler, the posterior shield, and an intervening region consisting of a varying number of bands made of "movable" osteoderms (fig. 10.3). The plates making up the shields are often called "immovable" osteoderms to contrast them with those of the

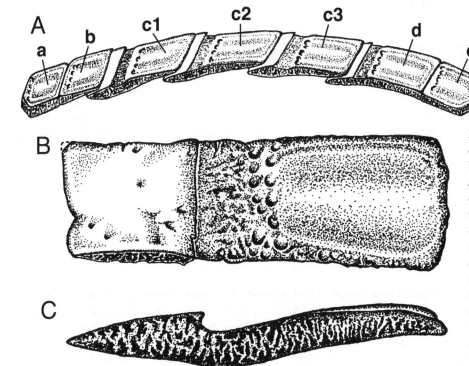

Fig. 10.3. *A*, relationship of osteoderms in the region of the movable bands in the giant armadillo *Holmesina*. Anterior is to the left. Osteoderms labeled *c1* through *c3* are the typical, complete movable osteoderms; *a* and *b* make up the last two rows of the anterior shield, and *d* and *e* the first two rows of the posterior shield. Below is a typical movable osteoderm in dorsal *(B)* and lateral *(C)* views of *H. septentrionalis* (UF 15136) from Coleman 3B, middle Pleistocene of Sumter County. About 1×. After Edmund (1985); reproduced by permission of the Texas Memorial Museum, Austin.

bands, but they are capable of slight flexing in life. The carapace of North American glyptodonts is a single unit without movable bands. Fossil evidence from primitive South American glyptodonts demonstrates that their carapaces had limited regions of movement.

The shape and surface morphology of individual osteoderms varies considerably depending on its location on the body. Nevertheless, characters of the osteoderms alone have often been used in systematic descriptions of cingulates. In dasypodids and pampatheres, each osteoderm was covered in life by a single scale made of keratinous material (like a fingernail) that does not fossilize. Each glyptodont osteoderm was covered by several such scales arranged in a characteristic rosette pattern. This pattern can be determined by the grooves on the osteoderms caused by the scales, much as those made by the keratinous scutes covering the bony shell of a turtle. The

general rosette pattern on a glyptodont osteoderm is a single central scale surrounded by a symmetrical series of peripheral scales (fig. 10.4A). Details of this pattern are used to distinguish species of glyptodonts. Immovable osteoderms of cingulates are most often irregularly hexagonal, although the number of sides varies from four to eight (fig. 10.4). Pampatheres also have some rectangular immovable osteoderms. The movable osteoderms of dasypodids and pampatheres are rectangular, with a raised anterior end over which moves the posterior end of the osteoderm of the preceding band (figs. 10.3, 10.4J). The last row of immovable scutes in the anterior shield and the first row in the posterior shield are also part of the flexing system. The function of the bands is to allow the animal to curl up into a ball, with as little exposure of its unprotected belly as possible. The number of bands is of some systematic value, although there is intraspecific

Fig. 10.4. Osteoderms of Florida fossil edentates. *A*, dorsal view of UF 52643, osteoderm of *Glyptotherium arizonae* from the Kissimmee River, Okeechobee County, late Pliocene. *B–C*, dorsal views of UF 125898 and 125881, two immovable osteoderms of the giant armadillo *Holmesina floridanus* from Haile 16A, Alachua County, early Pleistocene. *D–E*, dorsal views of two uncataloged UF specimens, immovable osteoderms of *Pachyarmatherium leiseyi* from Haile 16A. Round holes are openings for hair follicles. *F*, lateral view of the osteoderm shown in *E*. *G–H*, dorsal view of two uncataloged UF specimens, immovable osteoderms of the extinct armadillo *Dasypus bellus* from Haile 16A. *I*, lateral view of the osteoderm shown in *G*. Compare its thickness with that of *F*. *J*, dorsal view of UF 24978, a movable osteoderm of *D. bellus* from Haile 16A. All about 1×.

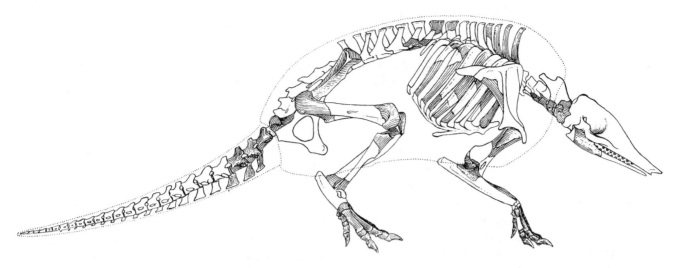

Fig. 10.5. Lateral view of UF 2478, a partial skeleton of the extinct armadillo *Dasypus bellus* from Medford Cave, Marion County, late Pleistocene. Recovered elements are indicated by shading; a considerable portion of the carapace was also preserved. About 0.15× (total length is an estimated 4 feet). After Auffenberg (1957); reproduced by permission of the Florida Academy of Sciences, Orlando.

variation in some dasypodids for this character state, and very complete fossil material is required to determine band number in an extinct taxon.

Dasypus bellus is the only described fossil dasypodid armadillo from the United States and ranges in age from late Pliocene to late Pleistocene with relatively little change (Downing and White, 1995). It differs only slightly from its modern relative *D. novemcinctus* except in terms of size. *D. bellus* tends to be twice as large as the modern *D. novemcinctus*. A relatively complete specimen about 1.2 meters long is known from Marion County (Auffenberg, 1957; fig. 10.5). Its carapace had nine movable bands, which is the typical number in northern populations of *D. novemcinctus*. The armadillo skull has a narrow, tubular snout and a wider but also tubular braincase. The teeth are simple and peglike, with each jaw having between seven and nine (fig. 10.6). *Dasypus* uses its long, muscular tongue to procure its main food items, beetles, ants, termites, millipedes, and other invertebrates. The modern species regularly includes amphibians and small lizards in its diet; *D. bellus* may have as well. The limbs and strong claws are specialized for digging.

A relatively newly recognized and still somewhat mysterious member of Florida's xenarthran fauna is *Pachyarmatherium leiseyi*. So far it is known only from the late Pliocene and early Pleistocene of Florida and the late Pliocene of South Carolina (Downing and White, 1995). It was similar to *Dasypus bellus* in size. *Pachyarmatherium* was first recognized as different from

other cingulates because of its small, thick osteoderms (fig. 10.4D–F). They have a rosette pattern (but one that differs in detail from that of glyptodonts), hair follicle pits, and no movable band osteoderms like those of dasypodids or pampatheres. Some of these osteoderm features at first suggested greater affinity with glyptodonts than armadillos. An associated skeleton of *P. leiseyi* with a carapace from Charlotte County (unfortunately sold to an overseas collection) allied it with the armadillo clan. The recovered portions of the mandible were delicate and edentulous. There were strong claws for digging, and the ulna also had adaptations for digging. This all suggests that the diet of *Pachyarmatherium* consisted mostly of ants and/or termites, which it would have dug out of their nests. The carapace turned out to be made of two solid portions, each made up of hundreds of small,

Fig. 10.6. Lateral views of UF 16698, partial right dentary and maxilla of *Dasypus bellus* from Haile 15A, Alachua County, late Pliocene. About 1×.

polygonal osteoderms. Those along the edge were extremely thick and pointed. The two parts of the carapace joined at a hinge that permitted limited mobility, but there were no bands of movable osteoderms. Downing and White (1995), who described this new beast, were cautious about its phylogenetic relationships with South American cingulates, assigning it to the superfamily Dasypodoidea, and left its familial status open awaiting further study.

Pampatheres were larger than typical armadillos, with the late Pleistocene representative *Holmesina septentrionalis* reaching a length of two or more meters. The invalid generic names *Chlamytherium* and *Chlamydotherium,* and the family group names formed from them (for example, Chlamytheriidae), have sometimes been applied to North American pampatheres, but *Holmesina* is usually recognized as the appropriate name (Edmund, 1985, 1987). Pampatheres were frequently included in the Dasypodidae as an extinct subfamily but are now recognized as their own family (the Pampatheriidae) and are considered to be more closely related to the glyptodonts than dasypodids. This relationship is supported by shared possession of anteroposteriorly elongated, lobate teeth with a raised central island of very hard dentine; greater fusion of the cervical vertebrae; and the transverse orientation of the cranial-mandibular joint (Engelmann, 1985). Both families primitively had nine teeth in each jaw, and this number is retained in *Holmesina* (fig. 10.7). The posterior teeth of pampatheres are bilobate or dumbbell-shaped, while the anterior teeth are rounded or reniform (kidney-shaped). A progressive trend in pampathere evolution is for the anterior teeth to become more lobate. The teeth wear down to form flat occlusal surfaces with transverse striations. These indicate that food was chewed with a predominantly sideways motion. Unlike armadillos, pampatheres were herbivorous (Vizcaíno et al., 1998). The carapace of pampatheres has three movable bands made up of rings of rectangular osteoderms (fig. 10.3). The immovable osteoderms of *Holmesina* are mostly polygonal (fig. 10.4B, C) and differ from those of glyptodonts in being thinner with no central figure or rosette pattern. Instead there is a broad, porous central surface (which may have a slight ridge or keel) surrounded by a narrow, beveled, and pitted edge. The limb elements are similar to those of *Dasypus* but proportionally heavier.

Two species of *Holmesina* are recognized in North America, with both names based on Florida specimens. The smaller *H. floridanus* lived in the late Pliocene and early Pleistocene, while *H. septentrionalis,* its apparent descendant, ranges from the middle to late Pleistocene

(Hulbert and Morgan, 1993). It is more than twice the size of the older species (fig. 10.8). *H. floridanus* was originally placed in the South American genus *Kraglievichia,* but Edmund (1987) has shown that the latter is quite distinct from *Holmesina* in osteoderm morphology. Thus the only genus currently recognized in Florida is *Holmesina.*

North American glyptodonts were thoroughly reviewed by Gillette and Ray (1981), who recognized a single genus, *Glyptotherium.* Other generic names applied to Florida glyptodonts such as *Boreostracon* and *Brachyostracon* are now regarded as junior synonyms. Glyptodonts resemble giant tortoises or Volkswagen "beetle" automobiles in their overall shape and size (fig. 10.9A, B). Two species of *Glyptotherium* are recognized in Florida, *G. arizonae* in the late Pliocene and very early Pleistocene, and *G. floridanum* in the late Pleistocene. The latter is smaller, with its individual carapace osteoderms normally less than 4 centimeters in diameter and the central figure is proportionally smaller (less than one-third the diameter of the entire osteoderm). The number of peripheral scales per osteoderm is nine or less in *G. floridanum,* and their margins are less distinct than in *G. arizonae* (which typically has ten or more peripheral scales surrounding a larger central figure). *Glyptotherium* had eight teeth per jaw, for a total of thirty-two. Except for the anterior first and sometimes second tooth, they are all trilobate with a raised central island (figs. 10.2, 10.9C). The skull and mandible are unusually modified, and the animal may have had a short proboscis much like a tapir (fig. 10.9B). The tooth row length is very long relative to the length of the skull. Like many sloths, the zygomatic arch bore a large ventral process that served as a prominent point for jaw muscle attachment. Unlike pampatheres, the major direction of movement of the jaw during chewing was anterior and posterior, similar to that of many rodents. The limbs were especially adapted to the large size of the animal, being massive and stout. The pelvis and associated fused vertebrae forms an enormous structure that supports much of the weight of the carapace. Gillette and Ray (1981) speculated that *Glyptotherium* was semiaquatic, with similar ecological preferences as the capybara. *G. arizonae* was about 3 meters long and is estimated to have weighed about a ton.

Fossil Ground Sloths of Florida

Three families of sloths are represented in the fossil record of Florida: the Mylodontidae, Megatheriidae, and Megalonychidae. They have been distinct since the

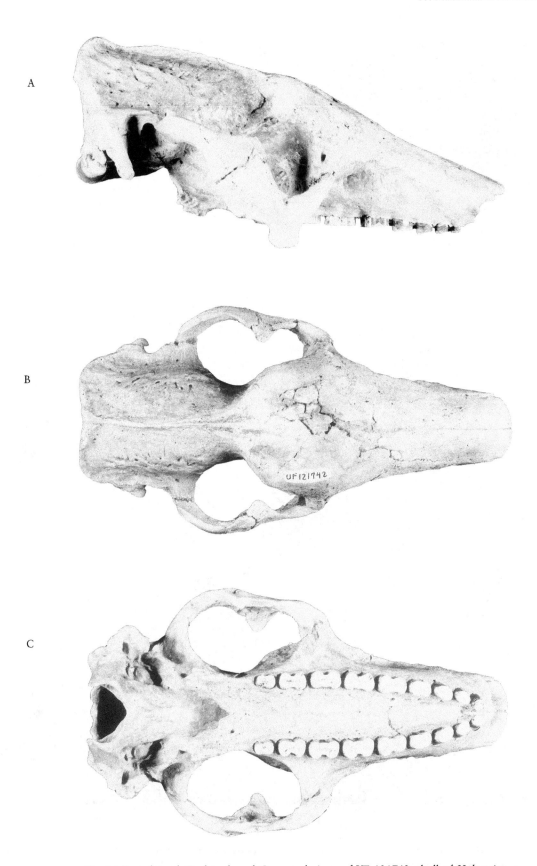

Fig. 10.7. *A,* lateral, *B,* dorsal, and *C,* ventral views of UF 121742, skull of *Holmesina floridanus* from Haile 7C, Alachua County, late Pliocene. About 0.5×.

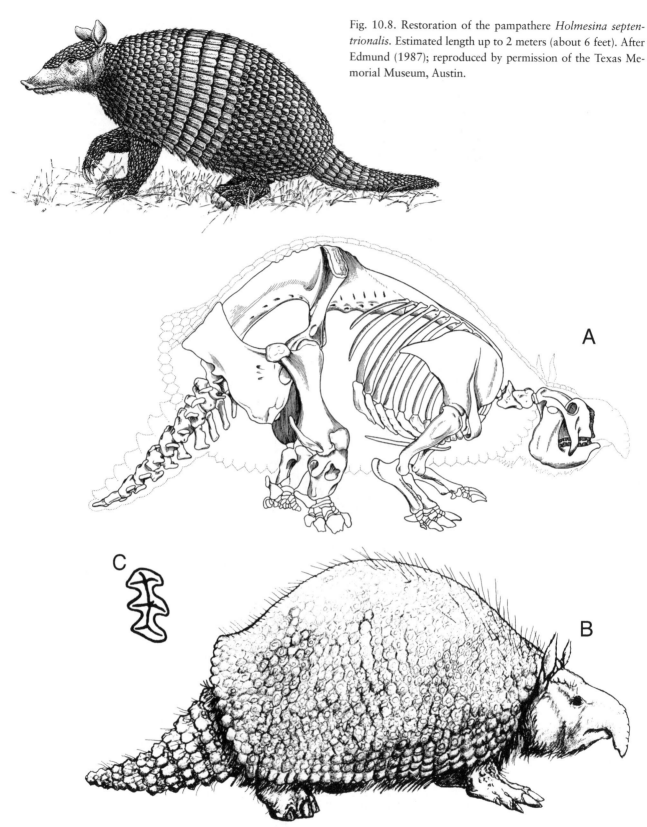

Fig. 10.8. Restoration of the pampathere *Holmesina septentrionalis*. Estimated length up to 2 meters (about 6 feet). After Edmund (1987); reproduced by permission of the Texas Memorial Museum, Austin.

Fig. 10.9. *A*, right lateral view of the skeleton of the Pliocene glyptodont *Glyptotherium arizonae*, with restored outline of body; *B*, restoration of possible external appearance of *G. arizonae*; note tapirlike proboscis and tail lacking terminal spines. *C*, right lower molariform tooth of *G. floridanum* from Seminole Field, Pinellas County, late Pleistocene. C about 1×. *A–B* after Gillette and Ray (1981); reproduced by permission of the Smithsonian Institution Press, Washington. C after Simpson (1929); reproduced by permission of the American Museum of Natural History, New York.

Paleogene, so almost all skeletal elements differ among them. All sloths found in Florida are relatively large mammals and are no doubt "ground" sloths as opposed to the much smaller living tree sloths. Megatheres and megalonychids have some anatomical characters associated with an arboreal lifestyle, and the smaller, early members of these two families probably spent some time climbing trees. Two genera representing different sloth families are the oldest recorded xenarthrans in Florida (and North America), *Pliometanastes* (a member of the Megalonychidae) and *Thinobadistes* (Mylodontidae). Both appeared about 9 Ma in the late Miocene, and their arrival is used to denote the beginning of the Hemphillian Land Mammal Age.

Relative to armadillos, sloths generally have short, deep skulls and mandibles. There are three to five large teeth per jaw. The zygomatic arch is elaborately developed anteriorly, with large processes for attachment of jaw muscles. The feet are extraordinarily large with huge claws. Ground sloths walked on the sides of their feet to keep the claws out of the way (except in megalonychids, which must have solved this problem differently). The tail is very strong and evidently helped the sloth form a tripodlike stance when it stood up on its hindlimbs.

Mylodontid ground sloths twice invaded Florida. The first invader was *Thinobadistes,* which is known from two species in the late Miocene. The second group of mylodonts arrived in the late Pliocene and lasted until the end of the Pleistocene. This group is also represented by two species, the smaller *Glossotherium chapadmalense* and the larger *Paramylodon harlani* (fig. 10.10). One very characteristic feature of mylodonts is their lobate or triangular molariform teeth. They also had large

upper and lower caniniforms, which sharpen by wear against one another. Mylodont claws are heavy, broad (not compressed), and relatively straight (not very curved). The forelimbs were short in proportion to the hindlimbs. Within the skin, there was a layer of irregular osteoderms, which are sometimes mistakenly discarded as representing only bone fragments. They can usually be found at any site that has other mylodont fossils. *Glossotherium* and *Paramylodon* are distinguished from *Thinobadistes* by having a single facet for articulation between the calcaneum and astragalus (Webb, 1990). A mounted skeleton of *Thinobadistes* is exhibited at the Florida Museum of Natural History.

Megalonychid ground sloths also appeared in Florida in the late Miocene (*Pliometanastes*) and became extinct at the end of the Pleistocene. *Megalonyx* is the common Pliocene and Pleistocene megalonychid genus (fig. 10.11). It was named by President Thomas Jefferson (meaning "large claw") in the belief that it belonged to some large carnivore, although he later recognized it as an xenarthran, one of the earliest attempts at vertebrate paleontology in North America. Four successively larger species of *Megalonyx* are recognized in Florida: *M. curvidens* (early Pliocene); *M. leptostomus* (late Pliocene and very early Pleistocene); *M. wheatleyi* (late early Pleistocene); and *M. jeffersoni* (late Pleistocene) (Hirschfeld and Webb, 1969; McDonald, 1977). *M. jeffersoni* is about the size of a large bear, which is actually rather small for a "giant" ground sloth. The molariform teeth of megalonychids are oval to subrectangular in shape, with concave occlusal surfaces. The caniniform is relatively small in *Pliometanastes* but large with a flat wearing crown in *Megalonyx*. The sharp claws are curved and

Fig. 10.10. Medial view of UF 83335, a dentary of the ground sloth *Paramylodon harlani* from the Leisey Shell Pit, Hillsborough County, early Pleistocene. About 0.5×.

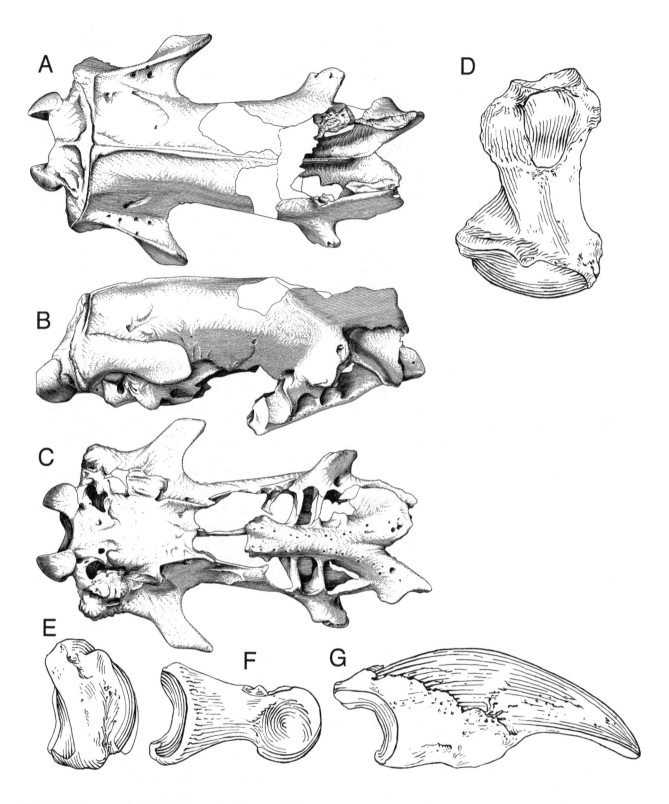

Fig. 10.11. The ground sloth *Megalonyx* from Texas and Florida. *A*, dorsal, *B*, right lateral, and *C*, ventral views of West Texas State University 1956, a partial skull of *Megalonyx leptostomus* from Randall County, Texas; *D–G,* medial views of AMNH 23423, left fourth pedal digit of *Megalonyx jeffersoni* from Sabertooth Cave, Citrus County; *D*, metatarsal 4; *E*, proximal phalanx; *F*, medial phalanx; *G*, distal phalanx (claw). *A–C* about 0.35×; *E–G* about 0.67×. *A–C* after Hirschfeld and Webb (1968); reproduced by permission of the Florida Museum of Natural History. *D–G* after Simpson (1928); reproduced by permission of the American Museum of Natural History, New York.

laterally compressed (fig. 10.11G). Unlike other sloths, megalonychids did not walk on the sides of their feet, but had a more normal plantigrade stance.

Two groups in the Megatheriidae independently reached Florida by the early Pleistocene. They are the gigantic Megatheriini (up to 6 meters long) and the much smaller Nothrotheriini (about 2.5 meters long). Some experts prefer to recognize them as distinct families and deny a close affiliation between the two; others classify them as two tribes in the same family. *Eremotherium* is the common North American genus of the megatherine tribe, and *Nothrotheriops* the common nothrotherine. Caniniforms are very reduced or absent in megatheres (McDonald, 1995). *Nothrotheriops* is known only from a few localities in the early Pleistocene, especially Leisey Shell Pit, where many skulls and mandibles are known

(McDonald, 1995; fig. 10.12). There are four upper and three lower molariform teeth, which are somewhat intermediate between *Megalonyx* and *Eremotherium* in morphology. They are subrectangular in shape, with the harder outer dentine forming two poorly defined transverse lophs or ridges. The mandibular symphysis forms an elongated spout (fig. 10.12B). The forelimb elements are long and slender, and the claws are laterally compressed like those of *Megalonyx* (fig. 10.12C, D). Although common during the late Pleistocene in the American Southwest, *Nothrotheriops* is only known from early Pleistocene in Florida.

Eremotherium is the largest of the giant ground sloths known from North America, weighing an estimated three tons in large males (females were much smaller). Its bulk was exceeded only by the largest of the probos-

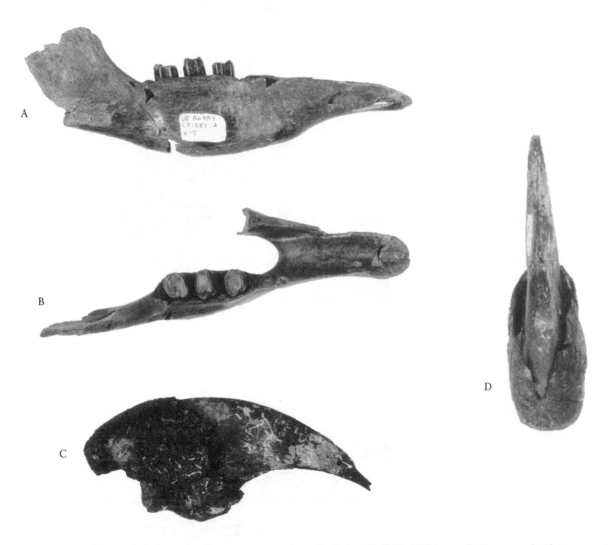

Fig. 10.12. The small ground sloth *Nothrotheriops texanum* from the Leisey Shell Pit, Hillsborough County, early Pleistocene. *A,* right lateral, and *B,* dorsal views of UF 86984, a right dentary with the spoutlike mandibular symphysis; *C,* lateral, and *D,* dorsal views of UF 87036, a distal phalanx of the third digit of the pes. Note the absence of caniform teeth and the laterally compressed claw, two characteristics of the genus. Both about 0.5×.

cideans. The large teeth (five uppers and four lowers) are typically square or rectangular (but the smaller last upper molariform is rounded) in cross section. The crown wears into a distinctive deep transverse, V-shaped valley that separates the anterior and posterior edges, which form high, sharp transverse ridges (fig. 10.13). With these interlocking teeth they must have consumed great quantities of leaves and twigs. Two species are recognized in Florida. The common late Pleistocene species is *Eremotherium laurillardi,* which has only a single claw on the hind foot and two on the front (Cartelle and De Iuliis, 1995). However, older megatheres with more than two claws on the front foot have been discovered in Florida (fig. 10.14) and have been given the name *Eremotherium eomigrans* (De Iuliis and Cartelle, 1999). This species is known from the late Pliocene and early Pleistocene. Like *Nothrotheriops,* the forelimb elements of

Eremotherium are slender and the tremendous claws are laterally compressed. The hindlimb elements and the pelvis are extremely massive to support the animal's enormous bulk. The species *Eremotherium rusconii* and *Eremotherium mirabile* are now considered junior synonyms of *E. laurillardi,* a species first described from Brazil (Cartelle and de Iuliis, 1995). It is also known from northern South America, Central America (a particularly good sample was collected in Panama), and the southeastern United States as far north as North Carolina. It is most abundant in coastal faunas, with the best sample deriving from Daytona Beach (Volusia County). An excellent mounted specimen of *E. laurillardi* is on display at the Daytona Beach Museum of Arts and Sciences. Several articulated skeletons of *E. eomigrans* were collected in the Haile quarries in western Alachua County.

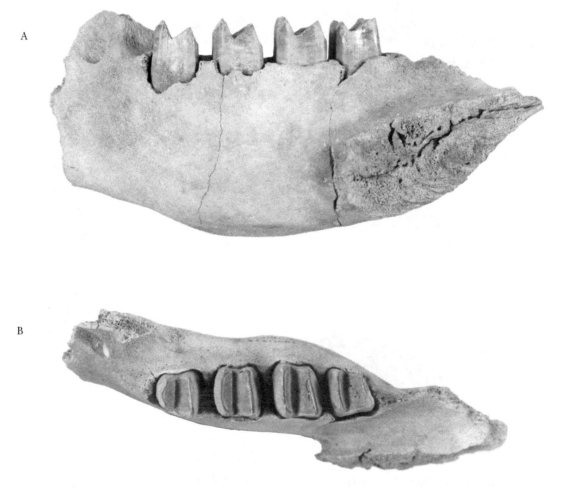

Fig. 10.13. *A,* medial, and *B,* occlusal views of UF 45475, a left dentary of a new species of the giant ground sloth *Eremotherium eomigrans* from Inglis 1A, Citrus County, late Pliocene. Note that this jaw is from a juvenile individual and that a fully grown adult could be twice as large. About 0.67×.

Fig. 10.14. Oblique lateral dorsal view of an articulated left manus of UF 121738, *Eremotherium eomigrans* from Haile 7C, Alachua County, late Pliocene. This species has four claws on its front foot (on digits 1 to 4), whereas the younger *E. laurillardi* only has claws on digits 3 and 4. Digits 1 and 2 are vestigial in *E. laurillardi*. About 0.2×.

REFERENCES

Auffenberg, W. 1957. A note on an unusually complete specimen of *Dasypus bellus* (Simpson) from Florida. Journal of the Florida Academy of Sciences, 20:233–37.

Cartelle, C., and G. De Iuliis. 1995. *Eremotherium laurillardi*: the panamerican late Pleistocene megatheriid sloth. Journal of Vertebrate Paleontology, 15:830–41.

De Iuliis, G., and C. Cartelle. 1999. A new giant megatheriine ground sloth (Mammalia: Xenarthra: Megatheriidae) from the late Blancan to early Irvingtonian of Florida. Zoological Journal of the Linnean Society, 127:495–515.

Downing, K. F., and R. White. 1995. The cingulates (Xenarthra) of Leisey Shell Pit 1A (Irvingtonian), Hillsborough County, Florida. Bulletin of the Florida Museum of Natural History, 37:375–96.

Edmund, G. 1985. The armor of fossil giant armadillos (Pampatheriidae, Xenarthra, Mammalia). Texas Memorial Museum Pierce-Sellards Series, 40:1–20.

———. 1987. Evolution of the genus *Holmesina* (Pampatheriidae, Mammalia) in Florida, with remarks on taxonomy and distribution. Texas Memorial Museum Pierce-Sellards Series, 45:1–20.

Engelmann, G. F. 1985. The Phylogeny of the Xenarthra. Pp. 51–64 *in* G. G. Montgomery (ed.), *The Evolution and Ecology of Armadillos, Sloths, and Vermilinguas*. Smithsonian Institution Press, Washington, D.C., 451 p.

Gillette, D. D., and C. E. Ray. 1981. Glyptodonts of North America. Smithsonian Contributions to Paleobiology, 40:1–255.

Hirschfeld, S. E., and S. D. Webb. 1968. Plio-Pleistocene megalonychid sloths of North America. Bulletin of the Florida State Museum, 12:213–96.

Hulbert, R. C., and G. S. Morgan. 1993. Quantitative and qualitative evolution in the giant armadillo *Holmesina* (Edentata: Pampatheriidae) in Florida. Pp. 134–77 *in* R. A. Martin and A. D. Barnosky (eds.), *Morphologic Change in Quaternary Mammals of North America*. Cambridge University Press, Cambridge.

Humphrey, S. R. 1974. Zoogeography of the nine-banded armadillo in the southeastern United States. Bioscience, 24:457–62.

McDonald, H. G. 1977. Description of the osteology of the extinct gravigrade edentate *Megalonyx* with observations on its ontogeny, phylogeny, and functional anatomy. Master's thesis, University of Florida, Gainesville, 328 p.

———. 1995. Gravigrade xenarthrans from the early Pleistocene Leisey Shell Pit 1A, Hillsborough County, Florida. Bulletin of the Florida Museum of Natural History, 37:345–73.

McKenna, M. C., and S. K. Bell. 1997. *Classification of Mammals above the Species Level*. Columbia University Press, New York, 631 p.

Robertson, J. S. 1976. Latest Pliocene mammals from Haile XVA, Alachua County, Florida. Bulletin of the Florida State Museum, 20:111–86.

Rose, K. D., and R. J. Emry. 1993. Relationships of Xenarthra, Pholidota, and fossil "edentates": the morphological evidence. Pp. 81–102 *in* F. S. Szalay, M. J. Novacek, and M. C. McKenna (eds.), *Mammal Phylogeny: Placentals*. Springer-Verlag, New York.

Simpson, G. G. 1928. Pleistocene mammals from a cave in Citrus County, Florida. American Museum Novitates, 328:1–16.

———. 1929. Pleistocene mammalian fauna of the Seminole Field, Pinellas County, Florida. Bulletin of the American Museum of Natural History, 56:561–99.

———. 1930. *Holmesina septentrionalis*, extinct giant armadillo of Florida. American Museum Novitates, 442:1–10.

Vaughn, T. A. 1972. *Mammalogy*. W. B. Saunders, Philadelphia, 463 p.

Vizcaíno, S. F., G. De Iuliis, and M. S. Bargo. 1998. Skull shape, masticatory apparatus, and diet of *Vassallia* and *Holmesina* (Mammalia: Xenarthra: Pampatheriidae): When anatomy constrains destiny. Journal of Mammalian Evolution, 5:291–322.

Webb, S. D. 1985. The interrelationships of tree sloths and ground sloths. Pp. 105–12 *in* G. G. Montgomery (ed.), *The Evolution and Ecology of Armadillos, Sloths, and Vermilinguas*. Smithsonian Institution Press, Washington, D.C.

Webb, S. D. 1990. Osteology and relationships of *Thinobadistes segnis*, the first mylodont sloth in North America. Pp. 469–532 *in* J. F. Eisenberg and K. Redford (eds.), *Advances in Neotropical Mammalogy*. Sandhill Crane Press, Gainesville, Fla.

Mammalia 3

Carnivorans

INTRODUCTION

The order Carnivora includes such familiar mammals as dogs, foxes, cats, bears, skunks, raccoons, and seals. It includes the top predators of most modern terrestrial ecosystems, although not all of them are strictly carnivorous. The word *carnivoran* is used to informally refer to members of the Carnivora; the word *carnivore* refers to any animal, regardless of its systematic classification, that is primarily a meat eater. Thus not all carnivorans are dietary carnivores (for example, giant panda, some bears, raccoon), nor are all carnivores carnivorans (for example, rattlesnakes, eagles, crocodiles). The ecology and behavior of living carnivorans are well known, especially the large species (for example, Gittleman, 1989, 1996; Macdonald, 1992). Despite numerous studies and a relatively good fossil record, the origin and evolutionary history of carnivorans remain controversial. Although some of Florida's fossil carnivorans are known from relatively complete specimens and large samples, many more are known by only a few, sometimes only one or two, specimens that may be no more than an isolated tooth or a fragmentary jaw. In these cases, information will be presented that is based on better samples from outside of Florida.

Despite its peripheral geographic location, Florida's fossil record includes most of the carnivoran families known to have lived in North America during the last 30 Ma (see chap. 3 for complete listing). A consensus of the evolutionary relationships among these families is shown in fig. 11.1. Most modern workers agree that the Carnivora is best divided into three major subgroups, the Aeluroidea, Cynoidea, and Arctoidea. There are four living aeluroid families, of which two are known from Florida, the Felidae (cats) and the Hyaenidae (hyenas), along with the Nimravidae, an extinct group of catlike carnivorans. The Cynoidea includes only a single family, the Canidae (dogs, foxes, and wolves). The Arctoidea is the most diverse group, containing six living families of

which five are known from Florida fossils: Ursidae, the bears; Phocidae, the true or earless seals; Odobenidae, the walrus; Procyonidae, the raccoon, coatis, and ringtail "cats"; and Mustelidae, which includes the otters, skunks, weasels, and badgers. Two important extinct arctoid groups, each accorded family status by McKenna and Bell (1997), are the Amphicyonidae and Hemi-

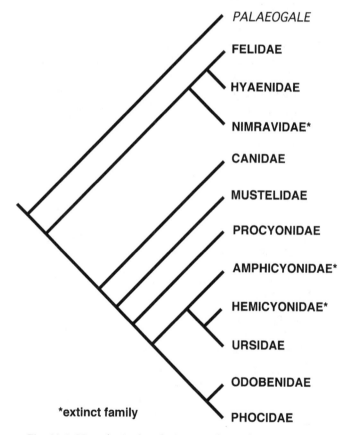

Fig. 11.1. Hypothesized evolutionary relationships among the eleven carnivoran families known from the fossil record of Florida, plus the genus *Palaeogale*. Primarily after Wyss and Flynn (1993). The position of the extinct Nimravidae relative to the others is especially controversial, and one of four possibilities is shown.

cyonidae. Fossils of both are known from Florida. Hemicyonids are often regarded as a subfamily in the Ursidae.

The Carnivora shares no known close evolutionary relationship with any other living group of placental mammals (fig. 9.1). The earliest known fossil carnivorans are about 64 Ma (early Paleocene), but the order probably originated in the late Cretaceous from an as yet unknown ancestor (Fox and Youzwyshyn, 1994). These earliest carnivorans were of small size and probably had a general (omnivorous) diet. The first large mammalian carnivores were *not* carnivorans but instead belong to two extinct lineages, the mesonychids and creodonts. Large, flightless, predaceous birds and terrestrial crocodiles also appeared in the early Paleogene. These four carnivorous groups were competing to succeed the theropod dinosaurs as the earth's dominant land-dwelling predators. In general, the creodonts were the most successful of the four and remained dominant through the Eocene Epoch (Janis et al., 1998). The mesonychids eventually became extinct on land, but their marine descendants, the whales, proved to be a long-term success (chap. 17).

Over the past 30 million years, the time period represented by land-dwelling vertebrates in Florida, there were four successive North American assemblages of medium- to large-sized carnivorous mammals (Hunt and Tedford, 1993; Janis et al., 1998). The first of these, termed the hyaenodont-nimravid association by Hunt and Tedford (1993), actually began about 40 Ma and lasted to about 22 Ma. During this interval, the hyaenodonts, the last creodonts, were the largest predators and were ecologically similar to modern wolves and hyenas. The nimravids filled a variety of cat "niches" and ranged in size from house cat to leopard; most had enlarged (saber-toothed) canines. Small canids and amphicyonids were the ecological equivalents of modern foxes and raccoons. This assemblage was replaced about 22 Ma by the "cat gap" assemblage, so-named because of the complete absence of either true cats (Felidae) or the catlike nimravids. From 22 to 18 Ma, most of the larger carnivorous mammals in North America were amphicyonids (filling a number of dog, hyena, and bear "niches"), while there was also a diverse array of fox- to coyote-sized canids. The entelodonts, a group of artiodactyls related to pigs and peccaries, may have been the bone-cracking scavengers of this interval (Joeckel, 1990). The middle to early late Miocene (18 to 9 Ma) carnivore guild, termed the amphicyonid-hemicyonid assemblage, included very large bear- and wolflike amphicyonids and hemicyonids and numerous medium to small canids. The borophagines, an extinct canid sub-

family, were especially diverse at this time. True cats and nimravids both dispersed from the Old World but generally remained uncommon. This was the first appearance of true felids in North America. While there had been earlier nimravids, they became extinct on this continent about 22 Ma. The final North American carnivore assemblage (9 Ma to present), the felid-ursid-canid association of Hunt and Tedford (1993) includes numerous canids (some with bone-cracking specializations), large bears, and both saber-toothed and normal-canined felids. Many dispersals are recorded from Asia, and it is this association that invaded South America in the Pliocene and effectively replaced its endemic mammalian carnivores.

CARNIVORAN SKELETAL AND DENTAL ANATOMY

Carnivorans have a wide range of body sizes (weasel to bear) and types of locomotion, from tree climbers to burrowers to runners to swimmers. Consequently there are a limited number of skeletal features shared by all members of the order. The paired lower limb elements (radius and ulna; tibia and fibula) always remain separate and never fuse together as in some ungulates, rodents, and xenarthrans. The length of the metapodials varies, being longest in good runners. Each digit bears a claw, which may have originally been used for climbing. In the felids and nimravids, the claws are enlarged and sharpened for use as weapons to augment the teeth. The primitive number of digits, five, is retained in most families, although commonly one is reduced in size. In the wrist, two of the carpal bones, the scaphoid and lunar, are fused into a single element called the scapholunar (fig. 11.2).

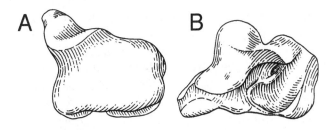

Fig. 11.2. *A*, proximal (dorsal), and *B*, distal (ventral) views of AMNH 108538, a left scapholunar of *Smilodon gracilis* from McLeod Limerock Mine, Levy County, early Pleistocene. All modern carnivorans have this fused element in their wrist, a combined scaphoid and lunar. About 1×. After Berta (1987); reproduced by permission of the Florida Museum of Natural History.

Rapid movement to close the jaws powerfully on live prey requires a rigid bony framework. There are heavy zygomatic arches along the sides of the skull and tall ridges on the braincase of the skull (the sagittal and nuchal crests) for attachment of massive jaw muscles. The mandible is relatively stout and typically bears a tall coronoid process, which has a deep fossa for attachment of jaw muscles.

Meat does not have to be thoroughly chewed to be easily digested the way plant matter does. Therefore, carnivorous mammals evolved teeth specialized to cut or slice flesh into swallowable pieces. This typically involves one or more pairs of upper and lower teeth that shear across each other when the mouth is closed, like the blades of a pair of scissors. Such specialized pairs of teeth are called carnassials. Although minor amounts of shear can occur on other teeth, in carnivorans the carnassial pair are the upper fourth premolar (P4) and the lower first molar (m1). In exclusively carnivorous carnivorans, such as felids, the carnassials are relatively long and the other cheek teeth are reduced or absent (fig. 11.3A–D). Canids, on the other hand, have a more generalized diet and need to chew and crush food as well as slice meat. They retain broad molars behind the carnassial pair (fig. 11.3E, F). The posterior molars are even larger in omnivorous carnivorans that eat a large amount of plant matter, and the slicing action of the carnassial pair is reduced or even lost (fig. 11.3I–K). The protocone of the P4 is primitively located anteriorly, farther forward than the paracone in carnivorans, and somewhat isolated from the shearing blade of the upper carnassial (fig. 11.3A, G, E). The protocone is moved posteriorly in omnivorous lineages and may even be joined by a hypocone (fig. 11.3I).

Fossil Canids of Florida

Recent systematic work has lead to a greatly improved understanding of canid evolution and has modified the family's traditional classification (Berta, 1988; Wang and Tedford, 1994; Wang, 1994; Tedford et al., 1995; Munthe, 1998; Wang et al., 1999). Unlike other carnivoran families, the Canidae was a North American group through most of its history, not dispersing to the Old World until about 5 Ma and to South America about 2.5 Ma.

The modern concept of canid evolution brought forth by the work of Tedford and Wang is that the family had three separate radiations that are formally recognized as subfamilies. Of these, only the Caninae survives, with its species subdivided into the fox and wolf tribes (Vulpini

and Canini, respectively). The other two canid subfamilies, the Hesperocyoninae and Borophaginae, are extinct, but both were quite abundant and diverse in the past, arguably more so than the modern Caninae. The Hesperocyoninae lived from the late Eocene to the middle Miocene (about 40 to 14 Ma) and were most abundant during the Oligocene, when ten genera are recognized (Wang, 1994). The Borophaginae lived from the early Oligocene to the late Pliocene (about 34 to 2 Ma) and were most diverse during the Miocene, when up to seven genera were alive at a time (Munthe, 1998). The oldest fossils of the Caninae are late Oligocene, but this lineage did not diversify until the late Miocene and Pliocene.

Canids have as yet not been reported from any of Florida's Oligocene localities; the state's oldest canids are early Miocene. These include late-surviving members of the Hesperocyoninae and early members of the Borophaginae. Morgan (1989) reported the oldest Florida canid as being from the White Springs fauna about 23 Ma. Other than small size (which is typical of early canids), it remains undescribed and not identified to subfamily. The slightly younger Live Oak and Buda localities both produced small samples of canids (Frailey, 1978, 1979). Most abundant were small species with large molars and reduced shear on their carnassials—adaptations for an omnivorous diet similar to the modern raccoon (fig. 11.4A, B). These are early members of the Borophaginae and are placed in the genera *Phlaocyon, Cormocyon,* and *Cynarctoides.*

The 18 Ma Thomas Farm site is the oldest locality in Florida to produce abundant samples of canids. Three species have been recovered here: the hesperocyonine *Osbornodon iamonensis* and the borophagines *Metatomarctus canavus* and *Euoplocyon spissidens. O. iamonensis* was originally described from a partial maxilla collected in Leon County just north of Tallahassee, but the sample of this coyote-sized canid from Thomas Farm is much larger. Olsen (1956) referred the Thomas Farm population to the species *O. iamonensis.* The genus *Osbornodon* was created by Wang (1994) for five hesperocyonine canids that ranged across much of the United States and had a combined chronologic range of early Oligocene to middle Miocene. Species of *Osbornodon* had long, slender muzzles, broad molars, slender premolars, robust skeletons, and short limbs (Wang, 1994; Munthe, 1998). In some of these features *Osbornodon* converged on characteristics later evolved independently by members of the Caninae, but the two are not closely related. The basic difference between *Osbornodon* and members of the Borophaginae and Can-

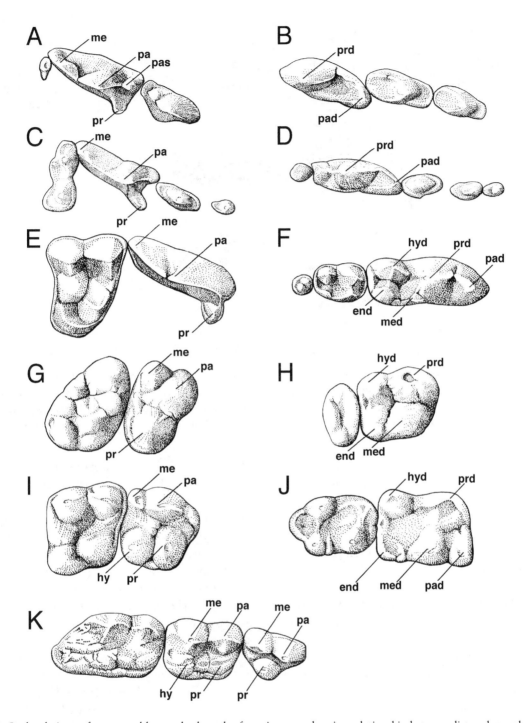

Fig. 11.3. Occlusal views of upper and lower cheek teeth of carnivorans, showing relationship between diet and morphology. Not to scale, abbreviations as in figure 1.11. *A*, right P3–M1, and *B*, left p3–m1 of the bobcat, *Lynx rufus;* note the reduced, buttonlike M1; *C*, right P2–M1, and *D*, left p2–m2 of the least weasel, *Mustela nivalis. Lynx* and *Mustela* are both nearly pure carnivores. Their carnassials (P4, m1) are thin and bladelike for shearing meat like a pair of scissors. The remaining taxa are more omnivorous and so retain more crushing ability in their cheek teeth; *E*, right P4–M1, and *F*, left m1–m3 of the coyote, *Canis latrans; G*, right P4–M1, and *H*, left m1–m2 of the sea otter, *Enhydra lutris* (must crush hard-bodied prey like sea urchins and crustaceans); *I*, right P4–M1, and *J*, left m1–m2 of the raccoon, *Procyon lotor; K*, right P4–M2 of the black bear, *Ursus americanus*. Figures from *Mammalogy* by Terry A. Vaughn, copyright © 1972 by Saunders College Publishing, reproduced by permission of the publisher.

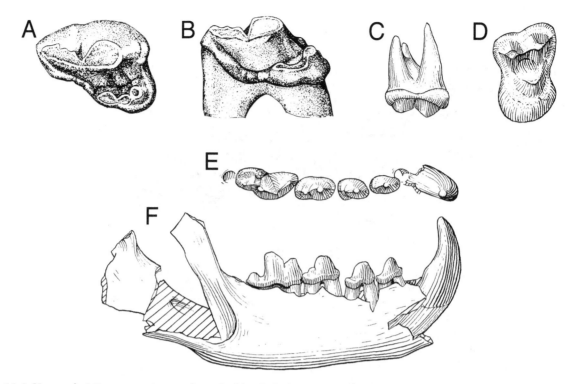

Fig. 11.4. Very early Miocene carnivorans from the Live Oak site, Suwannee County. *A*, occlusal, and *B*, lingual views of TRO 392, the right P4 of the borophagine canid *Phlaocyon leucosteus*; *C*, labial, and *D*, occlusal views of TRO 390, the right M1 of the amphicyonid *Mammocyon*; *E*, occlusal, and *F*, lateral views of UF 23928, right dentary with c, p2–m1 of the large mustelid *Paroligobunis frazieri*. A–B about 4×; C–F about 1×. After Frailey (1978), courtesy Natural History Museum, The University of Kansas.

inae is that the former lack a transverse ridge or crest connecting the entoconid and hypoconid on the m1 (fig. 11.5), while the m1 of members of the two other subfamilies have this feature (or lost it independently of *Osbornodon*). *O. iamonensis* was probably omnivorous, eating small vertebrates as well as carrion and fruits, and a good tree climber.

The two other Thomas Farm canids, *Metatomarctus canavus* and *Euoplocyon spissidens* (fig. 11.6), were similar in size to *O. iamonensis*. *M. canavus* is very abundant, almost as common as *O. iamonensis* (the two are by far the most abundant carnivorans at Thomas Farm), whereas *E. spissidens* is rare. *E. spissidens* is only known by lower jaws and teeth (Tedford and Frailey, 1976). Its borophagine classification is based on the presence of anterior and posterior stylids on all four lower premolars. The lower molars are more specialized for shearing than they are for crushing, so *E. spissidens* was evidently a purer carnivore than either *O. iamonensis* or *M. canavus*.

The common borophagine at Thomas Farm was originally named *Cynodesmus canavus*. Most subsequent authors listed it as *Tomarctus canavus* (for example, Olsen, 1956; Munthe, 1998). However, Wang et

al. (1999) made it the type species of their new genus *Metatomarctus*. It occupies an important position in the evolutionary history of borophagines, being intermediate between smaller forms such as *Desmocyon* and *Cormocyon* and more specialized scavengers like *Tomarctus*, *Aelurodon*, and *Epicyon* (Wang et al., 1999). In addition to Thomas Farm, the species is also known from Delaware, Nebraska, Wyoming, New Mexico, and California. The short muzzle and robust teeth suggest a more generalized diet than *Euoplocyon*, with more scavenging.

Epicyon is one of the more common North American carnivorans of the middle to late Miocene, and two species are recognized from Florida: *Epicyon saevus* and *E. haydeni* (figs. 11.7A, 11.8; Baskin, 1998b). *Epicyon* had longer limbs than earlier borophagines, such as *Metatomarctus*, that indicate it lived in more open, less wooded habitats. However, its skeleton does not include specializations for long-distance, high-speed pursuit. The skull of *Epicyon* had a short muzzle and a domed forehead, and its robust teeth show heavy wear indicating that it regularly chewed on hard objects. Apparently it was the first of the large borophagines to specialize as a bone-cracking scavenger. *E. haydeni* was the last and

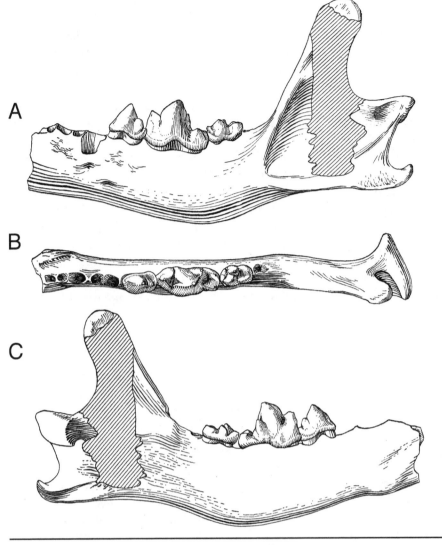

Fig. 11.5. *A*, lateral, *B*, occlusal, and *C*, lingual views of MCZ 3714, a left dentary with p4–m2 of the hesperocyonine canid *Osbornodon iamonensis* from the Thomas Farm site, Gilchrist County, early Miocene. This coyote-sized dog is the most abundant carnivoran at Thomas Farm. Shaded areas are broken and restored. About 1×. After Olsen (1956); reproduced by permission of the Museum of Comparative Zoology, Cambridge, Massachusetts.

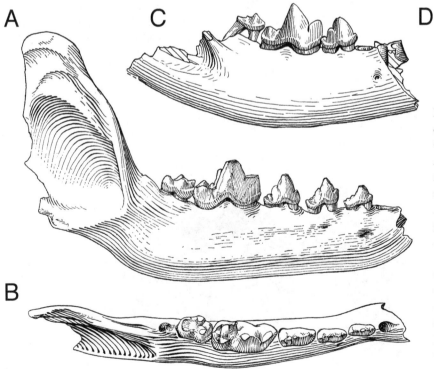

Fig. 11.6. Early Miocene borophagine canids from the Thomas Farm site, Gilchrist County. *A*, lateral, and *B*, occlusal views of MCZ 3628, a right dentary with p2–m2 of *Metatomarctus canavus*; *C*, lateral, and *D*, occlusal views of MCZ 7310, a partial left dentary with p4–m2 of *Euoplocyon spissidens*. Both about 1×. *A–B* after Olsen (1956); reproduced by permission of the Museum of Comparative Zoology, Cambridge, Massachusetts. *C–D* after Tedford and Frailey (1976); reproduced by permission of the American Museum of Natural History, New York.

largest species in the genus, exceeding the size of a modern lion (fig. 11.8). It is the largest canid species ever found.

Borophagus of the late Miocene and Pliocene shows more extreme bone-cracking adaptations than does *Epicyon*. The skull is extremely domed; the anterior premolars are reduced or lost; the carnassials are large; the tooth row is crowded because of the short muzzle; and the limbs are short, broad, and powerful. Species formerly in *Osteoborus* were transferred to *Borophagus* by Wang et al. (1999). *Borophagus* is first known in Florida from the 8 Ma Mixson site (Levy County), then coexisting with the very large *E. haydeni*. Fossils of a dwarf species of *Borophagus, B. orc*, were collected from the Withlacoochee River near Dunnellon (fig. 11.7C; Webb, 1969). Its age is about 6 million years old.

The youngest good samples of borophagines from Florida come from the early Pliocene Palmetto Fauna of the Bone Valley. At least four species are currently recognized: *Epicyon haydeni, Carpocyon limosus, Borophagus pugnator,* and *Borophagus hilli* (Wang et al., 1999). *Carpocyon* is much smaller and very different from the hyenalike *Borophagus*. It was a small omnivore with large molars for chewing plant matter (fig. 11.9). The larger Bone Valley borophagines have been usually regarded as a single species since the study by Webb (1969) and called *Osteoborus dudleyi* (fig. 11.7D). As part of their revision of the borophagines, Wang et al. (1999) recognized four large species in the Palmetto Fauna. The largest specimens are referred to the grizzly-sized *E. haydeni* and are the youngest known records of this species. The most common species is *Borophagus*

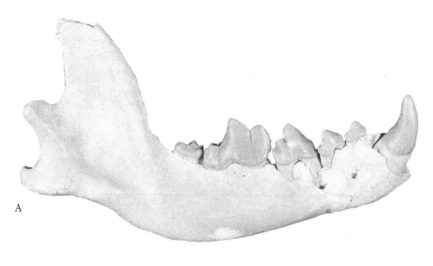

A

Fig. 11.7. Late Miocene and early Pliocene borophagine canids from Florida. *A*, lateral, and *B*, occlusal views of UF 37265, a right dentary with c, p2–m2 of *Epicyon saevus* from the Love site, Alachua County, late Miocene; *C*, occlusal view of UF 12313, a left maxilla with P2–M1 of *Borophagus orc* from the Withlacoochee River 4A site, Marion County, late Miocene; *D*, lateral view of UF 124521, a left dentary with p3–m1 of *Borophagus hilli* from the Palmetto Mine, Polk County, early Pliocene. *A* about 0.67×; *B–D* about 1×.

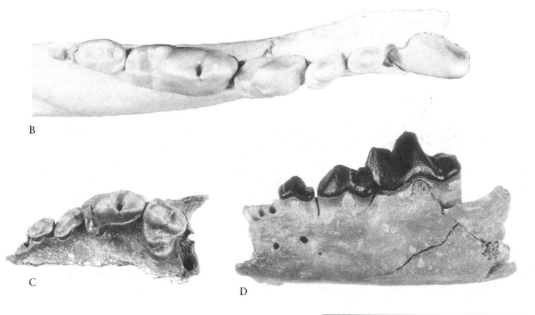

B

C

D

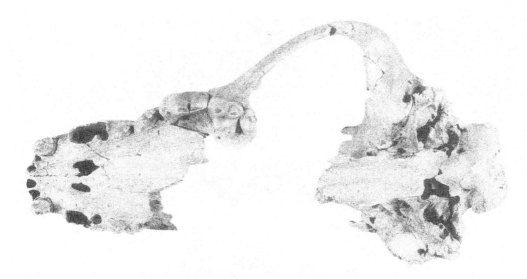

Fig. 11.8. Ventral view of UF 92000, a partial skull of *Epicyon haydeni* from Haile 19A, Alachua County, late Miocene. Cheek teeth preserved on the left side are the P2–M2. About 0.5×.

hilli, a widely distributed early Pliocene taxon. A more primitive species, *B. pugnator,* is also present. It is slightly smaller and has a longer, more slender mandible, and smaller P4 than *B. hilli.* The fourth species is *Borophagus dudleyi,* which Wang et al. (1999) recognized only from an edentulous partial skull. It is larger than would be expected for either *B. hilli* or *B. pugnator* and is morphologically different from *Epicyon.* Recovery of more complete specimens and larger sample sizes is needed to corroborate the presence of all four species.

The third canid subfamily, the Caninae, first appears in Florida in the late Miocene (fig. 11.10; Webb et al., 1981). At this time they were much less common than borophagines. To date, all records are of fox-sized canids. Representatives of both the red (*Vulpes*) and gray (*Urocyon)* fox lineages were present. Florida's oldest member of the wolf tribe, the Canini, is found in the Palmetto Fauna of the Bone Valley. Its name is *Eucyon davisi,* a species found across North America and in China (Tedford and Qiu, 1996). *Eucyon* can be viewed as the general ancestral group for both the wolf-dhole-hunting dog group (*Canis, Cuon,* and *Lycaon*) and the South American canid group (Berta, 1988; Tedford and Qiu, 1996).

Three major North American lineages are recognized within *Canis,* with their respective modern species being *Canis latrans,* the coyote; *Canis rufus,* the red wolf; and *Canis lupus,* the gray wolf. Representatives of all three are known from the Florida fossil record. The dog, *Canis familiarus,* the other living North American member of

Canis, is probably a domesticated offshoot of Asiatic *C. lupus.*

The earliest member of the coyote lineage recorded from Florida faunas is *Canis lepophagus.* It may be the direct ancestor of the living coyote, *C. latrans.* It differs from the latter in having a slightly larger and broader skull, deeper jaws, narrower premolars, and shorter metapodials. *C. lepophagus* is known from several late Pliocene sites, including the Santa Fe River and Macasphalt Shell Pit, but from very incomplete material. Undisputed remains referred to *C. latrans* have been record-

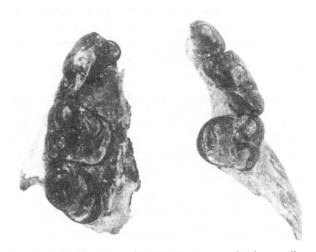

Fig. 11.9. Occlusal view of UF 12069, associated right maxilla with P4–M2 and left maxilla with P3–M1 of the omnivorous borophagine *Carpocyon limosus* from the Palmetto Mine, Polk County, early Pliocene. About 1×.

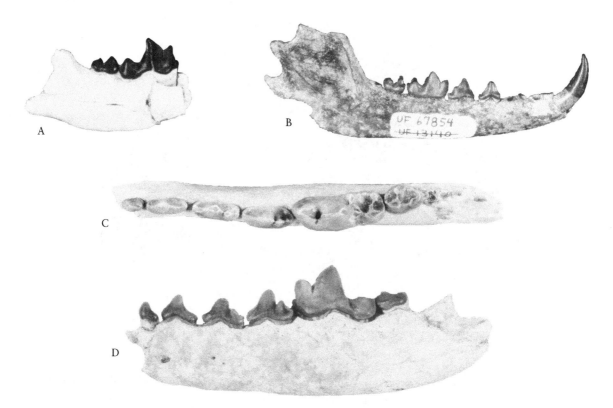

Fig. 11.10. Dentaries of fossil members of the subfamily Caninae from Florida. *A,* lateral view of UF 45884, a partial right dentary with m1–m2 of *Eucyon davisi* from the Fort Green Mine, Polk or Hardee County, early Pliocene; *B,* lateral view of UF 67854, a right dentary with c, p3–m2 of the extinct fox *Urocyon minicephalus* from Coleman 2A, Sumter County, middle Pleistocene; *C,* occlusal, and *D,* lateral views of UF 63175, a left dentary with p1–m2 of *Canis edwardii* from Haile 21A, Alachua County, early Pleistocene. All about 1×.

ed from several late Pleistocene sites in Florida, but it is not particularly common or well represented. Elsewhere it is known from the early Pleistocene, but not as yet from our state. *C. latrans* is the smallest living species of the genus in North America and ranges from Alaska as far south as Costa Rica, but it is most common in desert areas. Rodents and rabbits are the mainstay of the coyote diet, but it is far from choosy and will eat what it can catch or scavenge.

The red wolf lineage is thought to be more closely related to the coyote than to the gray wolf group, although its relationships are still being investigated. *Canis edwardii* is an early representative of the red wolf lineage (fig. 11.10C, D) and differs from the living species *C. rufus* in only a few details, such as a more prominent labial cingulum on the M1. Both are intermediate in size between *C. latrans* and *C. lupus,* with narrower skull proportions than either the gray or dire wolf. *C. rufus* is recorded from a few late Pleistocene sites in Florida, including Vero Beach and Devil's Den. It is much less common than the dire wolf. A mandible referred to *C. rufus* is also known from the early Holocene Nichol's

Hammock fauna of Dade County. Red wolves ranged over Florida in historic times only to be exterminated early in the twentieth century.

Although the modern gray wolf, *Canis lupus,* has apparently never lived in Florida, two closely related fossil species, *Canis armbrusteri* and *Canis dirus,* are well represented at a number of Pleistocene sites. *C. armbrusteri* was described on specimens collected from Cumberland Cave in Maryland (fig. 11.11). The species has also been found in Arizona, California, Florida, and Kansas. Several Florida faunas record its presence during the early to middle Pleistocene, including Leisey Shell Pit and Coleman 2A (Berta, 1995). Although *C. armbrusteri* is very similar to the gray wolf, it is regarded as a distinct species. It is distinguished from *C. lupus* by its narrower skull and muzzle, lower anterior premolars (p2 and p3) without posterior accessory cusps, and p4 with a small accessory cusp positioned on the heel of the tooth (fig. 11.11F). *C. armbrusteri* is most closely related to the Pliocene European wolf, *Canis falconeri.* The ancestry of the gray wolf lineage lies in the Old World, and *C. armbrusteri* represents its first appearance in North America.

It is last recorded in North America during the middle Pleistocene. *C. lupus* apparently represents a second dispersal of Eurasian wolves into North America that occurred late in the early Pleistocene.

Canis dirus, the dire wolf, was described in 1858 on the basis of a specimen collected near Evansville, Indiana. The dire wolf was the largest and most common canid in North America during the late Pleistocene. It has been reported from more than eighty localities across the continent, more than a quarter of these from Florida. It ranged as far north as Alaska and as far south as north-

ern South America. The largest sample of dire wolves is from the Rancho la Brea tar pits in Los Angeles, California. These have produced more than 1,600 individuals and hundreds of complete skulls. The most individuals ever recovered from a single site in Florida, 42, came from the Cutler Hammock site, a sinkhole in southern Dade County that probably represents a dire wolf den (Emslie and Morgan, 1995). The Aucilla River in Jefferson County has yielded the largest known specimens of dire wolves (Gillette, 1979). Greater size, wider skull proportions, exceptional heaviness of the mandible

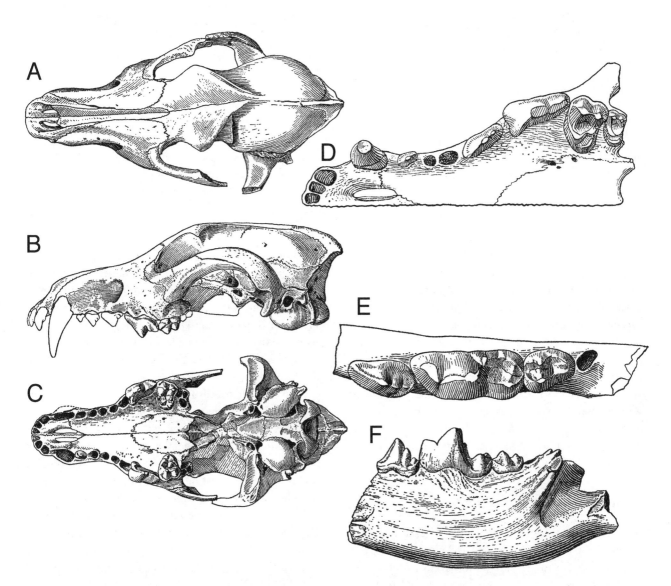

Fig. 11.11. Skull, dentary, and dentition of the early Pleistocene wolf *Canis armbrusteri*. These specimens are from Cumberland Cave, Maryland, but the same species is common in Florida. *A,* dorsal, *B,* left lateral, and *C,* ventral views of USNM 11885, nearly complete skull. Missing teeth are restored in outline in *B, D,* occlusal view of USNM 11883, showing left premaxilla and maxilla with C, P1, P3–M2; *E,* occlusal, and *F,* lateral views of USNM 7662, left dentary with p4–m2. Note small alveolus for missing m3. *A–C* about 0.4×; *D, F* about 0.67×; *E* about 1×. After Gidley and Gazin (1938); reproduced by permission of the Smithsonian Institution Press, Washington.

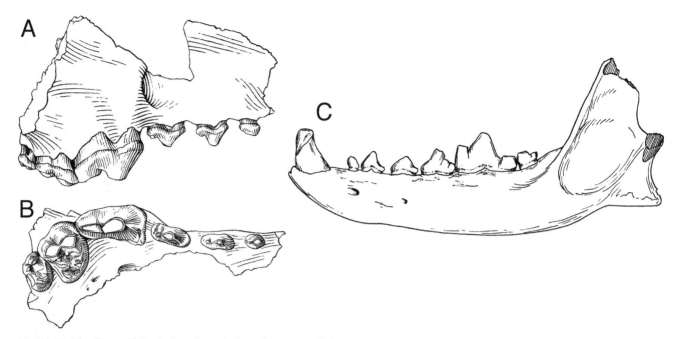

Fig. 11.12. The dire wolf *Canis dirus* from the late Pleistocene of Florida. *A*, lateral, and *B*, occlusal views of AMNH 23401, a right maxilla with P1–M2 from Seminole Field, Pinellas County; *C*, lateral view of UF/FGS 5692, a left dentary with c, p1–m2 from Hornsby Springs, Alachua County. *A–B* about 0.67×; *C* about 0.5×. *A–B* after Simpson (1929); reproduced by permission of the American Museum of Natural History, New York. *C* after Bader (1957); reproduced by permission of the Florida Museum of Natural History.

below the lower carnassial (fig. 11.12), extraordinary height and backward projection of the posterior portion of the skull, and enlarged carnassial teeth are all features that distinguish this species. Together these features permitted the large jaw muscles that this animal needed to capture large prey and provided some bone-cracking capabilities. *C. dirus* may have weighed as much as 20 percent more than the largest Alaskan wolves. The comparatively short limbs in relation to its overall weight and a massive head suggest that this animal was not as well adapted for running as are living wolves and coyotes. The origin of *C. dirus* is debated, with both *C. armbrusteri* and South American *Canis* being possible ancestors.

FOSSIL AMPHICYONIDS OF FLORIDA

Amphicyonids are an extinct family of arctoid carnivorans commonly called "beardogs," although they are truly neither dogs nor bears (Hunt, 1998). Among the living arctoids, they are most closely related to bears (fig. 11.1). Ecologically they were a diverse group and included species similar to modern canids, hyenas, bears, or some combination of the three. Genera with long-limbs and a digitigrade stance most often have cheek teeth well developed for shearing, while those with mas-

sive limbs and a plantigrade stance tend to have cheek teeth adapted for crushing. Sexual dimorphism is well developed in the family, with males larger than females. Much of the following discussion follows the recent review of Hunt (1998), who recognized three subfamilies in North America, the Daphoeninae, the Temnocyoninae, and the Amphicyoninae. Most of the amphicyonids from Florida belong to the latter group, although all three are known.

The earliest record of an amphicyonid from Florida comes from the Oligocene I-75 site near Gainesville. It was questionably referred by Patton (1969) to the genus *Daphoenus*, a common member of the Daphoeninae but never described. Hunt (1998) included this record in his text (203) but did not refer it to a genus and omitted it from his summary figure and appendix. Two amphicyonids are known from the very early Miocene of Florida, *Daphoenodon* and *Mammacyon*. *Mammacyon* is a member of the Temnocyoninae, a group characterized by a figure-eight-shaped M1 with a large, conical protocone (fig. 11.4C, D), and limb bones adapted for running. *Daphoenodon notionastes* is a lightly built, doglike amphicyonid, about the size of a large coyote. Overall, it is very similar to the well-known western species *Daphoenodon superbus* (fig. 11.13) except for smaller size and reduced molar talonids. Large burrows thought to be the dens of *D. superbus* have been excavated in western Nebraska.

The 18 Ma Thomas Farm site contains the oldest records of the subfamily Amphicyoninae from the state; two taxa are present: the very large, bearlike *Amphicyon longiramus* and the smaller *Cynelos caroniavorus* (figs. 11.14–11.17). Both genera evolved in Eurasia and dispersed to North America. *A. longiramus* was the top carnivore of the Florida early Miocene. Its skull was long relative to body size, but the brain was small. Incapable of running long distances, it must have hunted using the same stalking technique as the modern grizzly. *Cynelos caroniavorus* is the size of a small wolf (fig. 11.17) but with more robust, wolverinelike proportions. It is surprisingly similar to a species of *Cynelos* from the early Miocene of France and Germany, a resemblance that was first recognized by Tedford and Frailey (1976).

Amphicyonid numbers waned during the middle and late Miocene, and the family eventually became extinct. *Pliocyon robustus* is known from a single mandible from the middle Miocene of Polk County (fig. 11.18). The last North American amphicyonid, the bear-sized scavenger *Ischyrocyon* (fig. 11.19) is only known by a few fragments from the Love site in Alachua County. These large, bone-cracking forms were apparently replaced by borophagine canids in the late Miocene.

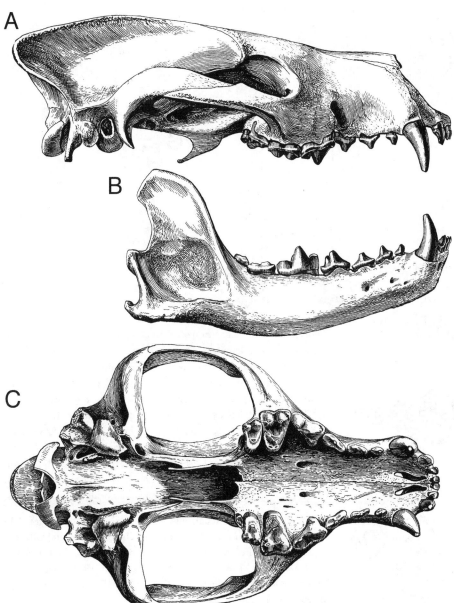

Fig. 11.13. *A*, lateral view of skull, *B*, lateral view of dentary, and *C*, ventral view of skull of the amphicyonid *Daphoenodon superbus* from the early Miocene of Nebraska. A similar, but smaller species, *D. notionastes*, is known from Florida by much less complete material. All about 0.5×. After Peterson (1906); reproduced by permission of the Carnegie Museum of Natural History, Pittsburgh.

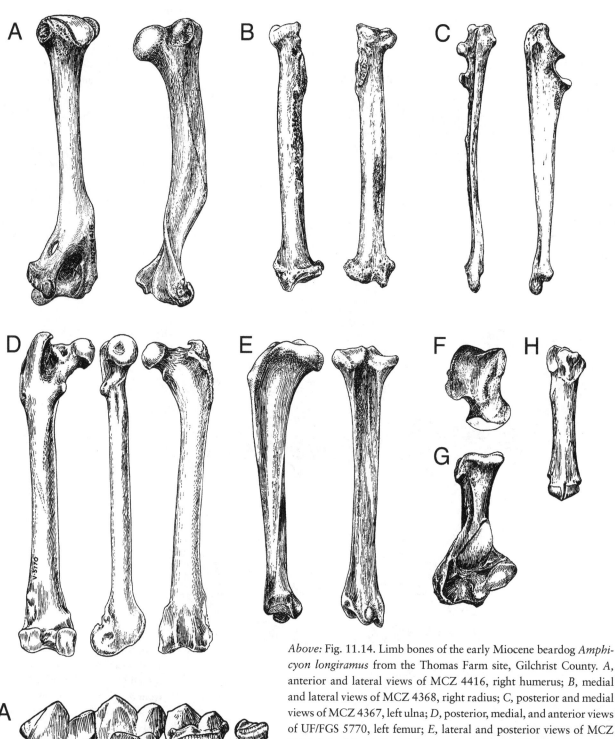

Above: Fig. 11.14. Limb bones of the early Miocene beardog *Amphicyon longiramus* from the Thomas Farm site, Gilchrist County. *A,* anterior and lateral views of MCZ 4416, right humerus; *B,* medial and lateral views of MCZ 4368, right radius; *C,* posterior and medial views of MCZ 4367, left ulna; *D,* posterior, medial, and anterior views of UF/FGS 5770, left femur; *E,* lateral and posterior views of MCZ 5886, left tibia; *F,* dorsal (or anterior) view of UF/FGS 5283, right astragalus; *G,* dorsal (or anterior) view of a right calcaneum; *H,* ventral (or posterior) view of UF/FGS 5774, left metatarsal 3. *A–E* about 0.25×; *F–H* about 0.5×. After Olsen (1960); reproduced by permission of the Museum of Comparative Zoology, Cambridge, Massachusetts.

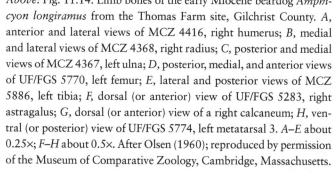

Left: Fig. 11.15. *A,* lingual, and *B,* occlusal views of the left P4–M3 of *Amphicyon longiramus* from the Thomas Farm site, Gilchrist County, early Miocene. A composite dentition comprised of MCZ 4060, 5833, and 7143. About 0.75×. After Olsen (1958c); reproduced by permission of the Museum of Comparative Zoology, Cambridge, Massachusetts.

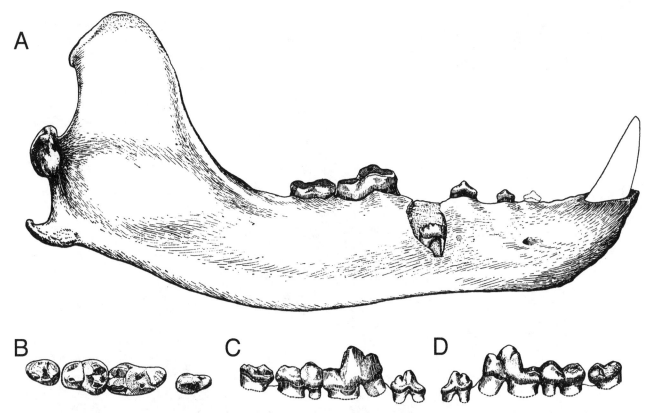

Fig. 11.16. Dentary and lower dentition of *Amphicyon longiramus* from the Thomas Farm site, Gilchrist County, early Miocene. *A*, lateral view of MCZ 3919, right dentary with p2–m2 (p4 embedded in jaw). The retention of the p4 in the jaw is a pathological condition. In normal individuals it would be erupted in a full adult such as this. *B*, occlusal, *C*, labial, and *D*, lingual views of a composite lower dentition (p4–m3) made from MCZ 5832 and 7141 and UF/FGS 5257. All about 0.5×. After Olsen (1958c); reproduced by permission of the Museum of Comparative Zoology, Cambridge, Massachusetts.

Fig. 11.17. *A*, lateral, and *B*, occlusal views of MCZ 7017, a left dentary with p2–m2 of *Cynelos caroniavorus* from the Thomas Farm site, Gilchrist County, early Miocene. About 1×. After Olsen (1958b); reproduced by permission of SEPM (Society for Sedimentary Geology).

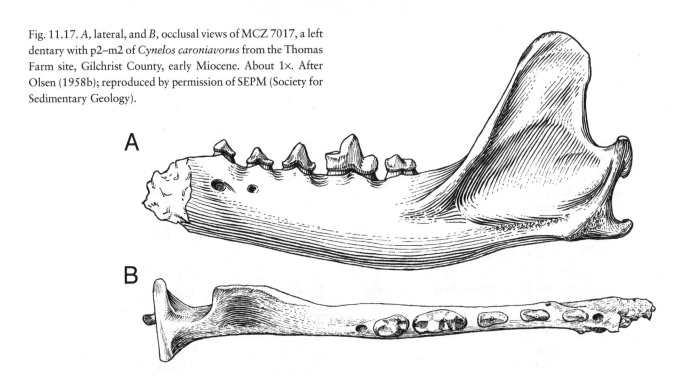

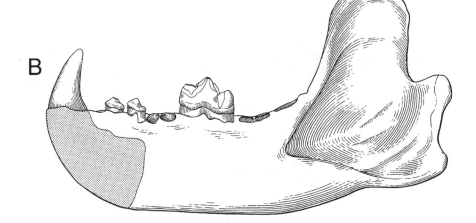

Fig. 11.18. *A,* occlusal, and *B,* lateral views of UF 24013, a left dentary with c, p2–p3, m1 of the rare amphicyonid *Pliocyon robustus* from Brewster Mine, Polk County, middle Miocene. About 0.5×. After Berta and Galiano (1984); reproduced by permission of the Society of Vertebrate Paleontology.

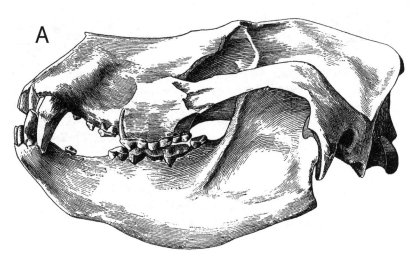

Fig. 11.19. The enormous, bone-crushing amphicyonid *Ischyrocyon gidleyi* from the middle to late Miocene. This widely distributed species is the last known amphicyonid in North America. Only a few fragments from the Love site indicate its presence in Florida. *A,* left lateral view of a skull and dentary from north Texas. *B,* occlusal view of UCMP 33853, left maxilla with P4–M3 and right maxilla with P3–M3 from Nebraska. *A* about 0.25×; *B* about 0.75×. *A* after Matthew (1902); reproduced by permission of the American Museum of Natural History, New York. *B* after Webb (1969a); reproduced by permission of the University of California Press, Berkeley.

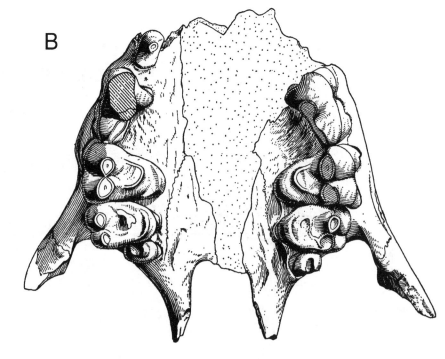

FOSSIL BEARS OF FLORIDA

The bear superfamily Ursoidea (Ursidae plus Hemicyonidae) is, like the Amphicyonidae, predominantly a Eurasian group that periodically dispersed representatives to North America during the Cenozoic. The earliest ursoids were small, generalized carnivorans, which evolved first into large, active predators and then into the very large, omnivorous bears of today. Much of the following discussion follows Hunt (1998), although the classification used is a combination of McKenna and Bell (1997) and Hunt (1998).

The Hemicyonidae resembled heavy-set dogs more than bears. They had a digitigrade stance, retained well developed carnassial teeth, and did not have the greatly enlarged posterior molars found in the Ursidae. *Phoberocyon johnhenryi* from Thomas Farm is the sole record of this genus in North America; otherwise, it is known only from Europe. It has elongate metapodials and a relatively slender mandible (fig. 11.20). Hunt (1998) moved all North American species formerly placed in *Hemicyon* to three other genera, *Phoberocyon, Plithocyon,* and an undescribed genus. Although hemicyonids persisted through the middle and early late Miocene in North America, specimens of this age have not been recovered from Florida.

After *Phoberocyon johnhenryi*, there are no recorded ursoids in Florida until the late Miocene, a gap of some 10 million years. Two very large bears dispersed from Asia to North America in the late Miocene, *Indarctos* at about 8 Ma and *Agriotherium* at about 6 Ma. The evolutionary origins of both of these are murky. According

to Hunt (1998), they represent independent lineages derived from the ancestral genus of the bear family, *Ursavus.* McKenna and Bell (1997) agreed with regards to *Indarctos* but placed *Agriotherium* in the Hemicyonidae. Despite some similarities, the two genera are easily distinguished with relatively complete fossils. *Indarctos* has four upper and lower premolars, while *Agriotherium* reduces its number of premolars; *Indarctos* has a longer M2, and *Agriotherium* has a short M2; *Indarctos* lacks a premasseteric fossa on the lateral side of the mandible, and *Agriotherium* has a premasseteric fossa. According to Hunt (1998), a single species of each is present in North America: *Indarctos oregonensis* and *Agriotherium schneideri.* The best remains of *Indarctos* from Florida come from the Withlacoochee River (Wolff, 1977; fig. 11.21A). Fossils of *Agriotherium* derive from the Palmetto Fauna of the Bone Valley (fig. 11.21C), where it is relatively rare.

Tremarctine bears are a North and South American subfamily distinguished by a relatively short muzzle, hence their common name of short-faced bears. The molars are long, with greatly wrinkled enamel, and the P4 is short. Three genera are recognized in North America, *Plionarctos, Tremarctos,* and *Arctodus. Tremarctos ornatus,* the spectacled bear, still lives in South America. Otherwise this subfamily is extinct. *Plionarctos* is the oldest and most primitive genus in the subfamily; its mandible lacks the premasseteric fossa found on the other two genera, and its M2 is slightly shorter (fig. 11.21B). *Plionarctos* is known from a few specimens from the Palmetto Fauna of the Bone Valley region of central Florida. *Arctodus* of the early Pleistocene and

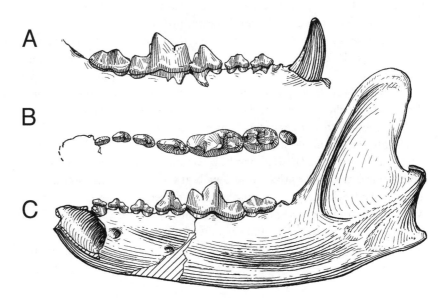

Fig. 11.20. The early Miocene ursoid *Phoberocyon johnhenryi* from the Thomas Farm site, Gilchrist County. *A,* lateral view of MCZ 4059, right lower dentition (canine, p2–m2); *B,* occlusal view of cheek teeth (p1–m2), and *C,* lateral view of AMNH 98608, a left dentary. About 0.5×. After Tedford and Frailey (1976); reproduced by permission of the American Museum of Natural History, New York.

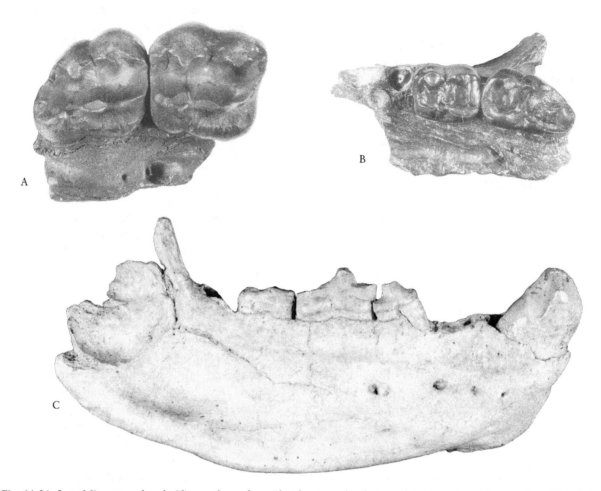

Fig. 11.21. Late Miocene and early Pliocene bears from Florida. *A,* occlusal view of UF 13792, a right maxilla with M1–M2 of *Indarctos* sp. from the Withlacoochee River 4A site, Marion County, late Miocene; *B,* occlusal view of UF 133945, a left maxilla with M1–M2 (also roots of the P4) of *Plionarctos* sp. from the Gardinier Mine, Polk County, early Pliocene; *C,* lateral view of UF/ FGS 652, a right dentary with c, p4–m2 of *Agriotherium schneideri* from the Bone Valley District, Polk County, early Pliocene. *A–B* about 1×; *C* about 0.5×.

Tremarctos of the late Pleistocene are more advanced tremarctines that evolved in North America, probably from *Plionarctos. Arctodus* is much larger than *Tremarctos* (figs. 11.22, 11.23) and has relatively longer metapodials.

The black bear, *Ursus americanus,* first appeared in Florida during the late Pleistocene and is much rarer in fossil deposits than *Tremarctos.* However, *U. americanus* still survives in the state, while the latter is long extinct. *Ursus* was the last Eurasian bear to range into North America, with the grizzly or brown bear (*U. arctos*) and the black bear representing separate dispersal events. *U. arctos* is not known as a fossil from Florida. The skulls of *Ursus* are longer and narrower than those of tremarctine bears, and have longer molars.

FOSSIL PINNIPEDS OF FLORIDA

Despite its miles of coastline, seals are a group of mammals one does not normally associate with Florida. However, the fossil record reveals their presence, although they have never been found in great abundance. Pinnipeds are classified into three families: the Otariidae, the fur seals and seal lions; the Phocidae, the true or earless seals; and the Odobenidae, the walrus. The evolutionary origin of the pinnipeds is thought to be from a bearlike ancestor and that odobenids and phocids share a closer common ancestor than either does with otariids. Studies supporting this concept include Berta et al. (1989) and Berta and Wyss (1994). All pinnipeds show similar modifications for an aquatic lifestyle. The limbs

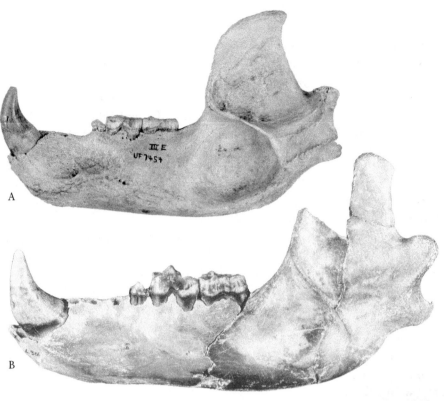

Fig. 11.22. Lateral views of dentaries with c, p4–m2 of Pleistocene tremarctine bears from Florida. *A*, UF 7454, *Tremarctos floridanus* from Devils Den, Levy County, late Pleistocene. The skull of this individual is shown in figure 11.23. *B*, UF 81692, *Arctodus pristinus* from Leisey Shell Pit, Hillsborough County, early Pleistocene.

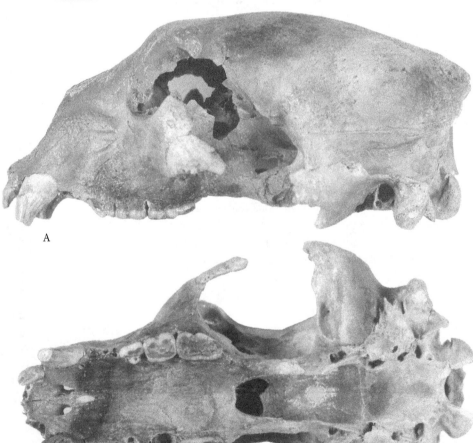

Fig. 11.23. *A,* left lateral, and *B,* ventral views of UF 7454, a skull of the short-faced bear *Tremarctos floridanus* from Devils Den, Levy County, late Pleistocene. Teeth present on the left side are I3, C, P4–M2. About 0.5×.

are modified into flippers. This means that the major limb elements are short and robust, especially the humerus and femur. The phalanges are elongated and flattened; those of the first digit are the longest. The dentition is simplified, with no demarcation between premolars and molars.

Otariids are mostly a Pacific group and are not known from Florida. Florida's fossil pinnipeds belong to the Odobenidae and Phocidae. The most common walrus fossil is the tusk, a modified upper canine with a diagnostic core of a type of dentine called globular osteodentine (fig. 11.24). Fossil walruses from Florida were at first reported to be Pleistocene in age (Ray, 1960), but subsequent discoveries in Florida and elsewhere and advances in stratigraphy and correlation strongly indicate an early to early-late Pliocene age for most occurrences of fossil walruses in the state. They have generally been referred to the species *Trichecodon huxleyi*. However, according to Deméré (1994), *T. huxleyi* is a Pleistocene species. Therefore, the Florida specimens need to be critically compared with Pliocene Atlantic walruses, such as *Prorosmarus alleni* and *Alachatherium cretsii*, a step no one has taken. Skulls, mandibles, and humeri are the most diagnostic elements and would be required for a positive identification.

After their first appearance about 15 Ma in the North Atlantic, phocid seals dispersed throughout much of the world's oceans. They are separated into two subfamilies, the phocines and monachines. Typical phocine seals are the harbor, ringed, harp, hooded, and gray seals. Monachine seals include the monk, Antarctic crabeater, Ross, leopard, and elephant seals. The only modern phocid native to Florida is the Caribbean monk seal (*Monachus tropicalis*), now unfortunately almost certainly extinct, but well known in and near Florida from living animals in historic time and from archaeological and a few Pleistocene fossil records.

Shallow marine deposits in central Florida record a diverse array of early Pliocene marine birds and mammals. While birds (chap. 8), whales (chap. 17), and sea cows (chap. 16) are abundant, remains of phocid seals are less well known. Representatives of both subfamilies, the phocines and monachines, are present, but they remain unstudied. More complete and abundant material of these seals is found in similar aged deposits from the Lee Creek (Aurora) Mine of North Carolina.

Fossil Procyonids of Florida

The Procyonidae is a group of small, primarily arboreal, mostly omnivorous arctoids. The raccoon, *Procyon*

lotor, is the best known modern example. While clearly members of the arctoid group (fig. 11.1), in some studies procyonids are thought to be most closely related to the Mustelidae, while in others they are allied with the bears and amphicyonids (Baskin, 1998a). They differ from mustelids by having enlarged the M2 and m2, a large m1 metaconid, and an m2 with a long talonid and tall hypoconulid (fig. 11.3I, J). The large molars and reduced carnassials are indicative of an omnivorous diet (except in *Bassariscus,* the modern ringtail "cat," which is more carnivorous).

Very little was known regarding procyonid evolution prior to the Pleistocene until the studies of Baskin (1982, 1989). He demonstrated the existence of a diverse array of Neogene procyonids, including several representatives from Florida. Two species of the large genus *Arctonasua* and a single species of the smaller *Paranasua* are known (figs. 11.25, 11.26). The latter is more closely related to the living *Nasua* and *Procyon*. Presumably the

Fig. 11.24. Lateral view of UF 3274, a tusk of the extinct walrus *Trichecodon huxleyi* from De Soto Lakes, Sarasota County, early Pliocene. About 0.3×.

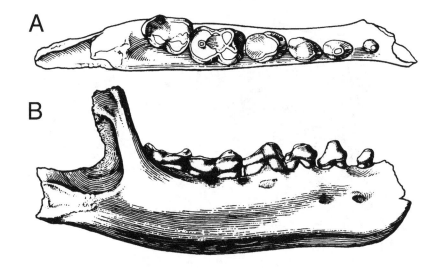

Fig. 11.25. *A*, occlusal, and *B*, lateral views of UF 32022, right dentary with p1–m2 of the procyonid *Arctonasua floridana* from the Love site, Alachua County, late Miocene. About 0.8×. After Baskin (1982); reproduced by permission of the Society of Vertebrate Paleontology.

two fossil genera were like modern procyonids in eating a wide range of small vertebrates, invertebrates (especially crayfish), and plants. *Procyon* is common in Florida Pliocene and Pleistocene deposits, with the modern species *P. lotor* first definitely recorded by the middle Pleistocene. There is some suggestion that early Pleistocene raccoons in Florida were significantly larger than the modern form, but more complete material is needed to firmly test this.

FOSSIL MUSTELIDS OF FLORIDA

The Mustelidae is a diverse family of medium- to very small-sized, predominantly northern hemisphere carnivorans. Mustelids usually have only one upper and two lower molars, and the P4 lacks a carnassial notch (fig. 11.3C, D). Most species have relatively short legs and long bodies. Many, especially among the subfamily Mustelinae, retain the primitive carnivoran habit of living and hunting in trees. During an early Miocene radiation, mustelids specialized into a number of different lifestyles. For example, there are ferrets, which hunt prey in their own burrows; otters, which are aquatic and feed on fish and invertebrates; and the large wolverines, which eat just about whatever they want to. The fossil record of mustelids is relatively poor until the Pleistocene, especially so in Florida. With a few exceptions, what specimens that are known have either not been studied at all or reviewed in great detail. Many of the species listed in chapter 3 are known from only a single specimen, or at best a handful, so they are difficult to characterize.

One relatively well-known fossil group is the Leptarctinae, an extinct subfamily of badgerlike mustelids (figs. 11.27, 11.28, 11.29A, B). They are known from the Miocene of Europe, Asia, and North America. A peculiar leptarctine feature is their double sagittal crests on the dorsal surface of the braincase (fig. 11.27B). The genus *Leptarctus* is present throughout the Miocene in

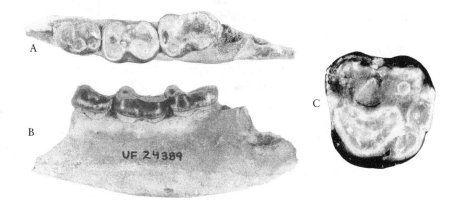

Fig. 11.26. The extinct procyonid *Arctonasua eurybates* from the Bone Valley District, early Pliocene. *A*, occlusal, and *B*, lateral views of UF 24389, a left dentary with p4–m2 from the Palmetto Mine. *C*, occlusal view of UF 60824, a left M1 from the Fort Green Mine. *A–B* about 1×; *C* about 2×.

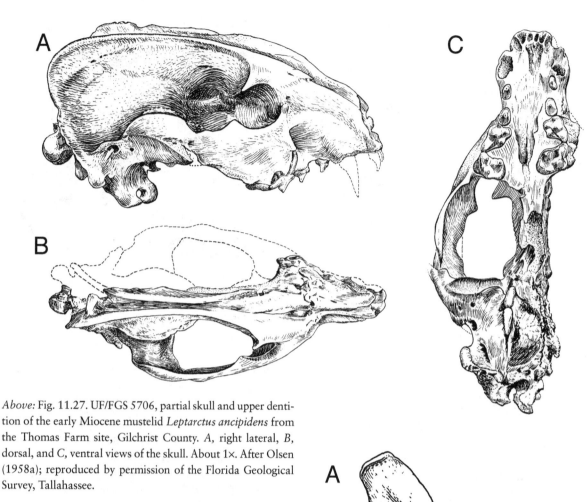

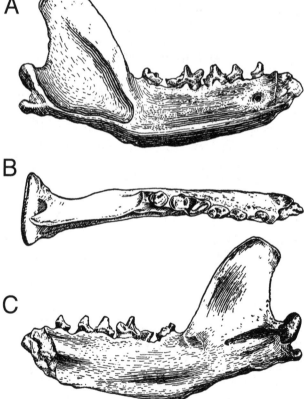

Above: Fig. 11.27. UF/FGS 5706, partial skull and upper dentition of the early Miocene mustelid *Leptarctus ancipidens* from the Thomas Farm site, Gilchrist County. *A,* right lateral, *B,* dorsal, and *C,* ventral views of the skull. About 1×. After Olsen (1958a); reproduced by permission of the Florida Geological Survey, Tallahassee.

Right: Fig. 11.28. *A,* lateral, *B,* occlusal, and *C,* lingual views of UF/FGS 5655, a right dentary with p2–m2 of *Leptarctus ancipidens* from the Thomas Farm site, Gilchrist County, early Miocene. About 1×. After Olsen (1957b); reproduced by permission of the American Society of Mammalogists.

Florida. Several larger, wolverinelike mustelids are also known from the Miocene, including *Oligobunis* and *Paroligobunis* early in the epoch (fig. 11.4E, F). *Plesiogulo,* a true wolverine related to the modern genus *Gulo,* dispersed into North America from Asia about 7 Ma and spread out across the continent (Harrison, 1981). It became extinct some three million years later, and subsequently *Gulo* came across the Bering Straits from Asia. This younger genus is adapted for cold climates and is not known from Florida. Both *Plesiogulo* and *Gulo* have large premolars, well-developed carnassials, broad molars, and relatively long limbs for mustelids. Their broad diet includes carrion, small vertebrates, and fruit. *Plesiogulo marshalli* is a relatively rare carnivoran in the Palmetto Fauna of the Bone Valley region (fig. 11.29F).

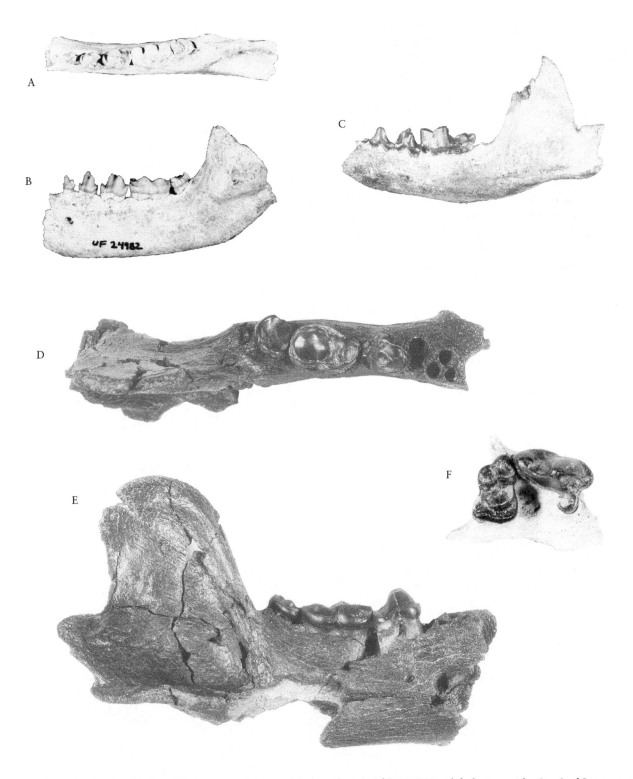

Fig. 11.29. Fossil mustelids from Florida. *A*, occlusal, and *B*, lateral views of UF 24982, a left dentary with p2–m2 of *Leptarctus* sp. from the Love site, Alachua County, late Miocene; *C*, lateral view of UF 124633, a left dentary with p3–m1 of the otter *Lutra canadensis* from the Leisey Shell Pit, Hillsborough County, early Pleistocene; *D*, occlusal, and *E*, lateral views of UF 125000, a right dentary with p4–m2 of the extinct otter *Enhydritherium terraenovae* from the Gardinier Mine, Polk County, early Pliocene; *F*, occlusal view of UF 19253, a right maxilla with P4–M1 of *Plesiogulo marshalli* from the Tencor Mine, Polk County, early Pliocene. All about 1×.

Otters, the subfamily Lutrinae, are semiaquatic carnivorans with especially broad molars for crushing (fig. 11.29C–E). The extinct otter *Enhydritherium terraenovae* first occurred in Florida in the late Miocene (Berta and Morgan, 1985). It is one of the best known of the pre-Pleistocene mustelids, as it is relatively common in the Palmetto Fauna of Polk County and a relatively complete skeleton was described from Marion County (Lambert, 1997). Relatives of *Enhydritherium* are known from Europe, South Africa, and India. It was apparently less fully aquatic than the modern sea otter of the Pacific Coast, *Enhydra lutris,* and made some incursions into fresh water based on its occurrence at the Moss Acres Racetrack site, an ancient pond. The large otter *Satherium* is known from Florida by a single late Pliocene humerus. It is related to the living giant otter of South America, *Pteronura brasiliensis.* The extant common otter of North America, *Lutra canadensis,* did not appear until the early Pleistocene (fig. 11.29C) and is another Asian immigrant. Fossils from the Leisey Shell Pit are some of the oldest records of *Lutra* in North America (Berta, 1995). *Lutra* is not uncommon as a fossil in some Florida river sites.

Skunks, the subfamily Mephitinae, are a familiar group of terrestrial, omnivorous mustelids. They modified the anal gland found in many carnivorans that is used to mark territories into a foul-smelling defensive weapon. Two skunks presently live in Florida: the small spotted skunk *Spilogale putorius* and the larger striped skunk *Mephitis mephitis.* Both are common in Pleistocene deposits (fig. 11.30C–E), but they are joined by an even larger variety, *Conepatus,* the hognosed skunk

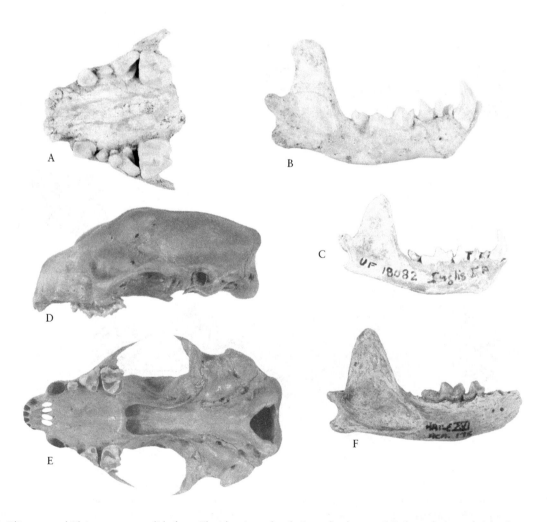

Fig. 11.30. Pliocene and Pleistocene mustelids from Florida. *A,* occlusal view of palate, and *B,* lateral view of right dentary of UF 22033, the extinct hog-nosed skunk *Conepatus robustus* from Haile 14B, Alachua County, late Pleistocene; *C,* lateral view of UF 18082, a right dentary of the spotted skunk *Spilogale putorius* from Inglis 1A, Citrus County, latest Pliocene; *D,* left lateral, and *E,* ventral views of UF 49232, a skull of the striped skunk *Mephitis mephitis* from Devils Den, Levy County, late Pleistocene; *F,* lateral view of UF 27509, a right dentary with p4–m1 of *Trigonictis cookii* from Haile 16A, Alachua County, early Pleistocene. All about 1×.

(fig. 11.30A, B). *Conepatus* today is a western species, ranging no further east than Texas, but at least two species have been recognized in the late Pleistocene of Florida: the living eastern hognosed skunk *C. leuconotus* and an extinct species, *C. robustus.* The latter is a larger variety, as the specific name implies.

Fossil Nimravids of Florida

The Nimravidae are an extinct family of catlike carnivorans that have been the subject of considerable study (for example, Baskin, 1981; Hunt, 1987; Bryant, 1988; Martin, 1998). In the past, some paleontologists thought that nimravids were the ancestors of living felids and classified them as a subfamily within the Felidae, but these hypotheses are not currently regarded as valid. In the dentition, nimravids lack the M2, M3, and m3, and the m2 is reduced or also absent. The more anterior premolars are also reduced or absent; only the P4 and p4 remain large and functional. The carnassials are large and very bladelike. The upper canines have serrations, and in contrast to saber-toothed felids, the deciduous upper canines are also very large and not shed until after maturity (Bryant, 1988). The mandible bears a flange on the anterior-ventral margin of the symphysis, which is greatly enlarged in forms with long canines such as *Barbourofelis* (fig. 11.31).

The most successful nimravid radiation, the subfamily Nimravinae, occurred in the late Paleogene and earliest Neogene. This subfamily is primarily Eurasian and North American, with a single entry into Africa. Size range in the Nimravinae was from that of the modern jaguar to the domestic cat; there were no really large (lion-size) members. Canine size also varied tremendously, although most fall into the "saber-tooth" category. Interestingly, those with the longest canines (the dirked-tooth type), such as *Eusmilus,* have similar cranial and postcranial features as true felids with very elongated canines. Other well-known genera of this subfamily are *Hoplophoneus* and *Dinictis.* In North America, these nimravids are well represented from the late Eocene and Oligocene White River Group badlands of South Dakota and Nebraska (Martin, 1998). This first nimravid radiation ended in the very early Miocene (about 22 Ma), when the subfamily Nimravinae became extinct in North America.

For a long time it was thought that the Nimravinae was not represented in Florida. The only report of a nimravine from Florida is based on a calcaneum from the very early Miocene Buda site. It can not be identified to genus but certainly represents a nimravid; based on the

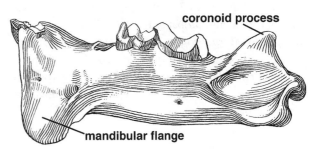

Fig. 11.31. Lateral view of UNSM 25546, a left dentary with p4–m1 of the nimravid *Barbourofelis whitfordi* from the middle Miocene of Nebraska. A few specimens of this species have been recovered from the Bone Valley District of Florida. Note the pronounced mandibular flange and the greatly shortened coronoid process. About 0.5×. After Schultz et al. (1970); reproduced by permission of the University of Nebraska State Museum, Lincoln.

age of the site, it is probably a nimravine. As was the case with hesperocyonine dogs, further discoveries of Oligocene faunas will probably add to the record of nimravines from Florida.

There is about an 11-million-year gap until the next record of a nimravid in Florida. This too is based on a single specimen, although a more diagnostic upper canine. It belongs to *Barbourofelis whitfordi,* a species best known from the middle Miocene of Nebraska and the Texas panhandle (fig. 11.31). The tooth was found in association with a middle Miocene fauna in Polk County. *Barbourofelis* is related to the middle Miocene *Sansanosmilus* of Eurasia, and its appearance in North America is a dispersal from Asia. *Barbourofelis* and *Sansanosmilus* are more advanced than the nimravines in their very long canines with lateral and medial grooves; loss of the P4 protocone; and presence of an extra anterior cusp on the P4, the ectoparastyle. Together they are classified in the subfamily Barbourofelinae. They had short legs and a plantigrade stance, which suggest they hunted from ambush. *Barbourofelis* differed from *Sansanosmilus* by having a postorbital bar (a unique character for carnivorans), reduced P3 and p3, and larger size. *B. whitfordi* is the oldest and smallest recognized species in the genus, about the size of a small leopard (fig. 11.31).

Barbourofelis lived in North America for a duration of approximately 5 million years, during which it greatly increased in size and relative canine length. The acme of nimravid evolution was the lion-sized *Barbourofelis fricki* of the late Miocene of Nebraska. A slightly smaller species, *Barbourofelis loveorum,* is well represented from the early late Miocene of Florida (Baskin, 1981;

Bryant, 1988; fig. 11.32). It is also known from Texas, California, and probably Kansas. The only mounted skeleton of *Barbourofelis* is on display at the Florida Museum of Natural History. *Barbourofelis* was the last of the nimravids. As its extinction coincides with the radiation of saber-toothed felids, competition may have played a role in its demise.

FOSSIL FELIDS OF FLORIDA

Felids originated in the Old World and dispersed to North America in the early Miocene, about 16 Ma. They are the only living aeluroid carnivorans (fig. 11.1) to successfully colonize North and South America. Felids resemble nimravids in reducing all of the cheek teeth with

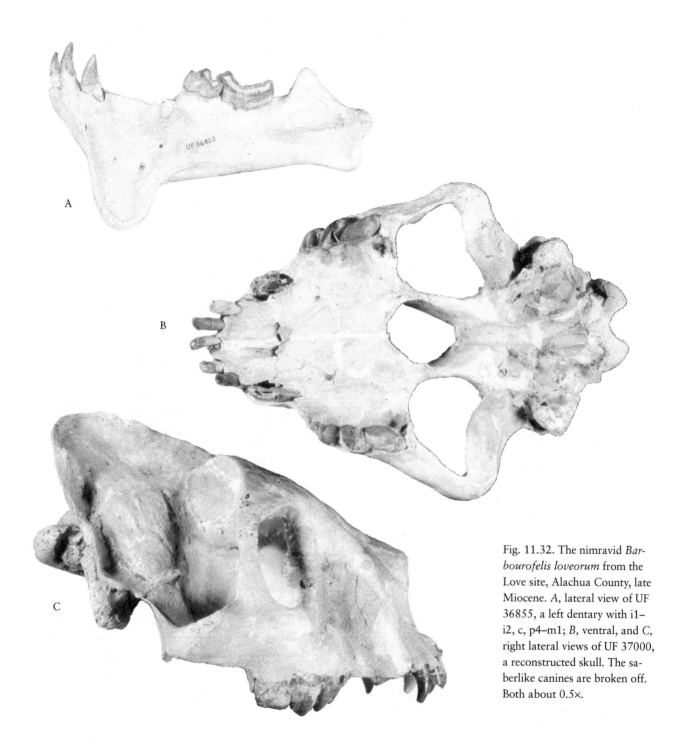

Fig. 11.32. The nimravid *Barbourofelis loveorum* from the Love site, Alachua County, late Miocene. *A,* lateral view of UF 36855, a left dentary with i1–i2, c, p4–m1; *B,* ventral, and *C,* right lateral views of UF 37000, a reconstructed skull. The saberlike canines are broken off. Both about 0.5×.

the exception of the carnassial pair, which are large and very well developed. The P1, p1, p2, M2, m2, M3, and m3 are absent, and the M1 is reduced to a rudimentary peg (fig. 11.3A, B). The talonid of the m1 is also lost, so all of the crushing ability of the molars is gone. The incisors are relatively small and chisel-like (except some saber-toothed cats have large incisors). The skull is short and broad, with an inflated braincase and large orbits. Five front and four hind toes have sharp retractile claws that are used to climb trees and as weapons. However, it is the teeth that are used to kill their prey. Smaller animals are usually killed with a bite into the neck vertebrae, while larger game are bitten in the throat and suffocated.

The upper canines of felids come in three basic sizes: normal, long, and extra long. Normal-sized canine teeth for a felid are large relative to body size compared to those of other carnivorans. To distinguish them from saber-toothed cats, those with normal-sized canines are called conical-toothed cats. Their canines are round in cross section, and the lower canine is relatively large, only slightly smaller than the upper canine. Two morphologically distinct groups of saber-toothed cats are recognized: the scimitar-toothed and dirk-toothed cats. Scimitar-toothed felids have shorter and broader canines (although longer and narrower than those of conical-toothed cats) that are usually coarsely serrated. Examples are *Nimravides, Machairodus, Xenosmilus,* and *Dinobastis.* Dirk-toothed cats have very long, narrow canines that may or may not be finely serrated. Examples are *Megantereon* and *Smilodon* among the Felidae, while the nimravid *Barbourofelis* and the South American marsupial *Thylacosmilus* are morphological equivalents. Until the Pleistocene, the larger felids (those the size of a lion or tiger) were all saber-toothed cats, while conical-toothed cats were jaguar size or smaller.

Four felid subfamilies are recognized in North America, the Felinae, Pantherinae, Acinonychinae, and the Machairodontinae (McKenna and Bell, 1997). All four have been identified from the Florida fossil record (chap. 3), where they range from the late Miocene to present, about the past 9 million years. Their fossils are normally rare, and good specimens are highly prized.

The oldest known felid from Florida is *Nimravides galiani* (Baskin, 1981; fig. 11.33A). The name of this genus is confusingly spelled similar to that of *Nimravus,* which is a nimravid, not a felid. The two are quite different and should not be confused. *N. galiani* is a member of the Felinae, but unlike most members of its subfamily it has large canines (it falls into the scimitar-toothed category). *Nimravides* does not show the cranial adap-

tations of most other saber-tooth cats (members of the Machairodontinae). The coronoid process of the mandible is high, the upper carnassial is less bladelike, and the lower canine is relatively large. *Nimravides* was a long-limbed cat about the size of a mountain lion or jaguar. The most complete specimens of this felid were found at the Love site.

About 7 Ma, the large Eurasian scimitar-toothed cat *Machairodus* appeared in North America. Remains of *Machairodus* have been found in the Withlacoochee River and in the Bone Valley region of central Florida (fig. 11.33B). It has long, somewhat flattened upper canines, and the mandible has a very slight flange. The p3 is large and double-rooted in contrast to *Homotherium* or advanced *Smilodon,* in which it is extremely reduced or absent. The limbs of *Machairodus* are long and relatively slender.

A short-limbed relative of *Homotherium* and *Dinobastis* is *Xenosmilus,* described by Martin et al. (2000) from the Haile 21A site (fig. 11.33C, D). Most scimitar-toothed felids have long limb bones and were apparently good runners. *Xenosmilus* in contrast had stout, robust limbs, giving it a more bearlike appearance and making it an ambush predator (Martin et al., 2000). The late Pleistocene *Dinobastis serus* is known from only a few sites. This species is sometimes referred to *Homotherium,* but *Dinobastis* has much shorter canines than *Homotherium* (Berta and Galiano, 1983; Berta, 1987). The upper canine of *Dinobastis* is very flattened and has razor-sharp, serrated cutting edges. The carnassial teeth are more bladelike than in *Smilodon* and are serrated. *Dinobastis* also has long, powerful front legs. The preferred prey of these large scimitar-toothed cats was probably large ungulates and sloths. *Machairodus, Homotherium, Xenosmilus,* and *Dinobastis* are members of the tribe Machairodontini within the Machairodontinae.

The other, more famous saber-toothed cat lineage is the tribe Smilodontini, which includes the genera *Megantereon* and *Smilodon.* At present the oldest known record of the Smilodontini is the very early Pliocene occurrence of *Megantereon hesperus* from the Palmetto Fauna of the Bone Valley region (Berta and Galiano, 1983; fig. 11.34), although this identification was disputed by Turner (1987). This genus is otherwise known from the middle Pliocene through middle Pleistocene of Eurasia and Africa and the late Pliocene of North America outside Florida. *M. hesperus* is a leopard- or puma-sized saber-toothed cat, much smaller than its contemporary *Machairodus.* The canines are not serrated in this genus (or only weakly so), and the p3 is well devel-

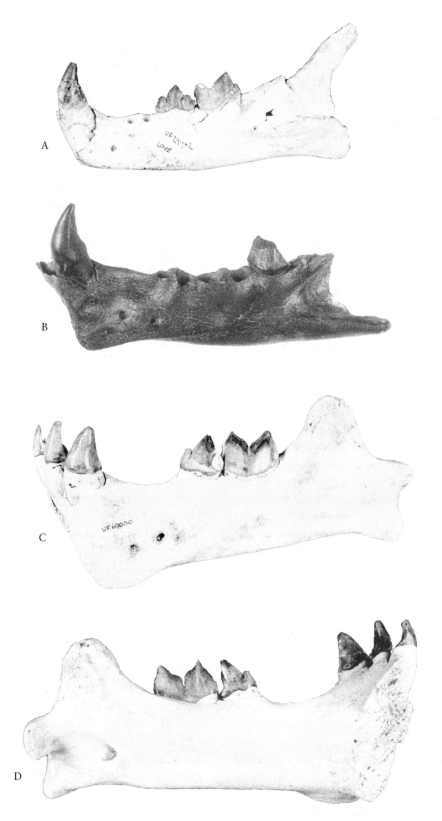

Fig. 11.33. Left dentaries of three large fossil cats from Florida. *A*, lateral view of UF 24462, *Nimravides galiani* from the Love site, Alachua County, late Miocene. This is the oldest true felid known from Florida; *B*, lateral view of UF 133905, *Machariodus* sp. from the Withlacoochee River 4A site, Marion County, late Miocene. Although the p3 and p4 have fallen out of this jaw, their alveoli indicate both were large and two-rooted. The posterior half of the m1 is also missing. *C*, lateral, and *D*, medial views of UF 60000, *Xenosmilus hodsonae* from Haile 21A, Alachua County, early Pleistocene. All about 0.5×.

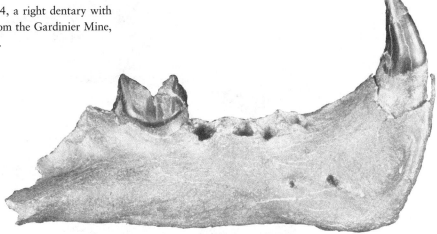

Fig. 11.34. Lateral view of UF 124634, a right dentary with c and m1 of *Megantereon hesperus* from the Gardinier Mine, Polk County, early Pliocene. About 1×.

oped relative to *Smilodon,* with a secondary posterior cusp and two roots. The mandible bears an anterior flange (fig. 11.34). The species is as yet not known from very complete cranial material from Florida, but several mandibles and postcranial elements have been recovered.

In the late Pliocene of Florida, *Megantereon* was replaced by the similar-sized but more advanced dirk-toothed cat, *Smilodon gracilis.* Unlike many of the cases discussed previously, this apparently represents an evolutionary event that took place in North America rather than a dispersal from another continent. *S. gracilis* is best represented in latest Pliocene and early Pleistocene faunas in Florida, especially the Inglis 1A and Leisey Shell Pit sites (Berta, 1987, 1995; fig. 11.35). It differs from *Megantereon hesperus* by having longer, slightly more curved, very finely serrated upper canines, a more reduced P3 and p3, and a larger ectoparastyle on the P4.

During the middle Pleistocene, a much larger, lion-sized species of *Smilodon* apparently evolved from *S. gracilis.* Several specific names for this larger taxon are presently used by different paleontologists. One view holds that all middle to late Pleistocene *Smilodon* from both North and South America belong to a single species, for which the oldest available species name is *Smilodon populator* (Berta, 1985). Another approach is to distinguish the populations from the two continents as distinct species and to regard all North American middle to late Pleistocene *Smilodon* as a single species. Under this scheme its name is *Smilodon fatalis,* based on a specimen from Texas. The presence of *Smilodon* in Florida was first noted a century ago on a partial skull found in a fissure-fill deposit near Ocala. This lion-sized species was the largest of the dirked-toothed cats. In addition to

its much greater size, it differs from *Smilodon gracilis* by its more curved upper canines with better developed serrations; the p3 is absent or rudimentary; the P4 protocone is lacking or very reduced; and the mandibular flange is greatly reduced (figs. 11.36, 11.37).

Fossils of late Pleistocene *Smilodon* are not particularly common in Florida. Instead, dire wolves and jaguars are the most abundant large carnivorans of that time. In addition to the Ocala locality, *S. fatalis* is known from the Ichetucknee and Aucilla Rivers, Vero Beach, Bradenton, Newberry, Cutler site, and Warm Mineral Springs. Most records are isolated teeth or skeletal elements, but a partial skeleton (unfortunately lacking the skull) was recovered from Arredondo, a subdivision southwest of Gainesville. The extinction of *Smilodon* and *Dinobastis* about 11,000 years ago brought an end to the era of saber-toothed cats in North America.

Five conical-toothed felid lineages are represented in the Pliocene and Pleistocene of Florida: the lynxes (*Lynx*); the large roaring cats (*Panthera*); the American cheetah (*Miracinonyx*); the mountain lion (*Puma*); and the small American spotted cats (*Leopardus*). Lynxes are small to medium-sized felines, typically with short faces, elongated limbs, relatively small canines, and a relatively large P3. The modern lynx and bobcat both lack the P2. *Lynx rexroadensis* is the oldest (Pliocene) member of the genus from Florida (fig. 11.38A). It is a medium-sized cat (similar to the Canadian lynx) with a somewhat larger P3 than modern lynxes. *Lynx rufus,* the bobcat, is a common element of the Pleistocene carnivoran fauna of Florida (fig. 11.38B). It is distinguished from the Canadian lynx *(Lynx canadensis)* by smaller size, relatively broader P3, and a large P4 paracone. Some early late Pleistocene bobcats from Florida were the size of *L.*

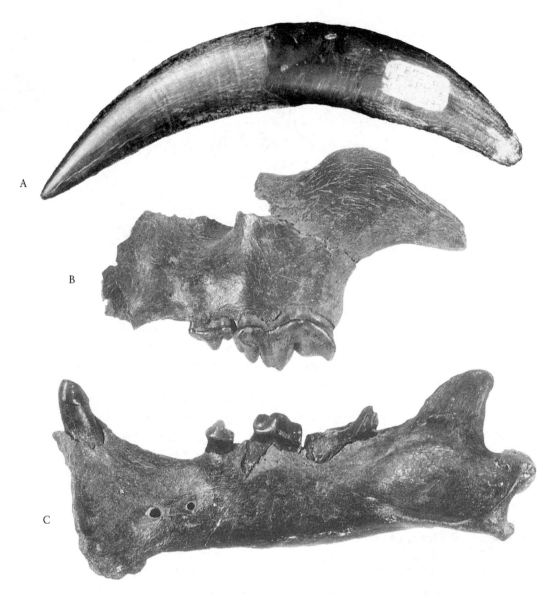

Fig. 11.35. The saber-toothed cat *Smilodon gracilis* from the Leisey Shell Pit, Hillsborough County, early Pleistocene. *A,* lateral view of UF 87276, an upper canine; *B,* lateral view of UF 87238, a left maxilla with p3–p4; *C,* lateral view of UF 81724, a left dentary with c, p3–m1. The m1 is heavily worn. All about 1×.

canadensis and have been considered a distinct subspecies, *L. rufus koakudsi.* By the very late Pleistocene, bobcats of modern proportions are known from across the state.

The largest modern felids are placed in the genus *Panthera* (fig. 11.39). Examples are the lion (*P. leo*), tiger (*P. tigris*), leopard (*P. pardus*), and jaguar (*P. onca*). Only the latter lives in the New World; the remainder live in Eurasia and/or Africa. In addition to their large size, members of *Panthera* have modifications to their hyoid apparatus that allows them to roar. In the early

Pleistocene, the first representative of this genus to enter North America from Asia was the jaguar (Seymour, 1993). Pleistocene jaguars are larger than living individuals (by 15 to 25 percent), and in the early Pleistocene had relatively longer limbs. The modern *P. onca* has adapted to live in forests and is relatively stocky. *P. onca* ranged widely in the New World during the Pleistocene, but its distribution contracted during the Holocene so that its present northernmost limit is the southwestern United States. In Florida, *P. onca* is the most common large cat of the middle and late Pleistocene (fig. 11.39C,

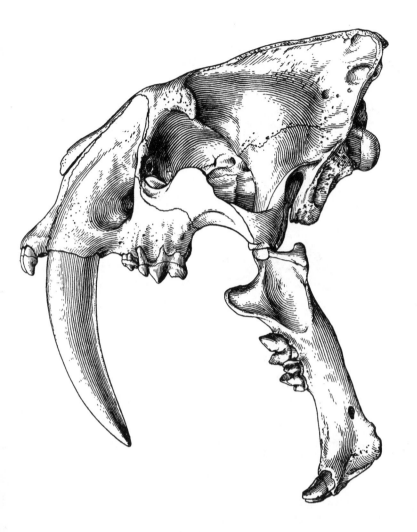

Fig. 11.36. Left lateral view of the skull and dentary of the late Pleistocene saber-toothed cat *Smilodon fatalis* from the Rancho la Brea tar pits, Los Angeles, California. About 0.3×. After Matthew (1910); reproduced by permission of the American Museum of Natural History, New York.

Fig. 11.37. Dentition of *Smilodon fatalis* from the Rancho la Brea tar pits, Los Angeles, California. *A*, Labial views of the P4 and m1; *B*, Occlusal view of left upper cheek teeth (P3–M1); note that the P4 makes up most of the postcanine dentition; *C*, lateral view of left dentary; note absence of p3. After Matthew (1910); reproduced by permission of the American Museum of Natural History, New York.

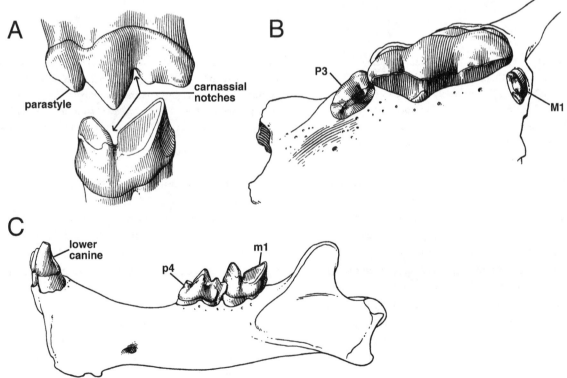

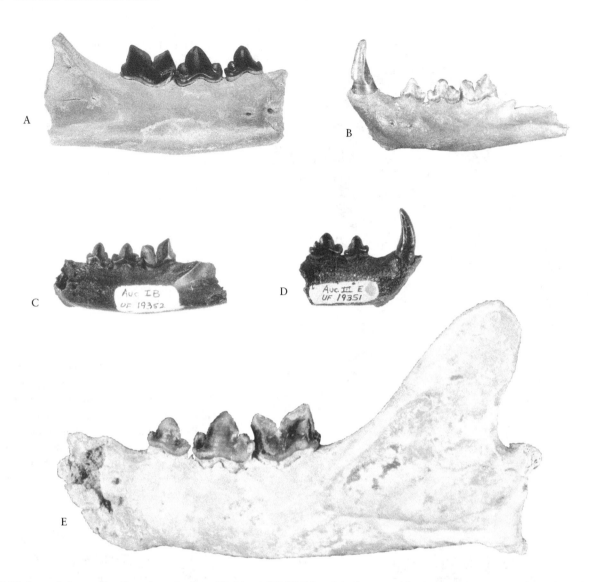

Fig. 11.38. Lateral views of fossil cat dentaries from Florida. *A,* UF 58308, a right dentary with p3–m1 of *Lynx rexroadensis* from the Fort Green Mine, Polk County, early Pliocene; *B,* UF 124631, a left dentary with c, p3–m1 of the bobcat *Lynx rufus* from the Leisey Shell Pit, Hillsborough County, early Pleistocene; *C,* UF 19352, a left dentary with p3–m1 of *Leopardus amnicola* from the Aucilla River 1B site, Jefferson County, late Pleistocene; *D,* UF 19351, a right dentary with c, p3–p4 of *L. amnicola* from the Aucilla River 3E site, Jefferson County, late Pleistocene; *E,* UF 21604, a left dentary with p3–m1 of the extinct American cheetah *Miracinonyx inexpectatus* from Inglis 1A, Citrus County, late Pliocene. All about 1×.

D), with excellent samples from Coleman 2A, Reddick, and the Cutler Hammock site.

The much larger *Panthera atrox,* a relative of the African lion, appeared very late in the Pleistocene and is rare in the state. However, a beautifully preserved skull and associated lower jaws were recovered from the Ichetucknee River (fig. 11.39A, B). Some authorities regard *P. atrox* as only a subspecies of *P. leo,* giving it the largest known range for any wild terrestrial mammal (Africa, Europe, Asia, North America, and northern South America). *P. atrox* was significantly larger than the mod-

ern African lion, and it even exceeded its saber-toothed contemporary *Smilodon* in size.

Although cheetahs (*Acinonyx jubatus*) are commonly thought of as African, their historical distribution included the Middle East and India, and *Acinonyx* was widespread in Eurasia during the Pliocene and Pleistocene. Two fossil species from North America are now regarded to be cheetahs, or at the very least to have many of the same physical and ecological attributes of cheetahs, *Miracinonyx inexpectatus* (late Pliocene-Pleistocene) and the smaller *M. trumani* (latest Pleistocene).

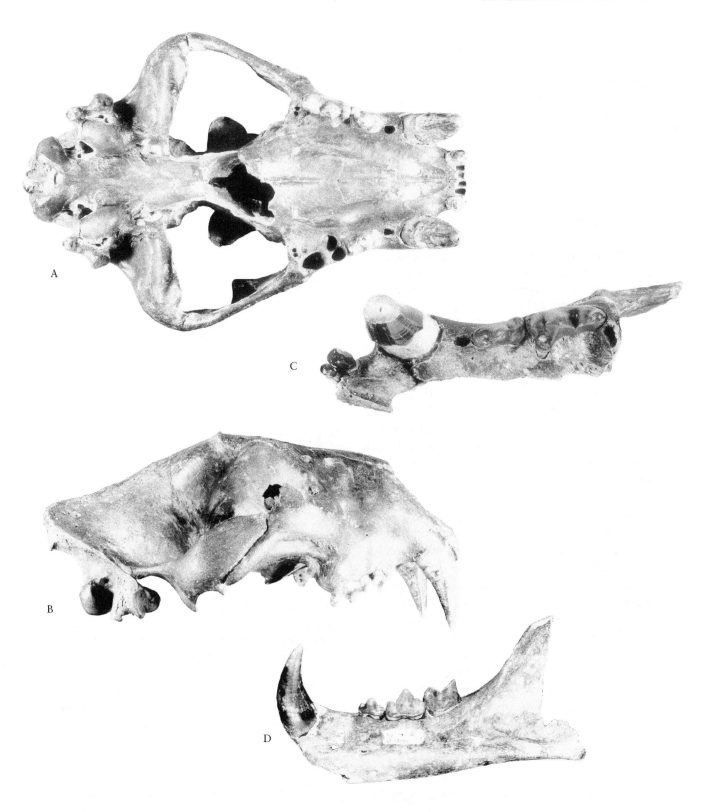

Fig. 11.39. Large Pleistocene conical-toothed cats from Florida. *A*, ventral, and *B*, right lateral views of UF 9076, a skull of the American lion *Panthera atrox* from the Ichetucknee River, Columbia County, late Pleistocene. *C–D*, the jaguar *Panthera onca* from Coleman 2A, Sumter County, middle Pleistocene; *C*, occlusal view of UF 12164, a left maxilla and premaxilla with I2–I4, C, P3–P4. Note single alveoli for small P1 and M1; *D*, lateral view of UF 12145, a left dentary with c, p3–m1. *A–B* about 0.33×; *C* about 0.67×; *D* about 0.5×.

Like Old World *Acinonyx*, these two have long, slender limb bones, small canines, and a reduced P4 protocone (figs. 11.38E, 11.40). The New World cheetahs differ by retaining fully retractile claws. *Miracinonyx* is sometimes regarded only as a subgenus of *Acinonyx*, while others regard the two as distinct but closely related. An alternate hypothesis is that the two are not closely related and that their similarities are the result of convergent evolution (Morgan and Seymour, 1997). *M. inexpectatus* has been definitely identified from five Florida localities (Van Valkenburgh et al., 1990; Berta, 1995; Morgan and Seymour, 1997), with the most specimens produced by the Inglis 1A site in Citrus County. A second Citrus County locality, Lecanto 2A, also produced fossils that likely represent *M. inexpectatus*, but, if so, it is one of the youngest known records of the species (Morgan and Seymour, 1997).

Although it is the lone surviving large cat from the Pleistocene, the panther *Puma concolor* is only known from relatively few fossils in Florida, mostly limb elements. Morgan and Seymour (1997) recently reviewed its fossil record in the state and illustrated the best specimens. *P. concolor* is only known from the latest Pleistocene and only from sites in peninsular Florida (not from the panhandle, although that is probably an artifact). It is represented by few specimens at most localities where it has been found. Fossils recovered show no significant differences with modern individuals of the species still persisting in small numbers in southern Florida.

The last group of felines known from Florida as fossils are the small Neotropical cats, whose living representatives include the ocelot, margay, and jaguarundi. The presence of the ocelot *(Leopardus pardalis)* in Florida is solely based on a well-preserved mandible from Reddick (fig. 11.41A, B). About the size of a bobcat, the specimen is referred to *L. pardalis* because of its short diastema between the c and p3 and the long, broad p4. No other records of this species have come to light since it was first described in the 1960s. An even smaller spotted cat has long been known from the late Pleistocene of Florida and was often considered to represent the modern jaguarundi, *Herpailurus yagouaroundi*. The records of this small cat were summarized by Gillette (1976), who described it as a new species, "*Felis*" *amnicola*. It is now thought to be more closely related to the margay, *Leopardus wiedii*, than the jaguarundi, and one study even placed it as a subspecies of *L. wiedii* (Werdelin, 1985). *L. amnicola* has a rather deep and robust mandible for such a small cat (figs. 11.38C, D, 11.41C, D). The mandible is the best-represented element for this species. It has a tall lower canine and a long p4. All

Fig. 11.40. Posterior view of UF 45353, a left femur of the American cheetah *Miracinonyx inexpectatus* from Inglis 1A, Citrus County, late Pliocene. The great length and slenderness are characteristic of an excellent runner. About 0.5×.

records of *L. amnicola* are very late Pleistocene, including specimens collected from the Aucilla, Ichetucknee, and Waccasassa Rivers and from Rock Springs.

FOSSIL HYENAS OF FLORIDA

The hyenas (family Hyaenidae) are the aeluroid ecological equivalents of canids: large, predaceous, pursuing carnivores with a highly complex social organization and a penchant for scavenging. As in other aeluroids, hyaenids have little or no crushing capabilities remaining in their molars. To compensate, hyaenid middle premolars (especially the P2, P3, p3, and p4) are large and

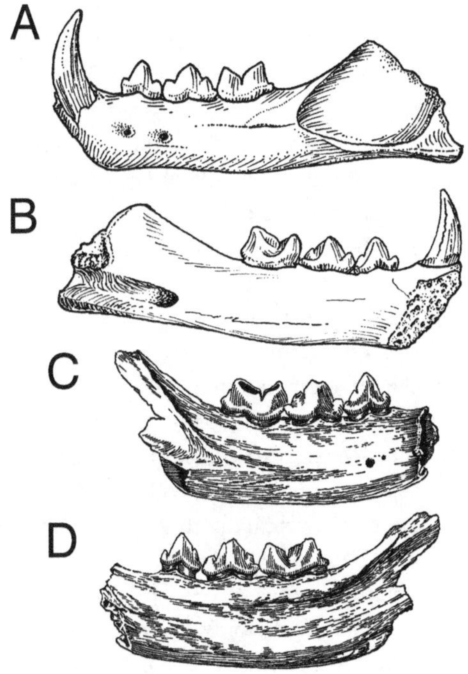

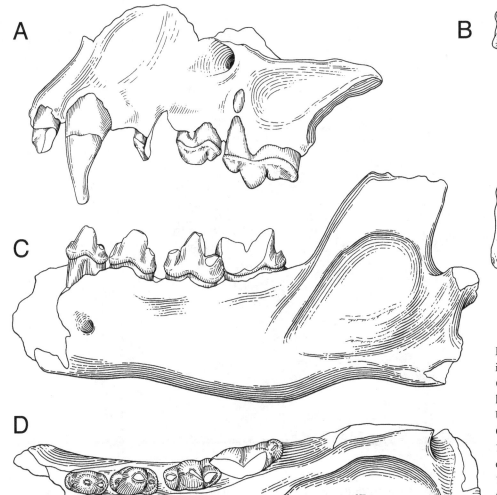

Fig. 11.41. Dentaries of two small cats from the late Pleistocene of Florida. *A,* lateral, and *B,* labial views of UF 3858, left dentary with c, p3–m1 of the ocelot *Leopardus pardalis* from Reddick, Marion County; *C,* lateral, and *D,* labial views of UF 4522, a partial right dentary with p3–m1 of *L. amnicola,* a close relative of the extant margay. From Rock Springs, Orange County. *A–B* about 0.75×; *C–D* about 1×. *A–B* after Ray et al. (1963); *C–D* after Ray (1964); reproduced by permission of the American Society of Mammalogists.

used for crushing. They typically have a high middle cusp that wears bluntly, along with anterior and posterior accessory cusps (fig. 11.42). The carnassial pair is well developed, with only a small protocone on the P4 and a large parastyle. The talonid basin of the m1 is reduced. The skull is broad with a high sagittal crest and massive zygomatic arches. Hyaenids are unique among carnivorans in having longer front legs than hind; nevertheless, they are good runners. The claws are not retractile,

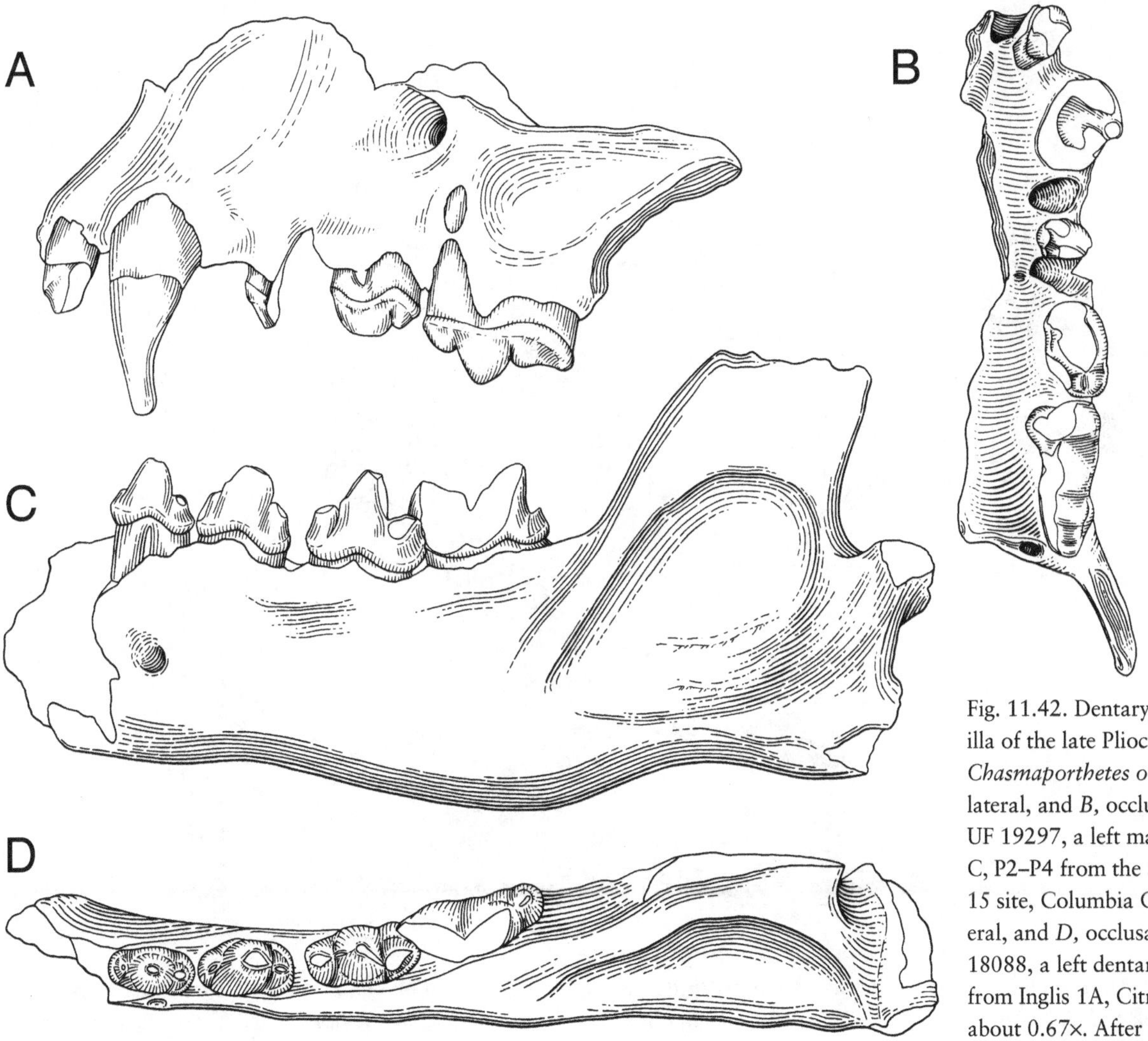

Fig. 11.42. Dentary and maxilla of the late Pliocene hyena *Chasmaporthetes ossifragus. A,* lateral, and *B,* occlusal views of UF 19297, a left maxilla with I3, C, P2–P4 from the Santa Fe River 15 site, Columbia County; *C,* lateral, and *D,* occlusal views of UF 18088, a left dentary with p2–m1 from Inglis 1A, Citrus County. All about 0.67×. After Berta (1981); reproduced by permission of the Society of Vertebrate Paleontology.

and each foot has four digits. The family is primarily Old World, with only three extant genera. Hyaenids were much more diverse and important in the Neogene, with different lineages specializing either for scavenging or pursuit hunting. It is the latter group, the pursuit-hunting hyenas (now a completely extinct radiation), that entered the New World.

When borophagine canids were first discovered late in the nineteenth century, they were initially thought to be true hyenas. This is reflected in the generic name given to one of these dogs, *Hyaenognathus* (meaning "hyena-jaw"; it is now regarded as a junior synonym of *Borophagus*). As more complete material of these animals was recovered, their canid affinities became clear. During the first decades of the twentieth century, it was assumed that no hyenas had ever lived in North America and that their ecological role as bone-cracking scavengers had been taken up by other groups. Thus in 1921, when O. P. Hay described a poorly preserved mandible from

Arizona as a true hyena, no one believed him. Hay's reputation as a taxonomic "splitter" and one who proposed names on very incomplete specimens served him poorly in this situation. It was not until the recovery twenty years later of a complete mandible with well-preserved teeth from Texas that the existence of hyenas in North America was substantiated and Hay's identification confirmed.

The North American hyena belongs to the genus *Chasmaporthetes,* a group of long-legged hyenas with relatively slender teeth (Kurtén and Werdelin, 1988). This widespread genus is also known from the Pliocene and early Pleistocene of Africa, Europe, and China. The Florida remains are questionably referred to *C. ossifragus,* the species Hay originally described from Arizona (Berta, 1981, 1998). They date from the late Pliocene and come from the Santa Fe River, Inglis 1A, and De Soto Shell Pit. They include the only known upper dentition and limb bones of the genus in North America (figs.

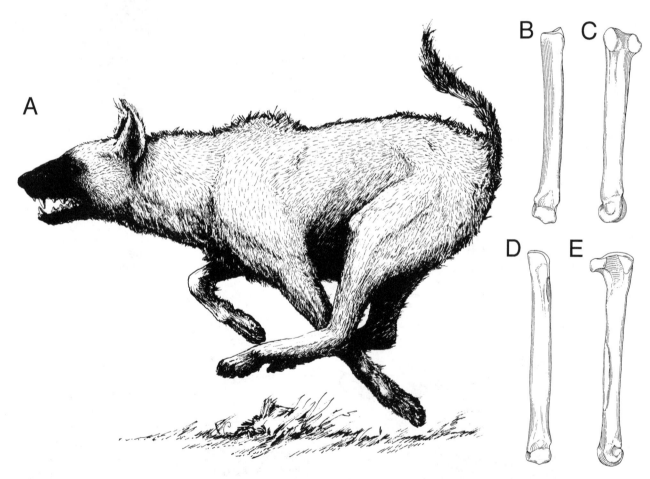

Fig. 11.43. The late Pliocene cursorial hyena *Chasmaporthetes ossifragus. A,* a restoration of how the species might have appeared; *B,* anterior, and *C,* lateral views of UF 27374, a left metatarsal 2; *D,* anterior, and *E,* lateral views of UF 27372, right metatarsal 4. Both metapodials are from Inglis 1A, Citrus County. The slender metapodials of *C. ossifragus* are characteristic of a carnivoran well adapted for running. *B–E* about 0.5×. After Berta (1981); reproduced by permission of the Society of Vertebrate Paleontology.

11.42, 11.43). The limbs are long but also robust, so the species was evidently very muscular. *Chasmaporthetes* is thought to have been a pursuit hunter, similar to the African hunting dog and the gray wolf, not a pure scavenger (Kurtén and Werdelin, 1988). The microscopic pattern of its tooth enamel is not as specialized for bone cracking as in the modern spotted or striped hyenas or the extinct canid *Borophagus* (Ferretti, 1999). It became extinct in North America during the early Pleistocene (Berta, 1998).

REFERENCES

Bader, R. S. 1957. Two Pleistocene mammalian faunas from Alachua County, Florida. Bulletin of the Florida State Museum, 2:55–75.

Baskin, J. A. 1981. *Barbourofelis* (Nimravidae) and *Nimravides* (Felidae), with a description of two new species from the late Miocene of Florida. Journal of Mammalogy, 62:122–39.

———. 1982. Tertiary Procyoninae (Mammalia: Carnivora) of North America. Journal of Vertebrate Paleontology, 2:71–93.

———. 1989. Comments on New World Procyonidae (Mammalia: Carnivora). Journal of Vertebrate Paleontology, 9:110–17.

———. 1998a. Procyonidae and Mustelidae. Pp. 144–73 *in* C. Janis, K. M. Scott, and L. L. Jacobs (eds.), *Evolution of Tertiary Mammals of North America*. Volume 1, *Terrestrial Carnivores, Ungulates, and Ungulatelike Mammals*. Cambridge University Press, Cambridge.

———. 1998b. Evolutionary trends in the late Miocene hyena-like dog *Epicyon* (Carnivora, Canidae). Pp. 191–214 *in* Y. Tomida, L. J. Flynn, and L. L. Jacobs (eds.), *Advances in Vertebrate Paleontology and Geochronology*. National Science Museum Monographs, No. 14, Tokyo.

Berta, A. 1981. The Plio-Pleistocene hyaena *Chasmaporthetes ossifragus* from Florida. Journal of Vertebrate Paleontology, 1:341–56.

———. 1985. The status of *Smilodon* in North and South America. Natural History Museum of Los Angeles County Contributions in Science, 370:1–15.

———. 1987. The sabercat *Smilodon gracilis* from Florida and a discussion of its relationships (Mammalia, Felidae, Smilodontini). Bulletin of the Florida State Museum, 31:1–63.

———. 1988. Quaternary evolution and biogeography of the large South American Canidae (Mammalia: Carnivora). University of California Publications in the Geological Sciences, 132:1–149.

———. 1995. Fossil carnivores from the Leisey Shell Pits, Hillsborough County, Florida. Bulletin of the Florida Museum of Natural History, 37:463–99.

———. 1998. Hyaenidae. Pp. 243–46 *in* C. Janis, K. M. Scott, and L. L. Jacobs (eds.), *Evolution of Tertiary Mammals of North America*. Volume 1, *Terrestrial Carnivores, Ungulates, and Ungulatelike Mammals*. Cambridge University Press, Cambridge

Berta, A., and H. Galiano. 1983. *Megantereon hesperus* from the late Hemphillian of Florida with remarks on the phylogenetic relationships of machairodonts (Mammalia, Felidae, Machairodontinae). Journal of Paleontology, 57:892–99.

———. 1984. A Miocene amphicyonid (Mammalia: Carnivora) from the Bone Valley Formation of Florida. Journal of Vertebrate Paleontology 4:122–25.

Berta, A., and G. S. Morgan. 1985. A new sea otter (Carnivora: Mustelidae) from the late Miocene and early Pliocene (Hemphillian) of North America. Journal of Paleontology, 59:809–19.

Berta, A., C. E. Ray, and A. R. Wyss. 1989. Skeleton of the oldest known pinniped, *Enaliarctos mealsi*. Science, 244:60–62.

Berta, A., and A. R. Wyss. 1994. Pinniped phylogeny. Pp. 33–56 *in* A. Berta and T. A. Deméré (eds.), *Contributions in Marine Mammal Paleontology Honoring Frank C. Whitmore, Jr.* San Diego Natural History Society, San Diego.

Bryant, H. N. 1988. Delayed eruption of the deciduous upper canine in the sabertoothed carnivore *Barbourofelis lovei* (Carnivora, Nimravidae). Journal of Vertebrate Paleontology, 8:295–306.

Deméré, T. A. 1994. The family Odobenidae: a phylogenetic analysis of fossil and living taxa. Pp. 99–123 *in* A. Berta and T. A. Deméré (eds.), *Contributions in Marine Mammal Paleontology Honoring Frank C. Whitmore, Jr.* San Diego Natural History Society, San Diego.

Emslie, S. D., and G. S. Morgan. 1995. Taphonomy of a late Pleistocene carnivore den, Dade County, Florida. Pp. 65–83 *in* D. W. Steadman and J. I. Mead (eds.), *Late Quaternary Environments and Deep History*. Mammoth Site of Hot Springs, Scientific Papers, vol. 3, Hot Springs, S.Dak.

Ferretti, M. P. 1999. Tooth enamel structure in the hyaenid *Chasmaporthetes lunensis lunensis* from the late Pliocene of Italy, with implications for feeding behavior. Journal of Vertebrate Paleontology, 19:767–70.

Fox, R. C., and G. P. Youzwyshyn. 1994. New primitive carnivorans (Mammalia) from the Paleocene of western Canada, and their bearing on relationships of the order. Journal of Vertebrate Paleontology, 14:382–404.

Frailey, C. D. 1978. An early Miocene (Arikareean) fauna from northcentral Florida (the SB-1A local fauna). Occasional Papers of the Museum of Natural History, University of Kansas, 75:1–20.

———. 1979. The large mammals of the Buda local fauna (Arikareean: Alachua County, Florida). Bulletin of the Florida State Museum, 24:123–73.

Gidley, J. W., and C. L. Gazin. 1938. The Pleistocene vertebrate fauna from Cumberland Cave Maryland. United States National Museum Bulletin, 171:1–99.

Gillette, D. D. 1976. A new species of small cat from the late Quaternary of southeastern United States. Journal of Mammalogy, 57:664–76.

———. 1979. The largest dire wolf: late Pleistocene of northern Florida. Florida Scientist, 42:17–21.

Gittleman, J. L. 1989. *Carnivore Behavior, Ecology and Evolution*. Cornell University Press, Ithaca, N.Y., 620 p.

———. 1996. *Carnivore Behavior, Ecology and Evolution*. Volume 2. Cornell University Press, Ithaca, N.Y., 640 p.

Harrison, J. A. 1981. A review of the extinct wolverine, *Plesiogulo* (Carnivora: Mustelidae), from North America. Smithsonian Contributions to Paleobiology, 46:1–27.

Hunt, R. M. 1987. Evolution of the aeluroid Carnivora: significance of auditory structures in the nimravid cat *Dinictis*. American Museum Novitates, 2886:1–74.

———. 1998. Ursidae and Amphicyonidae. Pp. 174–227 *in* C. Janis, K. M. Scott, and L. L. Jacobs (eds.), *Evolution of Tertiary Mammals of North America*. Volume 1, *Terrestrial Carnivores, Ungulates, and Ungulatelike Mammals*. Cambridge University Press, Cambridge

Hunt, R. M., and R. H. Tedford. 1993. Phylogenetic relationships within the aeluroid Carnivora and implications of their temporal and geographic distribution. Pp. 53–73 *in* F. S. Szalay, M. J. Novacek, and M. C. McKenna (eds.), *Mammal Phylogeny: Placentals*. Springer-Verlag, New York.

Janis, C., J. A. Baskin, A. Berta, J. J. Flynn, G. F. Gunnell, R. M.

Hunt, L. D. Martin, and K. Munthe. 1998. Carnivorous mammals. Pp. 73–90 in C. Janis, K. M. Scott, and L. L. Jacobs (eds.), *Evolution of Tertiary Mammals of North America*. Volume 1, *Terrestrial Carnivores, Ungulates, and Ungulatelike Mammals*. Cambridge University Press, Cambridge.

Joeckel, R. M. 1990. A functional interpretation of the masticatory system and paleoecology of entelodonts. Paleobiology, 16:459–82.

Kurtén, B. 1965. The Pleistocene Felidae of Florida. Bulletin of the Florida State Museum, 9:215–73.

———. 1966–67. Pleistocene bears of North America. 1. Genus *Tremarctos*, spectacled bears. 2. Genus *Arctodus*, short-faced bears. Acta Zoologica Fennica, 115:1–120; 117:1–60.

Kurtén, B., and L. Werdelin. 1988. A review of the genus *Chasmaporthetes* Hay, 1921 (Carnivora, Hyaenidae). Journal of Vertebrate Paleontology, 8:46–66.

Lambert, W. D. 1997. The osteology and paleoecology of the giant otter *Enhydritherium terraenovae*. Journal of Vertebrate Paleontology, 17:738–49.

Macdonald, D. 1992. *The Velvet Claw: A Natural History of the Carnivores*. BBC Books, London, 256 p.

MacFadden, B. J., and H. Galiano. 1981. Late Hemphillian cat (Mammalia, Felidae) from the Bone Valley Formation of central Florida. Journal of Paleontology, 55:218–26.

Martin, L. D. 1989. Fossil history of the terrestrial Carnivora. Pp. 536–68 in J. L. Gittleman (ed.), *Carnivore Behavior, Ecology, and Evolution*. Cornell University Press, Ithaca, N.Y.

———. 1998. Nimravidae and Felidae. Pp. 228–42 in in C. Janis, K. M. Scott, and L. L. Jacobs (eds.), *Evolution of Tertiary Mammals of North America*. Volume 1, *Terrestrial Carnivores, Ungulates, and Ungulatelike Mammals*. Cambridge University Press, Cambridge.

Martin, L. D., J. P. Babiarz, V. L. Naples, and J. Hearst. 2000. Three ways to be a saber-toothed cat. Naturwissenschaften, 87:41–44.

Martin, R. A. 1978. A new late Pleistocene *Conepatus* and associated vertebrate fauna from Florida. Journal of Paleontology, 52:1079–85.

Matthew, W. D. 1902. A skull of *Dinocyon* from the Miocene of Texas. Bulletin of the American Museum of Natural History, 16:129–36.

———. 1910. Phylogeny of the Felidae. Bulletin of the American Museum of Natural History, 28:289–316.

McKenna, M. C., and S. K. Bell. 1997. *Classification of Mammals above the Species Level*. Columbia University Press, New York, 631 p.

Morgan, G. S. 1989. Miocene vertebrate faunas from the Suwannee River Basin of north Florida and south Georgia. Pp. 26–53 in G. S. Morgan (ed.), *Miocene Paleontology and Stratigraphy of the Suwannee River Basin of North Florida and South Georgia*. Southeastern Geological Society Guidebook No. 30.

Morgan, G. S., and K. L. Seymour. 1997. Fossil history of the panther *(Puma concolor)* and the cheetah-like cat *(Miracinonyx inexpectatus)* in Florida. Bulletin of the Florida Museum of Natural History, 40:177–219.

Munthe, K. 1998. Canidae. Pp. 124–43 in C. Janis, K. M. Scott, and L. L. Jacobs (eds.), *Evolution of Tertiary Mammals of North America*. Volume 1, *Terrestrial Carnivores, Ungulates, and Ungulatelike Mammals*. Cambridge University Press, Cambridge.

Olsen, S. J. 1956. The Caninae of the Thomas Farm Miocene. Breviora, 66:1–12.

———. 1957a. Leptartines from the Florida Miocene (Carnivora, Mustelidae). American Museum Novitates, 1861:1–7.

———. 1957b. The lower dentition of *Mephititaxus ancipidens* from the Florida Miocene. Journal of Mammalogy, 38:452.

———. 1958a. The skull of *Leparctus ancipidens* from the Florida

Miocene. Florida Geological Survey Special Publications, 2(2):1–11.

———. 1958b. Some problematic carnivores from the Florida Miocene. Journal of Paleontology, 32:595–602.

———. 1958c. The fossil carnivore *Amphicyon intermedius* from the Thomas Farm Miocene. Part I, skull and dentition. Bulletin of the Museum of Comparative Zoology, 118:157–72.

———. 1960. The fossil carnivore *Amphicyon longiramus* from the Thomas Farm Miocene. Part II, postcranial skeleton. Bulletin of the Museum of Comparative Zoology, 123:1–44.

Patton, T. H. 1969. An Oligocene land vertebrate fauna from Florida. Journal of Paleontology, 43:543–46.

Peterson, O. A. 1906. The Miocene beds of western Nebraska and eastern Wyoming and their vertebrate faunae. Annals of the Carnegie Museum, 4:21–72.

Ray, C. E. 1960. *Trichecodon huxleyi* (Mammalia: Odobenidae) in the Pleistocene of southeastern United States. Bulletin of the Museum of Comparative Zoology, 122:129–42.

———. 1964. The jaguarundi in the Quaternary of Florida. Journal of Mammalogy, 45:330–32.

Ray, C. E., E. Anderson, and S. D. Webb. 1981. The Blancan carnivore *Trigonictis* (Mammalia: Mustelidae) in the eastern United States. Brimleyana, 5:1–36.

Ray, C. E., S. J. Olsen, and J. H. Gut. 1963. Three mammals new to the Pleistocene fauna of Florida, and a reconsideration of five earlier records. Journal of Mammalogy, 44:373–95.

Schultz, C. B., M. R. Schultz, and L. D. Martin. 1970. A new tribe of saber-toothed cats (Barbourofelini) from the Pliocene of North America. Bulletin of the University of Nebraska State Museum, 9:1–31.

Seymour, K. L. 1993. Size change in North American Quaternary jaguars. Pp. 343–72 in R. A. Martin and A. D. Barnosky (eds.), *Morphological Change in Quaternary Mammals of North America*. Cambridge University Press, Cambridge.

Simpson, G. G. 1929. Pleistocene mammalian fauna of the Seminole Field, Pinellas County, Florida. Bulletin of the American Museum of Natural History, 56:561–99.

Tedford, R. H., and D. Frailey. 1976. Review of some Carnivora (Mammalia) from the Thomas Farm local fauna (Hemingfordian: Gilchrist County, Florida). American Museum Novitates, 2610:1–9.

Tedford, R. H., and Qiu Z. 1996. A new canid genus from the Pliocene of Yushe, Shanxi Province. Vertebrata PalAsiatica, 34:27–40.

Tedford, R. H., B. E. Taylor, and X. Wang. 1995. Phylogeny of the Caninae (Carnivora: Canidae): the living taxa. American Museum Novitates, 3146:1–37.

Turner, A. 1987. *Megantereon cultridens* (Cuvier) (Mammalia, Felidae, Machairodontinae) from Plio-Pleistocene deposits in Africa and Eurasia, with comments on dispersal and the possibility of a New World origin. Journal of Paleontology, 61:1256–68.

Van Valkenburgh, B., F. Grady, and B. Kurtén. 1990. The Plio-Pleistocene cheetah-like cat *Miracinonyx inexpectatus* of North America. Journal of Vertebrate Paleontology, 10:434–54.

Vaughn, T. A. 1972. *Mammalogy*. W. B. Saunders, Philadelphia, 463 p.

Wang, X. 1994. Phylogenetic systematics of the Hesperocyoninae (Carnivora: Canidae). Bulletin of the American Museum of Natural History, 221:1–207.

Wang, X., and R. H. Tedford. 1994. Basicranial anatomy and phylogeny of primitive canids and closely related miacids (Carnivora: Mammalia). American Museum Novitates, 3092:1–34.

Wang, X., R. H. Tedford, and B. E. Taylor. 1999. Phylogenetic systematics of the Borophaginae (Carnivora: Canidae). Bulletin of the American Museum of Natural History, 243:1–391.

Webb, S. D. 1969a. The Burge and Minnechaduza Clarendonian

mammalian faunas of north-central Nebraska. University of California Publications in Geological Sciences, 78:1–191.

———. 1969b. The Pliocene Canidae of Florida. Bulletin of the Florida State Museum, 14:273–308.

Webb, S. D., B. J. MacFadden, and J. A. Baskin. 1981. Geology and paleontology of the Love Bone Bed from the late Miocene of Florida. American Journal of Science, 281:513–44.

Werdelin, L. 1985. Small Pleistocene felines of North America. Journal of Vertebrate Paleontology, 5:194–210.

Wolff, R. G. 1977. Function and phylogenetic significance of cranial anatomy of an early bear (*Indarctos) from Pliocene sediments of Florida. Carnivore, 1(3):1–12.*

Mammalia 4

Rodents and Lagomorphs

INTRODUCTION

The order Rodentia is extremely diverse and successful. It includes about fifty living families and many familiar mammals such as squirrels, beavers, rats, mice, hamsters, guinea pigs, and porcupines. Most rodents are relatively small in size and herbivorous, although many will eat the occasional invertebrate if they can catch it. Except for large rodents like beavers and capybaras, their fossils are rarely recovered without specifically screen-washing matrix and looking for them. Before the widespread use of this strategy (about 1950), rodent evolution was poorly known and based mostly on the study of living species. Now many paleontologists specialize in their study, and the group is extremely important in studies of evolution and biochronology. Because of their small size, rodent fossils are usually observed using a low-power microscope, as the detailed features of their minute teeth cannot otherwise be distinguished. The order Lagomorpha, in contrast to the Rodentia, contains only two living families, the Leporidae, the hares and rabbits, and the Ochotonidae, the pikas. Lagomorphs too are relatively small herbivores, although on average larger in size than most rodents. The fossil record of leporids is surprisingly poor in Florida, with few or no specimens known for many ages. This could represent either a bias in the record, an accurate portrayal of their relative past abundance, or some combination of the two.

During the nineteenth century, a close evolutionary relationship between lagomorphs and rodents was frequently proposed. Lagomorphs were often classified in the Rodentia as the suborder Duplicidentata. Throughout much of the twentieth century, resemblance between the two was regarded as convergent, and they were generally considered as being only very distantly related, if at all. Recently, paleontologists have begun to reexamine the possibility of a close lagomorph-rodent relationship.

Early Paleogene mammals from China include the oldest known lagomorphs, the most primitive rodent, and some intermediate taxa (Korth, 1994). The rodents, lagomorphs, and the elephant "shrew" are classified together in the grandorder Anagalida (McKenna and Bell, 1997).

RODENT ANATOMY

The primitive rodent dentition consists of one pair of upper and lower incisors, one lower and two upper premolars (the p4, P3, and P4), and all three molars. The upper and lower incisors are hypselodont, curved into semicircles, and worn into chisel-shaped occlusal surfaces (fig. 12.1). Enamel is present on only the anterior side of the incisors. The incisors are primarily used for gnawing, but some burrowing species also use them to dig by literally chewing through soil. Canine teeth are completely absent. There is a long diastema between the incisor and the cheek teeth. Some or all of the premolars are lost in certain lineages, but most rodents retain all three molars. If present, the premolars are molariform in all but the most primitive species. The earliest rodents' cheek teeth were brachyodont and bunodont. Such teeth are retained in a few living groups, most notably squirrels. Other rodent lineages have evolved lophodont and/or hypsodont cheek teeth, and hypselodont cheek teeth have probably evolved more times among rodents than in all other mammalian groups combined. The arrangement of tooth cusps and lophs varies between groups (fig. 12.2) and is used for classification. In many cases, convergence in dental anatomy is a major problem in unraveling evolutionary relationships.

The principal direction of movement of the jaw during chewing is back and forth (anterior-posterior). The development and arrangement of the muscles that move the jaw vary between major groups of rodents and have long been used to classify them. These in turn effect the morphology of the mandible and skull, so these muscle

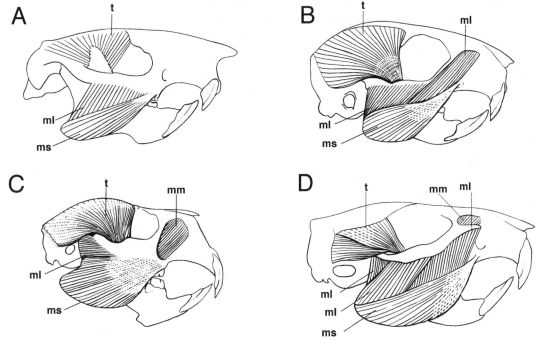

Fig. 12.1. Patterns of jaw muscles that form the major basis of rodent classification. Abbreviations: *ml*, lateral masseter muscle; *mm*, medial masseter (the deepest portion of this muscle); *ms*, superficial masseter; *t*, temporalis muscle. *A*, the primitive or protrogomorphous condition where all of the masseter muscles originate on the lateral side of the zygomatic arch. Among living rodents, this pattern is found only in the mountain beaver *Aplodontia* from the northwestern United States (not a relative of the true beaver). *B*, the sciuromorphous condition where parts of the lateral masseter originate on the anterior portion of the zygomatic arch, just in front of the orbit. *C*, the hystricomorphous condition in which the medial masseter originates on the rostrum and passes through an enlarged infraorbital foramen to insert on the dentary. *D*, the myomorphous condition in which the lateral masseter partially originates on the anterior part of the zygomatic arch and the deep medial masseter originates on the rostrum and passes through a narrow, slitlike infraorbital foramen to reach the jaw. Each major group of rodents shows one of these four types of muscle patterns. Figures from *Mammalogy* by Terry A. Vaughn, copyright © 1972 by Saunders College Publishing, reproduced by permission of the publisher.

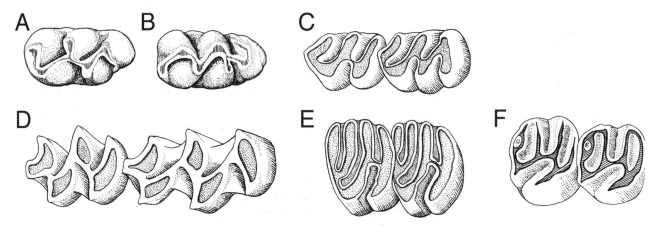

Fig. 12.2. Examples of some of the various types of molar teeth found in rodents, at greatly enlarged scales. *A*, right M1, and *B*, left m1 of the murid *Reithrodontomys*. Other common sigmodontine rodents having similar brachyodont teeth are *Peromyscus*, *Oryzomys*, and *Copemys*. *C*, right M1–M2 of the lophodont murid *Sigmodon*. *D*, right M1–M2 of the arvicoline *Microtus* showing the prismatic enamel triangles characteristic of voles and lemmings. *E*, right M1–M2 of the beaver *Castor*. *F*, right M1–M2 of the porcupine *Erethizon*. *E* and *F* are also of the lophodont type. In *C–E*, the unshaded portion of the occlusal surface is enamel, the stippled areas exposed dentine. In *F*, the dentine is represented by cross hatching. Figures from *Mammalogy* by Terry A. Vaughn, copyright © 1972 by Saunders College Publishing, reproduced by permission of the publisher.

patterns can be determined on fossils. Two basic types of rodent mandibles are recognized: sciurognathous and hystricognathous. Sciurognathous mandibles are primitive and represent rodents with the angle of the jaw (the posterior-ventral corner) aligned in the same general plane as the incisor (fig. 12.3A). Sciurognathous mandibles are found in squirrels, beavers, rats, and mice. The derived condition, a hystricognathous mandible, is found primarily in rodents native to Africa and South America. In these species the angle of the jaw is shifted laterally relative to the incisor (fig. 12.3B). This allows a longer pterygoideus muscle and strengthens jaw movement. This derived state is believed to have evolved once and is used to unite these rodents in the suborder Hystricognatha.

On the skull, differing arrangements of jaw muscles (principally the masseter muscle) manifest themselves on the zygomatic arch and in the size and shape of the infraorbital foramen. In the most primitive condition, called protrogomorphous, all of the masseter muscle originates on the lateral surface of the zygomatic arch (fig. 12.1A). The infraorbital foramen is very small. This is also the primitive mammalian state. Only a single protrogomorphous rodent is still living, the "mountain beaver" (*Aplodontia*) of the northwestern United States. Three other states, all derived, are believed to have evolved independently several times within the Rodentia. In the sciuromorphous condition, some of the masseter muscle has moved forward to originate on the anterior

portion of the zygomatic arch (fig. 12.1B). The infraorbital foramen is not enlarged. Sciuromorphous rodents include the squirrels, beavers, kangaroo rats, and pocket gophers. In the hystricomorphous condition, deeper portions of the masseter pass through a very enlarged infraorbital foramen to originate on the side of maxilla (fig. 12.1C). Hystricomorphous rodents include most, but not all, of the hystricognathous forms, as well as a few sciurognathous groups. Porcupines, capybaras, and guinea pigs are examples of hystricomorphous rodents. The most derived condition is called myomorphous and is basically a combination of the two previously described states. A thin portion of the masseter passes through the infraorbital foramen (which is slitlike or keyhole shaped), while most of the muscle originates on the anterior margin of the zygomatic arch (fig. 12.1D). It is thought that this condition evolved from the hystricognathous state. Myomorphous rodents include the murids (rats, mice, voles) and some dormice.

The postcranial skeleton of rodents is rather generalized, and is little studied by paleontologists. Exceptions to this are highly modified diggers (for example, geomyids), jumpers (for example, dipodids), and gliders (for example, flying squirrels). The digits bear claws and the fibula is often fused with the tibia.

FOSSIL RODENTS OF FLORIDA

As a whole, the fossil rodents of Florida are relatively poorly studied. Only a handful of papers have described pre-Pleistocene taxa. This is not the result of a lack of material (as recent collecting efforts have demonstrated an abundance of Miocene specimens), but more from a preference of most resident paleontologists toward larger mammals. There seems little doubt that the number of described pre-Pleistocene species will more than double in the future. The list in chapter 3 hints at where some of this diversity will come from. Eleven rodent families are presently recognized in the fossil record of Florida; the most diverse groups are the Sciuridae (at least ten genera and seven named species), Sigmodontinae (at least seven genera and fourteen recognized species), and the Arvicolinae (five genera and thirteen recognized species).

The Mylagaulidae is an extinct family of moderate-size burrowing rodents distantly related to the living *Aplodontia*. They are known from the late Oligocene to early Pliocene of North America and the Miocene of Asia. Their ecology was presumably like that of modern pocket gophers (geomyids): living underground in tunnels and eating roots and tubers. The mylagaulid skull

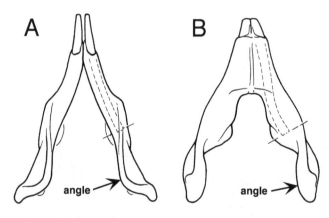

Fig. 12.3. The two basic types of mandibles found in rodents, with the primary difference in the formation of the angle. *A*, a sciurognathous mandible with the angle located directly in line with the incisor. This type of mandible is found in squirrels, beavers, rats, and geomyids. *B*, a hystricognathous mandible with the angle located lateral to the incisor, a result of a longer pterygoid muscle. This type is found in porcupines, capybaras, and nutrias. Not to scale; after Jacobs (1984); reproduced by permission of the Paleontological Society.

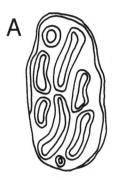 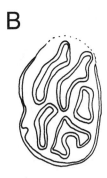

Fig. 12.4. Occlusal pattern of teeth in the late Miocene rodent *Mylagaulus elassos* from the Love site, Alachua County. *A*, UF 24192, right p4. *B*, UF 24196, left P4. About 5×. After Baskin (1980); reproduced by permission of the American Midland Naturalist.

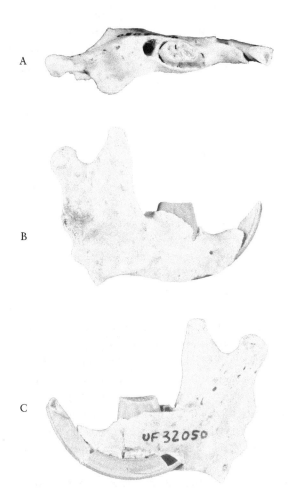

Fig. 12.5. *A*, occlusal, *B*, lateral, and *C*, medial views of UF 32050, a right dentary with i1 and p4 of *Mylagaulus elassos* from the Love site, Alachua County, late Miocene. About 1.5×.

was flat, broad, and protrogomorphous. Some had a pair of horns on the nasal bones—the only known rodents to have evolved horns. However, none of the species from Florida had horns. Mylagaulids have a very unusual dental morphology. The major cheek teeth are the hypsodont P4 and p4, which are oval and bear numerous enamel lakes (figs. 12.4, 12.5). The diameter of the P4 and p4 increase with wear, so as they erupt the molars are forced out. In very old individuals the fourth premolars are the only cheek teeth present. The molars are small, buttonlike teeth, with little functional significance.

The oldest known mylagaulids from Florida are late early Miocene, from the Midway and Brooks Sink localities. They have been referred to the common early Miocene genus *Mesogaulus,* but have yet to be thoroughly studied (Morgan and Pratt, 1988). *Mesogaulus* is smaller than the later *Mylagaulus,* with simpler teeth. In western North America, *Mylagaulus* displays a number of trends during the Miocene, including increased size, increased hypsodonty, and increased dental complexity (in terms of the number of lakes in each fourth premolar). These sorts of trends are common in many lineages of herbivorous mammals. Interestingly, the pattern of evolution observed in Florida is the reverse of several of these trends (Baskin, 1980). Here, *Mylagaulus* got smaller through time and evolved simpler cheek teeth. One hypothesis to explain these reversals is that Florida populations adapted to environmental instability by decreasing body size to increase fecundity and the rate of maturation. Simplification of complex structures often occurs as a secondary effect in this type of evolutionary process. Florida is thought to have a less stable environment (from the point of view of a burrowing rodent) than

other parts of North America because its low-lying terrain is adversely affected by even small changes in sea level, which in turn affects the level of water tables.

The Sciuridae, or squirrels, are among the most familiar rodents, as several varieties have adapted well to living in urban and suburban areas. Their cheek teeth are brachyodont and bunodont, with variable, usually slight, development of crests or lophs (best developed in the flying squirrels, the subfamily Petauristinae). The P3, if present, is small and vestigial. The molariform P4 and p4 are only slightly smaller than the M1 and m1. Tree squirrels have specialized ankle joints that can be rotated 180°. This allows them to climb down trees head first, with the hind claws catching into the bark. This adaptation was perfected by the Oligocene, as seen in the genus *Protosciurus,* and since then the squirrel's postcranial skeleton has not changed notably. Squirrels fall into three ecological categories: the arboreal tree squirrels such as *Sciurus* (gray and fox squirrels) and *Tamiasciurus* (red squirrels); the flying squirrels, which glide using a membrane of skin that they stretch between their front and hind legs (for example, the small, local *Glaucomys* and the much larger Eurasian *Petaurista*); and the ground squirrels, which often live in burrows. All three types are found as fossils in Florida (figs. 12.6, 12.7). Today no ground squirrels live in the state except for rare records of chipmunks from the far northwestern panhandle. Perhaps the most interesting of the fossil squirrels from Florida are the large flying squirrels *Petauristodon* and *Miopetaurista* from the Miocene and Pliocene, respectively (Pratt and Morgan, 1989). They certainly indicate well-forested habitats. The gray squirrel *Sciurus carolinensis* is first recorded at the middle Pleistocene Coleman 2A site and is common in the late Pleistocene, as is the native small flying squirrel *Glaucomys volans*.

The Castoridae are today represented only by a single genus, *Castor*, the beaver. The family comprises medium- to very large-sized rodents that feed on bark, leaves, and other plant matter. Not all fossil beavers were aquatic and felled trees like the modern genus. Some lived in deep, corkscrew-shaped burrows that are preserved in parts of northern Great Plains. Castorids have four upper and lower cheek teeth that decrease in size in a posterior direction. The teeth are lophodont and often hypsodont (fig. 12.8), and one lineage evolved hypselodont teeth. The small beaver *Eucastor* was present in the middle and late Miocene of Florida (fig. 12.8A). Two castorid lineages are well represented from the Pleistocene, the gigantic, hypselodont *Castoroides*, and the modern genus *Castor*, which has rooted teeth. Fossils of *Castor* are particularly common in northern Florida rivers, such as the Santa Fe, Ichetucknee, Chipola, St. Marks, and Waccasassa (fig. 12.8D). The bear-sized *Castoroides* is a widely distributed Pleistocene taxon (fig.

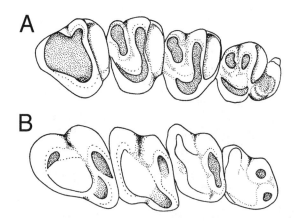

Fig. 12.7. Cheek teeth of the ground squirrel *Spermophilus* sp. from Haile 14A, Alachua County, late Pleistocene. *A*, occlusal view of UF 12366, right P3–M3 (anterior to right); *B*, occlusal view of UF 12364, left p4–m3. About 8×. After Martin (1974b); reproduced by permission of the University Press of Florida.

12.9). Its large incisors with their peculiar grooved enamel are easily recognized, even from fragments. *Castoroides* is thought to have fed on swamp vegetation rather than trees. Two species of giant beaver are recognized, the early Pleistocene *Castoroides leiseyorum* and the late Pleistocene *Castoroides ohioensis*. The two are differentiated mostly on the basis of characters on the base of the skull. The teeth, jaws, and limbs are similar.

The Eutypomyidae is a relatively small family of extinct rodents most closely related to beavers. They ranged from the Oligocene to the Miocene and were primarily North American in their distribution. They had a sciuromorphous skull. They differed from the castorids in having an elongated snout and more complex enamel pattern on the cheek teeth. The youngest genus referred to the family, *Anchitheriomys*, is known from a single tooth from the Bone Valley region (fig. 12.8B). Elsewhere, this genus is known from the middle Miocene of the Great Plains, Asia, and Europe.

Geomyids are moderate-sized burrowing rodents commonly called pocket gophers, or, in some places in Florida, salamanders. This is thought to be a corruption of the term *sandy mounders*, a reference to the mounds of sand found on the surface above the burrows of one of these rodents. Where soil conditions are to its liking, the eastern pocket gopher, *Geomys pinetis*, is quite common in northern and central Florida. Geomyid adaptations for burrowing include small eyes, strong forelimbs with large claws, and a stocky build. Two genera are recognized in the Pleistocene, *Geomys* and *Thomomys*. The latter is now restricted to western North America. The extinct species *Thomomys orientalis* was named from a

Fig. 12.6. Lateral view of TRO 401, right dentary with c, p4 of *Protosciurus* sp. from the Live Oak site, Suwannee County, early Miocene. About 2×. After Frailey (1978), courtesy Natural History Museum, The University of Kansas.

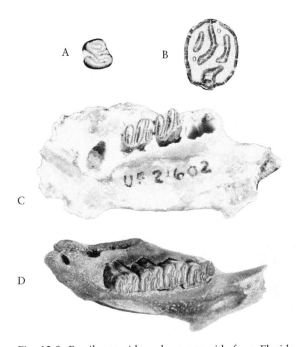

cave deposit in Citrus County but has subsequently been found at several other central Florida localities (Wilkins, 1985). Both genera have massive mandibles with simple, hypselodont teeth (fig. 12.10). The p4 and P4 are large and figure-eight shaped. The molars are simple ovals. *Geomys* differs from *Thomomys* in lacking the anterior enamel band on its lower molars and having two distinctive anterior grooves in the upper incisor (fig. 12.10). In *Thomomys*, a slight medial groove may or may not be visible on the upper incisor. The two genera have been found together at several fossil sites, although dissimilarities in soil preferences probably mean that the two did not actually live side by side. Wilkins (1984) demonstrated several trends in the evolution of *Geomys* in Florida during the past two million years, including increased depth of the retromolar fossa in the mandible (for a greater biting force) and increased cheek tooth width (for more efficient chewing).

Fig. 12.8. Fossil castorids and eutypomyids from Florida. *A,* occlusal view of UF 91017, a right M3 of *Eucastor* sp. from the Phosphoria Mine, Polk County, middle Miocene; *B,* occlusal view of UF 27503, a P4 of *Anchitheriomys* sp. from the Tiger Bay Mine, Polk County, middle Miocene; *C,* occlusal view of UF 21602, a left dentary with m1–m2 of *Dipoides* sp., Palmetto Mine, Polk County, early Pliocene; *D,* occlusal view of UF 14063, a left dentary with p4–m3 of *Castor canadensis* from the Ichetucknee River, Columbia County, late Pleistocene. *A* and *C* about 2×, *B* about 1.6×, *D* about 1.5×.

Fig. 12.9. *A,* lateral, *B,* medial, and *C,* occlusal views of UF 115965, a left mandible with i1, p4–m3 of the giant beaver *Castoroides leiseyorum* from Leisey Shell Pit, Hillsborough County, early Pleistocene. About 0.4×.

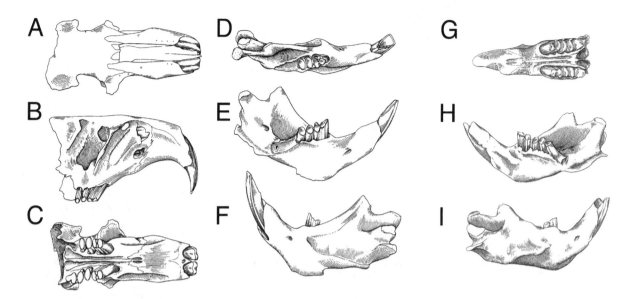

Fig. 12.10. Fossil pocket gophers (Geomyidae) from Florida. *A–F, Geomys propinetis* from Inglis 1A, Citrus County, late Pliocene; *G–I, Thomomys orientalis. A,* dorsal, *B,* right lateral, and *C,* ventral views of UF 46001, partial skull; *D,* occlusal, *E,* lingual, and *F,* lateral views of UF 46021, left dentary with i1, p4–m2; *G,* occlusal view of UF 46567, partial skull from Williston 3B, Levy County, early Pleistocene; *H,* lingual, and *I,* lateral views of UF 46572, right dentary with i3, p4–m3 from Rock Springs, Orange County, late Pleistocene. About 1.5×; after Wilkins (1984, 1985). *A–F* reproduced by permission of the Society of Vertebrate Paleontology; *G–I* reproduced by permission of the Biological Society of Washington.

Heteromyid rodents are today especially well known from the arid American southwest, although those of the subfamily Heteromyinae (the spiny pocket mice) live in the tropical forests of Central America. However, during the Oligocene and early Miocene, heteromyids and their extinct relatives, the eomyids, were more widespread and diverse. It appears that they occupied many of the small rodent niches that today are filled by rats and mice. Heteromyids are the most common rodents of the early Miocene in Florida; their small, cuspidate teeth outnumber all others. Only two species have been described, *Proheteromys floridanus* and the larger *Proheteromys magnus* (Black, 1963), and a third tentatively identified as a species from California, *Perognathus minutus*. A number of undescribed species have been recovered from a number of sites (Morgan and Pratt, 1988; Morgan, 1989; Bryant, 1991). They are thought to have eaten mostly seeds and were quadrupedal (not bipedal as the modern kangaroo rats). During the middle Miocene, heteromyids were replaced by the sigmodontines, a transition that is rather well documented in the Florida fossil record, and are subsequently absent.

Eomyids are an extinct family related to heteromyids and geomyids. Together the three make up the infraorder Geomorpha. Eomyids are first known from the Eocene and dispersed to the Old World in the Oligocene. An undescribed large eomyid is known from the late Oligocene and very early Miocene of Florida, with a smaller genus present later in the early Miocene. Representative eomyid teeth are shown in figure 12.11.

The Muridae, as currently composed, is the largest mammalian family in terms of numbers of species and genera. There are about 260 living murid genera and 1,135 species. Vast numbers of fossil taxa have also been described. In the past, murid rodents were divided between two major families, the Muridae for Old World rats and mice and the Cricetidae for Old World hamsters, New World rats and mice, and the Holarctic voles and lemmings. Carleton and Musser (1984) combined these two families (along with some smaller groups) into a single, enormous family. In their subfamilial partitioning of this group, the Old World hamsters, the Cricetinae in the strict sense, and the New World rats and mice (the Sigmodontinae) are separated and not thought to be especially closely related. Thus many of the North American rodents that mammalogists and paleontologists have for years been calling cricetids must be given another name, sigmodontines. Almost all of the scientific literature on Florida rodents published before 1984 (and much afterward out of force of habit) will generally refer to these species as "cricetids." Korth (1994) used "cricetid" in its traditional sense (and referred North American

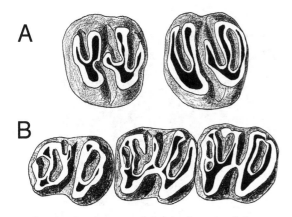

Fig. 12.11. Occlusal views of the cheek teeth of the extinct eomyid rodent *Pseudotheridomys hesperus* from the early Miocene of Colorado. Eomyids are now known from the late Oligocene and early Miocene of Florida but remain poorly studied. *A*, right P4 and M1; *B*, left p4–m2. About 25×. After Engesser (1979); reproduced by permission of the Carnegie Museum of Natural History, Pittsburgh.

species to the family), but most modern classifications, such as McKenna and Bell (1997), deny this and follow Carleton and Musser (1984).

Features found in murids are the myomorph pattern of the skull (fig. 12.1D) and loss of the P4 and p4. The M1 and m1 are usually the largest teeth. Primitively, the teeth are brachyodont and bunodont, but this pattern is altered in many lineages. Most murids are relatively small; that is, mouse to rat size. The biogeography and evolutionary history of this huge family has been and continues to be a field of intense study.

The oldest murid rodent is from Asia and is middle Eocene in age. The family dispersed widely and is known by the late Eocene in North America, early Oligocene in Europe, and early Miocene in Africa. Members of two relatively successful, early radiations of the Muridae are known from Florida: the subfamilies Paracricetodontinae and Cricetodontinae. Both are primarily Eurasian groups, but the paracricetodontines dispersed to North America in the early Oligocene and the cricetodontines in the middle Miocene. Several paracricetodontines are known from the late Oligocene to early Miocene in Florida (about 24 to 20 Ma), including the common genus *Leidymys* (Morgan and Pratt, 1988; Morgan, 1989; fig. 12.12A). Their fossils are much rarer than contemporaneous heteromyid and eomyid rodents.

Cricetodontines replaced the older paracricetodontines in the Miocene about 16 Ma and proved to be more successful in North America than their older relatives. Many species have been described, most in the genus *Copemys,* which is now known from several middle to

late Miocene sites in Florida (fig. 12.12B), including the Willacoochee Creek fauna of the panhandle (Bryant, 1991) and the Love site (Baskin, 1986).

Sigmodontine rodents are presently the most successful murids in the New World. They originated in the late Miocene, probably from a cricetodontine species, and radiated rapidly in the Pliocene and Pleistocene. Eleven tribes are recognized in this subfamily (McKenna and Bell, 1997), many from Central and South America, where the group had its greatest success. But many of the native rats and mice of the temperate and subtropical regions of the United States also belong to this subfamily. Two sigmodontine genera, *Oryzomys* and *Sigmodon,* are members of tribes that are primarily South and Central American but are common as fossils in Florida and are still living here. The modern species are *Oryzomys palustris,* the rice rat, and *Sigmodon hispidus,* the cotton rat. *Sigmodon* lives in grassy areas. The teeth are fairly low crowned but lophodont (fig. 12.12D, E). A number of late Pliocene and Pleistocene species have been described that vary in size and details of the dentitions (Martin, 1979). Their rapid evolution makes them very useful in biochronology.

The remaining sigmodontine rodents found in Florida are principally North American rather than Neotropical in their distribution. Most important of these are *Peromyscus* and *Neotoma*. *Peromyscus* includes many of the small mice native to the United States and Mexico. Two species have been identified from Florida fossils: the cotton mouse, *Peromyscus gossypinus,* and the very small beach mouse, *Peromyscus polionotus.* Species of *Neotoma* are called woodrats in eastern North America and packrats in the West. They are large for sigmodontines with high-crowned, rooted, lophodont teeth. The living species *Neotoma floridana* is a common late Pleistocene fossil. The Florida mouse, *Podomys floridanus,* is endemic to the state, where it survives in scrub habitats. It resembles a large *Peromyscus* and was formerly placed in that genus. Its fossil record is also limited to Florida.

Arvicoline rodents are characterized by molars that in occlusal view resemble a series of triangles (fig. 12.13). The triangles usually alternate along the right and left sides of the tooth (fig. 12.13A), but in some, such as *Atopomys* (fig. 12.13H), the triangles are directly across from one another. Primitive arvicolines have rooted cheek teeth, but many lineages evolved hypselodont molars. Lemmings are principally thought of as arctic animals, but an extinct bog lemming (*Synaptomys australis*) is known from Florida and other parts of the South (Olsen, 1958; Morgan and White, 1995). Lemmings have different-sized alternating triangles on the

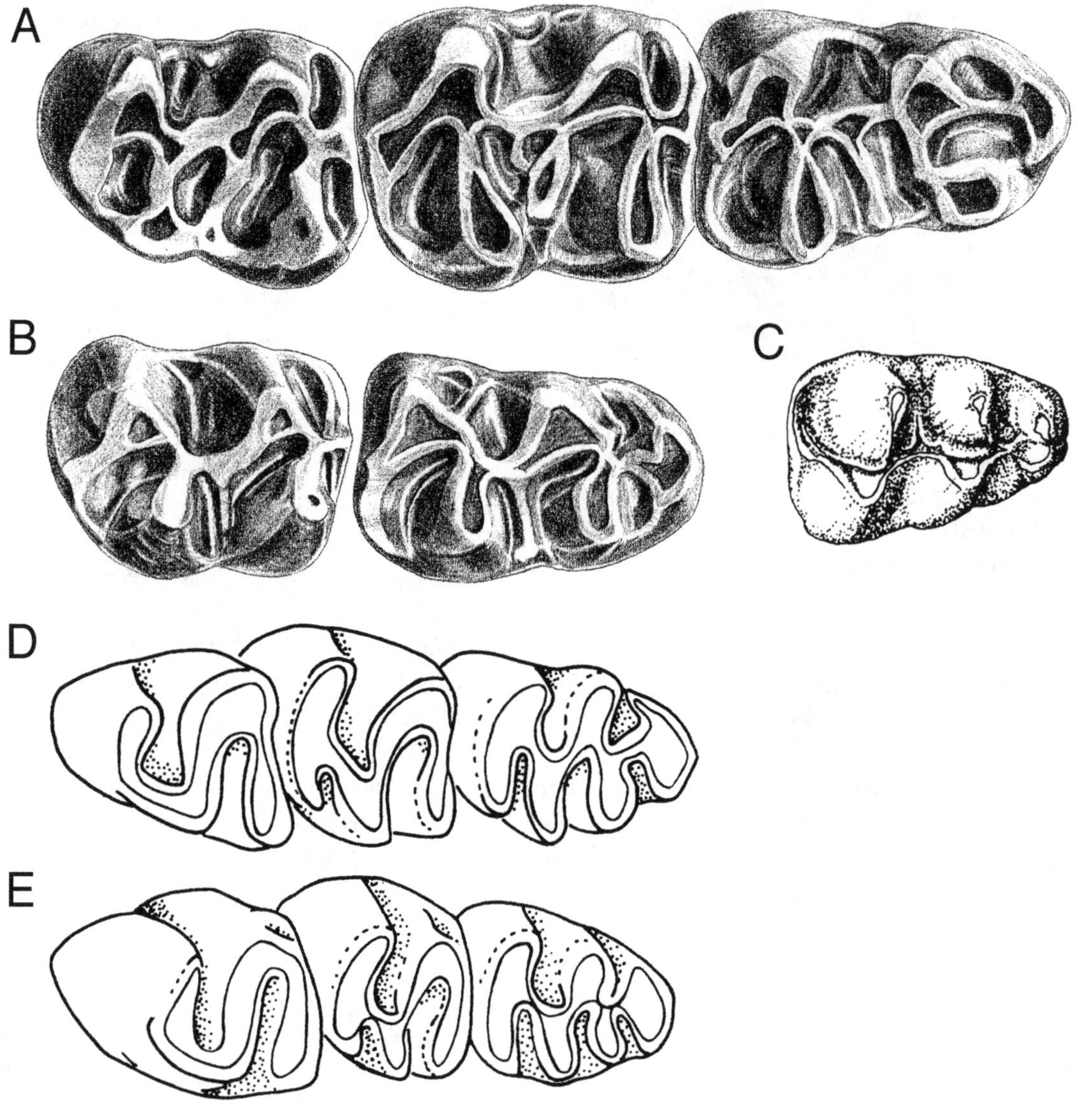

Fig. 12.12. Occlusal views of some fossil and Recent murid teeth. *A*, left m1–m3 (anterior to right) of *Leidymys alicae* from the Oligocene of Montana. This genus is now known from the very early Miocene of Florida. *B*, left m1–m2 of *Copemys barstowensis* from the middle Miocene of California. Species of *Copemys* or related forms are widely thought to have given rise to later sigmodontine rodents. *C*, UF 61335, right m1 of *Abelmoschomys simpsoni* from the Love site, Alachua County, late Miocene. *D*, UF 11700, left m1–m3 of *Sigmodon bakeri* from Coleman 2A, Levy County, middle Pleistocene. *E*, UF(M) 549, left m1–m3 of *Sigmodon hispidus* from Brevard County, Recent. *A–C* about 25×; *D–E* about 15×. *A–B* after Engesser (1979); reproduced by permission of the Carnegie Museum of Natural History, Pittsburgh. *C* after Baskin (1986); reproduced by permission of the Department of Geology and Geophysics, University of Wyoming, Laramie. *D–E* after Martin (1974a); reproduced by permission of the University Press of Florida.

labial and lingual sides of their molars (fig. 12.13E) that distinguish them from other arvicolines in which the triangles are of approximately the same size. *S. australis* is a large species for the genus and is related to the living southern bog lemming *Synaptomys cooperi,* which today ranges as far south as the mountains of Tennessee and North Carolina. *S. australis* was first described from a cave site near Lecanto in Citrus County. Its presence

suggests that the summers in Florida during the Pleistocene were not as hot as they are today.

Voles (tribe Arvicolini) are among the most abundant mammals in cool temperate to arctic regions in North America, Asia, and Europe. Their evolutionary history is complex and, as so often is the case with diverse groups, is subject to considerable debate among experts. The classification used here largely follows R. Martin (1987,

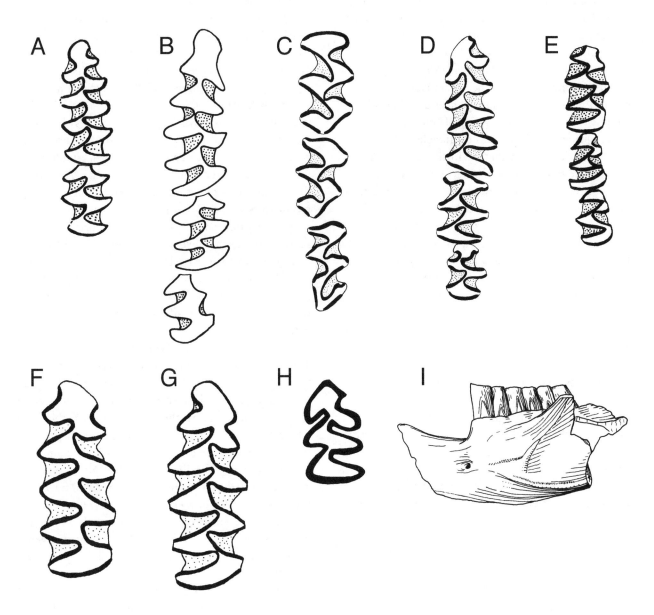

Fig. 12.13. Examples of fossil arvicolid (or microtine) teeth, in occlusal view (except *I*). *A*, UF 3586, left m1–m2 of *Microtus pennsylvanicus* from Arredondo 1A, Alachua County, late Pleistocene; *B*, UF 11685, right m1–m3 of *Microtus (Pitymys) aratai* from Coleman 2A, Sumter County, middle Pleistocene; *C*, right m1–m3, and *D*, right M1–M3 of *Ondatra annectens* from the early Pleistocene of Texas (UMMP 46174); *E*, AMNH 23440, right m1–m3 of *Synaptomys australis*, Sabertooth Cave, Citrus County, late Pleistocene; *F*, AMNH 98972B, right m1 of *Neofiber leonardi* from McLeod Limerock Mine, Levy County, middle Pleistocene; *G*, UF 22093, right m1 of *Neofiber alleni* from the Ichetucknee River, Columbia County, late Pleistocene; *H*, CM 20040, left m1 of *Atopomys salvelinus* from Trout Entrance Locality, West Virginia, middle Pleistocene; *I*, lateral view of UF 69235, left dentary of *A. salvelinus* from Haile 16A, Alachua County, early Pleistocene. *A–B, I* about 10×; *C–E* about 6×; *F–G* about 10×; *H* about 15×. *A–B* after Martin (1968, 1995); *A* reproduced by permission of the American Society of Mammalogists; *B* reproduced by permission of the Society of Vertebrate Paleontology. *C–D* after Hibbard and Dalquest (1968); reproduced by permission of the University of Michigan Museum of Paleontology, Ann Arbor. *E* after Simpson (1928); reproduced by permission of the American Museum of Natural History, New York. *F–G* after Frazier (1977); reproduced by permission of the American Society of Mammalogists. *H* after Zakrzewski (1975); reproduced by permission of the Carnegie Museum of Natural History, Pittsburgh. *I* after Winkler and Grady (1990); reproduced by permission of the Society of Vertebrate Paleontology.

1995). He combined the North American voles into one genus, *Microtus,* within which he recognized a number of subgenera. Two of these are living in Florida, each represented by a single species: *Microtus (Pitymys) pinetorum,* the pine vole, and *Microtus (Microtus) pennsylvanicus,* the meadow vole. The former is the more common of the two and has a much wider distribution. It is our lone representative of a group of voles characterized by having an m1 with three closed triangles (fig. 12.13B). Voles are well represented in the fossil record of Florida from the Pleistocene, with four recognized species. The oldest of these, *M. (Pedomys) australis* is known from the early Pleistocene at Haile 16A and Leisey Shell Pit. *M. (Microtus) pennsylvanicus* is a member of a more advanced grade, with five closed triangles on their m1s (fig. 12.13A). It is only known from the late Pleistocene in Florida, from both the northern peninsula and panhandle. Long thought to have gone extinct in the state, a relict population was discovered living in salt marshes around the Gulf Coast town of Cedar Key. The meadow vole is widespread today through the northern United States.

The largest arvicolines are the semiaquatic muskrats. Two distinct evolutionary lineages are recognized, the tribes Ondatrini and Neofiberini. There is a single living species in each, *Ondatra zibethicus* and *Neofiber alleni,* respectively. The muskrat *Ondatra* does not currently live in Florida, nor does its modern range overlap with the smaller but ecologically similar round-tailed muskrat *Neofiber.* The latter is today restricted to Florida and

Fig. 12.14. *A,* right lateral, and *B,* ventral views of UF 127677, a nearly complete skull of the round-tailed muskrat *Neofiber alleni* from Surprise Cave, Alachua County, late Pleistocene. Because they are so fragile, small rodent skulls are rarely this complete. About 1×.

southern Georgia (figs. 12.13, 12.14). However, both ranged much wider during the Pleistocene, and they overlapped extensively at that time. Both are known by fossils from Florida. *Ondatra* is distinguished by its larger size and rooted cheek teeth in older adults.

The common group of Old World rats and mice, the subfamily Murinae, is not native to North America. The house mouse, *Mus musculus,* and two species of rat, *Rattus norvegicus* and *R. rattus,* were introduced by Europeans onto the continent and are now well established in urban, farming, and other disturbed areas. The molars of these murines are distinguished from similar-sized sigmodontines such as *Peromyscus* by having three rows of cusps rather than two. Any remains of murines found in Florida are less than five hundred years old and are not considered to be fossils. Their recognition at archaeological sites is of importance, as they must postdate the arrival of the Spanish. *Mus* is distinguished from *Rattus* on the basis of much smaller size and a relatively longer m1.

Caviomorph rodents evolved in South America. There they first appear in the fossil record in the early Oligocene, when that continent was isolated from all other landmasses. Their ancestors must have dispersed to South America across the open ocean, probably from Africa. Caviomorphs are both hystricognathous and hystricomorphous, and most are large by rodent standards. Two caviomorph families appeared in North America about 2.5 Ma: the porcupine (family Erethizontidae) and the capybara (Hydrochoeridae). Both have long fossil records in South America, so it is clear that they represent dispersals, presumably following the Pliocene emergence of the Panamanian Isthmus. Both families are known from Florida.

The porcupine genus *Erethizon* is first recognized in the late Pliocene in Florida (Frazier, 1982; Hulbert, 1997). Through the mid-1960s, only fragments of porcupine material had been found, such that some authorities doubted its presence in the state. Since then, definitive mandibles and maxillae have been recovered, much of it from the late Pliocene and early Pleistocene, when the genus seems to have been more common. Three species are recognized, the late Pliocene *Erethizon poyeri* (fig. 12.15); the distinctively small, latest Pliocene *Erethizon kleini;* and the living species *Erethizon dorsatum.* The large, heavily rooted, lophodont cheek teeth are easily distinguished from those of all other rodents. The P4 and p4 are the largest cheek teeth, although those of *E. poyeri* are relatively small. *E. dorsatum* first appeared in Florida in the late early Pleistocene and was an inhabitant of the state until the end of the epoch (Morgan and

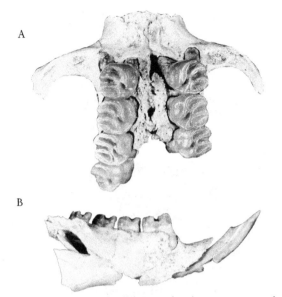

Above: Fig. 12.15. A late Pliocene fossil porcupine, *Erethizon poyeri,* from Haile 7C, Alachua County. *A,* occlusal view of UF 121740, palate with right P4–M3. *B,* lateral view of UF 121740, right dentary with i1, p4–m3. This specimen represents the oldest known porcupine in eastern North America. *A* about 1.5×, *B* about 1×.

White, 1995). Its modern range is well to the north of Florida.

Capybaras have long been recognized from fossils in Florida. They are the largest living rodents and are semi-aquatic. Their hypselodont cheek teeth are formed by a number of obliquely oriented plates held together by cement (fig. 12.16), somewhat reminiscent structurally of mammoth teeth, but on a much smaller scale. The third molars are very elongated and have many more plates

Below: Fig. 12.16. Comparison between the modern South American capybara *Hydrochaeris hydrochoerus (A–D)* and the Pleistocene capybara *H. holmesi (E–H)* from Saber-tooth Cave, Citrus County. *A,* lateral view of skull and dentary; *B,* ventral view of skull; *C,* occlusal view of right P4–M3; *D,* occlusal view of left p4–m3; *E,* occlusal view of AMNH 23434, left M3; *F,* occlusal view of AMNH 23434, left m1 (restored); *G,* anterior view of lower incisor; *H,* cross section of lower incisor at indicated point in *G.* The anterior groove is a characteristic of capybaras. *A–B* about 0.33×; *C–H* about 1×. *A–D* modified after Ellerman (1940). *E–H* after Simpson (1928); reproduced by permission of the American Museum of Natural History, New York.

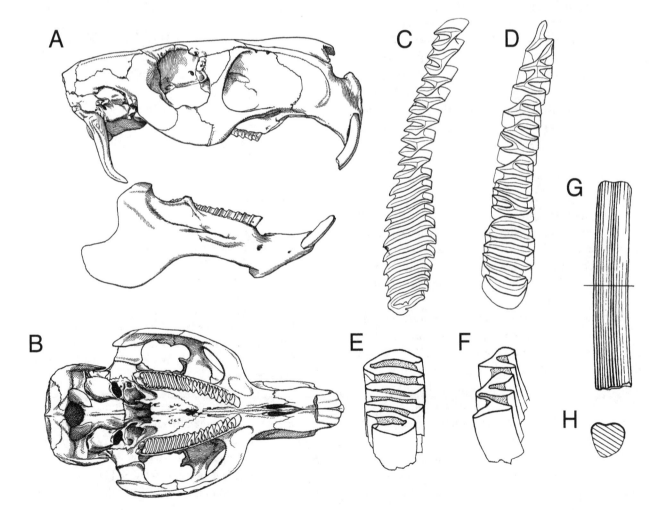

than the other teeth. The number of plates per tooth is important for their systematics, but most fossils are fragmentary and only identifiable to the family level. Two genera are usually recognized: the extant *Hydrochaeris* and the extinct giant form *Neochoerus*. The larger species of *Neochoerus*, *N. pinckneyi*, was about the size of a tapir and vies with *Castoroides* for the title of the largest rodent to have ever lived in North America. Remains of both genera are common in Pleistocene deposits associated with freshwater environments. A smaller species, *Neochoerus dichroplax*, is known from the late Pliocene of Arizona and Florida and is the oldest North American capybara.

MORPHOLOGY AND CLASSIFICATION OF LAGOMORPHS

Lagomorphs have three upper and two lower premolars. The third and fourth premolars are molariform in all but the most primitive genera. As in rodents there is a large pair of hypselodont incisors for gnawing, the canine is absent, and there is a long diastema (fig. 12.17A, B). However, a smaller, second pair of upper incisors are located posterior to the first pair. The large incisors bear single anterior grooves. The P3-M2 are nearly identical and isolated teeth cannot be identified to their original position. The same is true for the p4-m2. The third molars are reduced but rarely absent. The p3 is the most important tooth for lagomorph identification and systematics. Its pattern of cement-filled infoldings of enamel (called reentrant angles; fig. 12.18A) is usually diagnostic to the generic or even species level.

The skull is lightly built, and the preorbital region is fenestrated (figs. 12.17A, B, 12.19A). It fossilizes poorly. In the Leporidae (rabbits and hares), the only family known from Florida, the limb elements are long, slender, and hollow (fig. 12.19D, E). Of all mammals, leporid limb elements are the ones most often confused with those of birds. They are morphologically distinct from those of all other mammals.

Lagomorphs and rodents diverged from their common ancestor in the Paleogene, probably in China. The division between the two major lagomorph families, the Ochotonidae (pikas) and Leporidae occurred soon afterward. Ochotonids are principally an Old World group with a good fossil record in Europe. They entered North America in the middle Miocene but have never been recorded from the southeastern United States.

The Leporidae is a more cosmopolitan group. Their first North American record is in the Eocene, and they have been relatively common ever since. Their broad

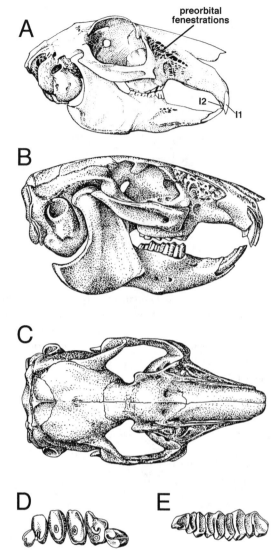

Fig. 12.17. Skulls, dentaries, and dentition of leporid lagomorphs. *A,* lateral view of the skull and dentary of the modern jackrabbit *Lepus;* note the openings (fenestrations) in the skull anterior to the orbit; *B,* lateral view of the skull and dentary of the Oligocene leporid *Palaeolagus; C,* dorsal view of the skull of *Palaeolagus; D,* right P2–M3 of *Palaeolagus; E,* right p3–m3 of *Palaeolagus.* Note the presence of a small I2 behind the larger I1 in *A* and *B,* a feature that distinguishes lagomorphs from rodents. *A* from *Mammalogy* by Terry A. Vaughn, copyright © 1972 by Saunders College Publishing, reproduced by permission of the publisher. *B–E* from *Vertebrate Paleontology and Evolution* by Carroll © 1988 by W. H. Freeman and Company. Used with permission.

ecological preferences and stereotyped morphology have apparently prevented them from becoming more diverse taxonomically. Three leporid subfamilies are usually recognized: the Paleogene Palaeolaginae, the primarily Neogene Archaeolaginae, and the living Leporinae. General trends observed in the evolution of the leporid fam-

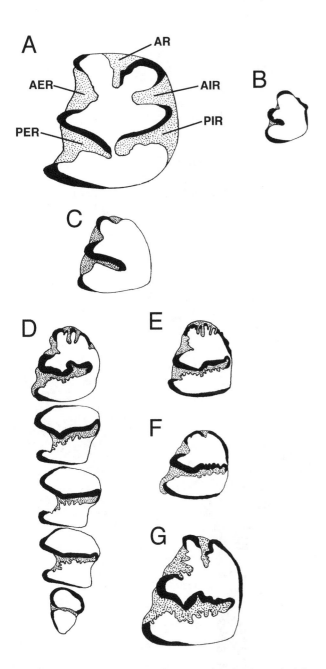

Fig. 12.18. Occlusal views of lower cheek teeth of fossil and modern leporids. All specimens are shown as from the left side; right teeth are shown reversed (mirror image) to allow easier comparison. *A*, left p3 showing names of reentrant angles used to identify species: *AER*, anterior external reentrant; *AIR*, anterior internal reentrant; *AR*, anterior reentrant; *PER*, posterior external reentrant; *PIR*, posterior internal reentrant. *B*, UF 23992, right p3 (reversed) of *Hypolagus*, probably *H. tedfordi*, from Nichols Mine, Polk County, middle Miocene. This species is characterized by small size and a shallow PER. *C*, UNSM 91154, left p3 of *Hypolagus ringoldensis* from the early Pliocene of Nebraska. *H. ringoldensis* is known from the contemporary Palmetto Fauna of the Bone Valley District. *D*, UF 51052, left p3–m3 of *Sylvilagus webbi* from Inglis 1A, Citrus County, late Pliocene. Note the multiple AR's on the p3. *E*, UF 95636, right p3 (reversed) of the marsh rabbit *Sylvilagus palustris* from Oldsmar Pit 1, Pinellas County, middle Pleistocene. *F*, UF(M) 22156, left p3 of *Sylvilagus floridanus* a modern specimen from Okaloosa County. *S. floridanus*, the eastern cottontail, is very common in Pleistocene sites in Florida. *G*, left p3 of a modern specimen of the jackrabbit *Lepus alleni*. Jackrabbits were present in Florida from the late Pliocene to the middle Pleistocene. Black is enamel, stippled areas are cement; white is exposed dentine. All about 6×. After White (1987, 1991a, 1991b); reproduced by permission of the Society of Vertebrate Paleontology.

ily include increased hypsodonty in the cheek teeth followed by attainment of hypselodonty (all archaeolagines and leporines are hypselodont), molarization of all premolars except the P2, primitive cuspidate tooth patterns replaced by simple transversely oriented lophs, and elongation of limb elements.

Fossil Lagomorphs of Florida

As previously noted, fossil lagomorphs are surprisingly rare in Florida, especially so in pre-Pleistocene deposits. However, members of all three leporid subfamilies are now recognized from the state. Palaeolagines are the most poorly known, and specifically diagnostic material

is wanting. A single upper molariform tooth referable to *Palaeolagus* was recovered from the White Springs fauna in north Florida (Morgan, 1989). Several Miocene records of the archaeolagine *Hypolagus* are now known. *Hypolagus* was common and widespread during the middle Miocene through late Pliocene in North America. The p3 of *Hypolagus* is distinguished from those of leporines by lack of internal reentrant angles, and the anterior reentrant angle is usually absent or shallow (fig. 12.18B, C). At least twelve species of *Hypolagus* are recognized in North America. Identifiable remains (that is, p3s) are rare, but at least two species have been identified from the Bone Valley region of Florida (White, 1987). *H. tedfordi* is a small middle Miocene species with a shallow posterior external reentrant angle on the p3 (fig. 12.18B). A much larger species, *H. ringoldensis,* is recognized from the early Pliocene. The p3 of *H. ringoldensis* has deep anterior and posterior external reentrant angles and, in about half the individuals, an anterior reentrant angle (fig. 12.18C).

Leporine rabbits are first recognized from Florida in the early Pliocene (White, 1991a). The oldest is a dentary of the extinct genus *Nekrolagus* from the Bone Valley region (fig. 12.19B). The extant genera *Sylvilagus* and *Lepus* are traditionally thought to have evolved from *Nekrolagus.* Fossils close in size and morphology to the widespread living species *Sylvilagus floridanus,* the eastern cottontail (figs. 12.18F, 12.19A), first occur in the late Pliocene. Fossils of this species are also relatively common at Pleistocene localities in Florida. The larger marsh rabbit, *Sylvilagus palustris,* does not appear until the early Pleistocene and is somewhat less common. Other than greater size, *S. palustris* differs from *S. floridanus* in having a more complex (often double or triple) anterior reentrant angle on the p3 (fig. 12.18E). *Sylvilagus webbi,* an extinct species close to the marsh rabbit is present in the late Pliocene and early Pleistocene (White, 1991b; fig. 12.18D). A supposed dwarf form of marsh rabbit, *Sylvilagus palustrellus* was described from Melbourne (very late Pleistocene). Referred specimens from the type locality and Vero Beach need to be critically compared with other species of *Sylvilagus* to substantiate the validity of this species.

Lepus, the jackrabbit, is presently not native to Florida, although an introduced population is thriving among the runways of the Miami airport (or at least did so at one time in the past). Fossil remains of two species are known from very late Pliocene and early Pleistocene sites in the state (figs. 12.18G, 12.19C–E). They were first referred to the living species *Lepus alleni,* the antelope jackrabbit, but recent work suggests that most belong to a different, probably extinct species. A second form from Leisey Shell Pit is smaller than the other and close to the white-tailed jackrabbit, *Lepus townsendi,* an extant species from the Great Plains (Morgan and White, 1995). *Lepus* is much larger than most species of *Sylvilagus* and has a relatively longer diastema. Ecologically, its presence indicates a dry climate and open grasslands.

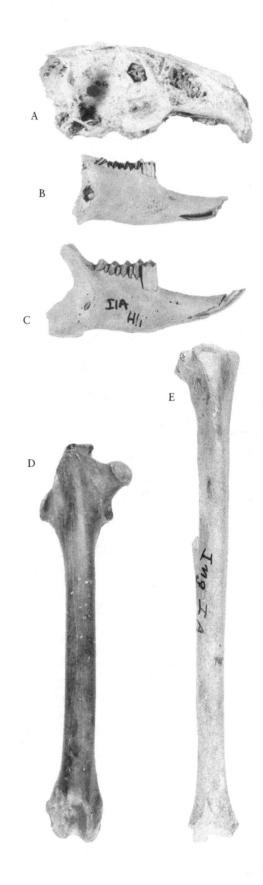

Fig. 12.19. Fossil leporids from Florida. *A,* right lateral view of UF 127676, a partial skull of the eastern cottontail *Sylvilagus floridanus* from Surprise Cave, Alachua County, late Pleistocene; *B,* lateral view of UF 12343, a right dentary with p3–m3 of *Nekrolagus* sp. from the Palmetto Mine, Polk County, early Pliocene; *C,* lateral view of UF 51054, a right dentary with i1, p3–m3 of the jackrabbit *Lepus* sp. from Inglis 1A, Citrus County, late Pliocene; *D,* anterior view of a right femur, and *E,* anterior view of a right tibia, both UF specimens of *Lepus* sp. from Inglis 1A. All about 1×.

REFERENCES

Baskin, J. A. 1980. Evolutionary reversal in *Mylagaulus* (Mammalia, Rodentia) from the late Miocene of Florida. The American Midland Naturalist, 104:155–62.

——. 1986. The late Miocene radiation of Neotropical sigmodontine rodents in North America. Pp. 287–303 *in* K. M. Flanagan and J. A. Lillegraven (eds.), *Vertebrates, Phylogeny, and Philosophy*. University of Wyoming Contributions to Geology, Special Paper 3.

Black, C. C. 1963. Miocene rodents from the Thomas Farm local fauna, Florida. Bulletin of the Museum of Comparative Zoology, 128:483–501.

Bryant, J. D. 1991. New early Barstovian (middle Miocene) vertebrates from the upper Torreya Formation, eastern Florida panhandle. Journal of Vertebrate Paleontology, 11:472–89.

Carleton, M. D., and G. G. Musser. 1984. Muroid rodents. Pp. 289–379 *in* S. Anderson and J. K. Jones (eds.), *Orders and Families of Mammals of Recent Mammals of the World*. J. Wiley & Sons, New York.

Carroll, R. L. 1988. *Vertebrate Paleontology and Evolution*. W. H. Freeman, New York, 698 p.

Ellerman, J. R. 1940. *The Families and Genera of Living Rodents*. Volume 1, *Rodents other than Muridae*. British Museum (Natural History), London, 689 p.

Engesser, B. 1979. Relationships of some insectivores and rodents from the Miocene of North America and Europe. Bulletin of the Carnegie Museum of Natural History, 14:1–68.

Frazier, M. K. 1977. New records of *Neofiber leonardi* (Rodentia: Cricetidae) and the paleoecology of the genus. Journal of Mammalogy, 58:368–73.

——. 1982. A revision of the fossil Erethizontidae of North America. Bulletin of the Florida State Museum, 27:1–76.

Hibbard, C. W., and W. W. Dalquest. 1966. Fossils from the Seymour Formation of Knox and Baylor Counties, Texas, and their bearing on the late Kansan climate of that region. Contributions from the Museum of Paleontology, University of Michigan, 21:1–66.

Hulbert, R. C. 1997. A new late Pliocene porcupine (Rodentia: Erethizontidae) from Florida. Journal of Vertebrate Paleontology, 17:623–26.

Jacobs, L. L. 1984. Rodents. Pp. 155–66 *in* P. D. Gingerich and C. E. Badgley (eds.), *Mammals: Notes for a Short Course*. University of Tennessee Department of Geological Sciences Studies in Geology, volume 8.

Korth, W. A. 1994. *The Tertiary Record of Rodents in North America*. Plenum Press, New York, 319 p.

Martin, R. A. 1968. Late Pleistocene distribution of *Microtus pennsylvanicus*. Journal of Mammalogy, 49:265–71.

——. 1974a. Fossil mammals from the Coleman IIA fauna, Sumter County, Pp. 35–99 *in* S. D. Webb (ed.), *Pleistocene Mammals of Florida*. University Presses of Florida, Gainesville.

——. 1974b. Fossil vertebrates from the Haile XIVA fauna, Alachua County. Pp. 100–113 *in* S. D. Webb (ed.), *Pleistocene Mammals of Florida*. University Presses of Florida, Gainesville.

——. 1979. Fossil history of the rodent genus *Sigmodon*. Evolutionary Monographs, 2:1–36.

——. 1987. Notes on the classification and evolution of some North American fossil *Microtus* (Mammalia; Rodentia). Journal of Vertebrate Paleontology, 7:270–83.

——. 1995. A new middle Pleistocene species of *Microtus (Pedomys)* from the southern United States, with comments on the taxonomy and early evolution of *Pedomys* and *Pitymys* in North America. Journal of Vertebrate Paleontology, 15:171–86.

McKenna, M. C., and S. K. Bell. 1997. *Classification of Mammals above the Species Level*. Columbia University Press, New York, 631 p.

Morgan, G. S. 1989. Miocene vertebrate faunas from the Suwannee River Basin of north Florida and south Georgia. Pp. 26–53 *in* G. S. Morgan (ed.), *Miocene Paleontology and Stratigraphy of the Suwannee River Basin of North Florida and South Georgia*. Southeastern Geological Society Guidebook No. 30.

Morgan, G. S., and A. E. Pratt. 1988. An early Miocene (late Hemingfordian) vertebrate fauna from Brooks Sink, Bradford County, Florida. Pp. 53–69 *in* F. L. Pirkle and J. G. Reynolds (eds.), *Heavy Mineral Mining in Northeast Florida and an examination of the Hawthorne Formation and post-Hawthorne clastic sediments*. Southeastern Geological Society Guidebook No. 29.

Morgan, G. S., and J. A. White. 1995. Small mammals (Insectivora, Lagomorpha, and Rodentia) from the early Pleistocene (Irvingtonian) Leisey Shell Pit local fauna, Hillsborough County, Florida. Bulletin of the Florida Museum of Natural History, 37:397–461.

Olsen, S. J. 1958. The bog lemming from the Pleistocene of Florida. Journal of Mammalogy, 39:537–40.

Pratt, A. E., and G. S. Morgan. 1989. New Sciuridae (Mammalia: Rodentia) from the early Miocene Thomas Farm local fauna, Florida. Journal of Vertebrate Paleontology, 9:89–100.

Simpson, G. G. 1928. Pleistocene mammals from a cave in Citrus County, Florida. American Museum Novitates, 328:1–16.

White, J. A. 1987. The Archaeolaginae (Mammalia, Lagomorpha) of North America. Journal of Vertebrate Paleontology, 7:425–50.

——. 1991a. North American Leporinae (Mammalia: Lagomorpha) from late Miocene (Clarendonian) to latest Pliocene (Blancan). Journal of Vertebrate Paleontology, 11:67–89.

——. 1991b. A new *Sylvilagus* (Mammalia: Lagomorpha) from the Blancan (Pliocene) and Irvingtonian (Pleistocene) of Florida. Journal of Vertebrate Paleontology, 11:243–46.

Wilkins, K. T. 1984. Evolutionary trends in Florida Pleistocene pocket gophers (genus *Geomys*), with description of a new species. Journal of Vertebrate Paleontology, 3:166–81.

——. 1985. Pocket gophers of the genus *Thomomys* (Rodentia: Geomyidae) from the Pleistocene of Florida. Proceedings of the Biological Society of Washington, 98:761–67.

Winkler, A. J., and F. Grady. 1990. The middle Pleistocene rodent *Atopomys* (Cricetidae: Arvicolinae) from the eastern and south-central United States. Journal of Vertebrate Paleontology, 10:484–90.

Zakrzewski, R. J. 1975. The late Pleistocene arvicoline rodent *Atopomys*. Annals of Carnegie Museum, 45:255–61.

13

Mammalia 5

Artiodactyls

INTRODUCTION

The remaining five chapters all deal with members of the diverse mammalian group Ungulata (fig. 13.1). Most modern ungulates are herbivorous, but more omnivorous, scavenging, and even carnivorous ungulates lived during the Paleogene Period. All but one of these meat-eating groups became extinct as the Carnivora (chap. 11) diversified and flourished. The lone surviving carnivorous ungulate lineage is the Cetacea, the whales, who evolved into marine predators (chap. 17). Ungulates are generally medium- to large-sized mammals and thus do not greatly overlap in size with the other dominant group of herbivorous mammals, the rodents (chap. 12). Large size can be considered a trademark of the group, as it includes the tallest (the giraffe) and largest (the African elephant) living terrestrial mammals, the largest land mammal that ever lived (the extinct Asian rhinoceros *Paraceratherium* [=*Baluchitherium*]), and the blue whale is the largest known vertebrate of any type. Many ungulate lineages evolved body sizes greater than one metric ton, the criterion for status as a giant mammal. These include the living elephants, sirenians, rhinos, hippos, giraffes, and whales, as well as extinct groups such as the titanotheres, embrithopods, desmostylians, and toxodonts. Not all ungulates are giants, of course, but the vast majority are rabbit size or larger. With their large size necessarily comes strong and robust bones and teeth, which in turn increases their chances for preservation as fossils. Larger fossils are easier to see and collect. As herbivores, they are normally more common than large predators. So for all of these reasons, the fossil record of ungulates is relatively good, both worldwide in general and in Florida specifically. The ungulates of Florida have also been very well studied, so their history is better understood than other groups of mammals. Even so, new finds occur annually that increase our knowledge of this group.

Artiodactyls are presently the most diverse of the ungulate orders and include the pigs, peccaries, hippos, camels, deer, giraffes, antelope, and cattle. The first artiodactyls appeared suddenly in the early Eocene in Europe, Asia, and North America. By the late Eocene, they had radiated extensively into many different adaptive types. Before the end of the Oligocene they were the numerically dominant terrestrial ungulates, and they have continued to diversify ever since. Artiodactyls are commonly called the "even-toed" ungulates, because two large digits per foot bear most of the body weight, while two smaller side toes may or may not be present. This name is intended to contrast them with the so-called odd-toed ungulates, the order Perissodactyla (chap. 13). However, some perissodactyls (for example, tapirs) have four digits on their front feet, and other ungulates also have an "even" number of toes. What makes the artiodactyl foot unique is not the number of toes, but the symmetry of the digits. In artiodactyls, the axis of symmetry passes between the third and fourth digits, which are equally developed. The second and fifth digits are

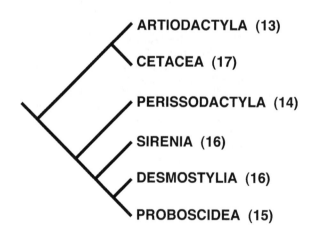

Fig. 13.1. Hypothesized evolutionary relationships among the major members of the mammalian group Ungulata. Numbers indicate the chapter in which that group is covered. There are other extinct groups not shown here that are not found in Florida.

smaller than the third and fourth, or are absent, while the first digit is reduced to a small, nonfunctional nubbin or is completely absent.

Artiodactyls are commonly divided into two major suborders: the Suiformes and Selenodontia. The Suiformes includes three living families, the Suidae (pigs), Tayassuidae (peccaries), and Hippopotamidae (hippos), as well as a number of extinct groups such as the Entelodontidae and Anthracotheriidae. Of the extant families, only the peccaries are native to North America. The remaining artiodactyls are placed in the suborder Selenodontia, which is in turn subdivided into two major infraorders, the Tylopoda and Ruminantia. The camels and llamas (Camelidae) are the only living tylopod family, but the group was appreciably more diverse in the past. The Ruminantia is currently more diverse and includes at least six extant families: the Tragulidae (mouse deer); Moschidae (musk deer); Cervidae (true deer, elk, moose); Antilocapridae (pronghorn "antelope"); Giraffidae (giraffe, okapi); and Bovidae (cattle, goats, true antelope). Within the Selenodontia, the Tylopoda is principally a North American group (although now extinct on that continent), and the Ruminantia is primarily an Old World group. However, many ruminant families dispersed to North America through the later half of the Cenozoic, and several of these underwent secondary radiations and were quite successful. The classification of artiodactyls used here for the most part follows McKenna and Bell (1997), with some modifications based on the artiodactyl chapters in Janis et al. (1998).

ARTIODACTYL SKELETAL AND DENTAL ANATOMY

In general, the artiodactyl skull is elongate, with the preorbital portion much longer than the postorbital region. The rear of the orbit is marked by either a strong postorbital process or a complete postorbital bar. The number of teeth varies greatly. Primitively, the basic placental mammal complement of 44 was present, and this state is retained in some suids and tylopods. Loss of teeth is a very common phenomenon in artiodactyl evolution and occurs most commonly to the upper incisors, upper canine, and anterior premolars. All but the most primitive artiodactyls have "squared-up" the upper molars with a posterior-lingual cusp. Although it is usually referred to as the "hypocone," it is actually the enlarged metaconule and is not equivalent to the true hypocone found in other ungulates.

Two basic morphologies are observed in artiodactyl molars: bunodont or selenodont. Bunodont cheek teeth (fig. 13.2A) have rounded cusps covered by heavy

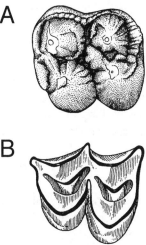

Fig. 13.2. Examples of the two major types of molar teeth found in artiodactyls, in occlusal view. *A*, UF 116822, a left upper molar of a peccary (Tayassuidae) from La Camelia Mine, Gadsden County, middle Miocene. This is an example of a bunodont tooth. *B*, UF 116823, a right M3 of an oreodont (Oreodontidae, *Ticholeptus*) from La Camelia Mine. This is an example of a selenodont tooth. *A* about 3×; *B* about 1.5×. After Bryant (1991); reproduced by permission of the Society of Vertebrate Paleontology.

enamel. Simple bunodont molars have four primary cusps with smooth enamel. Some lineages increase dental complexity by wrinkling the enamel and adding more cusps. Jaw movement during chewing is mostly up and down to produce a simple crushing action. Among bunodont artiodactyls, the process of molarizing the premolars varies from very little to quite advanced (for example, the peccary *Mylohyus*). Most bunodont artiodactyls also have brachyodont teeth, but some suids have hypsodont molars. Diets associated with bunodont teeth are typically omnivorous, as in the pig and peccaries. Others, like that of the hippopotamus, are strictly herbivorous.

Selenodont teeth are characterized by four (five in a few early examples) crescent-shaped lophs arranged in two pairs on each molar (fig. 13.2B). The upper enamel crescents are concave on their labial sides, whereas the lower enamel crescents are concave on their lingual sides. The functional significance of such teeth is evident when one sees a camel or cow chew. The lower jaw drops downward and outward, then it closes upward and lingually, bringing the concave enamel sides of the lower molars between the concave enamel sides of the corresponding upper molars. As the teeth come into occlusion, they shear food material trapped between the opposing concave surfaces. Chewing occurs on only one side of the mouth at a time. The great success of the many selenodont artiodactyl families is perhaps not surprising in view of their effective mechanism for chewing plant fibers. Selenodont artiodactyls include browsers, grazers, and mixed-feeders. While browsers usually retain brachyodont cheek teeth, increases in crown height are common in the other two feeding groups. Very hypsodont cheek teeth evolved in camelids, antilocaprids, and many bovids, but true hypselodonty is much rarer than in the rodents.

A characteristic feature of many artiodactyl groups (but not unique to them among the Mammalia) is the possession of horns, antlers, or similar structures that are collectively called cranial appendages. A few artiodactyl lineages (for example, camelids and peccaries) completely lack these structures. Their evolution and behavioral use are the subject of much investigation by biologists and paleontologists. Artiodactyls with cranial appendages evolved independently from hornless ancestors many times (more than six). This is evidenced by the different developmental pathways for horns and hornlike structures found in artiodactyls and by their evolutionary relationships with hornless forms. Artiodactyl cranial appendages are at least partially made of bone and thus preserve as fossils. Some groups also have a covering or sheath over the bony core made of organic proteins similar to those in fingernails (keratin). This sheath does not fossilize (except under extraordinary circumstances when other organic tissues are also preserved), so its morphology (or even existence) must be inferred from other evidence.

Four types of cranial appendages are found in living artiodactyls. Only bovids have true horns. These are bony outgrowths of the frontal bone which are never shed (nondeciduous) and are covered by a nondeciduous keratinous sheath. Bovid horns are located behind the orbits, do not branch or fork, but may curve or spiral. The living giraffids have what are called ossicones. They develop as separate centers of ossification in the skin that later fuse to the skull roof. They are also nondeciduous but are covered by skin rather than a sheath. Living and fossil antilocaprines have a nondeciduous bony core dorsal to the orbit that may or may not be branched or twisted into a spiral. In the living genus *Antilocapra*, this is covered by a forked keratinous sheath that is shed annually. Whether or not this was the case in extinct antilocaprines is uncertain but probable. The fourth type of cranial appendage is the antler of cervids. It is a bony, often branching structure that grows from a raised base on the frontal bone. The antler is initially covered by skin, which is then worn off. The deciduous antler later

breaks off at its base, to be grown again the following year.

Cranial appendages among extinct artiodactyls include varieties like those of extant groups, as well as some novel combinations of features. For example, cosorycine antilocaprids had a branched, supraorbital antlerlike structure that was apparently nondeciduous. Dromomerycines had long supraorbital and medial appendages that were nondeciduous, did not branch, and were skin covered. Whether or not they formed from separate ossification centers like those of giraffe ossicones is not known. The cranial appendages of protoceratids were structurally similar to those of dromomerycines, but some had a forked nasal appendage.

The primary function of all these bizarre structures is not for predator defense, but rather intraspecific behavior. They serve as signals of a male's age and status and, depending on the species, as weapons for intraspecific combat to determine dominance and mating rights in the social hierarchy. Hornless artiodactyls often have enlarged canines that are used in a similar manner. In many artiodactyl groups, the females either lack horns or have significantly smaller appendages than males. Because of their function as visual signals, cranial appendages often differ in morphology even between closely related species, more so than dentitions. Thus they have been used extensively in the classification and systematics of fossil artiodactyls, and the type specimens of many fossil species are isolated horns or antlers.

One of the two most characteristic features of the artiodactyl skeleton has already been described—the dual emphasis of the third and fourth digits, with the plane of symmetry of the foot passing between them. The other distinctive feature is in the astragalus (fig. 13.3). In addition to the usual condition of having a rotating joint between the astragalus and tibia (as expressed by a pulley-like articular surface on the proximal end of the astragalus), a similar, ventrally located joint is located between the astragalus and the navicular and cuboid. The double pulley system introduced great flexibility in the ankle joint, permitted freer running and springing, and faster acceleration. This system is seen in the earliest artiodactyls and presumably contributed to their evolutionary success. In advanced artiodactyls, the articulating surfaces of the astragalus have pronounced ridges that fit into grooves on the tibia. These permit only fore and aft movement of the foot relative to the hind limb.

Many artiodactyls, especially camelids and most ruminants, have specializations for running. The limb elements are elongated, especially the radius, tibia, and

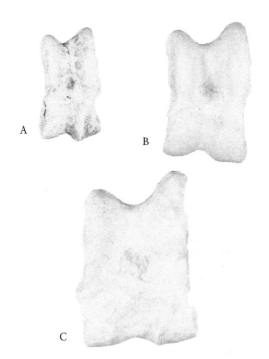

metapodials (fig. 13.4). The shafts of the ulna and fibula are reduced or lost, leaving only proximal and distal vestiges variously attached or fused onto the larger radius or tibia. The proximal end of the ulna, onto which the most powerful extensor muscles of the forelimb insert, fuses to the radius. Those groups with elongated metapodials have a tendency to tightly bind the third and fourth metapodials into a single functional element. This is carried to its most extreme condition possible, the fusing of the two into a single element, in some peccaries, advanced camelids, and most ruminants (figs. 13.4, 13.5). The name for such a structure is the cannon bone. In the feet, the second and fifth metapodials and phalanges are reduced to vestiges under the skin or are eventually lost. The toes of the third and fourth digits support the entire mass of the body. Each foot has an enlarged pair of hooves that are often (but inaccurately) called "a cloven hoof" in ruminants.

Fig. 13.3. Examples of artiodactyl astragali. The unique "double pulley" design is regarded as a key to the evolutionary success of the group. Shown are representatives of the three major artiodactyl groups found as fossils in Florida: *A* is a peccary (UF 29235, *Prosthenops* n. sp.); *B* is a ruminant (UF 38143, *Pediomeryx hamiltoni*); and *C* is a camel (UF 38833, *Procamelus grandis*). Only ruminants are still native to the state. All are right astragali from the Love site, Alachua County, late Miocene. All about 0.67×.

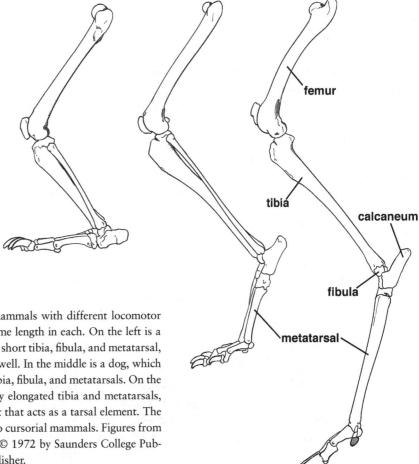

Fig. 13.4. Hind-limb proportions in three mammals with different locomotor specializations, drawn so the femur is the same length in each. On the left is a bear, which is plantigrade and has a relatively short tibia, fibula, and metatarsal, characteristics of a species that does not run well. In the middle is a dog, which is digitigrade, and has somewhat elongated tibia, fibula, and metatarsals. On the right is a deer, which is ungaligrade, has very elongated tibia and metatarsals, and has reduced the fibula to a small element that acts as a tarsal element. The deer and dog show specializations common to cursorial mammals. Figures from *Mammalogy* by Terry A. Vaughn, copyright © 1972 by Saunders College Publishing, reproduced by permission of the publisher.

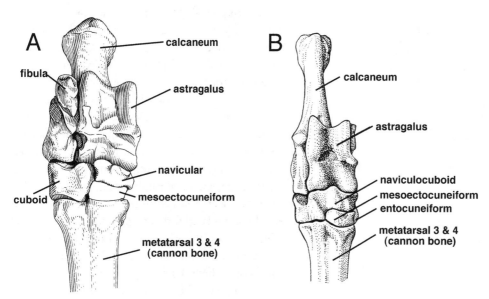

Fig. 13.5. Comparison of the tarsus of two cursorial artiodactyls. *A,* left tarsus of the Pleistocene camelid *Camelops. B,* right tarsus of the modern pronghorn *Antilocapra americana,* a ruminant. A major difference is that the navicular and cuboid are not fused in camelids as they are in ruminants. Both have a remnant of the fibula incorporated into the tarsus, although it is not depicted in *B. A* after Webb (1965), courtesy of S. David Webb, Florida Museum of Natural History. *B* from *Mammalogy* by Terry A. Vaughn, copyright © 1972 by Saunders College Publishing, reproduced by permission of the publisher.

FOSSIL BUNODONT ARTIODACTYLS OF FLORIDA

Two families of bunodont artiodactyls are known to have lived in Florida—the Entelodontidae and Tayassuidae, although neither is present today. Feral pigs (Suidae, Old World natives) are present in many parts of the state, and their bones are sometimes recovered from rivers and misidentified as fossils. Several other groups of bunodont artiodactyls occurred in North America during the Tertiary, but these are as yet not recognized from Florida.

The Entelodontidae is primarily an Oligocene group well known from Europe, Asia, and western North America. They are large, piglike animals, with many unique modifications of their own (fig. 13.6). In Florida, a few teeth from very early Miocene deposits represent the genus *Daeodon* (often called *Dinohyus,* a junior synonym), the largest and last of the entelodonts, best known from western Nebraska. The skull reached over a meter in length, and the body was more than three meters long. The cheek teeth were square with four or five large rounded cusps (fig. 13.6B). In general the skeleton resembled that of a very large pig. The incisors and canines are large and blunt. The anterior premolars are bladelike. The zygomatic arch bears a large flange sim-

ilar to that seen in some xenarthrans (chap. 10). Their diet is thought to have been omnivorous, while the wear patterns on the teeth suggest they regularly chewed on bones and other hard objects (Joeckel, 1990).

The peccaries (family Tayassuidae) originated in North America in the late Eocene (Wright, 1998). At the end of the Pleistocene the common tayassuid genera in North America became extinct along with many other herbivores. Tayassuids were a common component of Florida's mammalian fauna beginning in the Miocene. Tayassuids still live in the southwestern United States but are today principally a South and Central American group. The collared peccary, *Dicotyles tajacu,* ranges widely from the southern border of the United States (in Arizona, New Mexico, and Texas) south into temperate South America and lives in a broad range of habitats. The white-lipped peccary, *Tayassu pecari,* lives only in tropical forests from southern Mexico into Paraguay. The third living taxon, *Catagonus wagneri,* was found living in the scrub and dry forests of western Paraguay in the 1970s. It was previously known from fossils (Wetzel et al., 1975). The ancestors of these three living species dispersed into South America from North America.

Peccaries bear a general resemblance to pigs, and the two families diverged from a Eurasian common ancestor

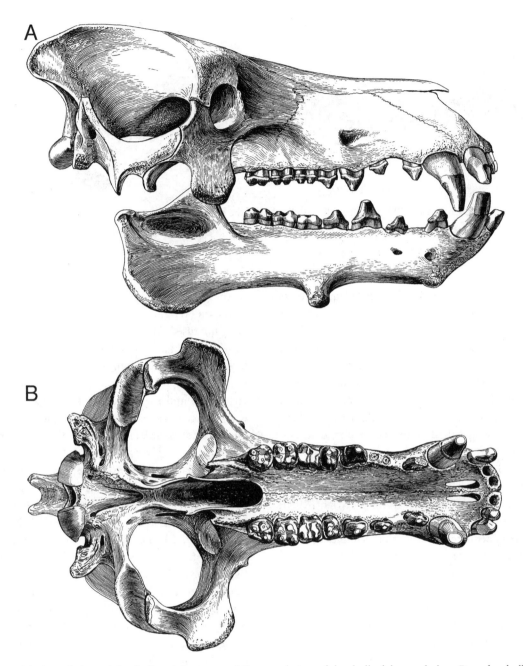

Fig. 13.6. *A*, right lateral view of the skull and dentary, and *B*, ventral view of the skull of the entelodont *Daeodon hollandi*. This particular specimen is from the early Miocene of Nebraska. A few, much less complete entelodont fossils are known from Florida. Some researchers believe entelodonts were primarily bone-cracking scavengers. About 0.2×. After Peterson (1906); reproduced by permission of the Carnegie Museum of Natural History, Pittsburgh.

near the end of the Eocene. In tayassuids, the triangular canine teeth stand vertically, with the posterior side of the lower canine occluding against the anterior surface of the corresponding upper. They occlude tightly and wear heavily against one another, producing continuously sharpened points. The long canines prevent peccaries from moving the lower jaw sideways, which is the chewing motion used in pigs and most other ungulates.

Peccaries chew with strictly vertical jaw motion. The broadly expanded cheekbone in peccaries provides the purchase for an enlarged masseter muscle, whose function is to pull the mandible strongly upward and slightly forward. These and many other features of the skull are related to their peculiar vertical jaw movement.

The peccaries of the Oligocene and very early Miocene in Florida have not yet been well described or un-

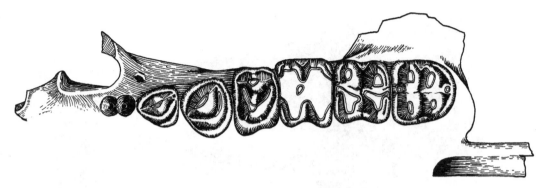

Fig. 13.7. Occlusal view of MCZ 4290, a left maxilla with P2–M3 of the rare peccary *Floridachoerus olseni* from the Thomas Farm site, Gilchrist County, early Miocene. About 1×. After White (1947); reproduced by permission of the Museum of Comparative Zoology, Cambridge, Massachusetts.

derstood, mainly because of a lack of adequate fossil material (Wright, 1998). A peccary of uncertain identity lived in the early Oligocene, as represented at the I-75 site. The early Miocene genus and species *Floridachoerus olseni* was described from Thomas Farm (fig. 13.7). It is closely related to the genus *Hesperhyus,* which is a common western form. It is an extremely rare species at Thomas Farm, and few additional specimens have been collected in recent years despite extensive excavations. Middle Miocene peccaries are represented by a single tooth (fig. 13.2A). Several late Miocene peccaries are known that resemble contemporary western taxa, but they have yet to be thoroughly described (Wright, 1993, 1998; fig. 13.8A, B).

The oldest adequate, well-described samples of tayassuids from Florida are early Pliocene. At least two, and possibly three, taxa are represented in the Palmetto Fauna of the Bone Valley region. The two best-known taxa are *Mylohyus elmorei* and *Catagonus brachydontus* (Wright and Webb, 1984; Wright, 1990, 1998), although in the past they were referred to the genera *Prosthenops* and *Desmathyus,* respectively. *C. brachydontus* is the larger and more common of the two (fig. 13.8C, D). In both, the zygomatic arches are large and inflated, thus differing from late Pleistocene and modern species of these genera, who have much smaller cheekbones. Large zygomatic arches are thought to be primitive for advanced Neogene tayassuids and independently lost in several lineages (Wright, 1993). The molars of *C. brachydontus* in general resemble those of *Platygonus* (described below), in which the cusps unite to form two transverse crests (bilophodont). The premolars are not molariform (fig. 13.8C). As suggested by the species name, the teeth are not high crowned. The reference to the living *Catagonus* (rather than *Platygonus*) is based

on the short diastema and details of the nasal and auditory region (Wright, 1990). The metapodials were apparently not fused. *Mylohyus elmorei* and younger members of *Mylohyus* have bunodont teeth, molariform premolars, and a long postcanine diastema (figs. 13.9, 13.10A–C). *M. elmorei* differs from its younger relatives in its larger size and expanded cheekbone. A possible third early Pliocene genus is *Platygonus,* but relatively complete cranial material is needed to distinguish it from the dentally similar *Catagonus.*

Two genera of peccaries are recognized in the late Pliocene and Pleistocene of Florida: *Platygonus* and, continuing from the early Pliocene, *Mylohyus.* They are readily distinguished by the structure of their cheek teeth (figs. 13.10, 13.11). The molars of *Platygonus* are bilophodont, whereas those of *Mylohyus* are bunodont (with four main cusps and several accessory cuspules). Both upper and lower molars show these differences. The lower premolars of *Platygonus* have long, narrow proportions and "stair-stepped" cusps (the front cusp being higher than the back). In *Mylohyus,* the lower premolars are nearly square and have low, flat crowns. The premolars of *Platygonus* are not molariform. Each upper premolar of *Platygonus* has only two high cusps, almost forming a transverse ridge, and in front of and behind these cusps, a wide cingulum. These provide space between the cusps of the upper premolars for the high cusps of the lower premolars to occlude (fig. 13.11).

Platygonus and *Mylohyus* also show significant differences in the skull. The snout and braincase was proportionally longer and narrower in *Mylohyus.* The bony buttress above and behind the upper canine teeth was much stronger in *Platygonus,* and the jugal bones flared down and out much more widely in *Platygonus.* The canine buttress and jugal flare were especially well de-

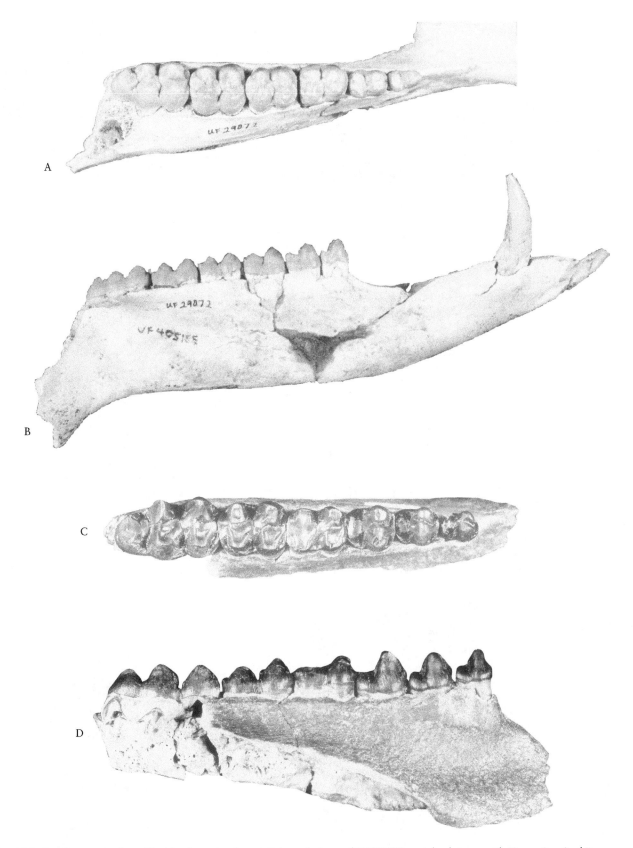

Fig. 13.8. Fossil peccaries from Florida. *A*, occlusal, and *B*, lateral views of UF 29072, a right dentary with i3, c, p2–m3 of *Prosthenops* n. sp. from the Love site, Alachua County, late Miocene; *C*, occlusal, and *D*, lateral views of UF 124190, a right dentary with p2–m3 of *Catagonus brachydontus* from the Gardinier Mine, Polk County, early Pliocene. *A, C, D* about 1×; *B* about 0.75×.

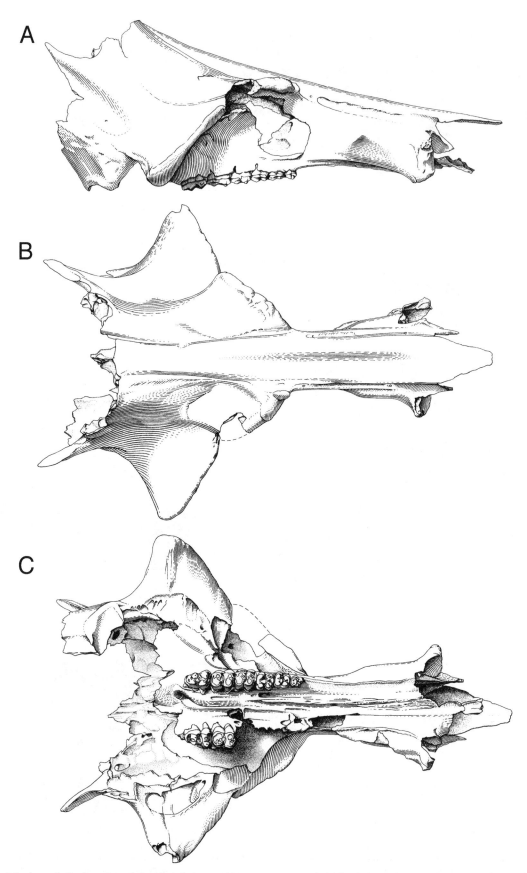

Fig. 13.9. *A,* right lateral, *B,* dorsal, and *C,* ventral views of UF 12265, partial skull of the peccary *Mylohyus elmorei* from the Palmetto Mine, Polk County, early Pliocene. About 0.33×. After Wright and Webb (1984); reproduced by permission of the Society of Vertebrate Paleontology.

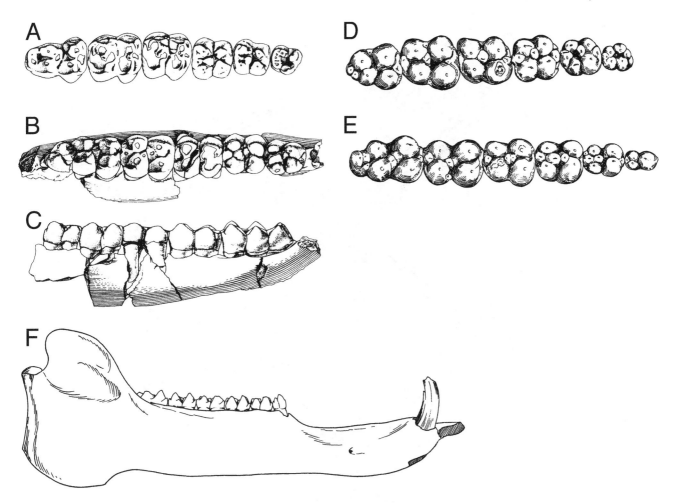

Fig. 13.10. Teeth and dentaries of *Mylohyus* from Florida. *A*, occlusal view of UF 12265, left P2–M3 of *M. elmorei* (same specimen as in fig. 13.9); *B*, occlusal, and *C*, lingual views of UF 49081, right p3–m3 of *M. elmorei* from Payne Creek Mine, Polk County, early Pliocene; *D–F*, UF/FGS 5691, *M. fossilis* from Arredondo 2, Alachua County, late Pleistocene; *D*, occlusal view of left P2–M3; *E*, occlusal view of right p2–m3; *F*, lateral view of right dentary. Note the very long postcanine diastema. *A–C*, about 0.8×; *D–E* about 1×; *F* about 0.5×. *A–C* after Wright and Webb (1984); reproduced by permission of the Society of Vertebrate Paleontology. *D–F* after Bader (1957); reproduced by permission of the Florida Museum of Natural History.

veloped in mature males of *Platygonus* (for example, fig. 13.11). Both *Mylohyus* and *Platygonus* had longer limbs than do modern peccaries. The metapodials were fused together to form a cannon bone, except in the manus of *Mylohyus*. There, as in the living tayassuids, the two bones are only flattened against each other. In *Platygonus*, the side toes are absent (fig. 13.12), while in *Mylohyus* they are reduced in front and absent behind. These and other features of the skeleton indicate that *Platygonus* and *Mylohyus* were swifter, more efficient runners than extant tayassuids.

In many parts of North America the common Pleistocene peccary is *Platygonus*. Dozens of skeletons often occur in a single site, especially in caves. At most late Pleistocene sites in Florida, however, *Mylohyus* predominates; for example, at Vero, Seminole Field, Reddick,

Arredondo, and the Cutler Hammock site. *Platygonus* is usually the more common of the two at late Pliocene through middle Pleistocene localities. The best samples are from Haile 21A (fig. 13.13), Leisey Shell Pit, and Coleman 2A. *Platygonus,* with its more hypsodont, bilophodont teeth, probably ate more cactus and other coarse vegetation. *Mylohyus,* with its low-crowned, bunodont cheek teeth presumably ate more fruits, nuts, and succulent vegetation. *Platygonus* is therefore thought to have lived in prairies and other open-country habitats, whereas *Mylohyus* is considered a forest species. The great abundance of *Mylohyus* and the relative rarity of *Platygonus* in late Pleistocene deposits in Florida suggest that wet, well-forested conditions then prevailed over much of the state. Earlier in the Pleistocene, however, drier periods with extensive prairies were represented, at

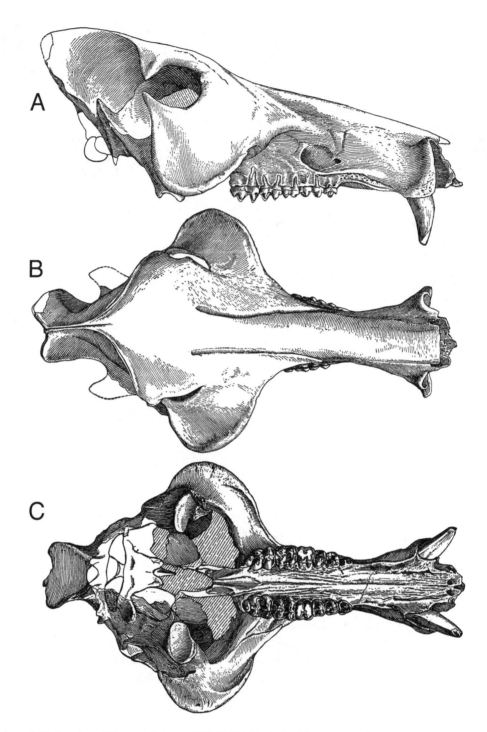

Fig. 13.11. *A*, left lateral, *B*, dorsal, and *C*, ventral views of USNM 8148, skull of the peccary *Platygonus cumberlandensis* from the middle Pleistocene of Maryland. This genus is common in Florida through the Pleistocene. About 0.33×. After Gidley and Gazin (1938); reproduced by permission of the Smithsonian Institution Press, Washington.

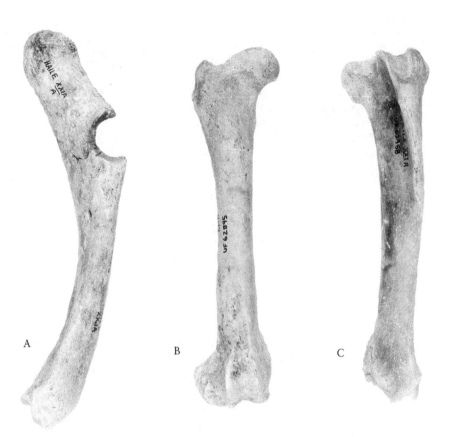

Fig. 13.13. Major limb bones of the early Pleistocene peccary *Platygonus vetus* from Haile 21A, Alachua County. *A*, lateral view of UF 62786, a radioulna (fused radius and ulna); *B*, anterior view of UF 62895, a right femur; *C*, anterior view of UF 62958, right tibia. About 0.5×.

Fig. 13.12. Anterior view of USNM 7690, left pes of the peccary *Platygonus cumberlandensis* from the middle Pleistocene of Maryland. Note the complete absence of lateral digits (digits 2 and 5). About 0.5×. After Gidley and Gazin (1938); reproduced by permission of the Smithsonian Institution Press, Washington.

least much of the time. The presence of other vertebrates, such as jackrabbits, ground squirrels, and hognose snakes, suggest a similar pattern.

FOSSIL OREODONTS OF FLORIDA

The Oreodontidae, or oreodonts, is a diverse group of selenodont, small- to medium-sized artiodactyls with body proportions similar to pigs from the late Eocene to Miocene of North America (Lander, 1998; fig. 13.14). They had four functional toes per foot and unfused, short metapodials. They are unusual for artiodactyls in lacking a diastema, or only having a very short one. Most retained the full complement of upper incisors and canines (fig. 13.15A). The evolutionary relationships and classification of oreodonts is uncertain; some regard them as Suiformes (for example, McKenna and Bell, 1997), others as Tylopoda (for example, Lander, 1998). Even the name of the family is disputed, with Merycoidodontidae vying with Oreodontidae. The generic and specific identifications of Florida oreodonts listed in

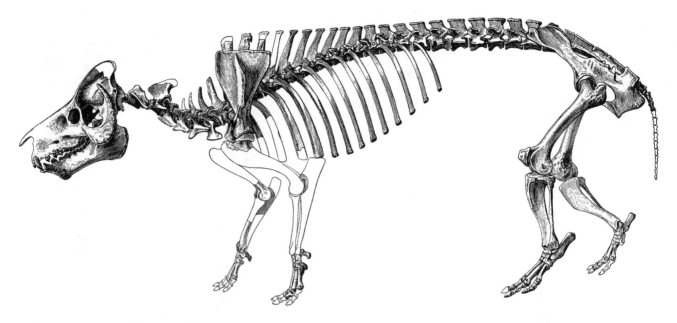

Fig. 13.14. Restored skeleton of the Miocene oreodont *Merycochoerus* from Nebraska. Despite their piglike body proportions, oreodonts had selenodont molars. Their relative evolutionary relationships with camels and pigs is presently open to debate. About 0.1×. After Peterson (1914); reproduced by permission of the Carnegie Museum of Natural History, Pittsburgh.

chapter 3 and used here follow the recent and somewhat controversial revisions of Lander (1998).

The rarity of oreodonts in Florida is in stark contrast with their abundance in western North America. They are known only from the early Oligocene to the early middle Miocene in Florida, despite persisting into the late Miocene on the Great Plains, and appear to be most common in older faunas. Lander (1998) recognized six species of oreodonts from Florida. The oldest are *Eporeodon occidentalis* and *Oreodontoides oregonensis* from the Oligocene. Morgan (1989) reported an associated skeleton of a very early Miocene oreodont collected on the bed of the Suwannee River (now on display at the Florida Museum of Natural History). This is the only locality in Florida at which oreodonts are very abundant. A partial maxilla found in Marion County (MacFadden, 1980; fig. 13.15C) belongs to the species *Merycoides harrisonensis,* which is also known from Nebraska, Wyoming, and Montana. A small form of *Merychyus elegans* is known from a few teeth and postcranial elements recovered from Thomas Farm (fig. 13.15B). The medium-sized *Ticholeptus* was reported from the early Barstovian of Gadsden County on the basis of a single tooth (fig. 13.2B; Bryant, 1991). Exactly why oreodonts were so uncommon in Florida remains a mystery.

Fossil Tylopods of Florida

Two families of tylopod artiodactyls are known from Florida: the Protoceratidae and Camelidae. The camelids are by far the more important and diverse group, judging from the fossil record. The Protoceratidae are medium-sized, selenodont artiodactyls limited to North America. They range from the late Eocene to the early Pliocene. The most notable protoceratid feature is their cranial appendages (found in males only), which are otherwise uncommon among the tylopods. In one subfamily, the Protoceratinae, three pairs of bony outgrowths are present on the top of the skull: one pair each from the maxilla, frontal, and parietal bones. In the other subfamily, the Synthetoceratinae, the posterior parietal pair is lacking, and the maxillary pair are fused to form a single anterior appendage (fig. 13.16A). Other characteristics of the family are unfused and relatively short metapodials; lateral metacarpals reduced but functional and bearing complete sets of phalanges; complete postorbital bar; very shallow mandible with a reduced coronoid process and masseteric fossa; upper incisors absent; upper canine saberlike and large in males; incisiform lower canine; and an anteriorly located, caniniform p1 (although this tooth is lost in *Synthetoceras*).

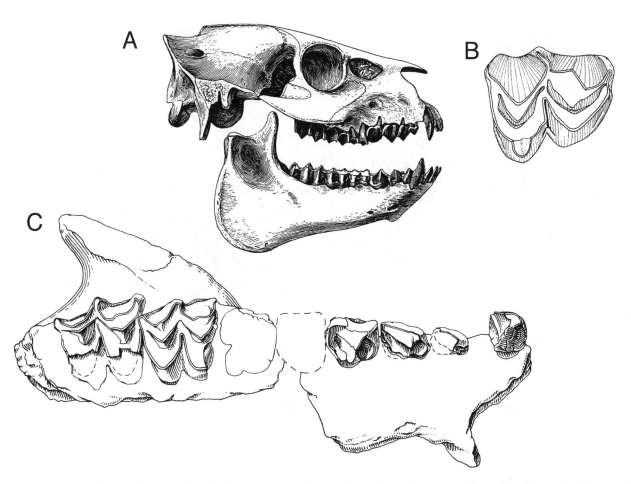

Fig. 13.15. *A*, right lateral view of the skull and dentary of a typical oreodont, *Merychyus elegans*, from the early Miocene of Nebraska. Note the full complement of incisors and the lack of a diastema between the canine and premolars; *B*, occlusal view of MCZ 7765, a left upper molar of *Merychyus elegans*, from the Thomas Farm site, Gilchrist County, early Miocene; *C*, occlusal view of UF 24202, a partial right maxilla with C, P1–P3, and M2–M3 of *Merycoides harrisonensis* from the Martin-Anthony Roadcut site, Marion County, early Miocene. *A* about 0.5×; *B* about 1.5×; and *C* about 1.3×. *A* after Peterson (1906); reproduced by permission of the Carnegie Museum of Natural History, Pittsburgh. *B* after Maglio (1966); reproduced by permission of the Museum of Comparative Zoology, Cambridge, Massachusetts. *C* after MacFadden (1980); reproduced by permission of SEPM (Society for Sedimentary Geology).

The earliest protoceratid from Florida is an undescribed sample from the early Oligocene I-75 site. On the basis of the known chronologic ranges of the two protoceratid subfamilies, it is more likely that these represent the only known specimens of the subfamily Protoceratinae from Florida. This is because that group is well represented in the Oligocene, while the Synthetoceratinae did not appear until the early Miocene.

The first well-represented protoceratid in Florida is the early Miocene *Prosynthetoceras texanus* (fig. 13.16), a species also known from the Texas Coastal Plain (Patton and Taylor, 1971). In this genus, and its close relative *Synthetoceras*, the nasal "horn" consists of a short, forked segment on top of a long base. The supraorbital

"horns" at first curve posteriorly, then arc anteriorly (fig. 13.16A). The teeth of *P. texanus* are simple and brachyodont. The next younger record of a protoceratid from the state is the late Miocene *Synthetoceras tricornatus*, a larger, more hypsodont taxon. As it is also known from the middle Miocene in Texas, it is likely that it also lived in Florida during this time as well. *S. tricornatus* has a longer diastema than *Prosynthetoceras*, as a result of losing the p1 and p2, and a taller nasal "horn" (fig. 13.17). The record from the McGehee Farm site is the youngest for this species. Until the discovery of *Kyptoceras*, this was thought to be the youngest record of the entire protoceratid family.

The largest, most hypsodont, and youngest proto-

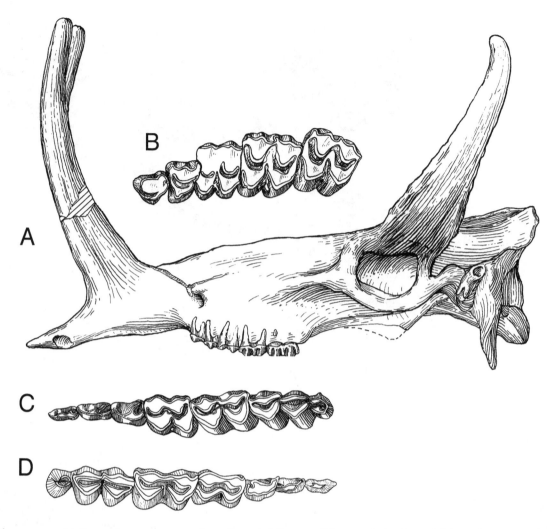

Fig. 13.16. The protoceratid *Prosynthetoceras texanus* from the early Miocene of Texas and Florida. *A*, left lateral view of AMNH 53500, skull from Walker County, Texas; *B*, occlusal view of the P3–M3 of AMNH 53500; *C*, occlusal view of AMNH 34181, left p2–m3 from Texas; *D*, occlusal view of right p2–m3 from the Thomas Farm site, Gilchrist County, early Miocene. Composite based on MCZ 3656, 3660, and 3654. *A* about 0.5×; *B–D* about 1×. *A–C* after Patton and Taylor (1971); reproduced by permission of the American Museum of Natural History, New York. *D* after Maglio (1966); reproduced by permission of the Museum of Comparative Zoology, Cambridge, Massachusetts.

ceratid was described by Webb (1981) from the early Pliocene Bone Valley deposits (Palmetto Fauna) of central Florida as a new genus and species, *Kyptoceras amatorum* (fig. 13.18). *Kyptoceras* differs from *Synthetoceras* and *Prosynthetoceras* in having a very short, broad-based nasal "horn" and long, anteriorly projecting frontal "horns." In these it more resembles the early Miocene genus *Syndyoceras* from Nebraska, with which Webb suggested a close evolutionary relationship. Despite some progressive features, like its hypsodont molars, *Kyptoceras* retained the short, unfused metapodials characteristic of the family.

The more common tylopod family from Florida is the Camelidae. Camelids were one of the most successful of the selenodont families in North America. Today's Old World camels and South American llamas are both derivatives of North American camelids. Thus it is a great irony for the camel family, as it is for the horses, that they did not survive on the continent where they had their greatest evolutionary success. In the early Oligocene, about four camelid genera were present, of which *Poebrotherium* is especially well known. Its morphology demonstrates the primitive characteristics of the family (fig. 13.19). It had a full set of forty-four teeth with very small spaces between the short incisors, canines, and anterior premolars (fig. 13.19C). The cheek teeth were very simple and low crowned. The face was short and shallow and had an incomplete postorbital bar behind

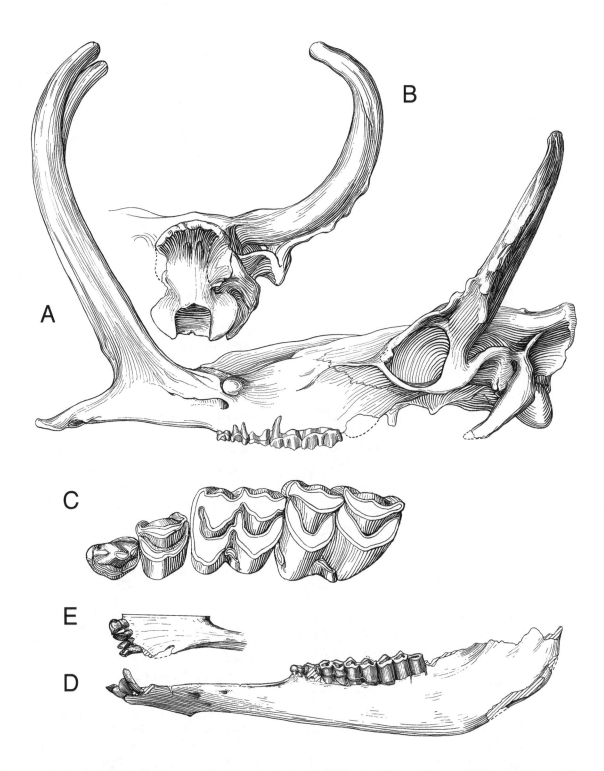

Fig. 13.17. The protoceratid *Synthetoceras tricornatus* from the late middle Miocene of Texas. This species is known from the late Miocene of Florida. *A*, left lateral, and *B*, posterior views of AMNH 33407, a male skull. *C*, occlusal view of AMNH 32467, right P3–M2. *D*, lateral view, and *E*, dorsal view of symphysis of AMNH 53444. *A, B, D, E* about 0.33×; *C* about 1×. After Patton and Taylor (1971); reproduced by permission of the American Museum of Natural History, New York.

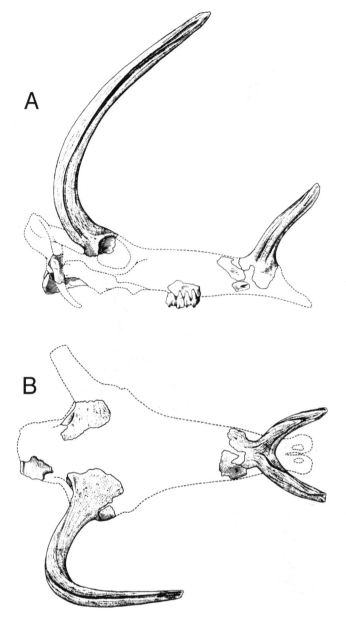

Fig. 13.18. *A,* right lateral, and *B,* dorsal views of UF 25711, a partial skull of the protoceratid *Kyptoceras amatorum* from Tiger Bay Mine, Polk County, early Pliocene. About 0.2×. After Webb (1981); reproduced by permission of the Society of Vertebrate Paleontology.

the eye socket. The mandible bears a sharp posterior hook (fig. 13.19C). The skeleton is only about 60 centimeters at the shoulder (the size of a sheep), yet it has proportionally very long limbs and neck. The feet, too, are notably elongate and show precocious reduction of the lateral (second and fifth) toes, as compared with any contemporaneous artiodactyl. The third and fourth metapodials are not fused, but the distal ends are splayed

laterally (fig. 13.19B), not oriented parallel as in other artiodactyls.

Three subfamilies of Neogene camelids are recognized in Florida. During the early Miocene, most known species belong to the subfamily Aepycamelinae, a group characterized by elongated metapodials and cervical vertebrae. Because of these specializations, they are commonly called "giraffe-camels." Another early Miocene group was the Floridatragulinae, with their long, low skulls. Later in the Miocene most of these early forms were replaced by members of the modern subfamily Camelinae.

Florida and the Gulf Coastal Plain of Texas provide an excellent record of early Miocene camelids. Although many of them have close relationships to genera from the mid-continent, some genera were apparently endemic to the Gulf Coast (for example, *Floridatragulus*). On the other hand, some common western forms, like the remarkable little gazelle camels (Stenomylinae), did not enter the Gulf coastal region, evidently because they were adapted to more arid habitats. The Miolabinae, another western group of camelids, although known from coastal Texas, has not yet been found in Florida.

Florida's early Miocene camelids are *Floridatragulus, Nothokemas, Oxydactylus,* and a very small form similar to *Gentilicamelus.* Most are rather small, with low-crowned molars and a full set of forty-four teeth. *Floridatragulus* has an unusually elongate skull and jaw (fig. 13.20), but in many structural details it is more primitive than the giraffe-camels. The very long rostrum has four caniniform teeth well separated from one another (fig. 13.20); the orbit is open posteriorly, as in *Poebrotherium;* and the posterior lobe of the m3 is peculiarly divided. Another distinctive feature is a diastema between the p2 and p3 (fig. 13.21).

Nothokemas was, in many ways, the most progressive early Miocene camelid. It had lost the p1, and the other lower premolars are modified by elongation and by addition of an anterior-lingual fold and a posterior-lingual cusp (fig. 13.22). The canines and incisors are very powerfully built, as is the mandibular symphysis. The molars are relatively complex, with well-developed styles, ribs, and intercolumnar tubercles. Two species are recognized, the very small, very early Miocene *N. waldropi* from near Live Oak (Frailey, 1978), and the more progressive *N. floridanus* from Thomas Farm. The only other records of the genus are from coastal Texas.

The correct generic assignment of a small, brachyodont camelid from the Buda locality (very early Miocene) is uncertain because of a lack of complete skulls or jaws.

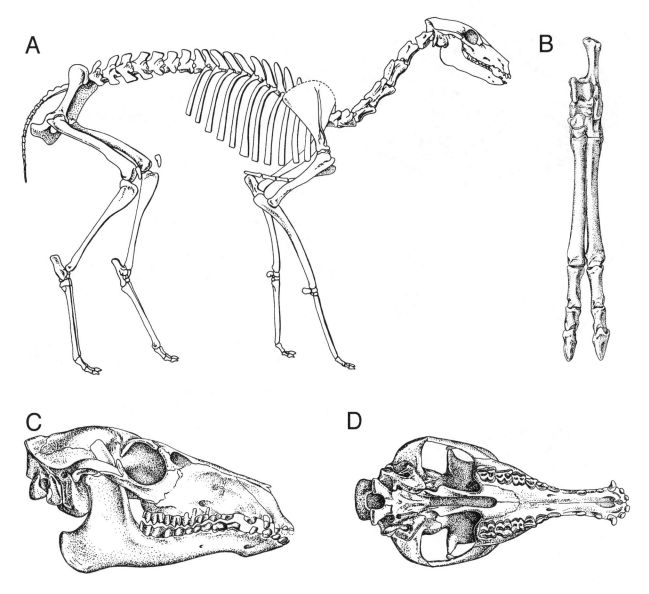

Fig. 13.19. The Oligocene camelid *Poebrotherium,* which retains many of the primitive features of the family, such as a full complement of forty-four teeth and unfused metapodials. *A,* lateral view of skeleton; *B,* anterior view of left pes; *C,* lateral view of skull and dentary; *D,* ventral view of skull. From *Vertebrate Paleontology and Evolution* by Carroll © 1988 by W. H. Freeman and Company. Used with permission.

It is the smallest known Miocene camelid, most similar to *Gentilicamelus,* a genus from Oregon. Frailey (1979) concluded that it represented a new genus but elected not to describe it. Its delicate teeth have strong ribs, stylar cusps, and intercolumnar tubercles. It is best distinguished by its short, broad p4 and its short, unfused metapodials (fig. 13.23).

By the late Miocene there had been a nearly complete turnover of camel genera in the Gulf coastal region, and most genera were more cosmopolitan, ranging into the mid-continent. The giraffe-camels persisted and were represented by their ultimate product, *Aepycamelus* (fig. 13.24). This genus was as tall as the living giraffe, standing more than 4 meters at the shoulder. The species *Aepycamelus major* is common at several late Miocene localities in north-central Florida. The rest of the late Neogene camels were advanced members of the subfamily Camelinae, which is divided into two tribes, the Camelini and Lamini. Members of this subfamily share an elongated rostrum, completely fused metapodials (acquired independently in *Aepycamelus*), a longer metacarpal than metatarsal, and loss of the I1.

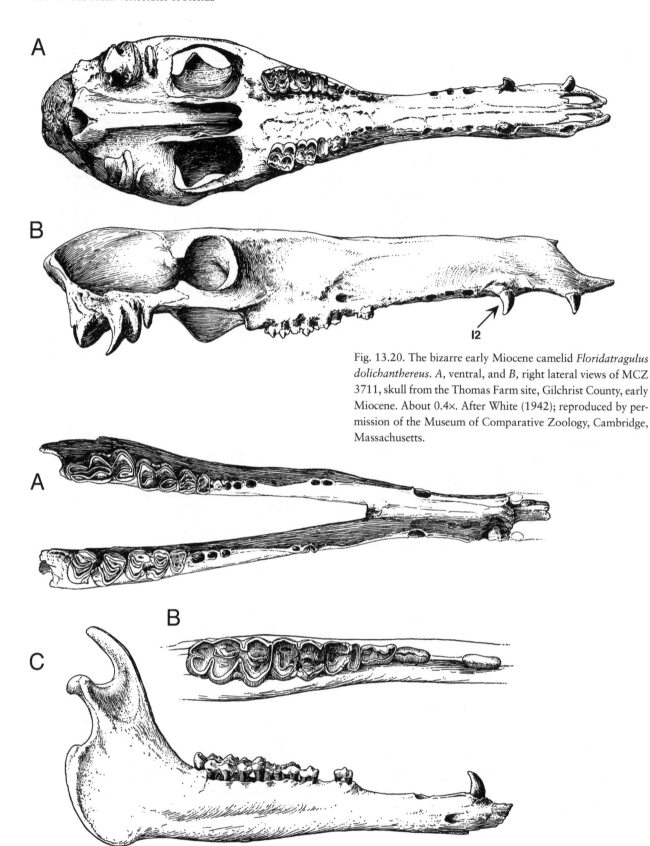

Fig. 13.20. The bizarre early Miocene camelid *Floridatragulus dolichanthereus*. *A*, ventral, and *B*, right lateral views of MCZ 3711, skull from the Thomas Farm site, Gilchrist County, early Miocene. About 0.4×. After White (1942); reproduced by permission of the Museum of Comparative Zoology, Cambridge, Massachusetts.

Fig. 13.21. Mandibles and lower teeth of the camelid *Floridatragulus* from the Thomas Farm site, Gilchrist County, early Miocene. *A*, occlusal view of MCZ 3635, mandible of *F. dolichanthereus*. *B*, occlusal view of cheek teeth (p2–m3), and *C*, lateral view of MCZ 4086, right dentary of *F. barbouri*. *A* about 0.8×; *B* about 1×; *C* about 0.5×. After White (1940, 1947); reproduced by permission of the Museum of Comparative Zoology, Cambridge, Massachusetts.

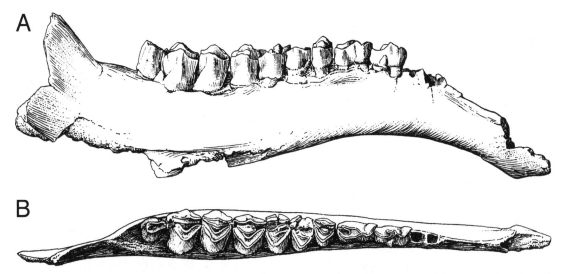

Fig. 13.22. The early Miocene camelid *Nothokemas floridanus* from the Thomas Farm site, Gilchrist County. *A*, lateral, and *B*, oc-clusal views of MCZ 3636, a right dentary with p3–m3. About 0.5×. After White (1947); reproduced by permission of the Museum of Comparative Zoology, Cambridge, Massachusetts.

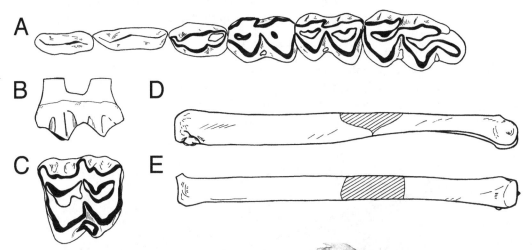

Fig. 13.23. The small camelid from the Buda Locality, Alachua County, very early Miocene. *A*, occlusal view of composite lower dentition (p2–m3); *B*, labial, and *C*, occlusal views of UF 18384, upper left molar; *D*, medial, and *E*, anterior views of UF 18367, a left metacarpal 3. *A*–*C* about 2.5×; *D*–*E* about 1×. After Frailey (1979); reproduced by permission of the Florida Museum of Natural History.

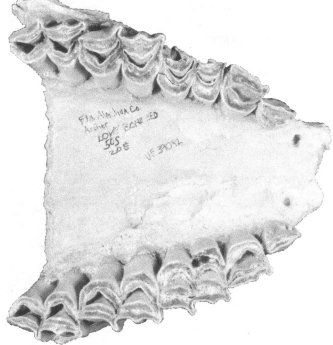

Right: Fig. 13.24. Occlusal view of UF 39042, a palate of the giant giraffe-camel *Aepycamelus major* from the Love site, Alachua County, late Miocene. About 0.5×.

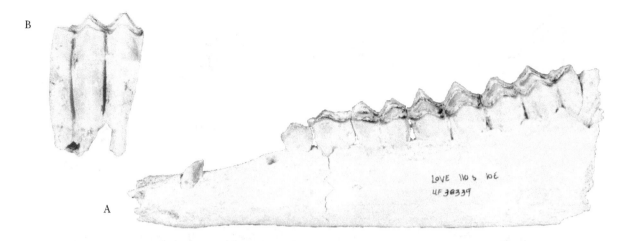

Fig. 13.25. The late Miocene camelid *Procamelus grandis* from the Love site, Alachua County, late Miocene. *A*, lateral view of UF 38339, a left dentary with caniniform p1, p3–m3; *B*, labial view of UF 38356, a right m3 showing how hypsodont the teeth are in this species. Both about 0.5×.

Procamelus was one of the more common middle and late Miocene camelids in North America. For a camel it is of moderately large size. It retained the P2 and the I2, although the latter is small and on the way to being lost. The p2 is reduced (figs. 13.25, 13.26), while the other premolars are narrow and long. The sample from the Love site, which is very similar to *Procamelus grandis* from South Dakota, has a moderately long diastema.

Procamelus-like teeth are also known from the middle Miocene of the Bone Valley in central Florida.

By the late Miocene, camelines were large and had very hypsodont teeth. The Florida representative was the gigantic *Megatylopus* (fig. 13.27). In *Megatylopus*, the p2 was lost or, in the early species, *M. primaevus* from Lapara Creek in Texas, functionless. The p1 and p3 were both reduced, but the canines were large and rounded in

Fig. 13.26. *A*, lateral view of the skull and dentaries, and *B*, ventral view of the skull of UCMP 32864, the camelid *Procamelus grandis* from the middle Miocene of South Dakota. This species is common in the late Miocene of Florida. About 0.25×. After Gregory (1942); reproduced by permission of the University of California Press, Berkeley.

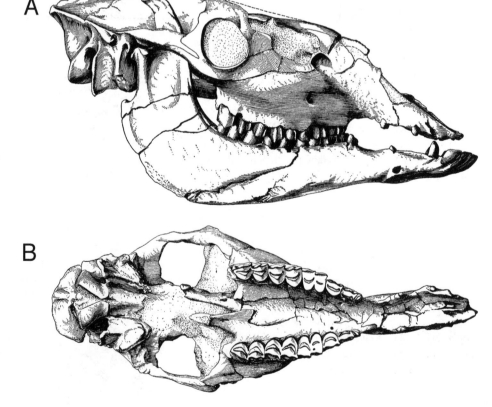

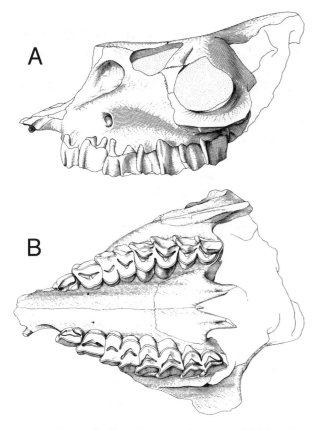

Fig. 13.27. *A*, lateral, and *B*, occlusal views of UCMP 31100, a partial skull with right and left P3–M3 of *Megatylopus matthewi*. This specimen is from Texas, but the species is also known from the early Pliocene of the Bone Valley District. About 0.25×. After Webb (1965), courtesy of S. David Webb, Florida Museum of Natural History.

cross section. *Megatylopus* is best represented in Florida from the early Pliocene Palmetto Fauna of the Bone Valley region; there its fossils consist mostly of isolated teeth and broken skeletal elements. It is the largest camelid in the Palmetto Fauna.

Lamine camelids are generally more common than camelines at southern latitudes in North America and are especially well represented in Florida. Members of the tribe Lamini are easily distinguished from camelines by the anterior-labial styles on their lower molars (the so-called llama buttresses, fig. 13.28). Lamines lack both the P2 and p2. In the late Miocene they are represented by several species, all of which are presently assigned to the genus *Hemiauchenia*. Some are about as tall as a modern camel, though lighter built, and others are the size of a modern llama, but with much longer limb proportions. *Hemiauchenia* is distinguished by its hypsodont, cement-covered cheek teeth, which have very an-

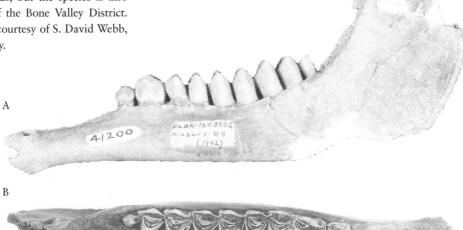

Fig. 13.28. Dentaries of the long-legged llama *Hemiauchenia* from Florida. *A*, lateral view of AMNH 41200, a left dentary with p3–m3 of *H. minima* from Mixson's Bone Bed, Levy County, late Miocene; *B*, occlusal, and *C*, lateral views of UF 104500, a right dentary with p3–m3 of *H. blancoensis* from the APAC (=Macasphalt) Shell Pit, Sarasota County, late Pliocene. *A* about 0.75×; *B–C* about 0.5×.

gular crescents and strong styles, and its very elongated metapodials and cervical vertebrae (figs. 13.28, 13.29B, D). *Hemiauchenia* ranges from the late Miocene to the late Pleistocene, with at least four recognized species in Florida. Two of these, *H. minima* from the late Miocene (fig. 13.28A) and an undescribed species from the early Pliocene, may more properly belong in another, undescribed genus. The large, late Pliocene *H. blancoensis* and the Pleistocene *H. macrocephala* are better known (figs. 13.28B, 13.29D). The latter underwent a profound decrease in size near the end of the Pleistocene. A catastrophic assemblage of a herd of mostly subadult *H. macrocephala* individuals was found at the Leisey Shell

Pit in 1986. The Leisey *Hemiauchenia* was referred to the species *H. seymourensis* by Webb and Stehli (1995), but it is included here in *H. macrocephala* for the reasons outlined by Webb (1974).

The Florida camelid fauna lost much of its diversity in the late Miocene, and after the early Pliocene it is restricted to lamines. Only *Hemiauchenia* is known during the late Pliocene. The genus *Palaeolama* first appears in Florida in the early Pleistocene, its earliest known occurrence. Only one species is presently recognized in Florida, *Palaeolama mirifica*. The relatively brachyodont dentition of *Palaeolama* with crenulated, uncemented enamel and a complex p4 are very distinctive (fig.

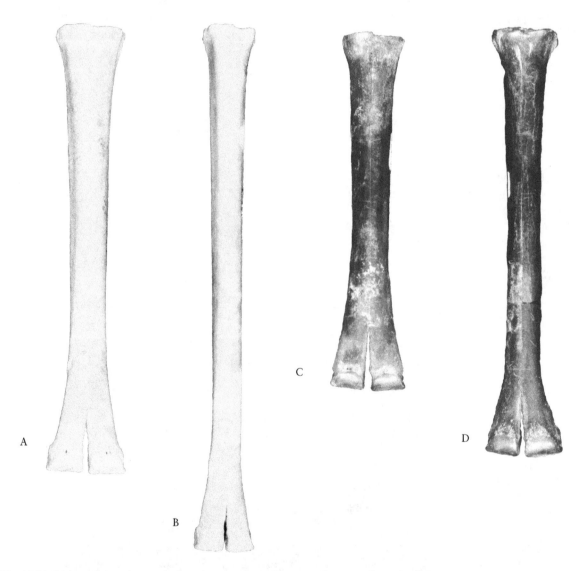

Fig. 13.29. Right metatarsals (cannon bone) in anterior views of camelids showing the difference in proportions between long-legged (B and D) and stout-legged (A and C) species. A, UF 38712, *Procamelus grandis* from the Love site, Alachua County, late Miocene; B, UF 39483, *Hemiauchenia minima* from the Love site; C, UF 65316, *Palaeolama mirifica* from the Leisey Shell Pit, Hillsborough County, early Pleistocene; D, UF 133908, *H. macrocephala* from the Leisey Shell Pit. All about 0.33×.

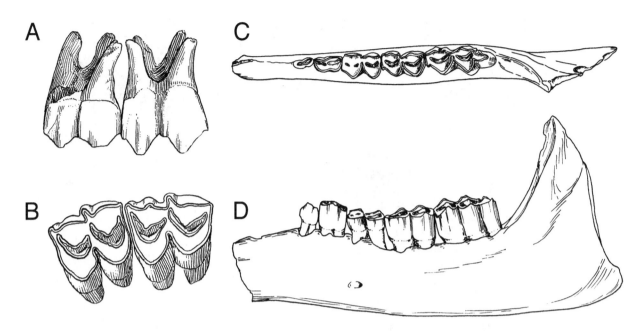

Fig. 13.30. Teeth of the common Pleistocene llama, *Palaeolama mirifica*. *A*, occlusal, and *B*, labial views of AMNH 23489, left M1–M2, from Seminole Field, Pinellas County; *C*, occlusal, and *D*, lateral views of UF 1093, left mandible with p3–m3 from Arrendondo 1B, Alachua County. The labial fold in the p4 is characteristic of this genus. *A–B* about 1×; *C–D* about 0.5×. *A–B* after Simpson (1929); reproduced by permission of the American Museum of Natural History, New York. *C–D* after Bader (1957); reproduced by permission of the Florida Museum of Natural History.

13.30), and its limbs are relatively stocky, especially compared to contemporaneous *Hemiauchenia* (figs. 13.29C, 13.31). It also lacks the caniniform p1 of the latter genus. Of the North American lamines, it is considered the most closely related to the living South American genus *Lama* (Webb, 1974; Honey et al., 1998). An alternative classification of lamines was proposed by Guérin and Faure (1999) in which *Palaeolama* and *Hemiauchenia* were regarded as subgenera within a more broadly defined *Palaeolama*. Both *Palaeolama* and *Hemiauchenia* persisted into the latest Pleistocene, only to go extinct along with many other large mammals about 11,000 years ago.

Fossil Ruminants of Florida

At least eight different ruminant families dispersed into the New World from Asia during the Cenozoic: the Hypertragulidae, Leptomerycidae, Gelocidae, Moschidae, Antilocapridae, Palaeomerycidae, Bovidae, and Cervidae. All eight are represented in Florida by fossils, although only bovids (bison) and cervids (deer) are historically native to the state. Ruminants are characterized by fusion of the cuboid and the navicular (two tarsal elements; fig. 13.5B), reduction or loss of the upper inci-

sors, and an incisiform lower canine. With the exception of the most primitive family, the Hypertragulidae, ruminants have also lost the P1, the trapezium, and the first metacarpal, and have a very reduced fibula.

The earliest ruminant fossils known in Florida are from the I-75 site and represent the Leptomerycidae and Hypertragulidae, the two oldest families of North American ruminants (Webb, 1998). To judge from the small samples of isolated teeth, the Florida species are very similar to those from the White River Group of South Dakota (fig. 13.32). Members of both families were hornless and retained brachyodont dentitions and small body size. They probably lived like their distant cousins, the living tragulids of tropical Africa and southeastern Asia.

The same two families of ruminants continued their residence in North America into the early Miocene (Webb, 1998). At the Buda Site in Alachua County, fairly extensive samples of dental elements and a few limb bones represent the hypertragulid *Nanotragulus loomisi*, a species described and better known from the Great Plains (fig. 13.33). This tiny hornless ruminant resembled the Asiatic mouse deer in many respects but had a more hypsodont dentition than any living tragulid. As in the living tragulids, it had very short front legs and

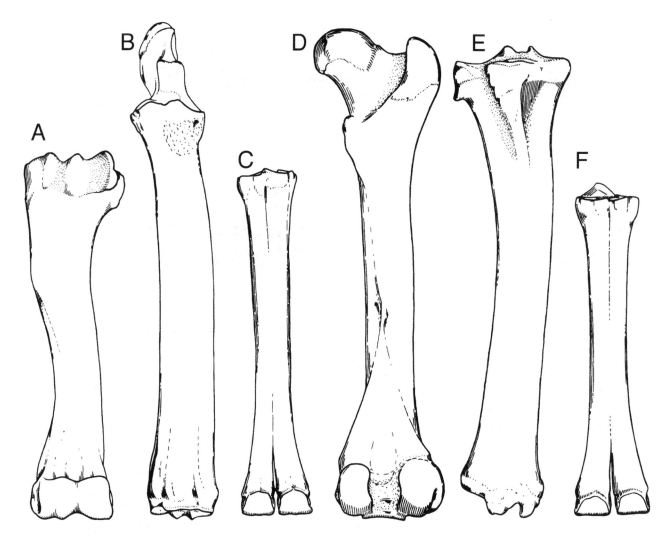

Fig. 13.31. Major limb bones of the short-legged llama, *Palaeolama mirifica*, from the Pleistocene of Florida. *A*, humerus; *B*, radio-ulna; *C*, metacarpal 3–4 (cannon bone); *D*, femur; *E*, tibia; *F*, metatarsal 3–4. *A, C, E, F* are anterior views; *B* and *D* posterior views. About 0.33×. After Webb (1974); reproduced by permission of the University Press of Florida.

evidently ran like a jackrabbit, with a flexible back, rather than like a deer. The primitive ruminant families Hypertragulidae and Leptomerycidae became extinct in the late Miocene.

Excluding the Hypertragulidae and Leptomerycidae, the remaining North America ruminants all belong to an advanced group called the Pecora. Pecorans are characterized by having compact, parallel-sided astragali, fused and elongated third and fourth metapodials, reduced or lost side toes, and, at least in all living members, a complex, four-chambered stomach. The Giraffoidea, the Bovoidea, and the Cervoidea compose the three great pecoran radiations, as well as some problematic fossil genera that do not fit in well with any of these groups, but which share derived pecoran character states. The Giraffoidea is entirely an Old World group with only two

living members, the giraffe and okapi. The Bovoidea (with the single family Bovidae) is primarily an Old World group but includes several immigrants to North America, such as mountain sheep, musk oxen, and bison. Bovids are characterized by their true horns with keratinous sheaths, upper canine small or absent, hypsodont cheek teeth, complete metapodial keels, and side toes absent. Cervoids, which include the antilocaprids, dromomerycines, and cervids, are characterized by a posterolingual fold of enamel running from the protoconid of the molars (secondarily lost in hypsodont taxa), and by the fusion of the remnant of metatarsal 2 to the proximal end of the canon bone. With the exception of a few extinct genera, all cervoids have complete distal metapodial keels (apparently acquired independently of bovids).

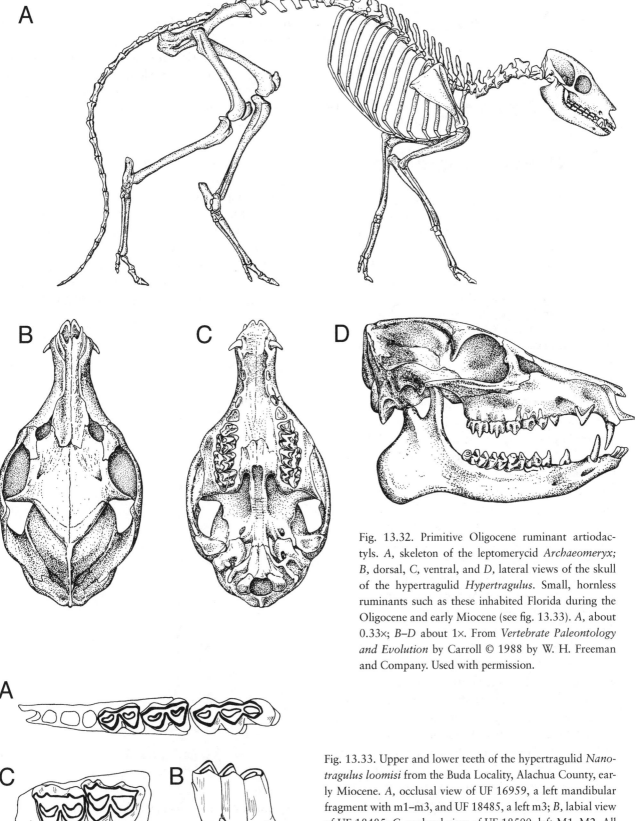

Fig. 13.32. Primitive Oligocene ruminant artiodactyls. *A*, skeleton of the leptomerycid *Archaeomeryx;* *B*, dorsal, *C*, ventral, and *D*, lateral views of the skull of the hypertragulid *Hypertragulus*. Small, hornless ruminants such as these inhabited Florida during the Oligocene and early Miocene (see fig. 13.33). *A*, about 0.33×; *B–D* about 1×. From *Vertebrate Paleontology and Evolution* by Carroll © 1988 by W. H. Freeman and Company. Used with permission.

Fig. 13.33. Upper and lower teeth of the hypertragulid *Nanotragulus loomisi* from the Buda Locality, Alachua County, early Miocene. *A*, occlusal view of UF 16959, a left mandibular fragment with m1–m3, and UF 18485, a left m3; *B*, labial view of UF 18485; *C*, occlusal view of UF 18500, left M1–M2. All about 2.5×. After Frailey (1979); reproduced by permission of the Florida Museum of Natural History.

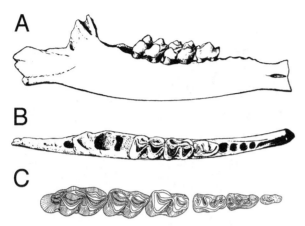

Fig. 13.34. The early Miocene moschid ruminants *Machaeromeryx* and *Parablastomeryx* from the Thomas Farm site, Gilchrist County. *A*, lateral, and *B*, occlusal views of MCZ 3651, a right dentary with p4–m2 of *M. gilchristensis*. *C*, occlusal view of composite left lower dentition (p2–m3) of *P. floridanus*. Composed of MCZ 7763 and 4349. *A–B* about 1.33×; C about 1.5×. *A–B* after White (1941); reproduced by permission of the Museum of Comparative Zoology, Cambridge, Massachusetts. *C* after Maglio (1966); reproduced by permission of the Museum of Comparative Zoology, Cambridge, Massachusetts.

In the early Miocene of North America, the moschids consisted of five genera (Webb, 1998). One genus, characterized by very small size, a long snout, and short premolars is *Machaeromeryx*. In contrast *Parablastomeryx* has a short snout and long premolars and was relatively large for a moschid. Fossils of both of these genera are known from Thomas Farm (figs. 13.34A, C, 13.35A, B). The Moschidae is also questionably represented by a few dental fragments and postcranial elements from older deposits in the Bone Valley region.

The subfamily Dromomerycinae includes three major lineages in the early Miocene of North America. The presence of a long, medial occipital "horn," in addition to the paired frontal "horns," characterizes the tribe Cranioceratini (fig. 13.36), the only members of the subfamily known from Florida. The neck and limbs were heavily built and relatively short. The body shape was roughly comparable to that of a moose. The entire subfamily became extinct very early in the Pliocene (Janis and Manning, 1998). Unfortunately, however, the earliest material representing dromomerycines in Florida has not been described and does not yet include horns or complete enough dentitions to determine its generic or tribal assignment. A few specimens of the moderate-sized, middle Miocene *Cranioceras* occur in the older beds of the Bone Valley region, while teeth referred to *Rakomeryx* and *Bouromeryx* have been found in Gads-

den County (Bryant, 1991; fig. 13.37). The best fossils of the Dromomerycinae in eastern North America come from the late Miocene Love site. There *Pediomeryx hamiltoni* is represented by an excellent sample of cranial, dental, and limb elements (Webb, 1983). The shallow depth of the mandible, the long diastema in front of the cheek teeth, and the low-crowned teeth with crenulated enamel are all features reminiscent of a giraffe jaw (fig. 13.38). Indeed dromomerycines were once classified in the Giraffoidea but are now considered cervoids (Janis and Scott, 1987). *Pediomeryx hemphillensis*, a smaller species, is known from a very late Miocene skull, mandibles, and partial skeleton from Marion County.

The hornless Gelocidae first appeared in North America in the late Miocene (Webb, 1998). This family of small, hornless ruminants is represented by sparse material of the genus *Pseudoceras* at about fifteen sites in all of North America. At the late early Hemphillian Withlacoochee River 4A site in central Florida, however,

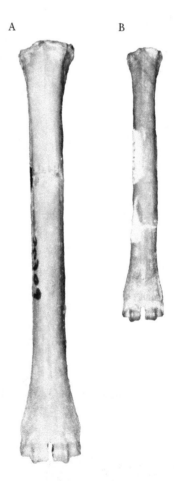

Fig. 13.35. Anterior views of the metacarpals of the two small ruminants from the Thomas Farm site, Gilchrist County, early Miocene. *A*, UF 20309, *Parablastomeryx floridanus*. *B*, UF/FGS 6493, *Machaeromeryx gilchristensis*. Both about 1×.

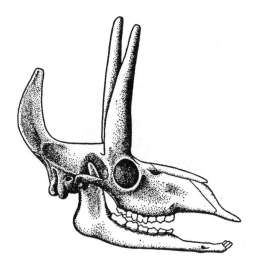

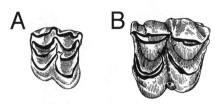

Fig. 13.37. Middle Miocene dromomerycine teeth from La Camelia Mine, Gadsden County. *A,* occlusal view of UF 116825, right upper molar of *Bouromeryx* sp., possibly *B. parvus; B,* occlusal view of UF 118530, left upper molar of *Rakomeryx* sp. Both about 1×; after Bryant (1991); reproduced by permission of the Society of Vertebrate Paleontology.

Fig. 13.36. Right lateral view of skull and dentary of the middle Miocene dromomerycine *Cranioceras.* Note the three "horns": two supraorbital and one posterior and medial. About 0.25×. Modified from Romer(1966); copyright © 1966 by The University of Chicago Press and reproduced by permission of the publisher.

an undescribed gelocid is the most common mammal, represented by hundreds of limb bones, dozens of jaws, and several skulls (fig. 13.39). An explanation for its unique abundance at this one site may be that, like the water chevrotain in equatorial Africa, it lived in dense forests near water, and that the green clay beds at Withlacoochee 4A represent a forest pond.

Pseudoceras dentally resembled both ruminants and camelids, but postcranially they were wholly similar to ruminants. They were once regarded as camelids because they had an upright lower canine and narrow premolars, and the postcranial skeleton was unknown. The large collection of mandibles from Withlacoochee 4A shows that only about half of the mature specimens had upright lower canines; the other half had incisiform canines as in other ruminants. Evidently the males had large, vertical canines for fighting, whereas the females had smaller, incisiform canines. The males also had long, bladelike upper canines (fig. 13.39), as do living tragulids. The narrow premolars are a primitive character for ruminants, still evident in Gelocidae as in Camelidae, but not retained in any higher ruminant groups. The preponderance of characters clearly indicates that *Pseudoceras* is a small hornless ruminant, but its relationships with other ruminants is uncertain, as is its actual affinity with the living Old World genus *Gelocus* (Webb, 1998).

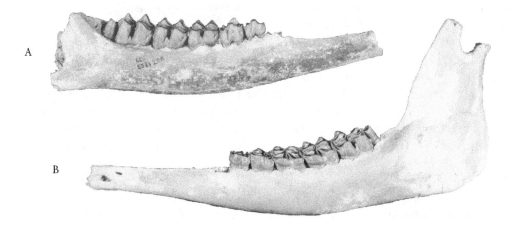

Fig. 13.38. Lateral views of dentaries of the dromomerycine *Pediomeryx hamiltoni* from the Love site, Alachua County, late Miocene. *A,* UF 27365, a right dentary with p3–m3 of a young adult; *B,* UF 27217, a left dentary with p3–m3 of an older adult with well-worn teeth. Both about 0.4×.

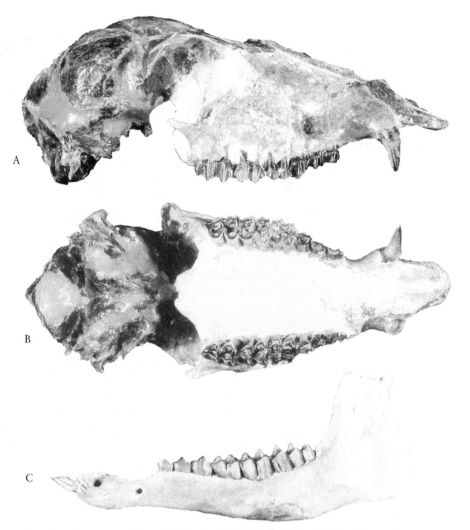

Fig. 13.39. The hornless ruminant *Pseudoceras* n. sp. from the Withlacoochee River 4A site, Marion County, late Miocene. *A*, right lateral, and *B*, ventral views of UF 19258–19260, a composite skull. This specimen actually consists of at least two individuals joined together to reconstruct the skull of this small forest dweller. The large canine indicates that it was a male. *C*, lateral view of UF 19395, a left dentary with i1, p2–m3. All about 1×.

In Florida, middle and late Miocene records of the family Antilocapridae (the family based on *Antilocapra,* the pronghorn antelope that lives in western North America) are extremely rare. Only a few specimens representing the older subfamily Cosorycinae have been found in the older parts of the Bone Valley. The more progressive subfamily Antilocaprinae is represented at the Love site by a small number of hypsodont teeth and isolated limb elements. Similar limited records come from two other late Miocene sites, McGehee Farm and Withlacoochee River 4A. These sparse records contrast strikingly with the rich samples from western sites, where extinct pronghorns often dominate the faunas in which they occur.

The two species of the Antilocapridae found in the early Pliocene Palmetto Fauna are both uniquely known

from those deposits. The larger and more common of the two is *Hexobelomeryx simpsoni,* a large six-horned antilocaprine often placed in its own genus, *Hexameryx.* This extraordinary animal has been taken as the symbol of the Florida Paleontological Society (fig. 13.40). Its closest relative is *Hexobelomeryx fricki* from the early Pliocene of northern Mexico, the only other known six-horned antilocaprid. As reconstructed by Webb (1973), each "horn" had a keratinous sheath that tapered slightly toward the tip but was probably not forked. Among *Hexobelomeryx* "horns" from the Bone Valley, two distinct size groups and two subtly different patterns of forking are evident (fig. 13.41A, B). Although these were originally recognized as separate species, Webb (1973) suggested that the smaller sets in which the middle "horn" more closely allied with the anterior

Fig. 13.40. Reconstructed head of the six-horned pronghorn *Hexobelomeryx simpsoni* in lateral and posterior views. After Webb (1973); reproduced by permission of the American Society of Mammalogists.

"horn" were females, whereas the larger ones with the middle "horn" more centrally placed were from males of the same species. The other Bone Valley antilocaprid is rarer than *Hexobelomeryx* and was first described as *Antilocapra (Subantilocapra) garciae*, a Pliocene subgenus of the living genus *Antilocapra* (Webb, 1973). *Subantilocapra* is now recognized as a distinct genus, although not disregarding the possibility of a close relationship with *Antilocapra*. The horn core of *Subantilocapra* consists of a flattened, dense plate of bone with short, unequally forked tips (fig. 13.41C).

Dentitions of both *H. simpsoni* and *S. garciae* are known from the Bone Valley. It is traditional to base the taxonomy of fossil antilocaprids primarily on horn core

morphology. Nevertheless, other parts of the skeleton are of importance as well. No teeth are actually associated directly with horn cores in the Bone Valley, but two distinct groups of dentitions can be securely assigned to the two species named on horn cores on the basis of size. The dentitions of *H. simpsoni* are about 20 percent larger than those of *S. garciae*. For their size, they are also less compressed transversely and somewhat less hypsodont (fig. 13.42). As in the living pronghorn, however, both species had hypsodont teeth that were well adapted to a diet of coarse vegetation.

The small antilocaprid *Capromeryx arizonensis* is extensively sampled in the late Pliocene Santa Fe River and Inglis 1A sites (fig. 13.43). These samples are the

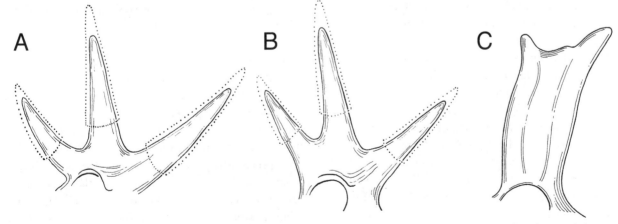

Fig. 13.41. Horn cores of the two early Pliocene antilocaprids of the Bone Valley region. *A*, male, and *B*, female *Hexobelomeryx simpsoni*. *C. Subantilocapra garciae*. *A–B* about 0.25×; *C* about 0.5×. After Webb (1973); reproduced by permission of the American Society of Mammalogists.

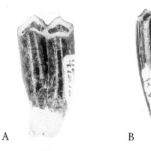

A   B

Fig. 13.42. Isolated teeth of the antilocaprid *Hexobelomeryx simpsoni* from the Bone Valley District, Polk County, early Pliocene. *A*, lingual view of UF 91215, a right M1 or M2 from the Mineco Mine. *B*, labial view of UF 18166, a left m3 from the Chicora Mine. Both about 1×.

only ones in Florida that seem to have preserved antilocaprid fossils abundantly, as one might expect of such herd animals. The remains of *C. arizonensis* are especially extensive at the Inglis site, a coastal plain sinkhole. *Capromeryx* is a four-horned antilocaprid. In *C. arizonensis,* the anterior and posterior horn cores are of approximately equal height and stand vertically. The Inglis record is the youngest known for antilocaprids in Florida. The family does not occur in any of the rich early Pleistocene sites in the state, such as Haile 16A or Leisey Shell Pit, so it apparently became locally extinct at the end of the Pliocene.

During the Pliocene and Pleistocene, North America received representatives of two new ruminant families: the Cervidae and the Bovidae. In fact, both families sent separate dispersals of several species from Asia. A large number of antlers, jaws, and some cranial and limb fragments from the Palmetto Fauna represent the oldest cervid in the New World, *Eocoileus gentryorum* (figs. 13.44A, 13.45A). Webb (2000) stated that this early Pliocene species represents a primitive member of the Odocoileinae, the cervid subfamily that includes the familiar white-tailed deer (*Odocoileus virginianus*), the only surviving member of the group in the eastern United States. Eight or more additional genera of this subfamily, however, occur in the Pleistocene and Holocene of North and South America. Much remains to be learned about the early fossil record of New World Cervidae.

The most common cervid in Florida late Pliocene and Pleistocene sites is *Odocoileus*, generally indistinguishable from *O. virginianus* living in the state today. Typical fossils of this deer are shown in figures 13.44 and 13.45. Occasionally, a very large antler fragment has been identified as that of an elk (*Cervus canadensis*). However, without the corresponding dentitions or nearly complete

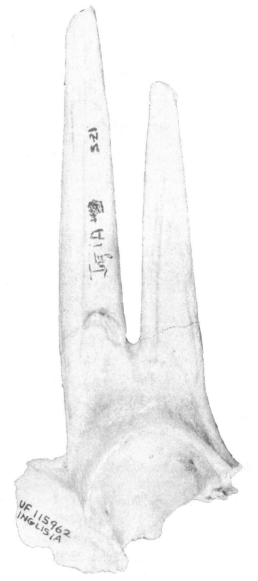

Fig. 13.43. Right lateral view of UF 115962, a skull fragment containing the orbit and the two supraorbital horn cores of the antilocaprid *Capromeryx arizonensis* from Inglis 1A, Citrus County, late Pliocene. About 1×.

antlers to show the characteristic branching pattern, this identification cannot be supported. It is more likely that some Pleistocene white-tailed deer grew to large size and had correspondingly very large antlers. A different type of cervid was named from Saber-tooth Cave in Citrus County, *Blastocerus extraneus* (fig. 13.46). The only living species of *Blastocerus* is *B. dichotomus*, the largest South American deer, which favors wetlands and ranges widely over parts of Uruguay, Brazil, and Argentina. Unfortunately, *Blastocerus extraneus* was based on a single mandible, whereas the large, multiple-forking antlers would be more distinctive. The fact that no further material has been recognized since its description more

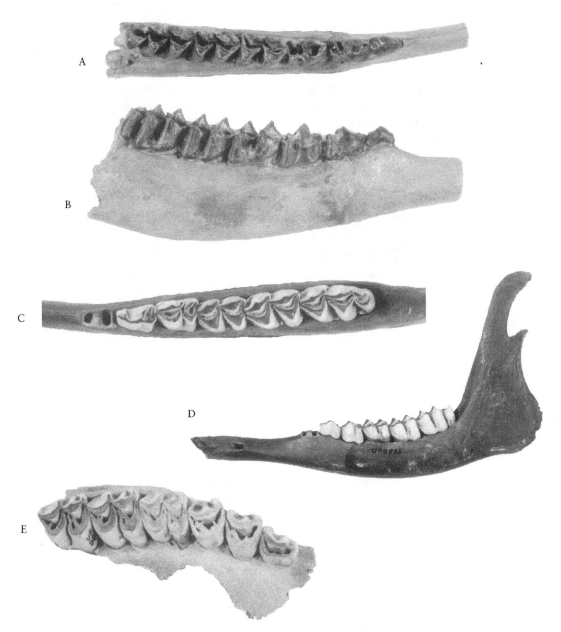

Fig. 13.44. Dentitions of fossil deer from Florida. *A*, occlusal, and *B*, lateral views of a MCZ specimen, a right dentary with p2–m3 of *Eocoileus gentryorum* from the Bone Valley District, early Pliocene. *C–E*, *Odocoileus virginianus* from Devils Den, Levy County, late Pleistocene. *C*, occlusal, and *D*, lateral views of UF 9835, a left dentary with p3–m3. *E*, occlusal view of UF 9824, a right maxilla with P2–M3. All are about 1× except *D*, which is about 0.5×.

than seventy years ago is worrisome but not conclusive one way or the other. The dentition of *B. extraneus* does seem to fall outside of the normal range of variation seen in *Odocoileus*.

It is interesting to note that samples of cervids are rarely very rich in Pliocene and early Pleistocene sites in Florida (Webb and Stehli, 1995). Cervids are usually less abundant than either camelids (for example, at Leisey Shell Pit) or *Capromeryx* (for example, at Inglis). Only in the latest Pleistocene and Holocene, during or after the

extinction of most of the other large ungulates, do *Odocoileus* numbers increase to modern levels of abundance.

The first record of the family Bovidae in Florida is late in the Pliocene and consists of fragmentary material from Macasphalt Shell Pit and Inglis 1A. These specimens are currently referred to the genus *Bison*, but not with certainty to any particular species (McDonald and Morgan, 1999). They represent a separate dispersal of Old World *Bison* to North America more than 1.5 million years prior to their well-established middle Pleistocene immi-

Fig. 13.45. Fossil deer antlers from Florida. *A*, UF 102048, *Eocoileus gentryorum* from the Fort Green Mine, Polk County, early Pliocene; *B*, UF/FGS 4934, *Odocoileus virginianus* from the Ichetucknee River, Columbia County, late Pleistocene. Both about 0.6×.

A

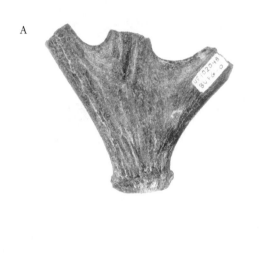

B

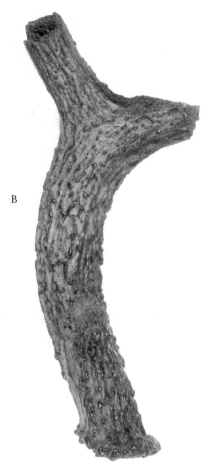

gration that marks the beginning of the Rancholabrean Land Mammal Age about 300 ka.

Bison rapidly became one of the characteristic large ungulates in North America in the middle Pleistocene, following their dispersal from Asia. Despite their status as newcomers, they managed to persist in large numbers into the Holocene, a claim that resident horses, llamas, or proboscideans cannot make. From the large, middle Pleistocene species *Bison latifrons*, through the late Pleis-

tocene *Bison antiquus*, to the modern *Bison bison*, bison evidently flourished in Florida. Many fossil species have been named, but only a few are generally regarded as valid. The most useful parts of the skeleton for species identification are the horn cores, but those of any one species are variable, especially between the sexes.

North American *Bison* migrated from Asia by way of the Bering Strait and extended its range over the entire continent in the late middle Pleistocene, about 300 ka.

A

B

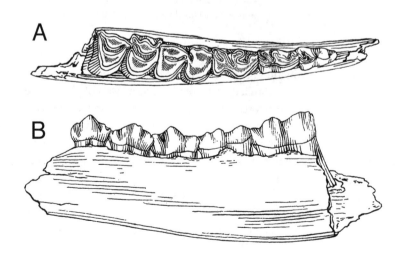

Fig. 13.46. *A*, occlusal, and *B*, labial views of AMNH 23457, a right dentary with p2–m2 of the cervid *Blastocerus extraneus* from Saber-tooth Cave, Citrus County, late Pleistocene. About 1×. After Simpson (1928); reproduced by permission of the American Museum of Natural History, New York.

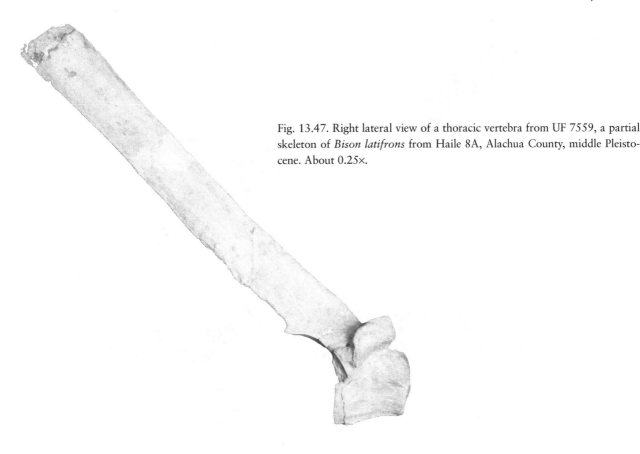

Fig. 13.47. Right lateral view of a thoracic vertebra from UF 7559, a partial skeleton of *Bison latifrons* from Haile 8A, Alachua County, middle Pleistocene. About 0.25×.

It was similar in appearance to the living *Bison* except for greater body size, a larger hump above the shoulders, and gigantic horns spanning more than 2 meters. The muscles necessary to support the extremely large and heavy horns were accommodated by the very high neural spines of the thoracic vertebrae (fig. 13.47). The hump is made up of these spines, attached muscles, and a covering of fat and skin. These large-horned animals may have been the ancestors of those with intermediate-sized horns (*B. antiquus* and relatives), which were also widespread in North America, or these could represent a second dispersal event from Asia.

The living *B. bison* probably evolved from those with intermediate-sized horns, such as *B. antiquus*. For that reason, *B. antiquus* is sometimes only considered a subspecies of *B. bison*. The morphological change leading to the living species involved further reduction of horn and hump size, as well as a decrease in overall body size. The surviving North American species is the smallest of all the *Bison*. Although the living *Bison* is generally regarded as a resident of the western United States, its presence in Florida, as in other parts of the eastern United States, was recorded by many early European explorers.

One species from each of the above mentioned groups was present in Florida. The large-horned variety, *B. latifrons* (figs. 13.47–13.50), is known from several sites in northern Florida. Three nearly complete skeletons and several isolated horn cores have been found. The skeletons had been little disturbed since the time of death. In one instance (at Haile 8A) the animal had fallen through the roof of a cave and was preserved in its entirety in a trap of its own making. Finds like these are rare, however, and most of the fossil *Bison* material from Florida consists of isolated teeth and limb bones. Excellent cranial material of *B. latifrons* is also known from near Bradenton.

A partial associated skeleton and several skulls of *B. antiquus* (figs. 13.48B, 13.49A) have been taken from north Florida rivers. An associated skeleton was found in place on the Santa Fe River bottom. Although river specimens are not usually associated, a lack of duplication of elements, as well as the extremely close agreement in measurements of left and right elements, is convincing evidence that these bones belonged to one individual. Most of the teeth and postcranial material of late Pleistocene *Bison* from Florida probably belong to *B. anti-*

Fig. 13.48. Fossil *Bison* horn cores from Florida. *A,* anterodorsal view of UF 3558, a right horn core of *B. latifrons* from the Bradenton 51st Street Locality, Manatee County, middle Pleistocene; *B,* posterior view of a skull of *B. antiquus* from the Aucilla River, Jefferson County, late Pleistocene. The horn cores of *B. latifrons* are much longer and less curved than those of *B. antiquus. A* about 0.14×; *B* about 0.17×.

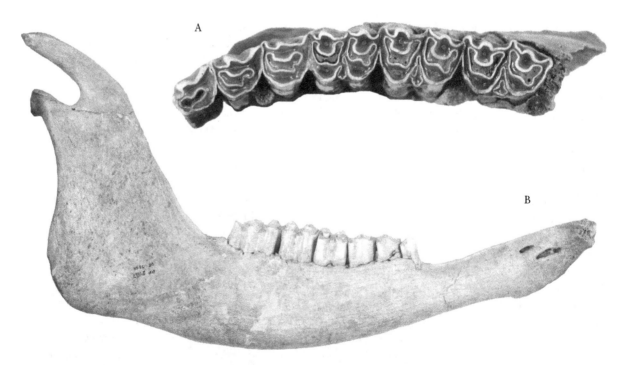

Fig. 13.49. *A,* occlusal view of UF 19376, a left maxilla with P2–M3 of *Bison antiquus* from the Ichetucknee River, Columbia County, late Pleistocene; *B,* lateral view of UF 7559, a right dentary with p2–m3 of *Bison latifrons* from Haile 8A, Alachua County, middle Pleistocene. *A* about 0.67×; *B* about 0.3×.

quus. One interesting specimen, from the Wacissa River, includes a partial skull with portions of a chert Paleoindian spear point imbedded in its frontal bone. Its has a carbon-14 date of 11.2 ka and provides direct evidence for human predation on *B. antiquus* in Florida.

The upper and lower molars of *Bison* have the accessory pillar characteristic of the family Bovidae (fig. 13.49A). Once familial identification has been established, the genus can be determined by tooth size and

enamel pattern. Positive specific identification of *Bison* from material other than horn cores and thoracic vertebrae is difficult. Isolated teeth and postcranial elements (which represent the bulk of the known material) can only be labeled "*Bison* sp." Postcranial material (fig. 13.50) can be distinguished from camelids and other bovids by key characters and size. Isolated teeth and postcranial material of the living species can be identified by their significantly smaller size, but at this point a fur-

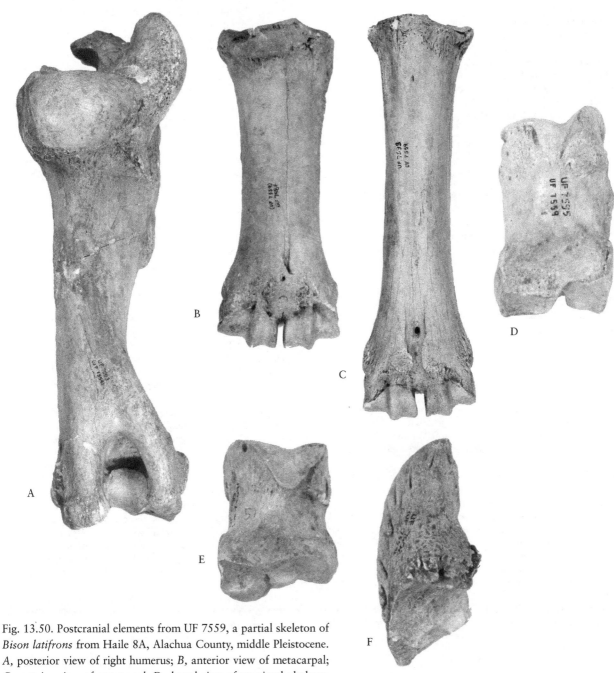

Fig. 13.50. Postcranial elements from UF 7559, a partial skeleton of *Bison latifrons* from Haile 8A, Alachua County, middle Pleistocene. A, posterior view of right humerus; B, anterior view of metacarpal; C, anterior view of metatarsal; D, dorsal view of proximal phalanx; E, dorsal view of medial phalanx; F, dorsal view of distal phalanx. A about 0.3×; B–C about 0.4×; D–F about 0.67×.

ther problem is encountered. The postcranial skeletons of cattle (*Bos*) are similar to those of *B. bison*. These can be separated only by a great deal of study and comparison (Olsen, 1960).

References

Bader, R. S. 1957. Two Pleistocene mammalian faunas from Alachua County, Florida. Bulletin of the Florida State Museum, 2:55–75.

Bryant, J. D. 1991. New early Barstovian (middle Miocene) vertebrates from the upper Torreya Formation, eastern Florida panhandle. Journal of Vertebrate Paleontology, 11:472–89.

Carroll, R. L. 1988. *Vertebrate Paleontology and Evolution*. W. H. Freeman, New York, 698 p.

Frailey, C. D. 1978. An early Miocene (Arikareean) fauna from northcentral Florida (the SB-1A local fauna). Occasional Papers of the Museum of Natural History, University of Kansas. 75:1–20.

———. 1979. The large mammals of the Buda local fauna (Arikareean: Alachua County, Florida). Bulletin of the Florida State Museum, 24:123–73.

Gidley, J. W., and C. L. Gazin. 1938. The Pleistocene vertebrate fauna from Cumberland Cave Maryland. United States National Museum Bulletin, 171:1–99.

Gregory, J. T. 1942. Pliocene vertebrates from Big Spring Canyon, South Dakota. University of California Publications, Bulletin of the Department of Geological Sciences, 26:307–446.

Guérin, C., and M. Faure. 1999. *Palaeolama (Hemiauchenia) niedae* nov. sp., nouveau Camelidae du Noreste brésilien et sa place parmi les Lamini d'Amérique de Sud. Geobios, 32:629–59.

Harrison, J. A. 1985. Giant camels from the Cenozoic of North America. Smithsonian Contributions to Paleobiology, 57:1–29.

Honey, J. G., J. A. Harrison, D. R. Prothero, and M. S. Stevens. 1998. Camelidae. Pp. 437–62 *in* C. M. Janis, K. M. Scott, and L. L. Jacobs (eds.), *Evolution of Tertiary Mammals of North America*. Volume 1, *Terrestrial Carnivores, Ungulates, and Ungulatelike Mammals*. Cambridge University Press, Cambridge.

Janis, C. M., and E. Manning. 1998. Dromomerycidae. Pp. 477–90 *in* C. M. Janis, K. M. Scott, and L. L. Jacobs (eds.), *Evolution of Tertiary Mammals of North America*. Volume 1, *Terrestrial Carnivores, Ungulates, and Ungulatelike Mammals*. Cambridge University Press, Cambridge.

Janis, C. M., and K. M. Scott. 1987. The interrelationships of higher ruminant families with special emphasis on the members of the Cervoidea. American Museum Novitates, 2893:1–85.

Janis, C. M., K. M. Scott, and L. L. Jacobs. 1998. *Evolution of Tertiary Mammals of North America*. Volume 1, *Terrestrial Carnivores, Ungulates, and Ungulatelike Mammals*. Cambridge University Press, Cambridge, 691 p.

Joeckel, R. M. 1990. A functional interpretation of the masticatory system and paleoecology of entelodonts. Paleobiology, 16:459–82.

Lander, B. 1998. Oreodontoidea. Pp. 402–425 *in* C. M. Janis, K. M. Scott, and L. L. Jacobs (eds.), *Evolution of Tertiary Mammals of North America*. Volume 1, *Terrestrial Carnivores, Ungulates, and Ungulatelike Mammals*. Cambridge University Press, Cambridge.

MacFadden, B. J. 1980. An early Miocene land mammal (Oreodonta) from a marine limestone in northern Florida. Journal of Paleontology, 54:93–101.

Maglio, V. J. 1966. A revision of the fossil selenodont artiodactyls from the middle Miocene Thomas Farm, Gilchrist County, Florida. Breviora, 255:1–27.

McDonald, J. N., and G. S. Morgan. 1999. The appearance of bison in North America. Current Research in the Pleistocene, 16:127–29.

McKenna, M. C., and S. K. Bell. 1997. *Classification of Mammals above the Species Level*. Columbia University Press, New York, 631 p.

Morgan, G. S. 1989. Miocene vertebrate faunas from the Suwannee River Basin of north Florida and south Georgia. Pp. 26–53 *in* G. S. Morgan (ed.), *Miocene Paleontology and Stratigraphy of the Suwannee River Basin of North Florida and South Georgia*. Southeastern Geological Society Guidebook No. 30.

Olsen, S. J. 1960. Post-cranial characters of bison and bos. Peabody Museum Papers, Harvard University, 35(4):1–15.

Patton, T. H. 1967. Revision of the selenodont artiodactyls from Thomas Farm. Quarterly Journal of the Florida Academy of Sciences, 29:179–90.

Patton, T. H., and B. E. Taylor. 1971. The Synthetoceratinae (Mammalia, Tylopoda, Protoceratidae). Bulletin of the American Museum of Natural History, 145:119–218.

Peterson, O. A. 1906. The Miocene beds of western Nebraska and eastern Wyoming and their vertebrate faunae. Annals of the Carnegie Museum, 4:21–72.

———. 1914. The osteology of *Promerycochoerus*. Annals of the Carnegie Museum, 9:149–219.

Prothero, D. R., E. M. Manning, and M. Fischer. 1988. The phylogeny of ungulates. Pp. 201–34 *in* M. S. Benton (ed.), *The Phylogeny and Classification of the Tetrapods*. Volume 2, *Mammals*. Clarendon Press, Oxford, England.

Robertson, J. S. 1974. Fossil *Bison* of Florida. Pp. 214–46 *in* S. D. Webb (ed.), *Pleistocene Mammals of Florida*. University Presses of Florida, Gainesville.

Romer, A. S. 1966. *Vertebrate Paleontology*. 3d ed. University of Chicago Press, Chicago, 468 p.

Simpson, G. G. 1928. Pleistocene mammals from a cave in Citrus County, Florida. American Museum Movitates, 328:1–16.

———. 1929. Pleistocene mammalian fauna of the Seminole Field, Pinellas County, Florida. Bulletin of the American Museum of Natural History, 56:561–99.

Vaughn, T. A. 1972. *Mammalogy*. W. B. Saunders, Philadelphia, 463 p.

Webb, S. D. 1965. The osteology of *Camelops*. Bulletin of the Los Angeles County Museum, 1:1–54.

———. 1973. Pliocene pronghorns of Florida. Journal of Mammalogy, 54:203–21.

———. 1974. Pleistocene llamas of Florida, with a brief review of the Lamini. Pp. 170–213 *in* S. D. Webb (ed.), *Pleistocene Mammals of Florida*. University Presses of Florida, Gainesville.

———. 1981. *Kyptoceras amatorum*, new genus and species from the Pliocene of Florida, the last protoceratid artiodactyl. Journal of Vertebrate Paleontology, 1:357–65.

———. 1983. A new species of *Pediomeryx* from the late Miocene of Florida, and its relationships with the subfamily Cranioceratinae (Ruminantia: Dromomerycidae). Journal of Mammalogy, 64:261–276.

———. 1998. Hornless ruminants. Pp. 463–76 *in* C. M. Janis, K. M. Scott, and L. L. Jacobs (eds.), *Evolution of Tertiary Mammals of North America*. Volume 1, *Terrestrial Carnivores, Ungulates, and Ungulatelike Mammals*. Cambridge University Press, Cambridge.

———. 2000. Evolutionary history of New World Cervidae. Pp. 38–64 *in* E. S. Vrba and G. B. Schaller (eds.), *Antelopes, Deer, and Relatives: Fossil Record, Behavioral Ecology, Systematics, and Conservation*. Yale University Press, New Haven.

Webb, S. D., and F. G. Stehli. 1995. Selenodont artiodactyls (Camelidae and Cervidae) from the Leisey Shell Pits, Hillsborough County, Florida. Bulletin of the Florida Museum of Natural History, 37:621–43.

Wetzel, R. M., R. E. Dubois, R. L. Martin, and P. Meyers. 1975. *Catagonus*, an "extinct" peccary alive in Paraguay. Science, 189:379–81.

White, T. E. 1940. New Miocene vertebrates from Florida. Proceedings of the New England Zoological Club, 18:31–38.

———. 1941. Additions to the Miocene fauna of Florida. Proceedings of the New England Zoological Club, 18:91–98.

———. 1942. The lower Miocene fauna of Florida. Bulletin of the Museum of Comparative Zoology, 92:1–49.

———. 1947. Additions to the Miocene fauna of north Florida. Bulletin of the Museum of Comparative Zoology, 99: 497–515.

Wright, D. B. 1990. Phylogenetic relationships of *Catagonus wagneri*: sister taxa from the Tertiary of North America. Pp. 281–308 *in* J. F. Eisenberg and K. Redford (eds.), *Advances in Neotropical Mammalogy*. Sandhill Crane Press, Gainesville, Fla.

———. 1993. Evolution of sexually dimorphic characters in peccaries (Mammalia, Tayassuidae). Paleobiology, 19:52–70.

———. 1995. Tayassuidae of the Irvingtonian Leisey Shell Pit local fauna, Hillsborough County, Florida. Bulletin of the Florida Museum of Natural History, 37:603–19.

———. 1998. Tayassuidae. Pp. 389–401 *in* C. M. Janis, K. M. Scott, and L. L. Jacobs (eds.), *Evolution of Tertiary Mammals of North America*. Volume 1, *Terrestrial Carnivores, Ungulates, and Ungulatelike Mammals*. Cambridge University Press, Cambridge.

Wright, D. B., and S. D. Webb. 1984. Primitive *Mylohyus* (Artiodactyla: Tayassuidae) from the late Hemphillian Bone Valley Formation of Florida. Journal of Vertebrate Paleontology, 3:152–59.

14

Mammalia 6

Perissodactyls

INTRODUCTION

The Perissodactyla, or "odd-toed" ungulates, includes only three living families and five living genera, not a very impressive total. The fossil record reveals that the group was once much more diverse and widespread. Perissodactyls are often very abundant at Florida fossil localities. Among the living ungulates, perissodactyls are most closely related to the elephants (Proboscidea), manatees (Sirenia), and hyraxes (Hyracoidea) (fig. 13.1). To reflect this relationship, these groups are classified together in the Altungulata (McKenna and Bell, 1997). However, they all originated prior to the Eocene (more than 55 Ma) and are now quite distinct from each other.

Five major perissodactyl groups are recognized: (1) the horses (family Equidae) and their extinct allies; (2) the chalicotheres, an extinct group of clawed perissodactyls; (3) the tapirs and their extinct allies; (4) the rhinoceroses and their extinct allies; and (5) the extinct brontotheres, a Paleogene group. As listed in chapter 3, only four perissodactyl families are recorded from Florida's fossil record (out of a possible total of about seventeen): the Equidae, Chalicotheriidae, Tapiridae, and Rhinocerotidae. The small number is because many perissodactyl families either became extinct before the early Oligocene (when Florida's terrestrial fossil record begins) or are only known from the Old World (or both). In fact, most of the state's perissodactyls derive from just two families, the Equidae and Rhinocerotidae (see list in chap. 3). Chalicotheres are as yet known from only a few localities, and tapirs, while they have a long record in the state, were never diverse.

PERISSODACTYL SKELETAL AND DENTAL ANATOMY

The perissodactyl skull tends to be elongated, like that of an artiodactyl, but the orbit is more centrally located. A postorbital bar is uncommon but is present in advanced equids (in the subfamily Equinae). The original placental complement of forty-four teeth was present in primitive perissodactyls, who had a lesser tendency for tooth loss than in artiodactyls. Chalicotheres and rhinoceroses had a general trend toward loss of incisors. The primitive perissodactyl molar was brachyodont and bunodont to slightly lophodont (fig. 14.1A). The individual cusps of the teeth were still recognizable in early perissodactyls. Their individuality became blurred and incorporated into lophs in younger taxa. In the upper molars, the primitive arrangement of the cusps and lophs resembled the Greek letter pi, (fig. 14.1A), with two transverse lophs and a labial ectoloph. Primitive lower molars were basically rectangular and bilophodont. Each of the perissodactyl groups evolved their cheek teeth from this basic morphology, emphasizing different structures depending on the particular ecological specializations of the group (fig. 14.1B–E).

Cranial appendages (horns and antlers), so common in artiodactyls, are rare features in perissodactyls. Only the brontotheres evolved true horns on their skulls. Some rhinoceroses also evolved "horns," but these are thickened masses of hair and organic fibers that are not true horns and do not preserve in fossils. Fortunately, the base for the attachment of such a rhino's "horn" on the skull is thickened and rugose. It is therefore possible to tell if an extinct rhino had or did not have a "horn." Surprisingly, although all living species have one or more "horns," most fossil rhinos did not have one (Prothero, 1998).

The skeletal elements of perissodactyls can usually be distinguished from those of artiodactyls with relative ease. The hind limb elements are especially characteristic. The femur bears a large process on the lateral side of the shaft (the third trochanter) for muscle attachment (absent in artiodactyls). The tibia has two oblique grooves for articulation with the astragalus on the distal end (straight grooves in artiodactyls). The distal surface of the astragalus, which articulates with the navicular, is partially concave (fig. 14.2) and lacks the second pulley

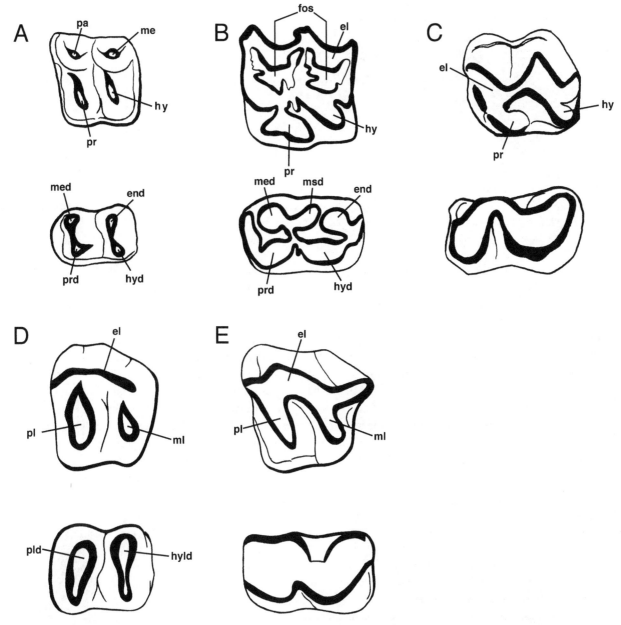

Fig. 14.1. Perissodactyl molars. Each pair contains an upper and lower left molar. For uppers, anterior is to the left and up is labial; for lowers, anterior is left and up is lingual. *A, Hyracotherium*, representing the basic pattern from which all the others evolved; *B,* the horse *Equus; C,* the chalicothere *Moropus; D,* the tapir *Miotapirus; E,* the rhino *Caenopus.* Abbreviations: *el,* ectoloph; *end,* entoconid; *fos,* fossettes; *hy,* hypocone; *hyd,* hypoconid; *hyld,* hypolophid; *ml,* metaloph; *msd,* metastylid; *mt,* metacone; *med,* metaconid; *pa,* paracone; *pl,* protoloph; *pr,* protocone; *pld,* protolophid; *prd,* protoconid. Figures from *Evolution of the Vertebrates,* 4th ed., by E. H. Colbert and M. Morales, copyright © 1991 by Wiley-Liss, Inc. Reprinted by permission of Wiley-Liss, Inc., a subsidiary of John Wiley & Sons, Inc.

of an artiodactyl astragalus. The metapodials never fuse together to form a true cannon bone. The third metapodial and its phalanges are enlarged, and the main weight-bearing axis of the foot passes through the third digit (this is secondarily modified in some chalicotheres). The first metacarpal and metatarsal and the fifth metatarsal are reduced. They eventually lose their respective

phalanges, while the metapodials are reduced to small, vestigial nubbins or are lost altogether in advanced perissodactyls. Primitively the fifth metacarpal and its phalanges are retained (so that the front foot has four functional digits), but these too are reduced and usually lost in progressive groups to form a foot with three functional digits (described as being tridactyl). Of the living

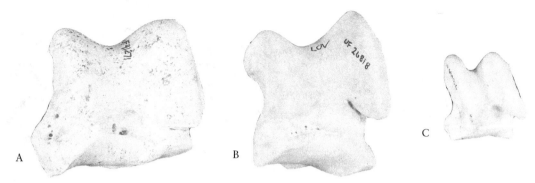

Fig. 14.2. Right astragali in anterior view of examples of the three common types of perissodactyls found in the fossil record of Florida. All are of late Miocene age. *A*, UF 41271, the rhinoceros *Aphelops malacorhinus*; *B*, UF 26818, the tapir *Tapirus simpsoni*; *C*, UF 69936, the horse *Nannippus aztecus*. All are about 0.67×.

perissodactyls, only the tapirs retain a front foot with four digits.

The perissodactyls most adapted for running are the equids, which show many of the same skeletal modifications found in other running specialists. The fibula and ulna are reduced, as are the lateral digits, while the metapodials and digits are elongated. In the most advanced horses each foot has only a single functional digit, the third. Movement of the limbs is restricted to the fore and aft direction by the configuration of the skeletal elements, and very little movement is possible within the wrist and ankle between the individual elements.

FOSSIL EQUIDS OF FLORIDA

From the oldest known terrestrial site (the I-75 fauna) through the latest Pleistocene, horses are usually among the most common mammals at Florida fossil localities. Their record from the middle early Miocene to the early Pliocene (about 18 to 4 Ma) is especially complete. The fossil record of horses has long been extolled as one of the best examples for observation of evolutionary patterns and trends, although in basic textbooks it is usually portrayed too simplistically. For general accounts of equid evolution, consult Simpson (1951), MacFadden (1992), and Hulbert (1996).

The oldest horse from Florida is known only from a very small sample of isolated, brachyodont teeth from the early Oligocene. At that time, two equid genera are recognized in North America, *Mesohippus* and *Miohippus* (MacFadden, 1998b). Primary differences between the two genera are that *Miohippus* had a deeper depression (fossa) on the side of the skull in front of the eye and a direct articulation between the cuboid and the third metatarsal (Prothero and Shubin, 1989). Unfortunately, neither of these can be observed on isolated teeth,

which is all we have so far from Florida. *Miohippus* upper cheek teeth usually had a larger, more developed hypostyle than those of *Mesohippus*. The stage of evolution of the hypostyle in the one good upper tooth in the I-75 sample better corresponds to that of the more progressive genus *Miohippus* rather than *Mesohippus* (fig. 14.3A). The limited sample precludes any degree of certainty about the identification, but it at least conforms to the early Oligocene age assigned to the fauna.

Equids were rare during the late Oligocene and very early Miocene (25 to 20 Ma) in Florida; artiodactyls were more abundant and diverse through this interval. Only a primitive, very small form of *Archaeohippus* is

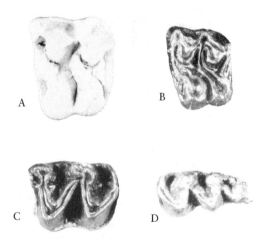

Fig. 14.3. Oligocene horse teeth from Florida. All in occlusal view, at about 1.5×. *A*, UF 171282, a left M1 or M2 of *Miohippus* sp. from the I-75 site, Alachua County, early Oligocene. *B–D*, *Archaeohippus* sp. from the Cowhouse Slough site, Hillsborough County, late Oligocene; *B*, UF 40120, a left M1 or M2; *C*, UF 40129, a right p3 or p4; *D*, UF 40127, a left m3.

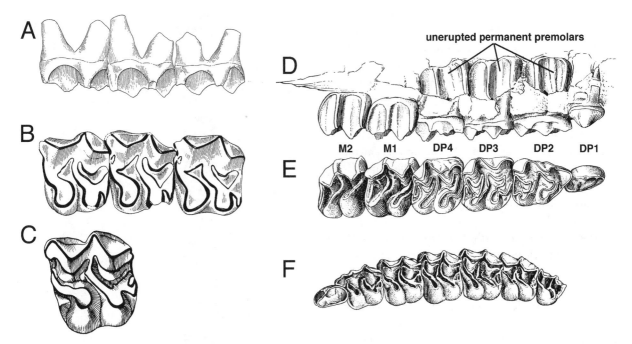

Fig. 14.4. Miocene browsing horses from Florida. *A*, labial, and *B*, occlusal views of UF 118529, left P3–M1 of *Anchitherium clarencei* from La Camelia Mine, Gadsden County; *C*, occlusal view of UF 32024, a left upper molar of *Hypohippus chico* from the Bone Valley District, Polk County; *D–F*, upper dentitions of the dwarf horse *Archaeohippus blackbergi* from the Thomas Farm site, Gilchrist County, early Miocene; *D*, labial, and *E*, occlusal views of MCZ 3840, a partial right maxilla with DP1–DP4 and M1–M2. The unerupted, permanent premolars are visible in *D*, where the bone has broken away. *F*, occlusal view of MCZ 3829, left DP1, P2–M3. In equids, as in most mammals, the DP1 (if present at all) is not replaced and remains with the adult dentition. All about 1×. *A–B* after Bryant (1991); reproduced by permission of the Society of Vertebrate Paleontology. *C* after MacFadden (1982); reproduced by permission of the Florida Academy of Sciences, Orlando. *D–F* after White (1942); reproduced by permission of the Museum of Comparative Zoology, Cambridge, Massachusetts.

known (fig. 14.3B–D). The most common large mammals at Thomas Farm (18 Ma) are equids, and this is often true at younger localities as well. Clearly some environmental or evolutionary factor must have changed to produce this dramatic shift in relative abundance. Members of the equid subfamilies Anchitheriinae and Equinae are both present at Thomas Farm. They are more advanced than *Miohippus* in having connected the metaloph to the ectoloph on the upper cheek teeth (fig. 14.4). Members of the Anchitheriinae retained the primitive brachyodont teeth of *Miohippus* and older horses. For this reason they are often called the "browsing horses" of the Miocene, as this is their inferred diet. This contrasts them with the Equinae, within which a tremendous increase in crown height (hypsodonty) occurred in many genera during the Miocene. It is commonly thought that this corresponded to an increased amount of grass (and associated grit) in the diet of these horses (MacFadden, 1992). Therefore, members of the Equinae are sometimes called "grazing horses" to contrast them with the anchitheres. Some advanced techniques have been developed to determine the diets of extinct herbi-

vores, and these have been applied to fossil horses (Hayek et al., 1991; MacFadden and Cerling, 1996; MacFadden et al., 1999). It is now clear that the transition from browsing to grazing took place within the Equinae and that many hypsodont species were *not* pure grazers, but instead were mixed-feeders to varying degrees. Grazing horses did not become common until the late Miocene and early Pliocene.

Two genera from the subfamily Anchitheriinae are recognized in Florida, *Anchitherium* and *Hypohippus,* while a third genus, *Megahippus,* might have been present. Anchitheres differ from members of the Equinae in lacking secondary labial ridges between the prominent stylar ridges of the upper cheek teeth. The third metatarsal proximally articulates with three tarsal elements, the ectocuneiform, cuboid, and mesoentocuneiform (this was later acquired independently by some members of the Equinae). The phalanges and metapodials of digits 2 and 4 were large and robust (fig. 14.5A); the proximal phalanx of the third digit was short, broad, and lacked the characteristic central V-scar of equines (fig. 14.5C); and the distal keel or ridge on the medial metapodial was incomplete and not present on the anterior side. *Anchitherium* was a very cosmopolitan genus and dispersed to Asia in the early Miocene and subsequently to Europe. *Anchitherium clarencei* is the best-known anchithere from Florida (fig. 14.4A, B). It was a relatively small anchithere, with an estimated mass of 290 pounds (132 kilograms; MacFadden, 1986a) and differed from other species in the genus by having a small M3. In addition to records from Thomas Farm, *A. clarencei* has also been found at a number of sites in the panhandle. Future study may show that it is the junior synonym of *Anchitherium navasotae* from the early Miocene of Texas (Forstén, 1975).

The second named anchithere from Florida is *Hypohippus chico,* which was described from a small sample of isolated teeth and a phalanx from the Bone Valley phosphate mines of Polk County (MacFadden, 1982; fig. 14.4C). It is also a rather small species by anchithere standards. The exact age of the deposits producing these specimens is unknown (although probably middle Miocene). A few specimens of larger anchitheres are also known from the Bone Valley and could represent *Hypohippus* or *Megahippus*. Presently available material is inadequate to differentiate between these two. The age of these specimens is also middle Miocene. No younger anchitheres are known from Florida, although in the Great Plains these large browsing horses persisted until the late Miocene, about 9 Ma.

A more common early Miocene horse in Florida is the diminutive *Archaeohippus,* for which the best sample comes from Thomas Farm. While it had primitive, brachyodont cheek teeth without cement (fig. 14.4D–F), *Archaeohippus* is a member of the subfamily Equinae instead of the Anchitheriinae because of its elongated phalanges, reduced lateral digits, and better developed distal keel on the slender third metapodials. Florida fossils ranging in age from 18 to 15 Ma are presently all referred to a species named from coastal Texas, *Archaeohippus blackbergi*. For several decades *A. blackbergi* was included in the genus *Parahippus*. *Archaeohippus* differs from *Parahippus* by having a much deeper fossa in front of the orbit, an incomplete postorbital bar, no strong V-scars on the proximal phalanges, and a much smaller size. *A. blackbergi* shares these features with other members of *Archaeohippus*. It was only as large as a medium-sized dog, with an estimated mass of about 95 pounds (MacFadden, 1986a).

Parahippus is a group of early to middle Miocene species characterized by a complete postorbital bar, shallow to very shallow fossa in front of the orbit, and elongated proximal phalanges with strong V-scars (figs. 14.5D,

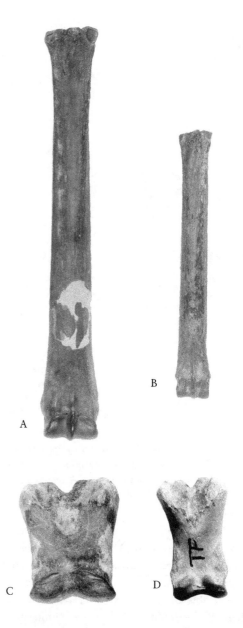

Fig. 14.5. Comparison between the metapodials and phalanges of anchitheriine and equine horses. *A,* posterior view of MCZ 7608, a metatarsal 3 of *Anchitherium clarencei; B,* posterior view of UF/FGS 6599, a metatarsal 3 of *Parahippus leonensis; C,* posterior/ventral view of UF 98400, a proximal phalanx, digit 3, of *A. clarencei; D,* posterior/ventral view of UF 176, a proximal phalanx, digit 3, of *P. leonensis*. All specimens are from the Thomas Farm site, Gilchrist County, early Miocene. Anchitheriines, represented by *A. clarencei,* tend to have stouter metapodials, relatively larger lateral metapodials and phalanges (not shown), shorter phalanges, and lack the V-shaped scar on the ventral surface of the proximal phalanx of the third digit, where the oblique sesamoid ligament attaches in equine horses. This V scar is evident in *D* on *P. leonensis. A–B* about 0.5×; *C–D* about 1×.

14.6). Most individuals show some development of a crochet and enlarged hypostyle on the upper cheek teeth (fig. 14.6B). The enamel has a thin to moderate coating of cement. Horses at the *Parahippus* grade of evolution are known from the early and early middle Miocene in Florida. The greatest known concentration of individuals of *Parahippus* in the world is the Thomas Farm sample of *Parahippus leonensis*. Fossils of many hundreds of individuals have been collected from this site alone. *P. leonensis* was first named from a small sample found in Leon County, Florida (hence the trivial name of the species), and was subsequently recovered from Texas and Nebraska (Hulbert and MacFadden, 1991). Its morphology is relatively advanced for a *Parahippus* and is very close to that expected in the ancestor of all more hypsodont horses. *P. leonensis* was a mixed-feeder, lived to a maximum of nine years, and averaged about 160 pounds (Hulbert, 1984; Janis, 1990). Sexual dimorphism was modest, as males were slightly larger on average than females, but this is more sexual dimorphism than is observed in modern horses.

Horses with more hypsodont teeth than *Parahippus* first appeared in North America about 17 Ma, although *Archaeohippus, Parahippus,* and anchitheres lingered on (Hulbert and MacFadden, 1991). Traditionally, the less advanced, more low-crowned of these horses have been lumped together into a single genus, *Merychippus.* Within this group of species is the initial radiation of "grazing" equids and the ancestors of all of the later Miocene genera. Recent work with this group has concentrated on understanding their evolutionary relationships and classifying them correctly. Only a few of the species traditionally placed in *Merychippus* actually form a natural group with the type species of the genus, *M. insignis,* and thus truly belong in the genus. Other species, like "*Merychippus*" *gunteri* from Florida, are not members of this natural group and do not truly belong in *Merychippus.* For some of these species, like "*M.*" *gunteri,* no generic name is available, so quote marks are placed around the generic name to signal that the species does not really belong in that particular genus.

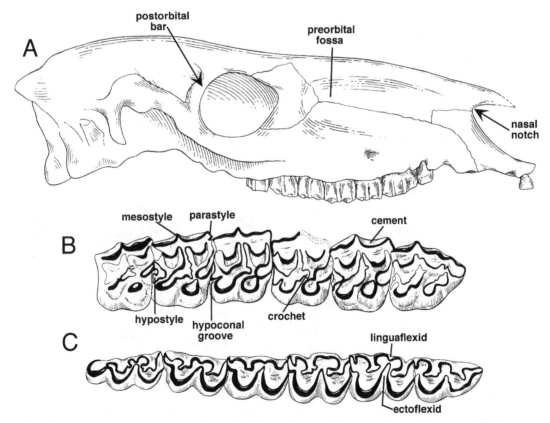

Fig. 14.6. Skull and dentition of the very common horse *Parahippus leonensis* from the Thomas Farm site, Gilchrist County, early Miocene. *A,* lateral view of reconstructed skull based on UF 56000 and 103753. The complete postorbital bar and very shallow depression in front of the orbit (preorbital fossa) are indicated. *B,* occlusal view of UF 100012, right P2–M3. *C,* occlusal view of UF/FGS 6449, right p2–m3. *A* about 0.5×; *B–C* about 1×. After Hulbert and MacFadden (1991); reproduced by permission of the American Museum of Natural History, New York.

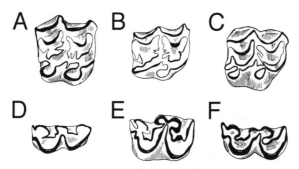

Fig. 14.7. Occlusal views of upper and lower cheek teeth of the primitive merychippine horse *"Merychippus" gunteri* from the Midway local fauna, Gadsden County, late early Miocene. *A*, UF/FGS 4114, right upper P3 or P4; *B*, UF/FGS 5045, left (reversed to look like a right) P4; *C*, UF/FGS 9952, right DP3 or DP4; *D*, UF/FGS 9937, right p2; *E*, UF/FGS 9942, right p3; *F*, UF/FGS 9959, right m1 or m2. All about 1×. After Hulbert and MacFadden (1991); reproduced by permission of the American Museum of Natural History, New York.

The heyday of *Merychippus*-like horses (or merychippines), the middle Miocene, is still poorly known in Florida, so no large samples of these interesting equids have been found. The most important population is probably that of *"Merychippus" gunteri* from the Torreya Formation of the panhandle (about 18 to 15 Ma). It is one of, if not the most, primitive of the merychippine horses (fig. 14.7), and in most respects not very different from *Parahippus leonensis* (Hulbert and MacFadden, 1991). Other merychippines were also present in Florida but are inadequately known (fig. 14.8). In the list in

chapter 3 they are referred to the western species that they most closely resemble, but larger sample sizes and complete skulls will be needed to confirm these identifications. *Merychippus westoni*, long thought to be another primitive merychippine from Florida, is now recognized as an early late Miocene *Nannippus* (Hulbert, 1993).

Three major groups of hypsodont horses evolved from a common ancestor similar to *"Merychippus" gunteri* about 17 Ma (Hulbert and MacFadden, 1991). These are the hipparionine, protohippine, and equine horses. Hipparionines were the numerically dominant horses in North America until very late in the Miocene and became extinct in the late Pliocene. They also migrated to the Old World in the middle Miocene about 11 Ma, where they underwent a successful radiation. Protohippines were a less diverse group (only two genera) that became extinct in the latest Miocene but are common in many Florida Miocene faunas. Their distribution is limited to North America. The third group, the equines, is the only one that survived, as it includes the sole living genus of horse, *Equus*. In the Miocene, this group was not especially abundant and was very rare in Florida. It was not until the very late Miocene and Pliocene that the equines truly became successful. *Equus* also dispersed to the Old World, probably several times, beginning in the late Pliocene.

Six hipparionine genera are recognized in Florida (see chap. 3 for listing of all species): true *Merychippus*, *Pseudhipparion*, *Neohipparion*, *Hipparion*, *Nannippus*, and *Cormohipparion*. Common examples are illustrated

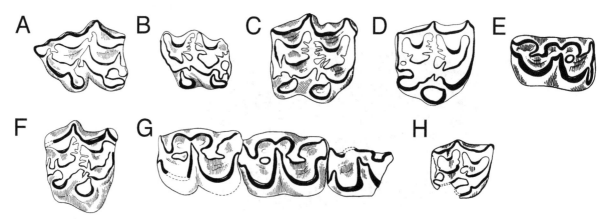

Fig. 14.8. Occlusal views of cheek teeth of merychippine horses from the middle Miocene of Florida. *A–B, Acritohippus isonesus,* Nichols Mine, Polk County; *C–E, "Merychippus" goorisi,* Phosphoria Mine, Polk County; *F, Merychippus brevidontus,* Phosphoria Mine; *G, A. isonesus,* Smith Mine, Gadsden County; *H, "Merychippus" primus,* La Camelia Mine, Gadsden County. *A,* UF 107056, left P2; *B,* UF 107057, a partial left M1; *C,* UF 93342, a right P3 or P4; *D,* UF 93296, a right M3; *E,* UF 93338, a right m1 or m2; *F,* UF 93287, a right P4; *G,* UF 65551, left p3–m1; *H,* UF 65560, right M3. All about 1×. After Hulbert and MacFadden (1991); reproduced by permission of the American Museum of Natural History, New York.

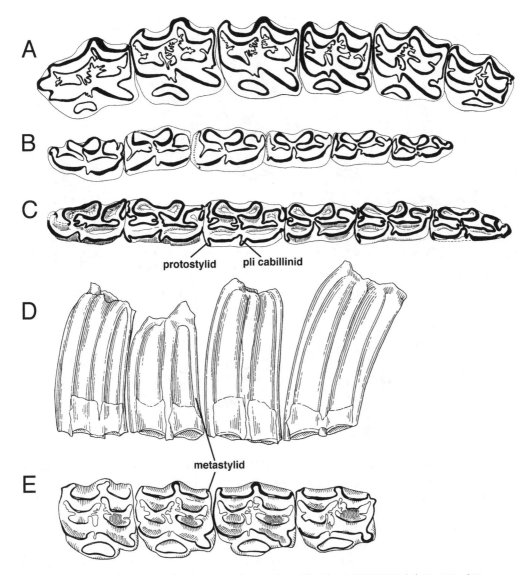

Fig. 14.9. Late Miocene and early Pliocene *Neohipparion* tooth rows from Florida. *A*, UF 27991, left P2–M3 of *N. trampasense* from the Love site, Alachua County; *B*, UF 17159, left p2–m3 of *N. trampasense* from McGehee Farm, Alachua County; *C*, UF 69969, right p2–m3 of *N. eurystyle* from Moss Acres, Marion County; *D*, labial, and *E*, occlusal views of UF 18318, left P3–M2 of *N. eurystyle* from the Payne Creek Mine, Polk County. All about 1×. *A–B* after Hulbert (1987); reproduced by permission of the Paleontological Society. *D–E* after MacFadden (1984); reproduced by permission of the American Museum of Natural History, New York.

in figures 14.9–14.19. Having a protocone that is isolated from the protoloph is characteristic of the P3 through M3 of most hipparionine horses (fig. 14.9A). The P2 may also have an isolated protocone, but more often it has a connected protocone. *Pseudhipparion* differed from the others in that its protocone typically connects to the protoloph in all of the cheek teeth, usually after about one-half of the crown of the tooth had worn away. Thus in young individuals of this genus, with slightly worn teeth, the protocone is isolated; in older individuals, with well-worn teeth, the protocone is connected (fig. 14.10). This pattern is actually the primitive one for hypsodont horses and is also seen in *Parahippus*

leonensis, some "*Merychippus,*" and *Protohippus.* The hypoconal groove in *Pseudhipparion,* the posterior-lingual indentation of enamel between the hypocone and hypostyle (fig. 14.10D), closes posteriorly to form a small isolated lake (fig. 14.10B) that eventually disappears with further wear (fig. 14.10R). The hypoconal groove usually remains open in other hipparionines, although it is very shallow in Pliocene *Nannippus.* The enamel fossette borders of hipparionines tend to have many infoldings, called plications. These sometimes are quite complex and numerous, especially in *Cormohipparion* (figs. 14.12–14.14). The fossette borders of protohippine and equine horses are usually simpler, with

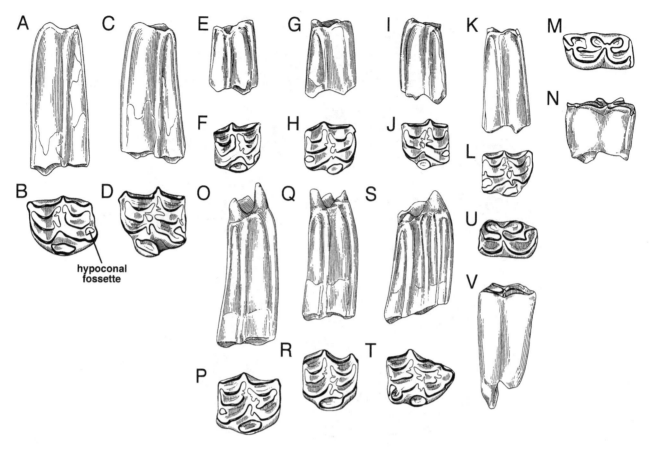

Fig. 14.10. Upper and lower cheek teeth of *Pseudhipparion* from Florida. *A–D, P. curtivallum* from Phosphoria Mine, Polk County, middle Miocene. *E–N, Pseudhipparion* sp. from Kingsford Mine, Polk and Hillsborough Counties, middle Miocene. *O–X, P. skinneri* from the Love site, Alachua County, late Miocene. *A,* labial, and *B,* occlusal views of UF 61302, left M1 or M2; *C,* labial, and *D,* occlusal views of UF 61301, a left P4; *E,* labial, and *F,* occlusal views of UF 61308, a right M1 or M2; *G,* labial, and *H,* occlusal views of UF 61306, right P3 or P4; *I,* labial, and *J,* occlusal views of UF 61309, a left M1 or M2; *K,* labial, and *L,* occlusal views of UF 61307, a right M1 or M2; *M,* labial, and *N,* occlusal views of UF 61310, a right dp3; *O,* labial, and *P,* occlusal views of UF 51951, a right M1 or M2; *Q,* labial, and *R,* occlusal views of UF 53631, a right P4; note connected protocone; *S,* labial, and *T,* occlusal views of UF 53623, a right P2; *U,* labial, and *V,* occlusal views of UF 64513, a left p3 or p4. All about 1×. After Webb and Hulbert (1986); reproduced by permission of the Department of Geology and Geophysics, University of Wyoming, Laramie.

fewer if any plications. Lower cheek teeth of hipparionines have strong protostylids (figs. 14.12B, 14.13B; except in the p2, and they are lost in advanced species of *Nannippus* [fig. 14.17Y–BB] and *Pseudhipparion*), large metaconids and metastylids, and variable development of enamel folds. All hipparionines are tridactyl, although the side toes are very reduced in some advanced species (fig. 14.19). The hooves are pointed with an anterior fissure or notch.

One of the more unusual aspects of horse evolution occurred in Florida populations of *Pseudhipparion* (Webb and Hulbert, 1986). *Pseudhipparion* became increasingly smaller and more hypsodont during the middle to late Miocene but retained the normal pattern of tooth development (fig. 14.10). The last "normal"

Pseudhipparion is known from the Withlacoochee River and several sites in Manatee County, about 6 Ma. A very unusual species of *Pseudhipparion, P. simpsoni,* occurs in the Palmetto Fauna of the Bone Valley region, one and a half million years later (fig. 14.11), and is also known from similar aged faunas from North Carolina, Texas, Oklahoma, and Kansas. *P. simpsoni* is the only known horse that began to evolve hypselodont teeth. The cheek teeth of *P. simpsoni* are rootless for the first few years of its life, and as the crown of its teeth are worn away, additional enamel and dentine are added at the root-end of the teeth, keeping it the same height. This process finally stopped, roots were added, and the teeth eventually wore out like those of other equids. Due to structural limitations, the portion added to the base of the tooth cannot

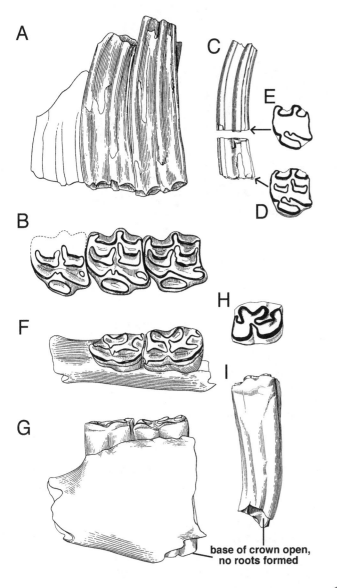

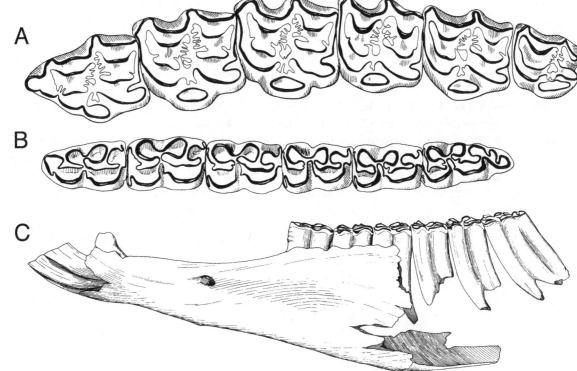

Fig. 14.11. Cheek teeth of the unusual horse *Pseudhipparion simpsoni* from the Palmetto Fauna of the Bone Valley District, early Pliocene. *A*, labial, and *B*, occlusal views of UF 12943, left P3–M1 from Fort Green Mine; *C*, labial, *D*, occlusal, and *E*, cross-sectioned views of UF 61347, a left M1 or M2; *F*, occlusal, and *G*, lateral views of UF 24791, partial left dentary with p2–p3 from Fort Green Mine; *H*, occlusal, and *I*, lingual views of UF 18313, a left m1 from the Palmetto Mine. Note lack of root formation and loss of fossettes (in *E*), diagnostic for this species. All about 1×. After Webb and Hulbert (1986); reproduced by permission of the Department of Geology and Geophysics, University of Wyoming, Laramie.

base of crown open, no roots formed

Below: Fig. 14.12. Cheek teeth and dentary of the horse *Cormohipparion plicatile* from the late Miocene of Florida. *A*, occlusal view of UF 32262, left P2–M3 from the Love site, Alachua County; *B*, occlusal view of cheek teeth, and *C*, lateral view of UF 69967, left dentary with i1–i3, c, p2–m3 from Moss Acres, Marion County. Note the very long diastema between the canine and the p2, a characteristic of *C. plicatile*, *C. ingenuum*, and *C. emsliei*, three common Miocene-Pliocene horses in Florida. *A–B* about 1×; *C* about 0.5×. After Hulbert (1988c); by permission of the Florida Museum of Natural History.

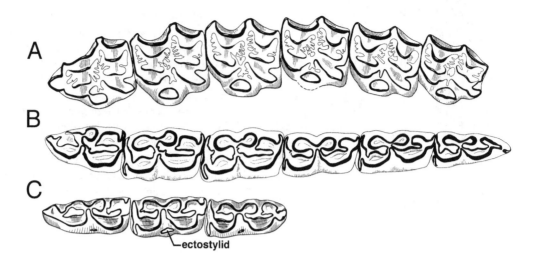

Fig. 14.13. Cheek teeth of the horse *Cormohipparion ingenuum* from the middle to late Miocene of Florida. *A*, occlusal view of UF 32300, left P2–M3 from the Love site, Alachua County; *B*, occlusal view of UF 98126, right p2–m3 from Hookers Prairie Mine, Polk County; *C*, occlusal view of UF 32294, left dp2–dp4 from the Love site. The prominent ectostylids in the deciduous premolars are a characteristic feature of this genus. All about 1×. After Hulbert (1988c); by permission of the Florida Museum of Natural History.

contain fossettes, so their presence or absence indicates the relative age of the individual (figs. 14.11D, E). Several incisors of *P. simpsoni* are known and are all rootless, suggesting that they are fully hypselodont. *P. simpsoni* became extinct following its appearance in the Bone Valley, but it is interesting to speculate whether given enough time it would have eventually evolved fully hypselodont cheek teeth. Unfortunately, this is an evolutionary experiment that nature apparently cut short.

Protohippus and *Calippus,* the two genera of protohippine horses, are common in the middle and early late Miocene of Florida (Hulbert, 1988b). They are characterized by having short, very broad muzzles and tridactyl feet. Their cheek teeth have connected protocones, relatively simple enamel patterns, and well-developed protostylids on the lower cheek teeth (figs. 14.20–14.22). *Calippus* includes very small-, small-, and medium-sized species, while *Protohippus* is larger in size. In *Calippus* the incisors form a nearly straight row, not the semicircle or arc found in other equids (fig. 14.21F). Such an adaptation was useful for feeding upon grass very close to the ground. The late Miocene species *Calippus elachistus* is one of the smallest hypsodont horses and is known only from Florida (fig. 14.20G–Q). It weighed about one hundred pounds. Advanced species of *Protohippus* paralleled the hipparionines in evolving an isolated protocone, but only when the teeth were slightly worn (fig. 14.22A). Both *Calippus* and *Protohippus* became extinct in the late Miocene, about 6 Ma. MacFadden (1998b) recorded the presence of two species of *Calippus* in the

early Pliocene Palmetto Fauna, but these are either misidentifications or represent fossils of Miocene, not Pliocene, age (Hulbert, 1988b).

Four genera belonging to the Equini are known from Florida: *Pliohippus, Astrohippus, Dinohippus,* and *Equus.* Primitive equines were tridactyl but had very reduced side toes and very slender but complete lateral metapodials. Advanced equines lost the side toes and reduced the lateral metapodials to splints, so they only have one functional digit per foot (monodactyl). Apparently this reduction evolved at least twice within the tribe. The cheek teeth of equines have connected protocones, very reduced or absent protostylids, and, except for advanced *Equus,* simple enamel patterns (fig. 14.23). *Pliohippus* and *Astrohippus* are each known from Florida by only a few isolated teeth but appear to resemble their respective western contemporaries. *Dinohippus* is somewhat better known in the late Miocene and early Pliocene (fig. 14.24) but is much rarer than its hipparionine contemporaries in Florida. Western horse faunas of this time are typically dominated by *Dinohippus* and/or *Astrohippus,* with hipparionines being rarer. This suggests that dry environmental conditions, which favor the monodactyl equines over tridactyl hipparionines, did not begin in Florida until the middle Pliocene, much later than in the western United States.

The oldest records of *Equus* from Florida are late Pliocene. The ancestry of *Equus* almost certainly lies within *Dinohippus* and not *Pliohippus,* as portrayed in many textbooks. Most Pliocene specimens belong to a

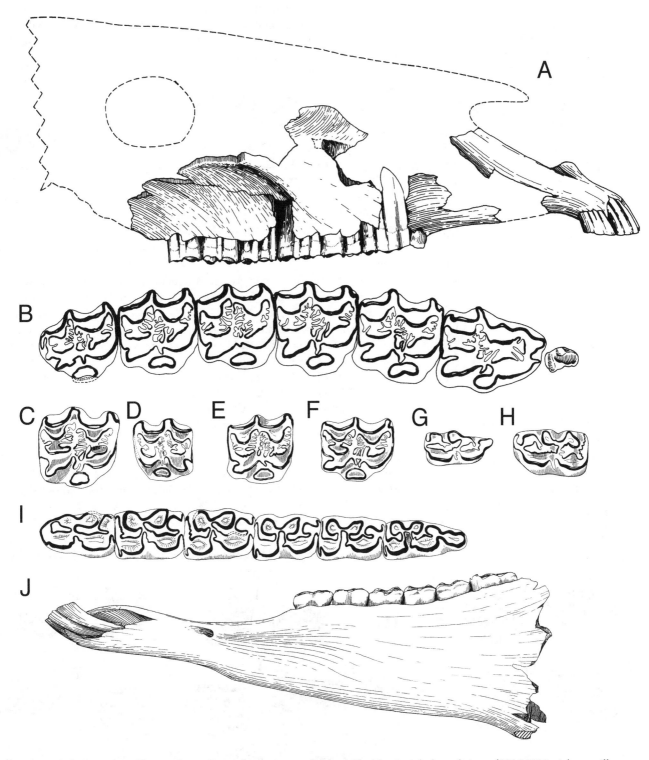

Fig. 14.14. The late Miocene-Pliocene horse *Cormohipparion emsliei* from Florida. *A*, right lateral view of UF 94700, right maxilla and premaxilla set in a reconstructed skull outline from the APAC (=Macasphalt) Shell Pit, Sarasota County, late Pliocene; *B*, occlusal view of right DP1, P2–M3 of UF 94700; *C–H*, occlusal views of cheek teeth from the Bone Valley District, early Pliocene; *C* from Fort Green Mine; *D* from Gardinier Mine; *E–H* from Palmetto Mine. *C*, UF 61981, right P3 or P4; *D*, UF 67976, left P3 or P4; *E*, UF 63642, right M1 or M2; *F*, UF 63641, right M1 or M2; *G*, UF 17152, right p2; *H*, UF 53495, right p3 or p4; *I*, occlusal view of p2–m3, and *J*, lateral view of UF 97259, left dentary from Moss Acres, Marion County, late Miocene. *A, J* about 0.5×; *B–I* about 1×. After Hulbert (1988a, 1988c); *A–H* reproduced by permission of the Society of Vertebrate Paleontology; *I–J* reproduced by permission of the Florida Museum of Natural History.

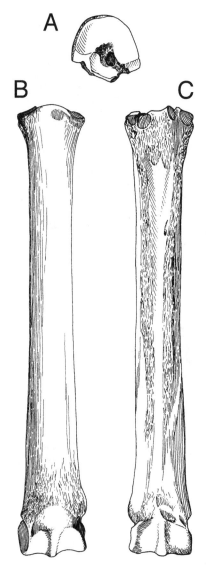

Left: Fig. 14.15. *A,* proximal, *B,* anterior, and *C,* posterior views of UF 94635, a right metatarsal 3 of *Cormohipparion emsliei* from the APAC (=Macasphalt) Shell Pit, Sarasota County, late Pliocene. About 0.5×. After Hulbert (1988a); reproduced by permission of the Society of Vertebrate Paleontology.

Above: Fig. 14.16. Occlusal view of UF 17205, left P2–M3 of the horse *Hipparion,* probably *H. tehonense,* from the Withlacoochee River site 4A, Marion County, late Miocene. This taxon is generally rare in middle and late Miocene faunas in Florida. About 1×. After Hulbert (1988c); by permission of the Florida Museum of Natural History.

Below: Fig. 14.17. Upper and lower cheek teeth of the small horse *Nannippus* from Florida in occlusal view. *A–H, Nannippus* sp., an undescribed middle Miocene species from the Agricola Fauna, Bone Valley District. *I–N, Nannippus westoni* from the Love site, Alachua County, late Miocene. *O–P, N. westoni* from the McGehee Farm site, Alachua County, late Miocene. *Q–T, Nannippus morgani* from the Withlacoochee River 4A site, Marion County, late Miocene. *U–CC, Nannippus aztecus* from the Palmetto Fauna, Bone Valley District, early Pliocene. This species was formerly known as *N. minor. A,* UF 98169, right P2; *B,* UF 98183, left P3 or P4; *C,* UF 28559, right M1 or M2; *D,* UF 28560, right M1 or M2; *E,* UF 98217, right p2; *F,* UF 98220, right p3 or p4; *G,* UF 98238, right m1 or m2; *H,* UF 28578, right m1 or m2; *I,* UF 60398, right P2; *J,* UF 62157, right P3 or P4; *K,* UF 62173, left M1 or M2; *L,* UF 62187, right M1 or M2; *M,* UF 63984, left p2; *N,* UF 64921, left p3 or p4; *O,* UF 95380, right p3 or p4; *P,* UF 95383, left m1 or m2; *Q,* UF 63648, right M1 or M2; *R,* UF 53520, right M1 or M2; *S,* UF 95372, left p3 or p4; *T,* UF 95367, right m3; *U,* UF 67980, right P2; *V,* UF 63628, right P3 or P4; *W,* UF 63978, right P3 or P4; *X,* UF 67981, right M1 or M2; *Y,* UF 57213, left p2–p4; *Z,* UF 53954, right p3 or p4; *AA,* UF 93235, right m1 or m2; *BB,* UF 65192, right m1 or m2; *CC,* UF 17275, right dp3 or dp4. All 1×. *I–T* after Hulbert (1993); reproduced by permission of the Society of Vertebrate Paleontology.

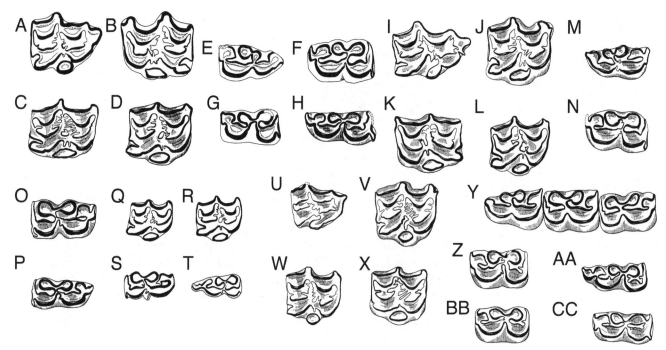

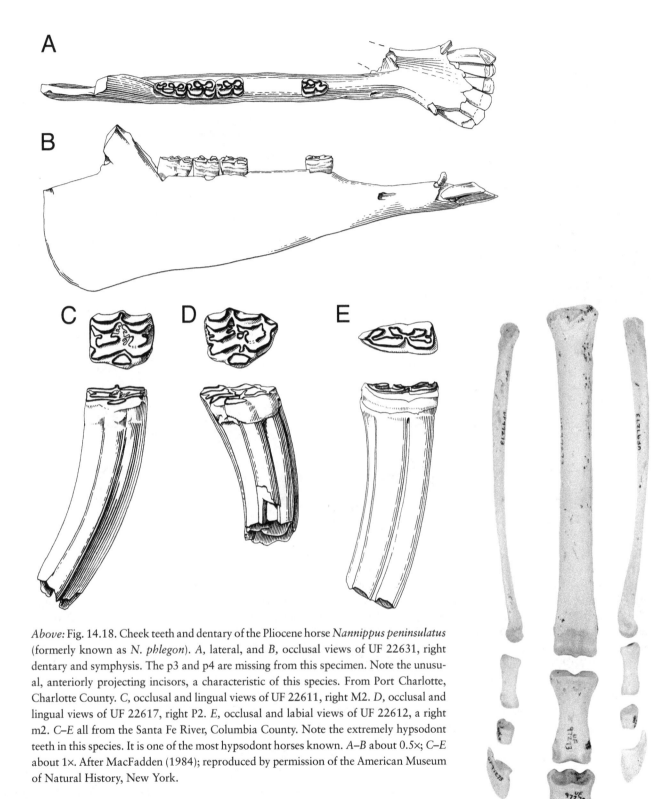

Above: Fig. 14.18. Cheek teeth and dentary of the Pliocene horse *Nannippus peninsulatus* (formerly known as *N. phlegon*). *A,* lateral, and *B,* occlusal views of UF 22631, right dentary and symphysis. The p3 and p4 are missing from this specimen. Note the unusual, anteriorly projecting incisors, a characteristic of this species. From Port Charlotte, Charlotte County. *C,* occlusal and lingual views of UF 22611, right M2. *D,* occlusal and lingual views of UF 22617, right P2. *E,* occlusal and labial views of UF 22612, a right m2. *C–E* all from the Santa Fe River, Columbia County. Note the extremely hypsodont teeth in this species. It is one of the most hypsodont horses known. *A–B* about 0.5×; *C–E* about 1×. After MacFadden (1984); reproduced by permission of the American Museum of Natural History, New York.

Right: Fig. 14.19. Hind foot of *Nannippus aztecus* from Moss Acres, Marion County, late Miocene. Note the slender but complete side toes. UF 97273, about 0.5×.

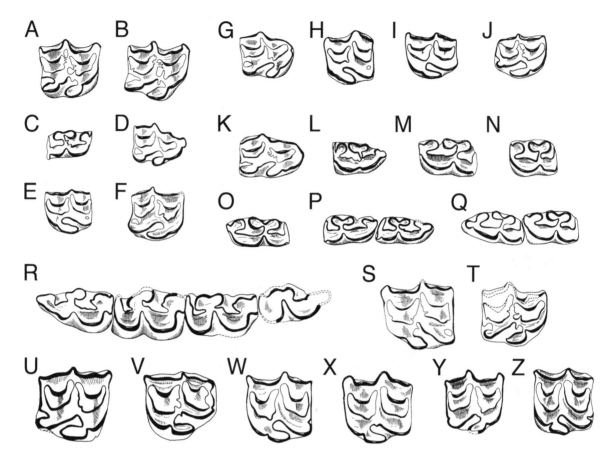

Fig. 14.20. Occlusal views of upper and lower cheek teeth of the protohippine horse *Calippus* from the Miocene of Florida. *A–B, C. proplacidus,* Phosphoria Mine, Polk County, middle Miocene; *C–F, Calippus* sp., Bone Valley District, middle Miocene (*C,* Kingsford Mine; *D–F,* Phosphoria Mine); *G–Q, C. elachistus,* the Love site, Alachua County, late Miocene; *R–W, C. martini,* Bone Valley District, middle Miocene (*R–S,* Phosphoria Mine; *T–U,* Hookers Prairie Mine); *V–W, C. cerasinus,* the Love site, Alachua County, late Miocene; *Y–Z, C. hondurensis,* late Miocene. *A,* UF 28469, right P3 or P4; *B,* UF 28421, right M1 or M2; *C,* UF 61343, right p3 or p4; *D,* UF 28421, right P2; *E,* UF 28680, left M1 or M2; *F,* UF 93201, right P3 or P4; *G,* UF 53431, right P2; *H,* UF 53436, left P3 or P4; *I,* UF 53448, left M1 or M2; *J,* UF 53576, right M1 or M2; *K,* UF 53620, right DP2; *L,* UF 53584, right p2; *M,* UF 68951, right p3 or p4; *N,* UF 53585, right p3 or p4; *O,* UF 92952, right dp3 or dp4; *P,* UF 53601, right m1–m2; *Q,* UF 32139, right m2–m3; *R,* UF 55886, left p2–m1; *S,* UF 28552, left P3 or P4; *T,* UF 98292, right P3 or P4; *U,* UF 98498, right M1 or M2; *V,* UF 98300, left M3; *W,* UF 60292, left M1 or M2; *X,* UF 60309, left P3 or P4; *Y,* UF 9506, right P3 or P4 from the McGehee Farm site; *Z,* AMNH 113643, left M1 or M2. All about 1×. After Hulbert (1988b); by permission of the Florida Museum of Natural History.

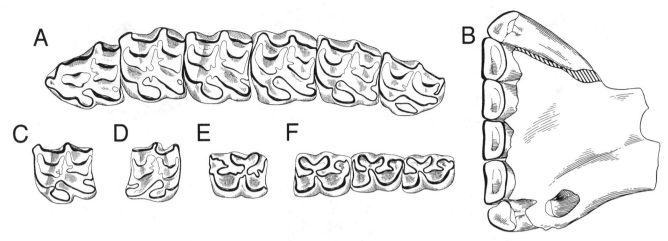

Fig. 14.21. The late Miocene protohippine horse *Calippus maccartyi* from Florida. *A*, occlusal view of left P2–M3, *B*, occlusal view of mandibular symphysis with right and left i1–i3, left c, and *F*, occlusal view of left p4–m2 of UF 69951 from Moss Acres, Marion County; *C*, occlusal view of UF 90299, left M1 or M2; *D*, occlusal view of UF 45536, right M1 or M2 from the Withlacoochee River 4A site, Marion County; *E*, occlusal view of UF 53462, right p3 or p4 from the Withlacoochee River 4A site. All about 1×. After Hulbert (1988b); by permission of the Florida Museum of Natural History.

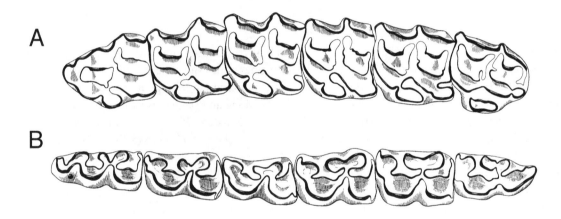

Above: Fig. 14.22. Upper and lower cheek tooth rows of the horse *Protohippus gidleyi* from the Love site, Alachua County, late Miocene. *A*, UF 32298, left P2–M3; *B*, UF 32173, right p2–m3. Both about 1×. After Hulbert (1988b); by permission of the Florida Museum of Natural History.

Right: Fig. 14.23. Upper cheek teeth of *Astrohippus stockii* from the Swift Mine, Polk County, early Pliocene. This species is the rarest of the six Upper Bone Valley horses. *A*, labial, and *B*, occlusal views of UF 18319, left P4; *C*, labial, and *D*, occlusal views of TRO 1920, right M2. About 1×. After MacFadden (1986b); reproduced by permission of the Paleontological Society.

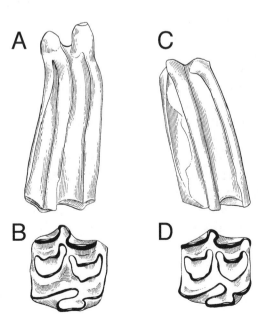

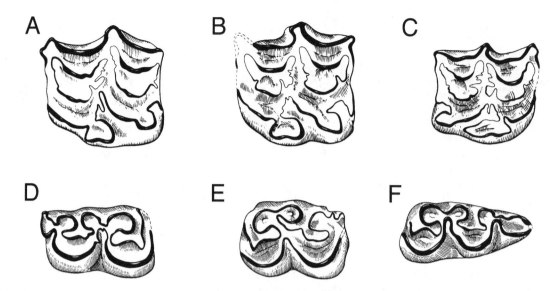

Fig. 14.24. Occlusal views of upper and lower cheek teeth of the monodactyl horse *Dinohippus* from Florida. *A–E, D. mexicanus* from the Bone Valley District, early Pliocene. *A*, UF 20056, left P3 or P4 from Payne Creek Mine; *B*, UF 53824, right P3 or P4 from Phosphoria Mine; *C*, TRO 1149, right M1 from Palmetto Mine; *D*, TRO 1174, right m1 from Palmetto Mine; *E*, UF 17230, left p3 or p4 from Chicora Mine; *F*, UF 61453, left m3 of *Dinohippus* sp. from Lockwood Meadows, Sarasota County, late Miocene. All about 1×. After MacFadden (1986b); reproduced by permission of the Paleontological Society.

moderately large species closely related to or conspecific with *Equus simplicidens,* a well-known western species. The protocone is fairly short in this species, the ectoflexid on the lower molars deep, and the enamel pattern relatively simple. Robertson (1976) concluded that a second, more advanced species of *Equus* was also present in the late Pliocene of Florida, as is the case in Texas, New Mexico, Kansas, and Nebraska. The evidence for the presence of this second species in Florida is weak and needs more complete specimens to substantiate it. Coexisting with *Equus* in Florida during the late Pliocene were two of the most advanced hipparionine horses, *Nannippus peninsulatus* and *Cormohipparion emsliei* (MacFadden, 1984; Hulbert, 1988a). This is the case, for example, at the famous Macasphalt (APAC) Shell Pit near Sarasota. These two are the youngest three-toed horses in North America.

Equus diversified in the latest Pliocene into several lineages. These can be distinguished by different cranial and metapodial proportions and by characters of the incisors and cheek teeth. However, the identification of a single specimen to a species is often difficult, even with complete material. Large samples of many individuals, preferably including skulls and metapodials, are usually needed to make a definitive identification. This is because of the great amount of intraspecific variation observed in *Equus*. A minimum of three species of *Equus* were present in Florida during the early Pleistocene

(Hulbert, 1995b). One was a slender-legged form belonging to the subgenus *Hemionus,* which today only lives in mountainous regions of Asia. Its limb bones are not as long and slender as those of *Equus calobatus,* an early Pleistocene *Hemionus* from northern Texas, and it is also smaller in size than *E. calobatus.* The Florida *Hemionus* probably represents a new species. The two other early Pleistocene species of *Equus* have shorter, stouter limb bones. One is a large species with compressed lower incisors; the other, a medium-sized species, is the most common of the three (Hulbert, 1995b).

Florida's late Pleistocene *Equus* is more poorly understood because large, complete samples from single quarries are unknown, but at least two species of different size were present until about 11,000 years ago (fig. 14.25). At that time, the genus became extinct in North and South America and only persisted in the Old World. Since feral *Equus* populations now thrive in many parts of the United States, the circumstances of its extinction have long perplexed scientists.

FOSSIL TAPIRS OF FLORIDA

Tapir teeth are very conservative and emphasize the transverse lophs and lophids (fig. 14.1D). The individual cusps making up the ectoloph remain distinctive, even in modern tapirs, while those of the transverse lophs are not. In tapir lower molars, the transverse lophids became

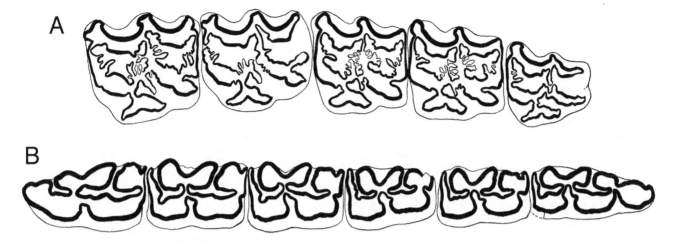

Fig. 14.25. Upper and lower cheek tooth rows of *Equus* sp. from Seminole Field, Pinellas County, late Pleistocene. *A*, left P3–M3; *B*, left p2–m3. The posterior fossettes on the P4 and M3 are not fully formed because these teeth are only slightly worn. About 1×. After Simpson (1929); reproduced by permission of the American Museum of Natural History, New York.

high crests. The m3 hypoconulid is lost in tapirids, so the tooth resembles the m2 (unlike equids). Tapir teeth are brachyodont, and tooth loss is minimal, as only the dp1 was lost. Unusual among perissodactyls (indeed among all mammals) is replacement of the DP1 by a true P1. In tapirids, the I3 is large and caniniform, while the true upper canine is reduced. The lower incisors are spatulate (shovel-shaped). The premolars were progressively molarized.

The oldest record of a tapir from Florida is a partial lower tooth from a very early Miocene (22 Ma) site near the town of Brooksville in Hernando County. This tooth is complete enough to establish the presence of the family in Florida at this time, but it cannot be identified to genus. It is from a relatively small tapir. Other early and middle Miocene faunas described from Florida completely lack tapirs, even the diverse Thomas Farm site. This is somewhat surprising considering that the environment was warm and well forested and so appears to have been very appropriate habitat for tapirs.

It is not until the late Miocene that tapirs became well established in Florida. These specimens belong to the living genus, *Tapirus* (figs. 14.26–14.28). *Tapirus* is a moderately large ungulate with a conspicuous nasal proboscis, a short, muscular trunk. It is used to grasp and move small branches and leaves into the mouth. *Tapirus* persisted in Florida until the end of the Pleistocene, when it became extinct along with *Equus* and many other large mammals. It lives today in the jungles of southeast Asia and Central and South America. At least four species of *Tapirus* lived in Florida (chap. 3), although only one or

at most two species were alive at any one time. The oldest species, *Tapirus simpsoni*, is actually very similar to the youngest, *Tapirus veroensis*, except for a few characters of the skull, even though the two are separated by about 9 million years. This demonstrates the conservative nature of tapir evolution. The two Pleistocene tapirs, *Tapirus haysii* (formerly called *Tapirus copei*) and *T. veroensis* are both well represented (figs. 14.27, 14.28), and are much more common in Florida than elsewhere in North America. Compared to modern *Tapirus*, these two had short muzzles and robust mandibles (Hulbert, 1995a). *T. veroensis* was about the size of living Neotropical *Tapirus*, but *T. haysii* was larger than any extant Neotropical tapir and about the size of the living Asiatic *T. indicus*.

A dwarf tapir was described from the Bone Valley region (Olsen, 1960; fig. 14.29), and was originally called *Tapiravus polkensis*. At the time its geologic age was uncertain. As other members of the genus *Tapiravus* are middle Miocene, this age was not unreasonably applied to the Florida specimens as well. However, more recently collected middle Miocene faunas from the Bone Valley have so far failed to provide any trace of this species (or of any tapir for that matter). Specimens of *T. polkensis* have instead been collected from the early Pliocene Palmetto Fauna and late Miocene Withlacoochee River sites. They indicate that the species is actually a dwarf member of *Tapirus* and does not belong in the older genus *Tapiravus*. Further study of the specimens and complete skulls are needed to demonstrate this conclusively.

Fig. 14.26. Dentitions of *Tapirus simpsoni* from the Love site, Alachua County, late Miocene. *A,* occlusal view of UF 26179, a right maxilla with P1–M2; *B,* dorsal view of UF 26191, mandible with right and left i1–i3, c, p2–m3. *A* about 0.67×; *B* about 0.35×.

Below: Fig. 14.27. Upper and lower cheek tooth rows of *Tapirus haysi* from Leisey Shell Pit 1A, Hillsborough County, early Pleistocene. *A,* medial view of UF 80973, left dentary with i1–i3, c, p2–m3; *B,* occlusal view of UF 80973, left p2–m3; *C,* occlusal view of UF 84190, palate with right and left P1–M3. *A* about 0.25×, *B* about 0.8×, *C* about 0.55×.

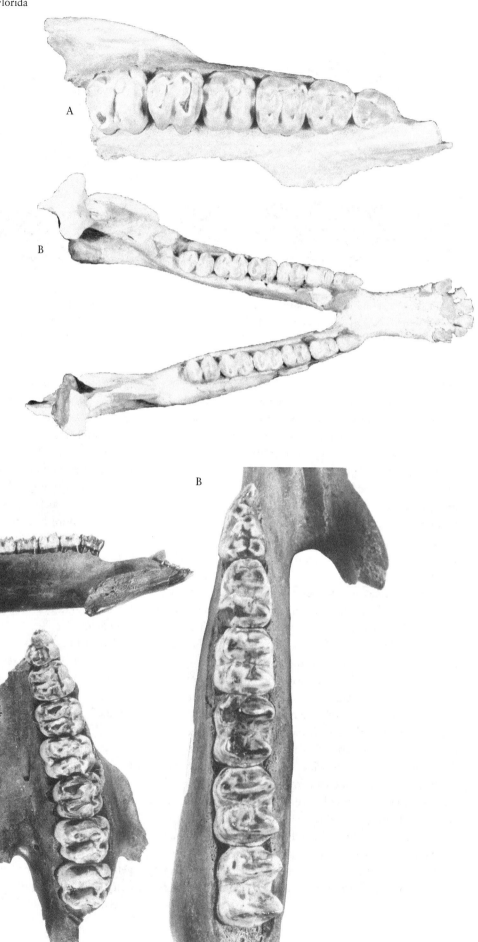

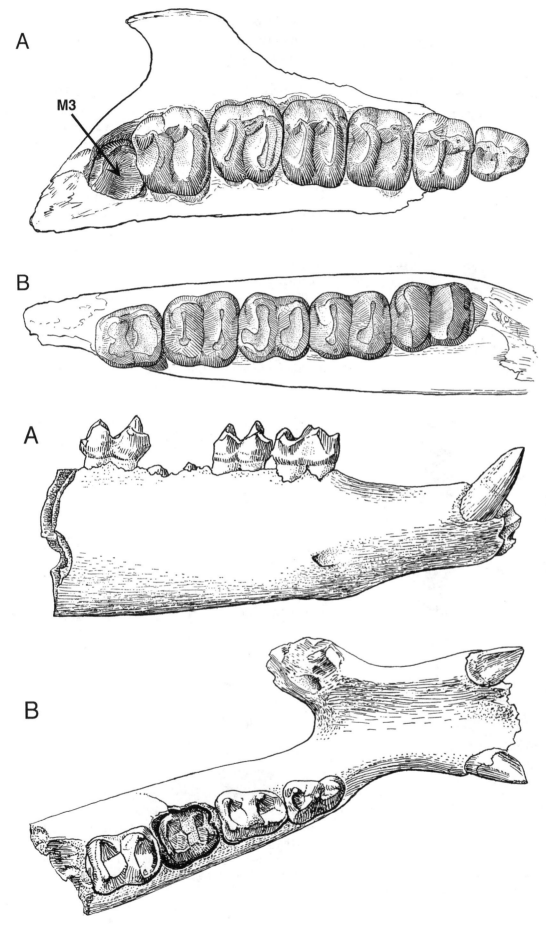

Fig. 14.28. Upper and lower cheek tooth rows of *Tapirus veroensis* from Seminole Field, Pinellas County, late Pleistocene. *A*, occlusal view of AMNH 23492, right maxilla with P1–M2 (the unerupted M3 is visible behind the M2); *B*, occlusal view of AMNH 23493, left dentary with p3–m3. Anterior to left in *B*. About 1×. After Simpson (1929); reproduced by permission of the American Museum of Natural History, New York.

Fig. 14.29. *A*, lateral, and *B*, occlusal views of UF/FGS 5941, right dentary and symphysis of *Tapirus polkensis* from a phosphate mine near Pierce, Polk County. The exact geologic age of this specimen has yet to be determined but is likely to be early Pliocene. Teeth visible in *A* are the right c, p2–p3, and m1. About 1×. After Olsen (1960); reproduced by permission of SEPM (Society for Sedimentary Geology).

FOSSIL RHINOCEROSES OF FLORIDA

The transverse lophs of the upper cheek teeth in rhinoceroses are strong, as in tapirs, but the labial ectoloph dominates the teeth (fig. 14.1E). The lower molars consist of two united, simple "L"-shaped lophs. The m3 hypoconulid is lost, independent of its loss in tapirids. Rhinocerotid cheek teeth are more highly crowned than those of tapirs, with true hypsodonty evolved by several lineages. All rhinoceroses from Florida belong to the family Rhinocerotidae. They are distinguished from the two other recognized rhino families (Amynodontidae and Hyracodontidae, both extinct) by having a very reduced or absent metacone and metastyle on the M3, so that the tooth is triangular rather than rectangular in outline (fig. 14.30E). The so-called "tusks" (if present) are composed of an enlarged I1 and i2, while all the other anterior teeth are either lost or reduced. While several successful rhinocerotid lineages existed in North America during the late Eocene and Oligocene, these were replaced in the Miocene by three groups that independently dispersed from Eurasia: the menoceratines, the aceratheres, and the rhinocerotines.

The lone North Amercan representative of the menoceratine group is *Menoceras* (Prothero, 1998). It was a relatively small rhino, with slender limbs and a pair of small "horns" on the anterior tip of its nasal bones (fig. 14.30). *Menoceras* is one of the best known of the North American rhinos, thanks to the large sample of *Menoceras arikarense* (often incorrectly called *Diceratherium cooki*) from the Agate Spring fauna of western Nebraska. Thomas Farm produced the largest sample of

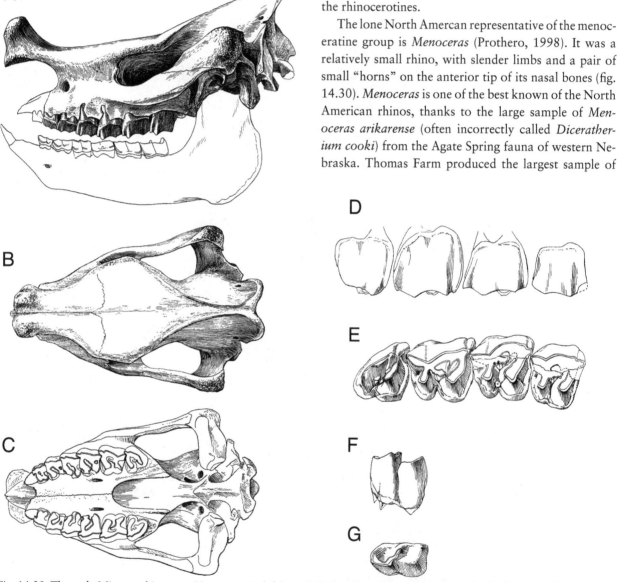

Fig. 14.30. The early Miocene rhinoceros *Menoceras*. *A*, left lateral, *B*, dorsal, and *C*, ventral views of a skull of *M. arikarense* from Nebraska. The protuberances on the nasals (visible in *A* and *B*) indicate that this genus bore two horns located side by side; *D–G, M. barbouri* from the Thomas Farm site, Gilchrist County; *D*, labial, and *E*, occlusal views of MCZ 4452, right P4–M3; *F*, labial, and *G*, occlusal views of MCZ 7445, left m3. *A* about 0.33×; *B–C* about 0.25×; *D–G* about 0.5×. *A–C* after Peterson (1906); reproduced by permission of the Carnegie Museum of Natural History, Pittsburgh. *D–G* after Wood (1964); reproduced by permission of the Museum of Comparative Zoology, Cambridge, Massachusetts.

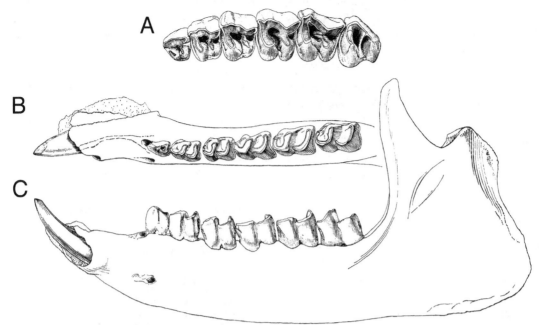

Fig. 14.31. Upper tooth row and dentary of the early Miocene rhinoceros *Floridaceras whitei* from the Thomas Farm site, Gilchrist County. *A*, occlusal view of MCZ 4046, left P2–M3; *B*, occlusal, and *C*, lateral views of MCZ 4435, left dentary with i2 (tusk), p2–m3. All about 0.25×. After Wood (1964); reproduced by permission of the Museum of Comparative Zoology, Cambridge, Massachusetts.

this genus from Florida, the early Miocene species *Menoceras barbouri* (fig. 14.30D–G), but it is actually uncommon at this locality. Better samples of this species come from Nebraska, Texas, and New Mexico. *M. barbouri* is slightly larger and longer limbed than *M. arikarense*.

More common than *Menoceras* at Thomas Farm is *Floridaceras whitei,* a much larger, hornless rhino belonging to the acerathere group (figs. 14.31, 14.32). Members of this group have four functional digits on the front foot (digits 2–5). Other rhinocerotids have either lost or have a vestigial fifth digit (Prothero, 1998). It is unclear whether the well-developed fifth digit of aceratheres represents the retention of a primitive state or the reacquisition of a previously suppressed character. The latter hypothesis was favored by Prothero (1998). *Floridaceras* differs from other members of the acerathere

Fig. 14.32. Lateral views of composite right forelimb and manus of the early Miocene rhinoceros *Floridaceras whitei* from the Thomas Farm site, Gilchrist County. The presence of the fifth metacarpal (and fifth digit) is characteristic of the rhino subfamily Aceratheriinae, which includes *Floridaceras* and *Aphelops*. About 0.125×. After Wood (1964); reproduced by permission of the Museum of Comparative Zoology, Cambridge, Massachusetts.

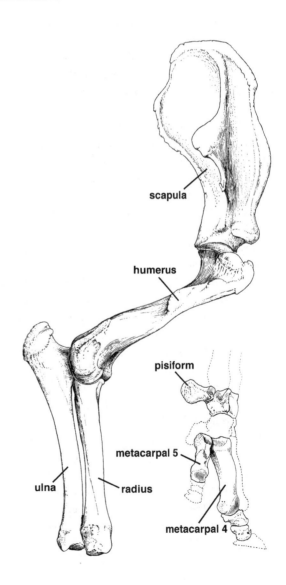

group by retaining an I1 and having a relatively shallow nasal incision that ends over the P2 or P3. *Floridaceras* is best known from the Thomas Farm site but is also known from Texas, Nebraska, and Oregon (Prothero, 1998).

Three genera of rhinocerotids are first recognized in North America beginning in the early Miocene. Two of them, *Aphelops* and *Peraceras,* are advanced aceratheres, while *Teleoceras* represents the rhinocerotine group. The former two were similar to *Floridaceras* in having a four-toed front foot, long limbs (fig. 14.33A), and relatively simple, low-crowned cheek teeth. They both differed *Floridaceras* in their deeper nasal incision (over the P4), loss of the I1 (no anterior upper teeth), and a greatly reduced premaxilla. The teeth of *Aphelops* and *Peraceras* are similar, and isolated specimens are difficult to distinguish. *Peraceras* has a shorter lower diastema and stronger lingual cingula on the lower molars. It also has short nasal bones with some anterior thickening indicating the presence of a "horn." *Aphelops* lacks any trace of a "horn." Middle Miocene rhino samples are sparse in Florida, but the species present appear to be similar to those described from the Texas Coastal Plain (Prothero and Manning, 1987; Prothero, 1998), and *Peraceras* seems to have been the more common acerathere during this time. It, however, became extinct about 9 Ma, while *Aphelops* continued into the early Pliocene.

Teleoceras represents a completely different type of rhino. The limb elements were very short and stout (fig. 14.33B), and the ribcage was enormous and rounded. The overall appearance was similar to the modern *Hippopotamus,* and a similar semiaquatic lifestyle has often been envisioned for this genus (Prothero, 1998). However, MacFadden (1998a) presented isotopic evidence that tended to discount this in favor of a purely terrestrial lifestyle. A mounted skeleton of this oddly proportioned rhino is on display at the Florida Museum of Natural History. The teeth of *Teleoceras* were relatively high crowned and more complex than those of aceratheres. They have strongly developed secondary folds in the enamel of the lophs (called crochets and antecrochets), and the lophs tend to merge to form isolated fossettes like those found in equids (figs. 14.34, 14.35). The anterior cheek teeth (DP1, P2, dp1, p2) are reduced or lost in adults. The upper tusk (I1) is large and bladelike, especially in males. Sexual dimorphism is pronounced both in tusk size (larger in males) and appearance of nasal horns (absent in females, small but present in males). The dental features of *Teleoceras* suggest a grazing diet, and this is supported by study of carbon isotopes found in the enamel of the teeth (MacFadden, 1998a). *Teleoceras* rep-

resents a very successful type of rhino that was abundant throughout North America, Asia, Europe, and Africa in the middle to late Miocene and very early Pliocene.

Rhinocerotid fossils are very abundant at many late Miocene localities in North America, and Florida is no exception. Excellent samples of both *Aphelops* and *Teleoceras* are known. The oldest of these in Florida, the Love site, contains large populations of the species *Aphelops malacorhinus* and *Teleoceras proterum* (figs. 14.34, 14.36A). The name *Aphelops longipes* is sometimes used instead of *A. malacorhinus.* These two species are also present at slightly younger sites such as Mixson's Bone Bed and McGehee Farm. In the very late Miocene,

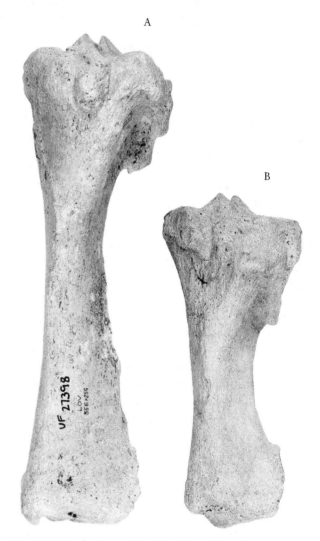

Fig. 14.33. Comparison of the left tibiae of Florida's two common rhinoceroses, the long-legged *Aphelops* and the short-legged *Teleoceras.* A, anterior view of UF 27398, *A. malacorhinus* from the Love site, Alachua County, late Miocene; B, anterior view of UF 59661, *T. proterum* from the McGehee Farm site, Alachua County, late Miocene. Both about 0.4×.

Fig. 14.34. *A*, left lateral view of the skull, and *B*, occlusal view of left P2–M3 of UF 9078, the rhinoceros *Teleoceras proterum* from the McGehee Farm site, Alachua County, late Miocene. *A* about 0.3×; *B* about 0.5×.

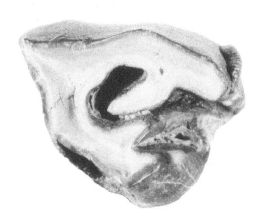

Fig. 14.35. Occlusal view of UF 14800, a left M3 of *Teleoceras hicksi* from the Palmetto Mine, Polk County, early Pliocene. About 1×.

about 6.5 Ma, a larger, more hypsodont species, *Aphelops mutilus*, replaced *A. malacorhinus*. The best sample of this species is from the Moss Acres Racetrack site (fig. 14.36B), which produced many skulls and jaws and a few nearly complete skeletons. Fossils of *Teleoceras* of this age are rare in Florida.

Aphelops is relatively rare in the earliest Pliocene sediments of the Palmetto Fauna in Polk and adjacent counties (MacFadden, 1998a). However, a large, progressive species of *Teleoceras* is relatively common (fig. 14.35). This sample is referred to *Teleoceras hicksi,* well known from the Great Plains and Mexico. The numerical dominance of *Teleoceras* over *Aphelops* in the early Pliocene is unique to Florida. In the Great Plains, *Aphelops* is much more common at this time (Prothero, 1998). Most rhino fossils found in the Bone Valley are either isolated teeth or limb elements, but a few complete jaws are known. This is the youngest sample of rhinos known from Florida, and one of the youngest from North America. The entire family became extinct in North America in the early Pliocene about 4.5 Ma.

FOSSIL CHALICOTHERES OF FLORIDA

Chalicotheres are primarily an Old World group that lived from the Eocene to the Pleistocene (Coombs, 1998). Their most unusual feature is that the distal phalanges bore claws rather than the hooves typical of ungulates (fig. 14.37). It is thought that the claws were used to grasp branches and tear them from trees, much as an elephant does with its trunk. Although very primitive chalicotheres are known from the middle Eocene of North America, they quickly became extinct on this con-

tinent. Then, at about the time of the Oligocene-Miocene boundary, about 24 Ma, a second group of chalicotheres dispersed into North America from Asia. Two genera represent this second phase in North America, *Moropus* and *Tylocephalonyx* (Coombs, 1998). Both are members of the family Chalicotheriidae and persisted until the late Miocene, when the group became extinct for good in North America. They had long necks and front limbs, giving the body the proportions of the African okapi (Coombs, 1983). Although widespread in North America, both genera are typically rare throughout their ranges.

The only published occurrence of chalicotheres from Florida is a limited sample (about ten specimens) from

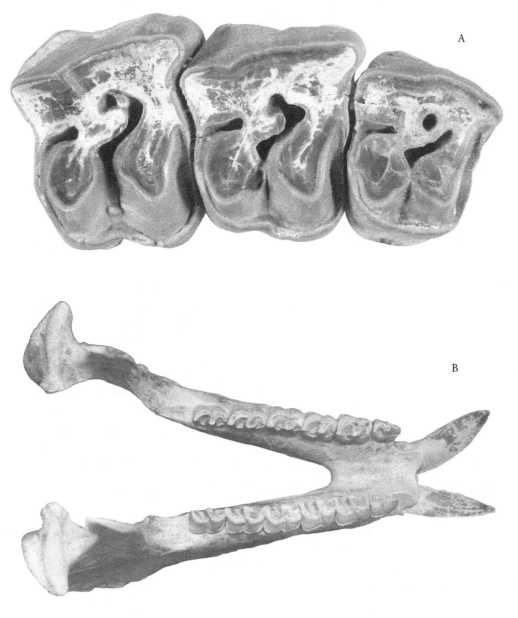

A

B

Fig. 14.36. Dentitions of the rhinoceros *Aphelops* from the late Miocene of Florida. *A,* occlusal view of UF 205641, an associated right P2–P4 of *A. malacorhinus* from the Love site, Alachua County; *B,* occlusal view of UF 135803, mandible with right and left i2, p2–m3 of *A. mutilus* from the Moss Acres Racetrack site, Marion County. *A* about 1×; *B* about 0.2×.

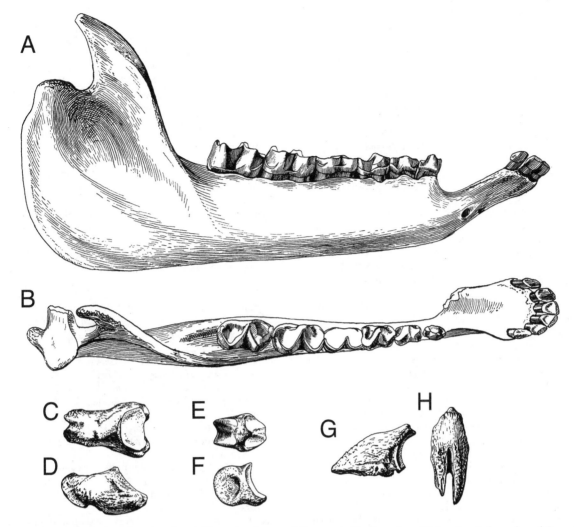

Fig. 14.37. The unusual, clawed perissodactyl *Moropus elatus*, a Miocene chalicothere from Nebraska. *A*, lateral, and *B*, occlusal views of a right dentary and symphysis with i1–i3, p2–m3 (Carnegie Museum 1758); *C*, dorsal, and *D*, lateral views of the right proximal phalanx of digit 2 of the pes; *E*, dorsal, and *F*, lateral views of the right medial phalanx of digit 2 of the pes; *G*, lateral, and *H*, anterior views of the right distal phalanx (claw) of digit 2 of the pes. The fissured claws are a diagnostic feature of the chalicothere family. Specimens shown in *C–H* are part of Carnegie Museum 1706A. Only a few chalicotheres are known from Florida. All about 0.33×. After Holland and Peterson (1914); reproduced by permission of the Carnegie Museum of Natural History, Pittsburgh.

the very early Miocene Buda site. Frailey (1979) referred them to the genus *Moropus,* but they could not be identified to species. Their small size is a primitive feature and in this respect similar to very early Miocene *Moropus* from Oregon and Texas (Coombs, 1998). The well-known species *Moropus elatus* from the Agate Spring fauna of Nebraska is much larger than the Florida specimens. Frailey (1979) noted that the Buda chalicothere lacked a facet for articulation between the fourth metatarsal and the ectocuneiform, which was otherwise found in *Moropus.* Thus the Buda sample represents a distinctive kind of chalicothere, possibly not even *Moropus.*

REFERENCES

Azzaroli, A. 1995. A synopsis of the Quaternary species of *Equus* in North America. Bollettino della Società Paleontologica Italiana, 34:205–21.

Bryant, J. D. 1991. New early Barstovian (middle Miocene) vertebrates from the upper Torrreya Formation, eastern Florida panhandle. Journal of Vertebrate Paleontology, 11:472–89.

Colbert, E. H., and M. Morales. 1991. *Evolution of the Vertebrates.* 4th ed. Wiley-Liss, New York, 470 p.

Coombs, M. C. 1983. Large mammalian clawed herbivores: a comparative study. Transactions of the American Philosophical Society, 73:1–96.

———. 1998. Chalicotherioidea. Pp. 560–68 *in* C. Janis, K. M. Scott, and L. L. Jacobs (eds.), *Evolution of Tertiary Mammals of North America.* Volume 1, *Terrestrial Carnivores, Ungulates, and Ungulatelike Mammals.* Cambridge University Press, Cambridge.

Forstén, A. 1975. The fossil horses of the Gulf Coastal Plain. Pierce-Sellard Series, Texas Memorial Museum. Number 22, 86 p.

Frailey, D. 1979. The large mammals of the Buda local fauna (Arikareean: Alachua County, Florida). Bulletin of the Florida State Museum, 24:123–73.

Hayek, L.-A., R. L. Bernor, N. Solounias, and P. Steigerwald. 1991. Preliminary studies of hipparionine horse diets as measured by tooth wear. Annales Zoologici Fennici, 28:187–200.

Holland, W. J., and O. A. Peterson. 1914. The osteology of the Chalicotheroidea. Memoirs of the Carnegie Museum, 3:189–406.

Hulbert, R. C. 1984. Paleoecology and population dynamics of the early Miocene (Hemingfordian) horse *Parahippus leonensis* from the Thomas Farm Site, Florida. Journal of Vertebrate Paleontology, 4:547–58.

———. 1987. Late Neogene *Neohipparion* (Mammalia, Equidae) from the Gulf Coastal Plain of Florida and Texas. Journal of Paleontology, 61:809–30.

———. 1988a. A new *Cormohipparion* (Mammalia, Equidae) from the Pliocene (latest Hemphillian and Blancan) of Florida. Journal of Vertebrate Paleontology, 7:451–68.

———. 1988b. *Calippus* and *Protohippus* (Mammalia, Perissodactyla, Equidae) from the Miocene (Barstovian-early Hemphillian) of the Gulf Coastal Plain. Bulletin of the Florida State Museum, 32:221–340.

———. 1988c. *Cormohipparion* and *Hipparion* (Mammalia, Perissodactyla, Equidae) from the Late Neogene of Florida. Bulletin of the Florida State Museum, 33:229–338.

———. 1993. Late Miocene *Nannippus* (Mammalia, Perissodactyla) from Florida, with a description of the smallest hipparionine horse. Journal of Vertebrate Paleontology, 13:350–66.

———. 1995a. The giant tapir, *Tapirus haysii*, from Leisey Shell Pit 1A and other Florida Irvingtonian localities. Bulletin of the Florida Museum of Natural History, 37:515–51.

———. 1995b. *Equus* from Leisey Shell Pit 1A and other Irvingtonian localities from Florida. Bulletin of the Florida Museum of Natural History, 37:553–602.

———. 1996. The ancestry of the horse. Pp. 11–34 *in* S. L. Olsen (ed.), *Horses through Time.* Roberts Rinehart Publishing, Boulder, Colo.

Hulbert, R. C., and B. J. MacFadden. 1991. Morphologic transformation and cladogenesis at the base of the radiation of hypsodont horses. American Museum Novitates. Number 3000, 61 p.

Janis, C. 1990. Correlation of cranial and dental variables with body size in ungulates and macropodoids. Pp. 255–99 *in* J. Damuth and B. J. MacFadden (eds.), *Body Size in Mammalian Paleobiology.* Cambridge University Press, Cambridge.

MacFadden, B. J. 1982. New species of primitive three-toed browsing horse from the Miocene phosphate mining district of central Florida. Florida Scientist, 45:117–25.

———. 1984. Systematics and phylogeny of *Hipparion, Neohipparion, Nannippus,* and *Cormohipparion* (Mammalia, Equidae) from the Miocene and Pliocene of the New World. Bulletin of the American Museum of Natural History, 179:1–196.

———. 1986a. Fossil horses from "Eohippus" *(Hyracotherium)* to *Equus*: scaling, Cope's Law, and the evolution of body size. Paleobiology, 12:355–69.

———. 1986b. Late Hemphillian monodactyl horses (Mammalia, Equidae) from the Bone Valley Formation of central Florida. Journal of Paleontology, 60:466–75.

———. 1992. *Fossil Horses: Systematics, Paleobiology, and Evolution of the Family Equidae.* Cambridge University Press, New York, 369 p.

———. 1998a. Tale of two rhinos: isotopic ecology, paleodiet, and niche differentiation of *Aphelops* and *Teleoceras* from the Florida Neogene. Paleobiology, 24:274–86.

———. 1998b. Equidae. Pp. 536–59 *in* C. Janis, K. M. Scott, and L. L. Jacobs (eds.), *Evolution of Tertiary Mammals of North America.* Volume 1, *Terrestrial Carnivores, Ungulates, and Ungulatelike Mammals.* Cambridge University Press, Cambridge.

MacFadden, B. J., and T. E. Cerling. 1996. Mammalian herbivore communities, ancient feeding ecology, and carbon isotopes: a 10 million-year sequence from the Neogene of Florida. Journal of Vertebrate Paleontology, 16:103–15.

MacFadden, B. J., and J. S. Waldrop. 1980. *Nannippus phlegon* (Mammalia, Equidae) from the Pliocene (Blancan) of Florida. Bulletin of the Florida State Museum, 25:1–37.

MacFadden, B. J., N. Solounias, and T. E. Cerling. 1999. Ancient diets, ecology, and extinction of 5-million-year-old horses from Florida. Science, 283:824–27.

McKenna, M. C., and S. K. Bell. 1997. *Classification of Mammals above the Species Level.* Columbia University Press, New York, 631 p.

Olsen, S. J. 1960. Age and faunal relationships of *Tapiravus* remains from Florida. Journal of Paleontology, 34:164–67.

Peterson, O. A. 1906. The Miocene beds of western Nebraska and eastern Wyoming and their vertebrate faunae. Annals of the Carnegie Museum, 4:21–72.

Prothero, D. R. 1998. Rhinocerotidae. Pp. 596–605 *in* C. Janis, K. M. Scott, and L. L. Jacobs (eds.), *Evolution of Tertiary Mammals of North America.* Volume 1, *Terrestrial Carnivores, Ungulates, and Ungulatelike Mammals.* Cambridge University Press, Cambridge.

Prothero, D. R., and E. Manning. 1987. Miocene rhinoceroses from the Texas Gulf Coastal Plain. Journal of Paleontology, 61:388–423.

Prothero, D. R., and N. Shubin. 1989. The evolution of Oligocene horses. Pp. 142–75 *in* D. R. Prothero and R. M. Schoch (eds.), *The Evolution of Perissodactyls.* Oxford University Press, New York.

Ray, C. E. 1964. *Tapirus copei* in the Pleistocene of Florida. Quarterly Journal of the Florida Academy of Sciences, 27:59–66.

Ray, C., and A. E. Sanders. 1984. Pleistocene tapirs in the eastern United States. Pp. 283–315 *in* H. H. Genoways and M. R. Dawson (eds.), *Contributions in Quaternary Vertebrate Paleontology: A Volume in Memorial to John E. Guilday.* Carnegie Museum of Natural History, Pittsburgh, Special Publication 8.

Robertson, J. S. 1976. Latest Pliocene mammals from Haile XV A, Alachua County, Florida. Bulletin of the Florida State Museum, 20:111–86.

Simpson, G. G. 1929. Pleistocene mammalian fauna of the Seminole Field, Pinellas County, Florida. Bulletin of the American Museum of Natural History, 56:561–99.

———. 1945. Notes on Pleistocene and Recent tapirs. Bulletin of the American Museum of Natural History, 86:33–82.

———. 1951. *Horses.* Oxford University Press, New York, 247 p.

Webb, S. D., and R. C. Hulbert. 1986. Systematics and evolution of *Pseudhipparion* (Mammalia, Equidae) from the Late Neogene of the Gulf Coastal Plain and the Great Plains. Pp. 237–72 *in* K. M. Flanagan and J. A. Lillegraven (eds.), *Vertebrates, Phylogeny, and Philosophy.* University of Wyoming Contributions to Geology, Special Paper 3.

White, T. E. 1942. The lower Miocene fauna of Florida. Bulletin of the Museum of Comparative Zoology, 92:1–49.

Wood, H. E. 1964. Rhinoceroses from the Thomas Farm Miocene of Florida. Bulletin of the Museum of Comparative Zoology, 130:361–86.

Mammalia 7

Proboscideans

INTRODUCTION

The living elephants and their extinct relatives make up one of the most spectacular groups of mammals, the Proboscidea. The fossil record of these large ungulates is quite extensive, and they are among the most common prehistoric mammals found in Florida. Among ungulates, proboscideans are most closely related to two groups of large, aquatic herbivores: the extinct Desmostylia and the living Sirenia (dugongs and manatees [see chap. 16 and fig. 13.1]). The collective name given to these three mammalian groups is the Tethytheria. The general consensus of studies on the evolutionary relationships among ungulates is that proboscideans are more closely related to desmostylians than to sirenians, and that, among living mammals, perissodactyls and hyraxes are closest to tethytheres (fig. 13.1). The classification of proboscidean families, subfamilies, and tribes used here follows McKenna and Bell (1997).

The most primitive proboscideans are known from the Eocene Epoch of southern Asia and northern Africa. In the late Paleogene and early Neogene Periods the group dispersed into southern Africa and throughout Eurasia, acquiring the characteristics most associated with the group: very large size; a manipulative proboscis (the trunk); the protruding tusks; molars with three or more transverse lophs; and a horizontal tooth replacement system (described below). The number of recognized proboscidean families varies among authorities, but the group was certainly much more diverse than now (Shoshani and Tassy, 1996; Lambert and Shoshani, 1998). During the middle Miocene (about 12 to 13 Ma), two kinds of proboscideans first spread out through North America, mastodonts and gomphotheres. Some evidence (based on tooth fragments from Nevada and California) indicates that mastodonts actually first arrived in North America somewhat earlier, about 17 Ma. If so, this family remained uncommon and limited to western North America until the middle Miocene. In all,

three families of proboscideans immigrated to North America from Asia, and several representatives of each are known from Florida (chap. 3).

A comment on common names. As they are so well known to the general public, a variety of common names are in use for groups of fossil and living proboscideans. The exact meaning of these names varies among authors. To eliminate any confusion, the common names used in this book are specifically defined as follows. The term *mastodont* is sometimes broadly used for all brachyodont proboscideans (that is, Mammutidae and Gomphotheriidae), but will be used here in a stricter sense to apply only to members of the Mammutidae. It is also seen spelled "mastodon," but *mastodont* appears more often in the scientific literature. Likewise *gomphothere* is often broadly applied but is limited here to members of the family Gomphotheriidae. Members of the tribe Amebelodontini are referred to either as *amebelodonts* or *shovel-tuskers* in reference to the shape of their lower tusks. The term *elephant* is applied only to members of the Elephantidae, not to proboscideans in general.

Proboscidean Skeletal and Dental Anatomy

Proboscideans are the largest living terrestrial mammals, with the African elephant *(Loxodonta africana)* weighing from 2,500 to 6,300 kilograms. Although most fossil proboscideans were somewhat smaller, they were all large mammals. Thus large size is especially characteristic of proboscidean skeletal elements (fig. 15.1), and they have many modifications related to supporting such a large mass. Not all large fossils represent proboscideans, of course. In Florida, bones of whales and the larger sloths can be mistaken for those of proboscideans unless attention is paid to their distinctive morphology. Proboscidean limb elements are heavy and massive (fig. 15.1). The humerus and femur are relatively long. The ulna is large and bears most of the weight in the forearm, unlike most ungulates, which have a larger radius and a reduced ulna. The fibula is also complete and separate from the tibia. The carpal and tarsal bones are dorsally and ventrally flattened (fig. 15.1). The feet have five digits that surround a dense heel pad that transmits most of the animal's mass to the ground. Although subtle morphological differences separate the postcranial skeletal elements of the four Florida proboscidean families, they can only be easily distinguished by direct association with cranial remains or by careful comparative analysis (for example, Olsen, 1972).

The proboscidean skull is primitively low and long (fig. 15.2A). The mandible is shallow and bears an elongated symphysis that houses the alveoli for the lower tusks (fig. 15.2B). A common trend in proboscidean evolution was to shorten the skull and jaw and reduce (or even lose) the lower tusks. The elephants, with their high, domed skulls (fig. 15.3), are among the most advanced in this respect. To reduce weight, much of the

bone from the skull roof contains numerous air passages and sinuses. Large chunks of bone with these air cavities are frequently found as fossils.

Proboscidean teeth can be divided into two categories, the tusks, which are enlarged, ever-growing incisors, and the cheek teeth. The primitive number of tusks is four, two uppers and two lowers (fig. 15.2). It is often stated that these are the right and left I2 and i2. Tassy (1987) suggested instead that they are the equivalent of the DI2 and di1 of other mammals, but Luckett (1996) found no embryological evidence to support this claim. The other incisors and the canines were lost very early in proboscidean evolution. Most of the tusk consists of laminated rings or sheets of a type of dentine called ivory. Viewed in cross section, ivory displays a unique crosshatched or herringbone pattern (fig. 15.4). This pattern allows even fragments of tusks to be identified as ivory. When the tusks first erupt in juveniles they have a short cap of enamel. Enamel deposition on the tusks stops, leaving bare ivory, except along one side of the upper tusk that bears a band of enamel (figs. 15.2A, 15.5). An enamel band is present on the upper tusk of primitive mastodonts and most gomphotheres, but it was independently lost in elephants, advanced mastodonts, and a few gomphotheres. A banded lower tusk evolved in one genus of North American gomphothere. Lower tusks tend to be smaller than uppers and are usually round or oval in cross section in the Mammutidae, Gomphotheriidae, and primitively the Elephantidae. Various lineages in all three of these families reduce and eventually lose the lower tusks. This was usually accompanied by shortening of the mandibular symphysis (compare the mandibles in figs. 15.2 and 15.3). Lower tusks were greatly flattened in the Amebelodontini.

Proboscidean cheek teeth are built upon a basic pattern of transverse lophs and lophids. The most primitive morphology was bilophodont, similar to the lower cheek teeth of tapirs. Very early in proboscidean evolution a third loph (and lophid) was added to the molars and fourth deciduous premolars by increasing the height of the posterior cingulum. This type of tooth is called trilophodont because of its three major transverse lophs (for example, the M1 and M2 in fig. 15.2C). The M3 and m3 have four, five, or even more lophs, so the last tooth in the jaw is much longer than the other cheek teeth. Some groups (for example, Mammutidae) retained trilophodont anterior molars, while various late Neogene species have four or even five lophs on the anterior molars. The number of lophs and lophids on the anterior molars is of systematic importance, while the number on the M3 and m3 tends to be more variable.

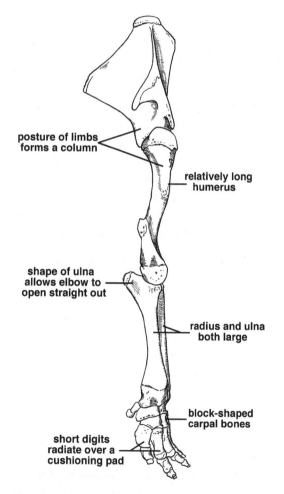

posture of limbs forms a column

relatively long humerus

shape of ulna allows elbow to open straight out

radius and ulna both large

block-shaped carpal bones

short digits radiate over a cushioning pad

Fig. 15.1. Special adaptations of the proboscidean limb to support its large mass. Figures from *Analysis of Vertebrate Structure* by Milton Hildebrand, copyright © 1974 by John Wiley & Sons, Inc. Reprinted by permission of John Wiley & Sons, Inc.

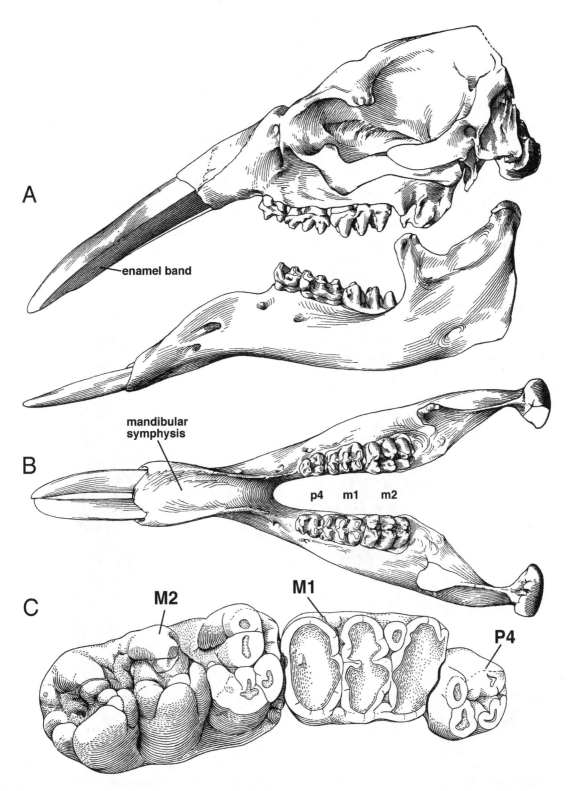

Fig. 15.2. Skull, mandible, and teeth of the middle Miocene proboscidean *Gomphotherium productum*. *A*, lateral view of skull and dentary, and *B*, occlusal view of mandible of AMNH 10673 from northern Texas. Note the elongated mandibular symphysis bearing tusks, the outer enamel band on the upper tusks, and the low skull; *C*, occlusal view of UCMP 35753, right P4–M2 (anterior to right) from Nebraska. Note the bilophodont P4 and trilophodont M1 and M2. *A–B* about 0.125×; *C* about 0.5×. *A–B* after Osborn (1936); reproduced by permission of the American Museum of Natural History, New York. *C* after Webb (1969); reproduced by permission of the University of California Press, Berkeley.

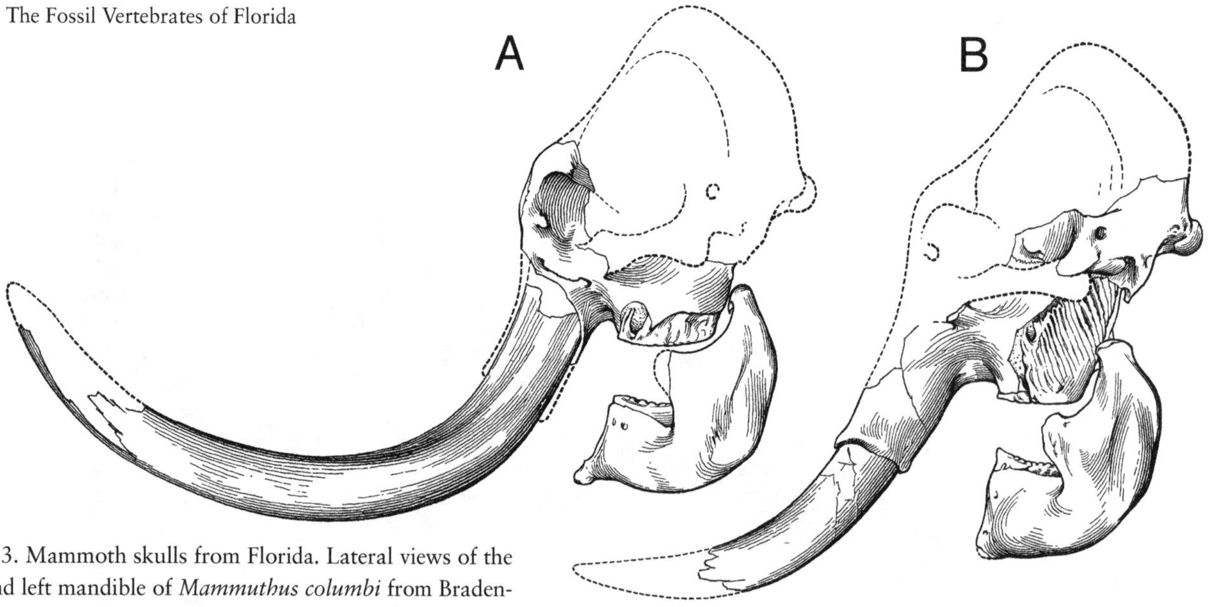

A

B

Fig. 15.3. Mammoth skulls from Florida. Lateral views of the skull and left mandible of *Mammuthus columbi* from Bradenton, Sarasota County, middle Pleistocene. *A*, AMNH 26821; *B*, AMNH 26820. In comparison with figure 15.2, note the short, domed skull, the much shorter mandibular symphysis without tusks, and the larger, more curved upper tusks without enamel bands. About 0.05×. After Osborn (1930); reproduced by permission of the American Museum of Natural History, New York.

Fig. 15.4. Cross section of a proboscidean tusk, probably Gomphotheriidae, in the UF/FGS collection from the Bone Valley District. This specimen clearly shows that proboscidean tusks are made of rings of specialized dentine called ivory and that the ivory forms a unique herringbone pattern that instantly allows even very small fragments to be identified as parts of a tusk. About 1×.

Fig. 15.5. Lateral view of an upper tusk of a juvenile gomphothere, probably *Rhynchotherium simpsoni*, from the Gardinier Mine, Polk County, early Pliocene. The more darkly preserved enamel once completely covered the tip of the tusk, up to about 2 inches back, but this cap is partially broken revealing the inner dentine core. Behind the enamel cap, the enamel is limited to a narrow lateral band that runs along the outer side of the tusk. About 0.75×.

The cheek teeth of elephantids are among the most specialized of the mammals. Elephants evolved from a gomphothere species with relatively simple enamel patterns on its cheek teeth. The original brachyodont, lophodont pattern was modified by greatly increasing crown height and the number of lophs per tooth, and by greatly narrowing each individual loph. In this family these lophs are referred to as "plates." Each plate consists of a core of dentine surrounded by a relatively narrow layer of enamel (fig. 15.6). The plates are fused at the base of each mature tooth but are really held together by cement. Upon wear, each plate becomes two long ridges of enamel alternating with softer regions of dentine and cement at the occlusal surface (fig. 15.6A). The plates of the occluding upper and lower teeth grind against each other as the animal chews, shearing vegetation into easily digested pieces. Elephants are generally thought to eat more grass than did the brachyodont mastodonts and gomphotheres. The latter two presumably ate mostly leaves, twigs, and bark (Lambert, 1992). Fossil mammoth teeth frequently break up into individual plates or groups of plates along the weaker planes of the cement layers.

Proboscidean cheek teeth erupt in an unusual manner called horizontal tooth replacement. In the normal mammalian eruption sequence (vertical replacement), the worn-out deciduous cheek teeth are replaced by permanent teeth that develop directly above or below them. The position of the permanent teeth stays relatively fixed in the jaw except for vertical movement (up or down). In a horizontal replacement system, as teeth wear out they move forward, pushed by younger, less worn teeth in the back of the jaw. Eventually the anteriormost tooth falls out or is shed, giving the more posterior teeth room to move forward. In an animal with this type of replacement system the complete tooth formula cannot be determined from any single individual. A series of individuals of differing ages (juvenile to mature adult) must be examined. Horizontal tooth replacement evolved once in the Proboscidea and independently in the manatee and the kangaroos.

Proboscidean tooth succession generally proceeds as follows, although some genera may take a slightly different approach. At or shortly after birth the animal has two small, usually bilophodont cheek teeth per jaw quadrant (the DP2, DP3, dp2, and dp3; see fig. 15.7). The much larger, trilophodont DP4 and dp4 next erupt behind these two. Then the first and second molars erupt sequentially from the rear of the jaw, pushing the premolars anteriorly. By the time the M2 and m2 have completely erupted and come into wear, the second and third deciduous premolars are lost, leaving three teeth in

Fig. 15.6. *A*, occlusal, and *B*, lateral views of UF 135733, a right lower third molar of *Mammuthus columbi* from the Aucilla River, Jefferson County, late Pleistocene. Mammoth teeth consist of a series of plates made of enamel with dentine cores held together with cement. As the anterior plates (to the right on this figure) wear away, they break off, and more posterior plates come into wear. A whole tooth may consist of up to twenty-five to thirty plates. About 0.33×.

the jaw (fig. 15.2). In some proboscideans, smaller permanent premolars may replace their deciduous counterparts in the normal vertical manner before they too are shed. In others, permanent premolars are completely suppressed or only the DP4/dp4 are replaced. Thus, in the young adult stage, the three teeth in the jaw of a mastodont will be the DP4, M1, and M2 (all trilophodont), while *Gomphotherium* has a bilophodont P4 with the trilophodont M1 and M2 (fig. 15.2C). As the M3/m3 begin to erupt and wear, first the DP4/dp4 (or P4/p4) are lost, and then coincident with the complete eruption of the last molar, the M1 and m1 are shed. This leaves only two teeth in the jaw during the typical mature adult phase in mastodonts and primitive gomphotheres (fig. 15.8). In these species, the second molars are not shed until very late in life, leaving only the very worn third molars. In advanced gomphotheres and amebelodonts, which have very long third molars, the second molars are shed at a somewhat earlier age, leaving a single tooth in the jaw for a longer period of time.

The tooth succession pattern in elephants is modified from the gomphothere system and is well understood because those of the living elephant species have been studied intensively. The age at death (in years) can be estimated for a mammoth with some degree of accuracy on the basis of studies on the closely related living elephants (Owen-Smith, 1988). As the individual teeth in elephants are longer than in similar-sized gomphotheres and mastodonts, two differences in the succession pattern emerge. Rarely is there room in the jaw for more

Fig. 15.7. Lateral view of UF 80286, a partial right dentary with dp2–dp3 of *Mammut americanum* from the Leisey Shell Pit, Hillsborough County, early Pleistocene. Note that these teeth are bilophodont in contrast to the trilophodont dp4 and first and second molars. About 0.5×.

than one and a half teeth to be in use at any one time. Second, heavily worn anterior pieces of teeth break off and are shed separately from the less worn posterior part of the same tooth. These pieces are sometimes recovered as fossils. The whole tooth is usually shed in other proboscideans. At birth and shortly thereafter, the very small buttonlike DP2/dp2 and the anterior half of the DP3/dp3 are the functioning teeth. The DP2/dp2 are shed at an age of about two years as the DP4/dp4 erupt. This is followed by shedding of the DP3/dp3 at age five years, the DP4/dp4 at eleven to twelve years, the M1/m1 at about twenty years, and the M2/m2 at about forty years. The third molar remains the only tooth in the jaw until the animal dies, which under normal conditions may be at

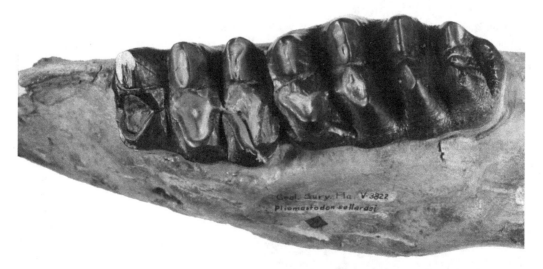

Fig. 15.8. Occlusal views of the tooth-bearing portion of UF/FGS 3822, a left dentary with m2–m3 of the mastodont *Mammut sellardsi* from the American Cyanamide Company Mine, Polk County, early Pliocene. The m2 is trilophodont; the m3 tetralophodont. About 0.5×.

an age of about fifty to sixty years. The great longevity of proboscideans is permitted by their horizontal tooth replacement system.

FOSSIL PROBOSCIDEANS OF FLORIDA

Three proboscidean families are recorded as fossils in Florida: Mammutidae, Gomphotheriidae, and Elephantidae (see chap. 3 for species listing). The two oldest proboscideans in the state represented by complete material are a mastodont, *Zygolophodon,* and a gomphothere, *Gomphotherium,* that were collected from the middle Miocene Bradley Fauna in Polk County (fig. 15.9). Both of these families persisted to the late Pleistocene, with the genera *Mammut* and *Cuvieronius* being their youngest respective genera. The mastodonts entered North America earlier than the gomphotheres, and a few

enamel fragments reported from the early middle Miocene of Gadsden County by Bryant (1991) may represent this family. Shovel-tuskers are well represented in the late Miocene of northern peninsular Florida by the genus *Amebelodon,* while less complete specimens of *Platybelodon* have also been found. The presence of either younger or older representatives of the shovel-tusker group is uncertain. The mammoth (genus *Mammuthus*), a member of the Elephantidae, appeared in North America during the early Pleistocene and quickly spread across the entire continent. *Mammuthus,* like *Mammut,* persisted until the latest Pleistocene in Florida.

Mastodonts (Mammutidae) are relatively small to moderate-sized for proboscideans and have relatively simple, brachyodont cheek teeth (figs. 15.8–15.11). The DP4, M1, M2, dp4, m1, and m2 are always trilophodont, while the third molars have between 3.5 and 5

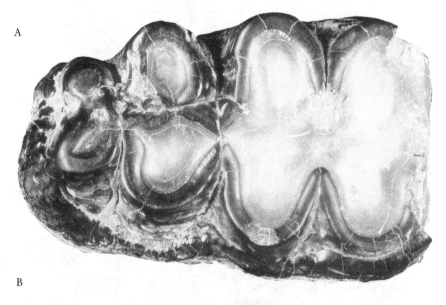

Fig. 15.9 Occlusal views of two middle Miocene proboscidean teeth from the Kingsford Mine, Bone Valley District. *A,* UF 116809, a right M3 of the mastodont *Zygolophodon* sp.; *B,* UF 116808, a left m3 of *Gomphotherium* sp. These are among the oldest proboscidean fossils from Florida and clearly distinguish the simple enamel patterns of mastodonts and the complex enamel of gomphotheres. Both about 0.9×.

lophs or lophids (most often four). In unworn teeth the lophs have sharp crests. The lower tusks are narrow and peglike, when present. Two mastodont genera are recognized in Florida, *Zygolophodon* (sometimes called *Miomastodon*) and *Mammut*. *Zygolophodon*, the early to middle Miocene mastodont, differed from *Mammut* by retaining a well-developed enamel band on the upper

tusk, having slightly narrower and lower crowned cheek teeth, and by its slightly smaller size (fig. 15.9A). The cheek teeth were slightly more complex in *Zygolophodon* as they often have accessory cuspules, but not to the degree of gomphothere cheek teeth. In Florida, it is currently known only from two middle Miocene teeth from the Bone Valley.

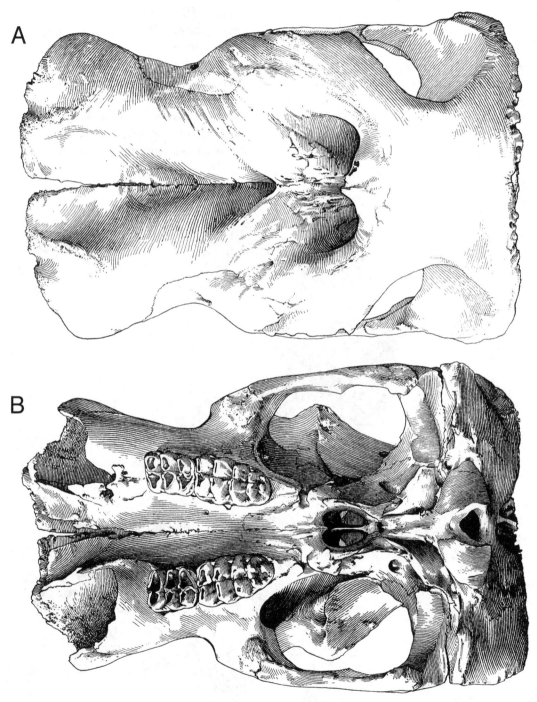

Fig. 15.10. *A*, dorsal, and *B*, ventral views of a nearly complete skull of the American mastodont *(Mammut americanum)* from New York, AMNH 9951. About 0.125×. After Osborn (1936); reproduced by permission of the American Museum of Natural History, New York.

Fig. 15.11. Occlusal view of UF 135709, a palate with right and left M1–M3 of *Mammut americanum* from the Aucilla River, Jefferson County, late Pleistocene. Compare with the specimen illustrated in figure 15.10. About 0.25×.

An 8-million-year gap separates *Zygolophodon* from the next record of mastodonts in Florida. Whether this is an artifact of preservational bias, or a true reflection of historical fact is uncertain. But the late Miocene is relatively well known in the state, and no localities of that age have as yet produced fossils of mastodonts. *Mammut sellardsi* is only known from the early Pliocene Palmetto Fauna of the Bone Valley region, but by much more complete material than *Zygolophodon* (fig. 15.8). This species was formerly placed in *Pliomastodon,* but that genus is no longer regarded as distinct from *Mammut* (Lambert and Shoshani, 1998). *Mammut americanum,* the American mastodont (figs. 15.7, 15.10, 15.11), first appeared in the late Pliocene and persisted until the end of the Pleistocene. *Mammut* is one of the best known of the extinct Pleistocene mammals, with relatively complete remains recovered from across North America and Eurasia. This species had more vestigial lower tusks (frequently absent in older adults), a shorter symphysis, and wider cheek teeth than late Miocene–early Pliocene *Mammut.* Many complete or nearly complete skeletons of *M. americanum* have been recovered from Florida, including individuals from the Aucilla River, Ichetucknee River, Palm Beach County, and Wakulla Springs (the mounted specimen displayed at the Museum of Florida History in Tallahassee). Partial jaws, isolated teeth, and postcranial elements of *M. americanum* are very common in riverbeds containing Pleistocene vertebrate

fossils. *Mammut* was apparently more of a forest animal than *Mammuthus,* with a diet principally of twigs, leaves, shrubs, fruits, pinecones and needles, and mosses (Webb et al., 1992).

The most diversified group of proboscideans are the gomphotheres (Lambert and Shoshani, 1998). Their teeth are readily distinguished from those of mammutids by their more rounded (rather than sharp-crested) lophs in unworn teeth and more abundant accessory conules (compare the two in fig. 15.9). When worn, the enamel pattern often forms a series of cloverleaf patterns, called "trefoils." The various types of gomphotheres are difficult to distinguish from one another on the basis of isolated cheek teeth or postcranial elements. Relatively complete skulls and jaws with the mandibular symphysis and tusks are needed for secure identifications. The relatively primitive genus *Gomphotherium* ranges from the middle Miocene to the late Pliocene in Florida (Lambert and Shoshani, 1998). It had trilophodont anterior molars; a long, fairly straight mandibular symphysis; a shallow mandible that is not expanded laterally; and long, rounded upper and lower tusks that are neither twisted nor spiraled (figs. 15.2, 15.9B). The upper tusk bore a prominent, straight enamel band. Specific designation of Florida *Gomphotherium* is difficult because adequate samples and studies are lacking. The largest sample is from the Love site (Alachua County). Florida appears to be the only region in North America in which *Gom-*

photherium survived into the Pliocene. The species name *Gomphotherium simplicidens* is apparently available for this small, brachyodont Pliocene gomphothere.

A more advanced Florida gomphothere is *Rhynchotherium*. *Rhynchotherium* is part of a group of New World gomphotheres with shortened mandibular symphyses. Other members of this group include *Cuvieronius*, *Stegomastodon*, and the South American gomphotheres. *Rhynchotherium* cheek teeth tended to be relatively simple for a gomphothere, with trilophodont

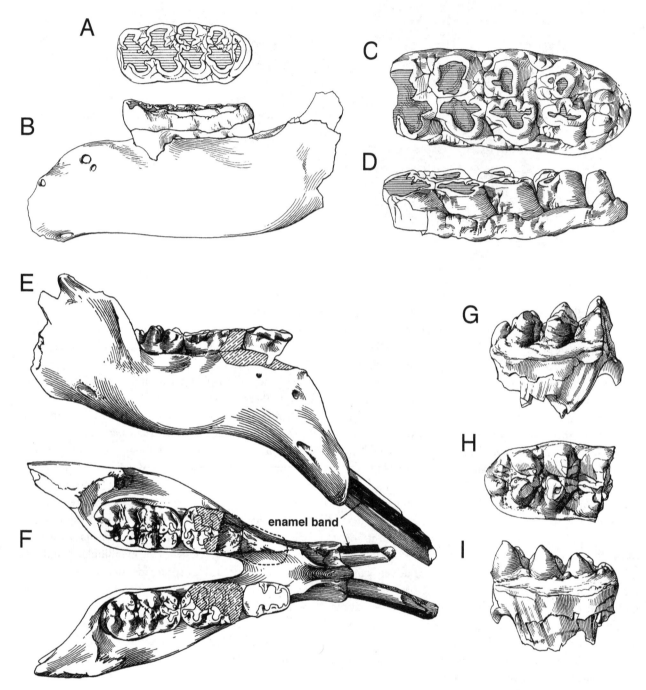

Fig. 15.12. Pliocene Gomphotheres from Florida and Mexico. *A–D, Rhynchotherium simpsoni* from the early Pliocene Palmetto Fauna, Bone Valley District. *A,* occlusal view of left m3, and *B,* lateral view of mandible of UF/FGS 77000; *C,* occlusal, and *D,* labial views of AMNH 1875, left m3; *E,* lateral, and *F,* occlusal views of AMNH 15550, mandible with m1–m3 of *Rhynchotherium browni* from Mexico; note the short, downturned mandibular symphysis and enamel bands on the lower tusks; *G,* labial, *H,* occlusal, and *I,* lingual views of AMNH 1907, partial right m3 of *Gomphotherium simplicidens,* early Pliocene, Polk County. *A–B* about 0.17×; *C, D, G–I* about 0.33×; *E–F* about 0.125×. After Osborn (1936); reproduced by permission of the American Museum of Natural History, New York.

anterior molars and third molars with four to five lophs/lophids. Accessory conules were usually limited to the lingual half of the upper cheek teeth and the labial half of the lower cheek teeth (fig. 15.12). The second molar tended to be lost earlier in life than in *Gomphotherium,* and the deciduous premolars were only rarely replaced by permanent premolars. The mandibular ramus below the cheek teeth was more massive and deeper than in *Gomphotherium* and tended to protrude or bulge laterally (fig. 15.12F). The mandibular symphysis was sharply turned ventrally. The enamel and dentine in the upper tusk grew in a spiral pattern, so that the enamel band wound around the tusk like the stripe of a barbershop pole. The lower tusk often, but not always, had a straight enamel band, a unique feature of *Rhynchotherium* (fig. 15.12E, F).

Rhynchotherium simpsoni was described from the Palmetto Fauna (early Pliocene) of Polk County based on an associated palate with upper tusks and the right and left mandibles with tusks (Olsen, 1957). The upper tusks are spiraled but lack an enamel band, as do the lowers. This is thought to reflect individual variation or postmortem loss (Webb and Tessman, 1968) and does not represent the morphology of the species as a whole. Unfortunately, additional complete material from the Bone Valley has not been described, so this hypothesis remains untested. A segment of a lower tusk of *Rhynchotherium* with an enamel band was recovered from the very late Miocene Manatee County Dam Site. Webb and Tessman (1968) suggested that all large late Neogene *Rhynchotherium,* including *R. simpsoni,* be synonymized into one species, *R. euhypodon.* Other classifications for the genus were proposed by Miller (1990) and Lambert and Shoshani (1998). *R. simpsoni* is tentatively retained here pending a comprehensive review. It is apparently the common large gomphothere in the Palmetto Fauna of the Bone Valley. Recent isotopic studies on its tooth enamel suggested that it was more of a grazer than is usually thought (MacFadden and Cerling, 1996). A more advanced species of *Rhynchotherium* (or perhaps a primitive *Cuvieronius*) is present in the late Pliocene of Florida; it is known from a mandible from the APAC Shell Pit and from a few teeth and tusk fragments from other southwestern Florida Pliocene sites. It apparently represents an undescribed species.

Junior synonyms of *Gomphotherium* and *Rhynchotherium* often seen in the older scientific literature include *Serridentinus, Trilophodon,* and *Ocalientinus.* In the 1920s, three species of gomphotheres were named on the basis of isolated molars of uncertain stratigraphic position from the Bone Valley region of Florida. These were *Serridentinus simplicidens, Serridentinus brewsterensis,* and *Serridentinus bifoliatus.* As noted above, *S. simplicidens* is probably referable to *Gomphotherium* and represents a small, primitive form. The other two likely represent the same species as *Rhynchotherium simpsoni,* but isolated teeth of gomphotheres are rarely diagnostic.

The youngest Florida gomphothere is *Cuvieronius tropicus,* which ranges from the late Pliocene to the middle Pleistocene. *Cuvieronius* is named after the great French anatomist of the early nineteenth century, Georges Cuvier. *Cuvieronius* resembled *Rhynchotherium* in having an upper tusk with a spiraled enamel band and a very short, downturned mandibular symphysis. It differed by lacking lower tusks, having a shorter symphysis, more complex cheek teeth (trefoils common on both sides of the molars), and greater development of a

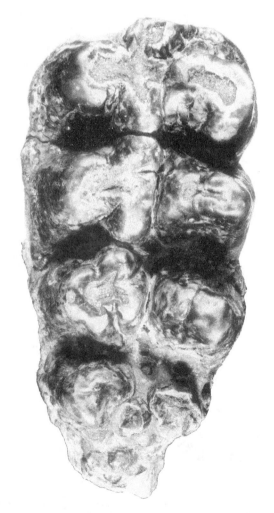

Fig. 15.13. Occlusal view of UF 129033, a left M3 of *Cuvieronius tropicus* from Leisey Shell Pit, Hillsborough County, early Pleistocene. About 0.5×.

fifth loph and lophid on the third molars. As it is the only gomphothere in the Pleistocene of Florida, specimens of that age can be confidently identified as *Cuvieronius*, even isolated teeth (fig. 15.13). A good sample of *Cuvieronius* was collected from the early Pleistocene Leisey Shell Pit in Hillsborough County.

Shovel-tuskers (tribe Amebelodontini) occurred in two basic forms, those with long, relatively narrow lower tusks (*Amebelodon* and *Serbelodon*) and those with short, relatively broad lower tusks (*Platybelodon* and *Torynobelodon*). In cross section, the lower tusk of an amebelodont was flat with a concave dorsal margin and a flat or squared-off medial margin. *Amebelodon* had normal, large upper tusks (with straight enamel bands), but those of *Platybelodon*, especially females, were reduced. Large samples of late Miocene *Amebelodon* with relatively complete material are known from Mixson's Bone Bed (Levy County) and the Moss Acres site (Marion County). The former is the type sample of *Amebelodon floridanus*, the first proboscidean species to be named in Florida, by Joseph Leidy in 1886. *A. floridanus* is a moderately large species with very complex, relatively high-crowned cheek teeth (fig. 15.14). The M3 had four well-developed lophs and a partial fifth

loph, while the m3 had five lophids and often the rudimentary beginnings of a sixth. The other molars and the DP4/dp4 were trilophodont. The DP3 and DP4 were replaced by small, irregularly shaped P3 and P4. In addition to the sample from Mixson's, referred specimens of *A. floridanus* are known from similar aged sites (late Miocene, about 9 to 7 Ma) in Marion and Alachua Counties. Specimens from the Bone Valley have also been referred to this species, but these identifications are based on isolated teeth and are therefore highly suspect.

An even larger, more advanced species of *Amebelodon*, *A. britti*, was described from the Moss Acres site (very late Miocene, about 6.5 Ma) (Lambert, 1990). The m3 is extremely long and has 6.5 to 7.5 lophids (fig. 15.15B, C), while the m2 was tetralophodont. An m3 similar to that of *A. britti* is known from the Bone Valley, but its age is uncertain. *A. britti* was one of the largest proboscideans known from Florida (fig. 15.15A), exceeded only by the largest Pleistocene mammoths. Some short, broad, flattened tusks from the Withlacoochee River Site 4A have been identified as *Platybelodon*. More complete material is needed for a specific identification. The large proboscidean sample from the Love site (early late Miocene, Alachua County) was provisionally

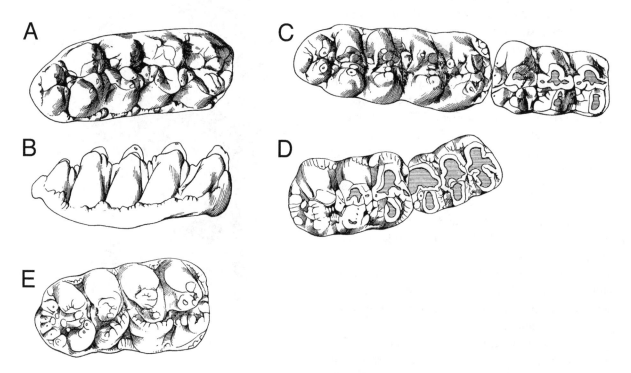

Fig. 15.14. Cheek teeth of the shovel-tusker *Amebelodon floridanus* from Mixson's Bone Bed, Levy County, late Miocene. *A*, occlusal, and *B*, labial views of USNM 3083, right m3; *C*, occlusal view of USNM 3066, left m2–m3; *D*, occlusal view of left M1–M2; *E*, occlusal view of right M3. All about 0.25×. After Osborn (1936); reproduced by permission of the American Museum of Natural History, New York.

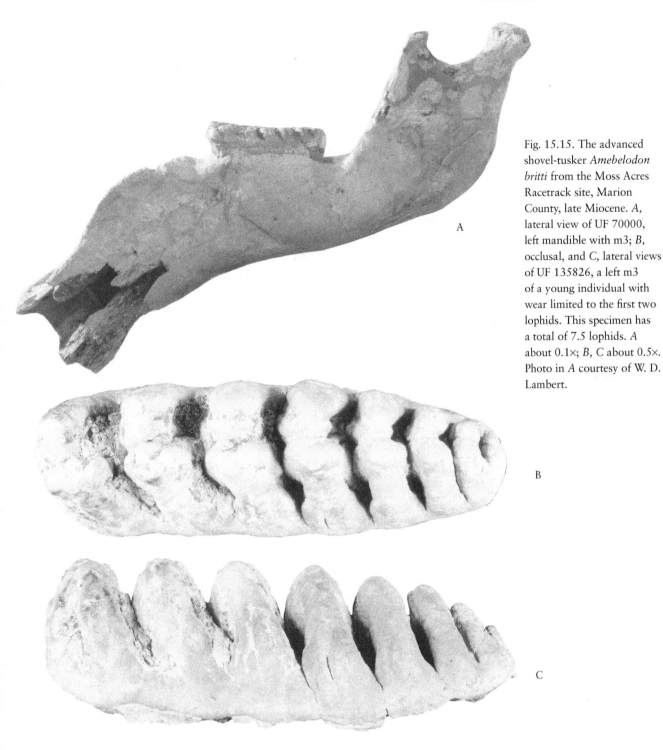

Fig. 15.15. The advanced shovel-tusker *Amebelodon britti* from the Moss Acres Racetrack site, Marion County, late Miocene. *A*, lateral view of UF 70000, left mandible with m3; *B*, occlusal, and *C*, lateral views of UF 135826, a left m3 of a young individual with wear limited to the first two lophids. This specimen has a total of 7.5 lophids. *A* about 0.1×; *B, C* about 0.5×. Photo in *A* courtesy of W. D. Lambert.

referred to *Amebelodon barbourensis*, a species often placed in *Serbelodon*. However, Lambert and Shoshani (1998) regarded the Love site specimens as belonging to *Gomphotherium* and not to a shovel-tusker.

The family Elephantidae evolved in Africa in the late Neogene (Maglio, 1973). The oldest elephants had lower tusks. These were lost in the common ancestor of the three Quaternary elephant genera, *Loxodonta, Elephas,* and *Mammuthus.* The oldest known species of each of these genera are from the early Pliocene of Africa. *Elephas* and *Mammuthus* dispersed to Eurasia between 3 and 4 Ma. These two are more advanced in their dental evolution than *Loxodonta,* having narrower plates and more plates per tooth. *Elephas* also remained common in Africa until the late Pleistocene; it eventually became extinct there and survived only in southern Asia. *Mammuthus* was common throughout Europe and northern Asia in the late Pliocene and Pleistocene. It twice invaded the New World, once very early in the Pleistocene and once in the late Pleistocene. The second invader, the wooly mammoth (*Mammuthus primigenius*), was adapted for cold climates and arctic steppe environments. Its range never progressed into the southern United States and it is not known from Florida.

Therefore, all of the mammoths from Florida represent the first immigrant lineage into North America, the *Mammuthus columbi* group. This lineage demonstrates considerable morphologic change through its chronologic range in North America, but little in the way of phylogenetic branching. So much disagreement exists in the literature that almost everyone who has studied this group has come up with a different scheme for classifying them. The classification used here follows Webb and Dudley (1995) in recognizing only two species, except *Mammuthus haroldcooki* is used instead of *Mammuthus hayi* as the proper species name for the early Pleistocene mammoth of North America (Webb, pers. comm.).

The oldest mammoths from Florida, *M. haroldcooki*, are early Pleistocene in age, about 1.3 to 1.5 million years old. The best known samples are from Leisey Shell Pit and Punta Gorda (Webb and Dudley, 1995). The cheek teeth of *M. haroldcooki* are relatively low crowned and have very thick enamel (2.5–3.0 millimeters) that is very crenulated on each plate in early wear. The M3 have

about sixteen to eighteen total plates; the m3 slightly more. The generic name *Archidiskodon* was applied to these early Pleistocene mammoths in the past, but recent work unites all mammoths in *Mammuthus* (for example, Maglio, 1973; Agenbroad, 1994; Lister and Bahn, 1994).

The common middle and late Pleistocene mammoth, *Mammuthus columbi,* differed from the older *M. haroldcooki* in having thinner (1.5–2.3 millimeters), less wrinkled or crenulated enamel, and much more closely spaced plates (figs. 15.3, 15.6, 15.16). Third molars had between 20 to 30 plates, with complete Florida specimens typically having 22 to 25. Specimens from Florida are relatively large in size and were described as a distinct species, *Mammuthus floridanus,* based on an excellent sample collected near Bradenton in the 1920s (Osborn, 1930). Most modern authorities treat it as a synonym of *M. columbi.* An obsolete generic name sometimes associated with this species is *Parelephas.* Specimens of *M. columbi* are common at most late Pleistocene deposits and have been found all over the state. *M. columbi* became extinct about 11,000 years ago, along with all other contemporary proboscideans in Florida. At that time, native Americans were actively hunting mammoths, mastodonts, and other "big game." Associated Paleoindian artifacts (some made of proboscidean ivory)

Fig. 15.16. Occlusal view of UF 14779, mandible with right and left m3 of *Mammuthus columbi* from the Aucilla River, Jefferson County, late Pleistocene. About 0.15×.

and butchered bone have been found from several sites in the state. Whether overhunting was a significant factor in the extinction of these large mammals is a much-debated topic among paleontologists and archeologists. Owen-Smith (1988) proposed that the extinction of proboscideans in the late Pleistocene (caused by human predation) resulted in far-reaching effects on patterns of vegetation and other mammals.

REFERENCES

Agenbroad, L. D. 1994. Taxonomy of North American *Mammuthus* and biometrics of the Hot Springs mammoths. Pp. 158–207 *in* L. D. Agenbroad and J. I. Mead (eds.), *The Hot Springs Mammoth Site.* Mammoth Site of South Dakota, Hot Springs.

Bryant, J. D. 1991. New early Barstovian (middle Miocene) vertebrates from the upper Torreya Formation, eastern Florida panhandle. Journal of Vertebrate Paleontology, 11:472–89.

Hildebrand, M. 1974. *Analysis of Vertebrate Structure.* John Wiley and Sons, New York, 710 p.

Lambert, W. D. 1990. Rediagnosis of the genus *Amebelodon* (Mammalia, Proboscidea, Gomphotheriidae), with a new subgenus and species, *Amebelodon (Konobelodon) britti.* Journal of Paleontology, 64:1032–40.

———. 1992. The feeding habits of the shovel-tusked gomphotheres: evidence from tusk wear patterns. Paleobiology, 18:132–47.

Lambert, W. D., and J. Shoshani. 1998. Proboscidea. Pp. 606–21 *in* C. Janis, K. M. Scott, and L. L. Jacobs (eds.), *Evolution of Tertiary Mammals of North America. Volume 1: Terrestrial Carnivores, Ungulates, and Ungulatelike mammals.* Cambridge University Press, Cambridge.

Lister, A., and P. Bahn. 1994. *Mammoths.* Macmillan, New York, 168 p.

Lucas, S. G., and G. S. Morgan. 1996. The Pliocene proboscidean *Rhynchotherium* (Mammalia: Gomphotheriidae) from south-central New Mexico. Texas Journal of Science, 48:311–18.

Luckett, W. P. 1996. Ontogenetic evidence for incisor homologies in proboscideans. Pp. 26–31 *in* J. Shoshani and P. Tassy (eds.), *The Proboscidea: Evolution and Palaeoecology of Elephants and Their Relatives.* Oxford University Press, Oxford.

MacFadden, B. J., and T. E. Cerling. 1996. Mammalian herbivore communities, ancient feeding ecology, and carbon isotopes: a 10 million-year sequence from the Neogene of Florida. Journal of Vertebrate Paleontology, 16:103–15.

Maglio, V. J. 1973. Origin and evolution of the Elephantidae. Transactions of the American Philosophical Society, 63:1–149.

McKenna, M. C., and S. K. Bell. 1997. *Classification of Mammals above the Species Level.* Columbia University Press, New York, 631 p.

Miller, W. E. 1990. A *Rhynchotherium* skull and mandible from southeastern Arizona. Brigham Young University Geology Studies, 36:57–67.

Olsen, S. J. 1957. A new beak-jaw mastodont from Florida. Journal of the Paleontological Society of India, 2:131–35.

———. 1972. Osteology for the archaeologist, 3: the American mastodont and the wooly mammoth. Papers of the Peabody Museum of Archaeology and Ethnology, Harvard University, 56:1–47.

Osborn, H. F. 1930. *Parelephas floridanus* from the upper Pleistocene of Florida compared with *P. jeffersoni.* American Museum Novitates, 443:1–17.

———. 1936. *Proboscidea.* Volume 1. American Museum Press, New York, 802 p.

Owen-Smith, R. N. 1988. *Megaherbivores.* Cambridge University Press, Cambridge, 369 p.

Shoshani, J., and P. Tassy. 1996. *The Proboscidea: Evolution and Palaeoecology of Elephants and Their Relatives.* Oxford University Press, Oxford, 472 p.

Tassy, P. 1987. A hypothesis on the homology of proboscidean tusks based on paleontological data. American Museum Novitates, 2895:1–18.

Tassy, P., and J. Shoshani. 1988. The Tethytheria: elephants and their relatives. Pp. 283–315 *in* M. S. Benton (ed.), *The Phylogeny and Classification of the Tetrapods.* Volume 2, *Mammals.* Clarendon Press, Oxford, England.

Webb, S. D. 1969. The Burge and Minnechaduza Clarendonian mammalian faunas of north-central Nebraska. University of California Publications in Geological Science, 78:1–191.

Webb, S. D., and N. Tessman. 1968. A Pliocene vertebrate fauna from low elevation in Manatee County, Florida. American Journal of Science, 266:777–811.

Webb, S. D., and J. Dudley. 1995. Proboscidea from the Leisey Shell Pits, Hillsborough County, Florida. Bulletin of the Florida Museum of Natural History, 37:645–60.

Webb, S. D., J. Dunbar, and L. Newsom. 1992. Mastodon digesta from North Florida. Current Research in the Pleistocene, 9:114–16.

16

Mammalia 8

Sirenians

INTRODUCTION

Sirenians (also called sea cows) are an almost extinct group of aquatic, herbivorous mammals. Only two genera are extant: *Trichechus* (manatee; three living species, one in Florida), which inhabits shallow bays of the Caribbean islands and fresh to brackish water of coastal rivers along the Atlantic coasts of southeastern North America, South America, and Africa, and *Dugong* (dugong, one living species), an inhabitant of shallow waters of western Pacific islands, the Indian Ocean, and the Red Sea. A third genus, *Hydrodamalis* (Steller's sea cow), lived in coastal waters near islands in the Bering Sea but was exterminated by hunting about two hundred years ago. Arguably sea cow remains are the most common mammalian fossils in Florida, but the rib fragments most often found are of little scientific value.

Sirenians diverged from related ungulate mammals early in the Paleogene. Their anatomy and fossil record demonstrate a close evolutionary relationship with the proboscideans and the extinct desmostylians (Savage et al., 1994; fig. 13.1). Like whales, they evolved from land-dwelling mammals and have lost all superficial resemblance to their ancestors. The oldest sirenians date back to the early Eocene of Jamaica and the middle Eocene of North Africa, Europe, Asia, Florida, and North Carolina, but even at that time the Sirenia were adapted to an aquatic life and their movement would have been limited on land.

The Sirenia is subdivided into four families, the living Dugongidae and Trichechidae, and the extinct Protosirenidae and Prorastomidae. The latter two are known only from the Eocene and represent the more primitive sea cows. The Dugongidae ranges from the middle Eocene to the present, with a moderately well-documented history. Fossil representatives occur on all continents (except Antarctica) but are best known in North Africa, Europe, and the United States. Thus its modern Indo-Pacific distribution is very relict compared to its once more cosmopolitan range. The Trichechidae, the manatees, are much less well known. The oldest known trichechids are from the Neogene of South America (Domning, 1982). An extinct Pliocene genus, *Ribodon* was the first trichechid to cross the Caribbean Sea and live in North America. It was followed in the early Pleistocene by *Trichechus*. They apparently dispersed northward following the demise of the dugongids in North American waters. Although sirenians achieved a worldwide distribution, they apparently were never very diverse. With the exception of the recently extinct *Hydrodamalis,* both living and fossil forms have been confined to warm waters. The acme of sirenian diversity was reached during the late Oligocene and Miocene. The maximum number of coexisting genera (three) was reached in the very late Oligocene of Florida. Florida is also the only area in the world with fossil evidence of continuous sirenian habitation from the middle Eocene to the present.

SIRENIAN SKELETAL AND DENTAL ANATOMY

Dense massive bones, especially the ribs, are characteristic of both fossil and living sirenians. All post-Eocene sirenians have a similar overall appearance resulting, in part, from their complete adaptation to water. The front limbs are modified into paddles for maneuvering, grasping the young, and manipulating food (fig. 16.1). External hind limbs were present in Eocene sirenians but lacking in modern species. A vestigial pelvis is present in modern species. Postcranial elements of sirenians, although common as fossils, are not very diagnostic to any particular family and have minor scientific importance (except in Paleogene taxa).

Sirenian bones are readily recognizable by their very dense, massive quality. Most completely lack marrow cavities or spongy bone. The weight of the bones helps keep the animal from floating by decreasing its buoyancy. The most common fossil elements are ribs, followed

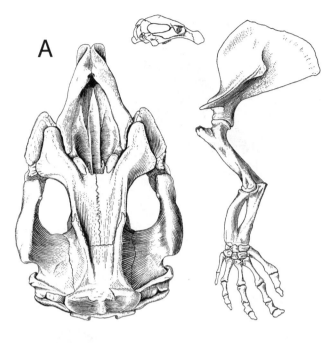

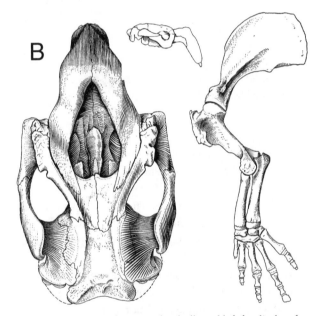

Fig. 16.1. Comparisons between the skulls and left forelimbs of the modern manatee, *Trichechus (A)*, and the dugong, *Dugong (B)*. Note the enlarged, downturned rostrum or snout of the dugong. After Howell (1930).

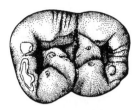

Fig. 16.2. Occlusal view of UF 114951, right M2 of *Metaxytherium crataegense* from the early Miocene of Florida. This specimen shows the basic tooth morphology of dugongs, being both bunodont and brachyodont. About 1.5×. After Bryant (1991); reproduced by permission of the Society of Vertebrate Paleontology.

than four or five cheek teeth in each jaw at one time, and often less in fossil forms. They have heavy, bunodont teeth with two transverse curved crests. The crests are greatly complicated by numerous small cusps (fig. 16.2). They somewhat resemble the teeth of gomphothere proboscideans. Advanced trichechids have numerous lophodont teeth that develop at the back of the jaw and migrate forward into position throughout the animal's life (horizontal tooth replacement). About seven teeth are present in each jaw at a time. Each tooth bears simple transverse crests and a posterior cingulum (fig. 16.3), while upper teeth also have anterior cingula.

Fossil Sirenians of Florida

Fossil sirenians in Florida range in age from Eocene to Pleistocene and are represented by all four known sea cow families (see chap. 3 for complete listing of genera and species). The most common sirenian fossils, ribs and vertebrae, cannot usually be identified to family if they are collected out of stratigraphic context and isolated from limb and cranial material. Sirenian bones are so strong and durable that they are often "reworked." In Florida river bottoms it is not uncommon to find fossils of Eocene and/or Miocene dugongs intermingled with those of Pleistocene and modern manatees. For these reasons, experts on sea cows tend to work mostly on complete cranial material or postcranial elements found in association with skulls. Sirenian remains in Florida are perhaps most common in late Oligocene and Miocene deposits, when a fortuitous combination of widespread shallow marine environments, the preferred habitat of sea cows, and a high diversity of sirenian taxa coincided. Sirenian fossils are particularly abundant in the Bone Valley phosphate mining region and in exposures of the Hawthorn Group in north-central Florida (Morgan, 1994). Some of the latter include numerous creek beds in

by vertebrae. Besides their great density, sirenian ribs are easily recognized by their rounded, bananalike shape, moderately large size, and, in cross section, by the absence of marrow cavities and the presence of distinct layers. The vertebrae are characterized by their density and their heart-shaped centra.

Dugongids and trichechids can be distinguished by differences in the dentition. Dugongids have no more

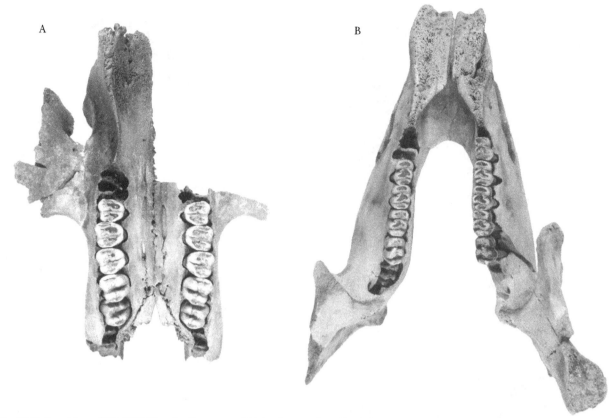

Fig. 16.3. Dentition of UF 123652, a fossil manatee, *Trichechus* sp., from Rock Springs, Orange County, late Pleistocene. *A*, occlusal view of right and partial left maxilla; *B*, occlusal view of nearly complete mandible. Note the heart-shaped, bilophodont upper cheek teeth and the trilophodont lowers. Both about 0.5×.

the Gainesville area; around the town of White Springs in Hamilton, Columbia, and Suwannee Counties; and in the fuller's earth mines of Gadsden County.

The oldest sea cows known from Florida were collected from late middle and late Eocene limestone in Columbia, Citrus, Levy, and Marion Counties. Some of the specimens were collected at limestone quarries, but most were found in the beds of rivers flowing over and eroding into Eocene bedrock. By far, the most specimens have been found in the Waccasassa River near the town of Gulf Hammock. Two partial atlases represent the most primitive (that is, the least aquatic) kind of sea cow, the Prorastomidae. This family is better known from much more complete fossils from the late early Eocene of Jamaica (Savage et al., 1994; fig. 16.4). Prorastomids were relatively small (compared to other tethytheres), amphibious herbivores with functional limbs. Savage et al. (1994) suggested that they lived along and in rivers, lakes, tidal marshes, and estuaries, feeding on floating and emergent aquatic plants. The atlas bears an unusual ventral knob on the large lateral wing, allowing the Florida specimens to be distinguished from other sirenians and referred to this ancient family.

The other Eocene sea cow family, the Protosirenidae, is better represented by specimens from Florida, but only by isolated teeth and fragmentary cranial and postcranial material (fig. 16.5). They have been referred to the genus *Protosiren* and are similar to the middle Eocene species *Protosiren fraasi* first described from Egypt (Domning et al., 1982). *Protosiren*, like all Eocene sea cows, had five premolars and three molars, as well as a full complement of incisors and canines. The molars of *Protosiren* were simple and bilophodont with the major crests running transversely. The mandibular symphysis was long, narrow, and not sharply deflected like younger sea cows. The specific status of Florida *Protosiren* remains uncertain pending recovery of a complete skull or skeleton. Since both Eocene sea cow families are so poorly known, any find of a relatively complete specimen of a sirenian embedded in Eocene limestone should be reported to professional paleontologists.

During the interval from the late Eocene to the late Oligocene (a period of about 15 million years), sea cow fossils are rare in Florida. Early Oligocene sirenians are known from ribs and even some skeletons, but these have yet to be properly described. Two basic types of

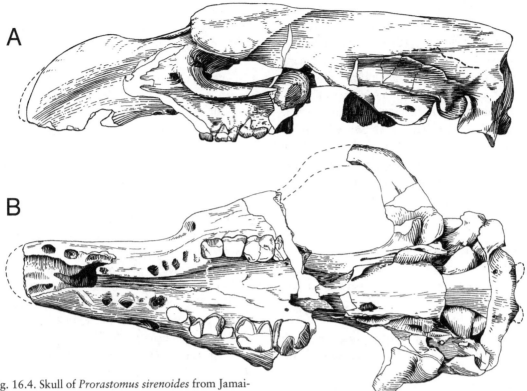

Above: Fig. 16.4. Skull of *Prorastomus sirenoides* from Jamaica, late early Eocene, the most primitive known sea cow. *A,* left lateral view; *B,* ventral view of British Museum (Natural History) 44897. Two specimens from Florida have been referred to this genus. About 0.5×. After Savage et al. (1994); reproduced by permission of the Society of Vertebrate Paleontology.

Right: Fig. 16.5. *A,* dorsal, *B,* ventral, and *C,* right lateral views of UF 55801, a parietal-supraoccipital skull cap of the early sirenian *Protosiren* sp. from the Cross-Florida Barge Canal, Citrus County, Eocene. About 1×.

dugongids were present in Florida during the latest Oligocene and Miocene, the subfamily Halitheriinae, represented by the cosmopolitan genus *Metaxytherium,* and the Dugonginae, a group characterized by larger upper tusks. Dugongids are distinguished from trichechids and protosirenids by their reduced dentition, enlarged and usually ventrally deflected snout and mandibular symphysis, upper tusks, and a strong bicipital groove on the humerus.

Two named species of *Metaxytherium* are recognized in Florida, the late early to early middle Miocene *Metaxytherium crataegense* (formerly known as *Hesperosiren crataegensis*), and the middle to late Miocene *Metaxytherium floridanum* (includes records of *M. ossivalense;* Domning [1988] provided a thorough review of this taxon). A third species of *Metaxytherium,* the oldest in the state, is known from the latest Oligocene from near the town of White Springs (Domning, 1997). *M.*

crataegense was smaller than *M. floridanum,* and had a less deflected rostrum; otherwise, the two were very similar (Domning, 1988; fig. 16.6). *M. floridanum* is the typical sirenian of the Bone Valley region, and many skulls and articulated specimens are known (figs. 16.7, 16.8). This species is also common in northern Florida, where it has been found at both middle and late Miocene sites including the Occidental Mine, McGehee Farm, and the Gainesville region. An articulated skeleton of *M. floridanum* was collected from a creek bed near Gainesville High School. The youngest records of *Metaxytherium* in Florida are from late Miocene sites along the Gulf Coast in Hillsborough and Manatee Counties (Morgan, 1994).

Individuals of *Metaxytherium* had six cheek teeth per jaw quadrant during their lifetime, three deciduous premolars that were never replaced by permanent premolars, and three molars. Tooth replacement proceeded somewhat like that of proboscideans (chap. 15). The anterior teeth were shed as they wore out and the more posterior teeth moved forward in the jaw. Adult dentitions typically consisted of only the three molars (figs. 16.7, 16.8; the deciduous premolars having been shed), or just the second and third molars in older individuals. The teeth were brachyodont and bunodont. The upper molars and the molariform DP5 had two to four major transverse lophs with gomphothere-like accessory conules. The upper tusk was small with a pointed crown 1.5 to 2.5 centimeters high when unworn (fig. 16.7B, C). The mandible did not have any anterior teeth (fig. 16.8).

The dugongid subfamily Dugonginae (formerly classified as the Rytiodontinae) is represented in Florida by three described genera: the latest Oligocene *Crenatosiren,* the late Oligocene to middle Miocene *Dioplotherium,* and the early Pliocene *Corystosiren.* Characteristic features of the subfamily are large upper tusks (blade-like in advanced species), enlarged, sharply downturned supraorbital processes, and a deep nasal opening (fig. 16.9). Dugongine cheek teeth are indistinguishable from those of *Metaxytherium. Crenatosiren olseni* (formerly referred to the Old World genus *Halitherium*) is the most primitive dugongine, with relatively short, conical tusks (Domning, 1997). It was based on a partial skeleton found on the banks of the Suwannee River near White Springs, as was a nearly complete skull of *Dioplotherium manigaulti.* The latter species was originally described from fragmentary material found near Charleston, South Carolina, in the 1870s. Compared to *C. olseni,* it has larger, laterally compressed, saberlike tusks (fig. 16.9) and a deeply concave frontal roof (Domning, 1989). A second species of *Dioplotherium, D. allisoni,* is

known from a single tusk fragment from the Bone Valley region. Domning (1989) hypothesized that the large tusks of dugongines functioned to cut up the thick but nutritious roots of sea grasses.

The youngest described dugongine, *Corystosiren varguezi,* is known from the Yucatan Peninsula of Mexico and Florida (Bone Valley and the Waccasassa River). Like *Dioplotherium* and *Rytiodus,* it had very large, flattened tusks, but differed in its broader, very thick dorsal roof of the skull (Domning, 1990). According to Morgan (1994), two Florida dugongines remain to be described, a small, middle Miocene species from the Bone Valley region and a latest Pliocene species from Charlotte County. The latter is the youngest known Atlantic dugongine and is close to the living genus *Dugong.*

The supposedly oldest record of manatees (family Trichechidae) from Florida is late Pliocene, from the Santa Fe River (Domning, 1982). However the Santa Fe deposits are known to contain a mixed fauna of late Pliocene and late Pleistocene fossils, so the age of the specimens is questionable. The oldest well-documented record of *Trichechus* in Florida is the early Pleistocene of Leisey Shell Pit, and it is subsequently known throughout the rest of the Pleistocene. However, most specimens are collected from unconsolidated river-bottom sediments and could also be of Holocene age. Their very dense structure makes it difficult to distinguish between fossil and modern manatee elements. Pleistocene specimens generally resemble the living species *Trichechus manatus,* except in the ventral surface of the mandibular symphysis (fig. 16.3). Modern manatees evidently evolved in isolation in the Amazon Basin during the Cenozoic and have only recently spread out from South America (Domning, 1982). Their horizontal tooth replacement system with a constant supply of new molars allows them to more successfully eat a diet of sea grasses than the brachyodont dugongids. To determine whether manatees out-competed dugongids in the Caribbean and drove them to extinction there, or if dugongids went extinct for other reasons (because of climatic change?) prior to the arrival of manatees, will require a more complete Pliocene record.

DESMOSTYLIANS IN FLORIDA?

Desmostylians are an odd group of extinct ungulate mammals well known from the middle Tertiary of the northern rim of the Pacific Ocean. They are believed to have been amphibious, wading out into shallow coastal waters to graze on sea grasses, algae, and seaweed. The cheek teeth of advanced desmostylians were composed

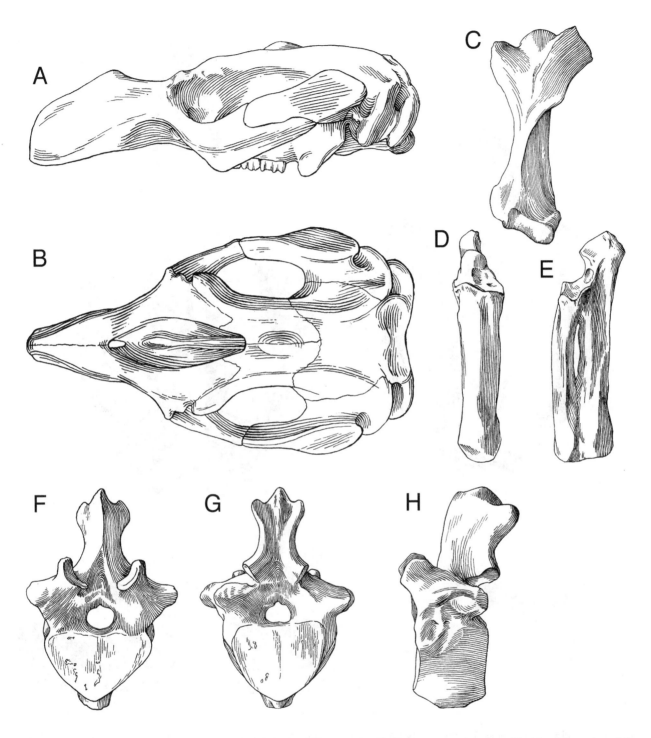

Fig. 16.6. Skull, vertebrae, and forelimb elements of the fossil dugong *Metaxytherium crataegense* from Florida. *A*, lateral, and *B*, dorsal views of the skull of AMNH 26838. The drawing is reconstructed to correct for crushing in the actual specimen; the snout was originally more downturned. *C*, anterior view of AMNH 26839, left humerus; *D*, anterior, and *E*, lateral views of AMNH 26839, left radioulna; *F*, anterior, *G*, posterior, and *H*, left lateral views of AMNH 26838, fourteenth thoracic vertebra. *A* and *B* 0.25×; *C*–*H* 0.33×. After Simpson (1932); reproduced by permission of the American Museum of Natural History, New York.

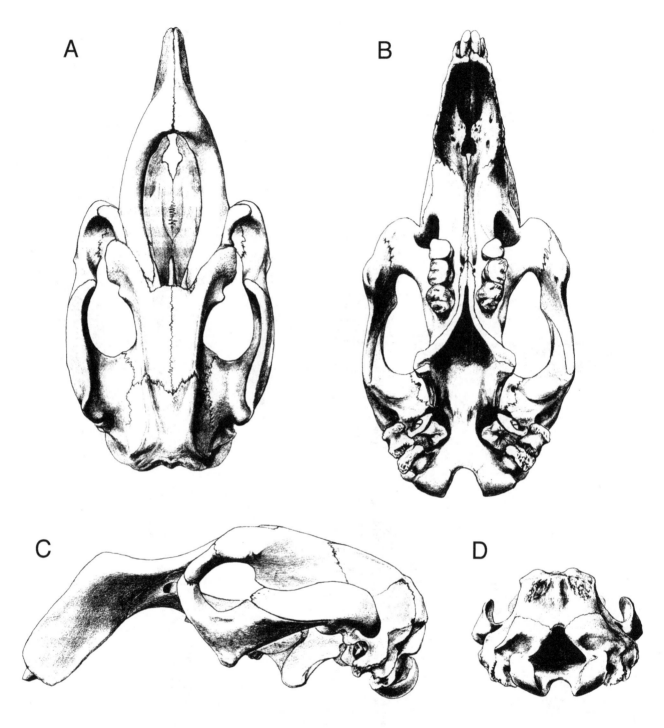

Fig. 16.7. Skull of the dugongid *Metaxytherium floridanum* in *A*, dorsal, *B*, ventral, *C*, left lateral, and *D*, posterior views. Based on USNM 377509 and 323193, both collected from the middle Miocene of the Bone Valley District. About 0.25×. After Domning (1988); reproduced by permission of the Society of Vertebrate Paleontology.

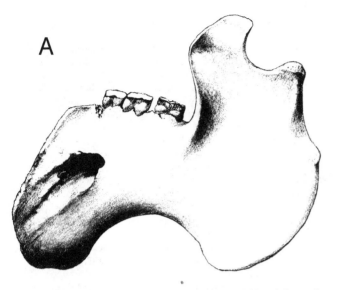

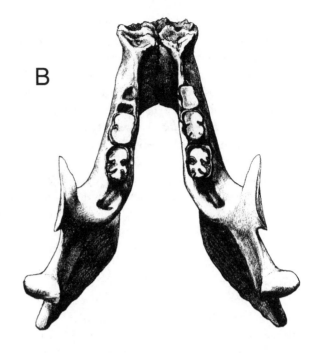

Fig. 16.8. Composite restoration of the mandible of the middle Miocene dugongid *Metaxytherium floridanum*, based on several specimens collected from the Bone Valley District. *A*, left lateral, and *B*, occlusal views. About 0.3×. After Domning (1988); reproduced by permission of the Society of Vertebrate Paleontology.

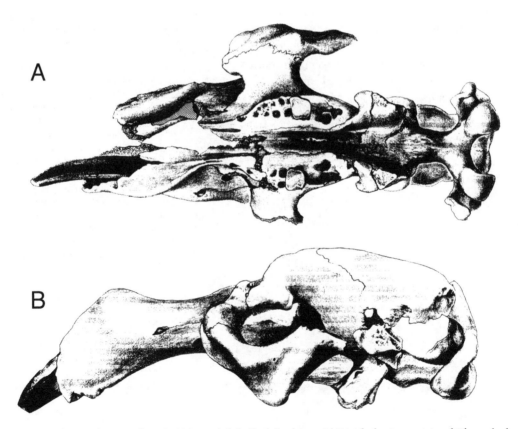

Fig. 16.9. *A*, ventral, and *B*, left lateral views of UF 95615, partial skull of the dugongid *Dioplotherium manigaulti* from the banks of the Suwannee River near White Springs, Hamilton County, very late Oligocene or very early Miocene. Compared with *M. floridanum* (fig. 16.7), *D. manigaulti* shows much stronger sagittal crests and larger, bladelike tusks. About 0.25×. After Domning (1989); reproduced by permission of the Society of Vertebrate Paleontology.

of multiple rounded pillars of thick enamel surrounding a narrow core of dentine. Reinhart (1976) reported on five mammalian tooth fragments supposedly collected in the Bone Valley region. These so resemble isolated pillars of broken desmostylian teeth that Reinhart concluded they were indeed desmostylian and even referred them to the genus *Desmostylus*. They are the only credibly reported remains of desmostylians from the Atlantic Coast.

As the Panamanian Isthmus was open during the Miocene, no land barrier prevented these creatures from crossing around Mexico and coming into the Gulf of Mexico. Thus their presence in Florida is a distinct possibility. However, their distribution in the Pacific indicates a strong preference for cool to cold water, with their southernmost record from Baja California. Furthermore, not a single complete fossil of a desmostylian has ever been identified from the extensive, fossiliferous marine deposits of the Atlantic Coastal Plain, only those five tooth fragments. Almost every element of the body is easily identifiable because whole skeletons are known from the Pacific. Thus, pending more compelling evidence, the conclusion must be drawn that desmostylians were not present in Florida. The tooth fragments identified by Reinhart are either proboscidean, or, more likely, actual desmostylian teeth that were collected in California or elsewhere on the West Coast, transported to Florida, and later mistakenly thought to have been found in Polk County (see Morgan, 1994).

REFERENCES

Aranda-Manteca, F. J., D. P. Domning, and L. G. Barnes. 1994. A new middle Miocene sirenian of the genus *Metaxytherium* from Baja California and California: relationships and paleobiogeographic implications. Pp. 191–204 *in* A. Berta and T. A. Deméré (eds.), *Contributions in Marine Mammal Paleontology Honoring Frank C. Whitmore, Jr.* San Diego Natural History Society, San Diego.

Bryant, J. D. 1991. New early Barstovian (middle Miocene) vertebrates from the upper Torreya Formation, eastern Florida panhandle. Journal of Vertebrate Paleontology, 11:472–89.

Domning, D. P. 1982. Evolution of the manatees: a speculative history. Journal of Paleontology, 56:599–619.

———. 1988. Fossil Sirenia of the West Atlantic and Caribbean Region. I. *Metaxytherium floridanum* Hay, 1922. Journal of Vertebrate Paleontology, 8:395–426.

———. 1989. Fossil Sirenia of the West Atlantic and Caribbean Region. II. *Dioplotherium manigaulti* Cope, 1883. Journal of Vertebrate Paleontology, 9:415–28.

———. 1990. Fossil Sirenia of the West Atlantic and Caribbean Region. IV. *Corystosiren varguezi*, gen. et sp. nov. Journal of Vertebrate Paleontology, 10:361–71.

———. 1994. A phylogenetic analysis of the Sirenia. Pp. 177–89 *in* A. Berta and T. A. Deméré (eds.), *Contributions in Marine Mammal Paleontology Honoring Frank C. Whitmore, Jr.* San Diego Natural History Society, San Diego.

———. 1997. Fossil Sirenia of the West Atlantic and Caribbean Region. VI. *Crenatosiren olseni* (Reinhart, 1976). Journal of Vertebrate Paleontology, 17:397–412.

Domning, D. P., G. S. Morgan, and C. E. Ray. 1982. North American Eocene sea cows (Mammalia: Sirenia). Smithsonian Contributions to Paleobiology, 52:1–69.

Howell, A. B. 1930. *Aquatic Mammals: Their Adaptations to Life in the Water.* Charles Thomas, Springfield, Illinois, 338 p.

Morgan, G. S. 1994. Miocene and Pliocene marine mammal faunas from the Bone Valley Formation of central Florida. Pp. 239–68 *in* A. Berta and T. A. Deméré (eds.), *Contributions in Marine Mammal Paleontology Honoring Frank C. Whitmore, Jr.* San Diego Natural History Society, San Diego.

Reinhart, R. H. 1976. Fossil sirenians and desmostylians from Florida and elsewhere. Bulletin of the Florida State Museum, 20:187–300.

R. J. G. Savage, D. P. Domning, and J. G. M. Thewissen. 1994. Fossil Sirenia of the West Atlantic and Caribbean Region. V. The most primitive known sirenian, *Prorastomus sirenoides* Owen, 1855. Journal of Vertebrate Paleontology, 14:427–49.

Simpson, G. G. 1932. Fossil Sirenia of Florida and the evolution of the Sirenia. Bulletin of the American Museum of Natural History, 59:419–503.

Mammalia 9

Whales and Dolphins

INTRODUCTION

Whales, dolphins, and porpoises compose the group of mammals formally called the Cetacea. Compared to other mammals, their anatomy is highly specialized for a completely aquatic mode of life. Some of the aquatic modifications found in modern cetaceans include paddle- or flipper-like forelimbs (fig. 17.1); absence of the external hind legs; a large tail with horizontal flukes for propulsion through the water; location of the external nostrils (the blowhole) on the top of the skull (fig. 17.2); modifications to the ear region for underwater hearing (and for echolocation in odontocetes); and an overall streamlining of the entire body, including an almost total loss of hair and formation of an insulating layer of blubber. The two major bony elements of the ear, the auditory bulla and petrosal (whale specialists usually refer to the petrosal element as the periotic) are heavily ossified and very dense. This makes them more durable than the remainder of the skull and prone to separation from it after death. Many fossil records of cetaceans consist solely of these dense, isolated ear bones. Although some of these anatomical modifications are found in other aquatic mammals (such as sea cows or seals), no mammal is so completely adapted for life in the sea as are the cetaceans. The Cetacea is traditionally divided into three subgroups: the extinct Archaeoceti and the living Odontoceti (toothed whales) and Mysticeti (baleen whales). The classification used here and in chapter 3 follows Fordyce and Barnes (1994) at the rank of superfamily and below, while the higher classification follows McKenna and Bell (1997).

All cetaceans are carnivorous, with principal prey items including invertebrates, fish, and other marine mammals. Thus, when looking for a terrestrial ancestor for whales, one would logically start with carnivorous mammals (such as the Carnivora, chap. 11) rather than the mostly herbivorous ungulates. Yet now a large body of morphological, paleontological, and molecular evidence indicates that whales are more closely related to ungulates than they are to the Carnivora (Thewissen, 1994, 1998; Fordyce and Barnes, 1994; O'Leary and Geisler, 1999). The basal group from which the ungulates evolved did include some omnivorous and carnivorous families. One of the latter, the Mesonychia, is widely regarded by paleontologists as the closest terrestrial group to the Cetacea (O'Leary and Geisler, 1999; O'Leary and Uhen, 1999). The oldest well-known cetacean, *Pakicetus inachus,* from the early Eocene of Pakistan, has numerous dental and cranial characters that are intermediate between mesonychids and later cetaceans, and it thus forms a good evolutionary link between the two groups. *Pakicetus* and other very early cetaceans were probably amphibious and bred on land like modern seals. Whales continued to have functional hind limbs through the middle Eocene (Hulbert et al., 1998).

The Archaeoceti are an extinct group of primitive toothed whales known only from the Eocene (Fordyce and Barnes, 1994). It is within the Archaeoceti that whales made the evolutionary shift from a terrestrial to a completely aquatic mode of life. Archaeocete whales retained many characteristics of land mammals and are distinguished from odontocete and mysticete cetaceans by a number of primitive skeletal and dental features. These include the following: (1) incomplete posterior migration of the external nasal opening (it is located about halfway between the tip of the snout and the top of the head); (2) absence of telescoping among the skull elements; (3) teeth differentiated into morphologically distinct incisors, canines, premolars, and molars; (4) a tooth formula similar to the numbers found in primitive placental mammals, three incisors, one canine, four premolars, and two to three molars in each jaw quadrant for a total of forty-two to forty-four teeth; and (5) presence of external hind limbs. Fossils of archaeocete whales have been found in Africa, England, Pakistan, India, New Zealand, and in the southeastern United States from North Carolina to Texas, including Florida.

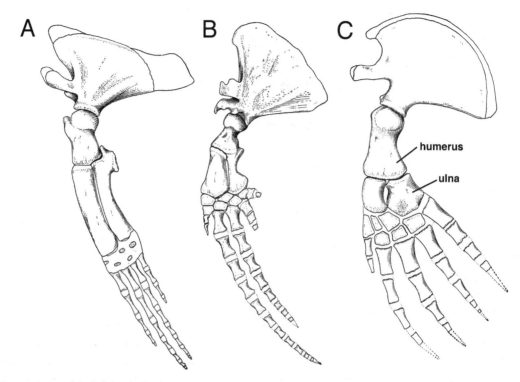

Fig. 17.1. Comparisons of the left forelimb of three cetaceans. Note the shortened humerus, blocky carpals, and elongated phalanges. Some digits have increased the number of phalanges beyond the normal mammalian three. In modern cetaceans, movement in the forelimb is restricted to the joint between the humerus and scapula. There is no movement at the elbow or wrist. *A*, a mysticete, the blue whale *Balaenoptera musculus*; *B*, an odontocete, the pilot whale, *Globicephala*; *C*, a river dolphin, *Platanista*. Not to scale; after Howell (1930).

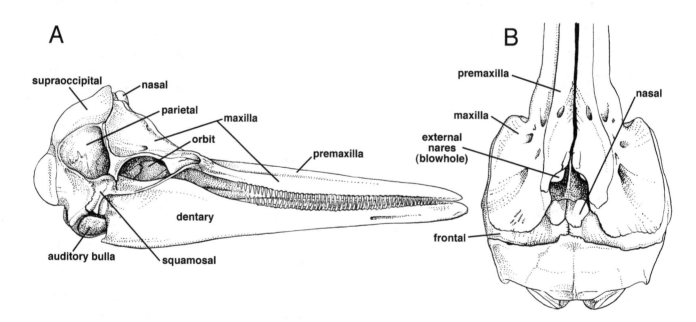

Fig. 17.2. *A*, right lateral, and *B*, dorsal views of modern dolphin skulls showing some of the modifications to the skull found in odontocete whales. *A* is *Delphinus bairdii*, *B* is *Tursiops truncatus*. The premaxilla and maxilla extend far posteriorly, overlapping the frontal (telescoping), and the external nares are located on the top of the skull. Also note the very thin zygomatic arch, which is rarely preserved in fossils because it is so fragile. The dentition is homodont. Figures from *Mammalogy* by Terry A. Vaughn, copyright © 1972 by Saunders College Publishing, reproduced by permission of the publisher.

Toothed whales (odontocetes) are more common and diverse than baleen whales. All of the smaller living cetaceans and a few of the larger species are odontocetes. Toothed whales include a wide variety of living forms, such as beaked whales, sperm whales, narwhales, belugas, pilot whales, orcas, and numerous species of smaller dolphins and porpoises. Odontocetes are characterized by the following: (1) the presence of teeth, which in living species are homodont (a few odontocetes are edentulous or nearly so because they prey upon soft-bodied animals such as squid and need no teeth). Most modern odontocetes have simple conical teeth. Many toothed whales have a much larger number of teeth in comparison to other mammals; for example, the La Plata River dolphin may have as many as 240 teeth; (2) blowhole with a single external opening; (3) skull bones are asymmetrical, especially in the region around the blowhole (fig. 17.2B). This asymmetry appears to be a relatively modern trend in the odontocetes, as most Oligocene and some Miocene toothed whales had symmetrical skulls, and asymmetry apparently evolved independently in various odontocete lineages; (4) maxilla extends posteriorly over the orbit to form a supraorbital process (Fordyce and Barnes, 1994); and (5) echolocation using high-frequency sound waves (sonar). Modifications to the skull to send and receive these sound waves can be detected in fossils. Thus it has been determined that even early odontocetes had this ability. Odontocetes first appeared in the Oligocene, and by the Miocene they had already become quite diverse. Several extinct families of toothed whales and a large number of extinct genera have been described (Fordyce and Barnes, 1994). In addition, fossil representatives of most of the modern families of odontocetes are known from the Miocene onward.

The mysticetes include the largest animal that ever lived, the blue whale, *Balaenoptera musculus*. Along with the toothed sperm whale, the baleen whales formed the basis for the commercial whaling industry. Mysticetes are characterized by the following: (1) complete lack of teeth in adults—mysticetes have long plates of baleen attached to the roof of their mouth, which filter planktonic crustaceans (krill), other small invertebrates, and fish from sea water; (2) blowhole with two external openings; (3) symmetrical skull bones; (4) posteriorly extended maxilla under the orbit; and (5) mandibles do not meet in a symphysis. Baleen whales first occurred in the early Oligocene (possibly the late Eocene), but their remains do not become common until the Miocene. The earliest mysticetes had large, multicusped teeth, similar to those of late Eocene archaeocetes. These functioned primarily to strain small prey from water taken into the mouth. This function was eventually taken over by the baleen plates, and some early mysticetes apparently had both teeth and baleen. Toothed mysticetes became extinct in the late Oligocene. The cetotheres, an extinct group of relatively small-bodied, toothless whales known worldwide, were the common mysticetes of the Miocene, but they continued through the Pliocene. About sixty cetothere species have been named (Fordyce and Barnes, 1994). Representatives of modern mysticete families first began to appear in the Miocene.

Fossil Cetaceans of Florida

Most of the fossil cetaceans from Florida are either Miocene or Pliocene, although specimens are known from the Eocene and Pleistocene epochs as well. The majority of fossil cetaceans found in Florida belong to some type of toothed whale, although mysticetes are also common. Most are not surprisingly found in the same marine sediments where shark teeth, ray mouth plates, and sea cow material are abundant, but fossils of whales tend to be less common than these other vertebrates. Isolated finds, especially vertebrae, teeth, partial mandibles, and the durable elements of the ear region (the auditory bulla and petrosal), are the most commonly recovered cetacean fossils. Much rarer and more important are complete skulls and associated skeletons. The portion of this chapter dealing with Oligocene and younger fossils is largely taken from Morgan (1994), which should be consulted for further details, extensive references, and more figured specimens.

Specimens of archaeocete whales from Florida have thus far been rare, in contrast to their fairly common occurrence in Eocene rocks in Alabama and Mississippi (Kellogg, 1936; Thewissen, 1998). In recent years, archaeocete fossils, including several nearly complete skeletons, have also been found in Georgia and South Carolina (Hulbert et al., 1998). The shoreline of the Eocene ocean extended across central Georgia, Alabama, and Mississippi, and it has been hypothesized that archaeocetes lived mostly in nearshore waters and seldom ventured far out to sea. It is also likely that archaeocete fossils are more common in Florida's Eocene limestone than their meager fossil record would indicate, and what is needed to find them are more exposures and outcrops. All archaeocete whales that have been found in Florida belong to the family Basilosauridae (figs. 17.3–17.5).

The very distinctive cheek teeth of basilosaurid whales were double-rooted, coarsely serrated, and laterally compressed so that the serrations were lined up in a row from the front to the back of the tooth (fig. 17.3). One serration or cusp in the center of the tooth is much larger than the others. The incisors and canines are similar to each other in appearance, all being basically conical in shape (fig. 17.3). They are long, somewhat curved,

laterally compressed, and have a distinct enamel crown ornamented with coarse grooves. The crown is about one-fourth of the tooth's total length.

Three genera of archaeocetes are known from Florida, each represented by a single species. One of these,

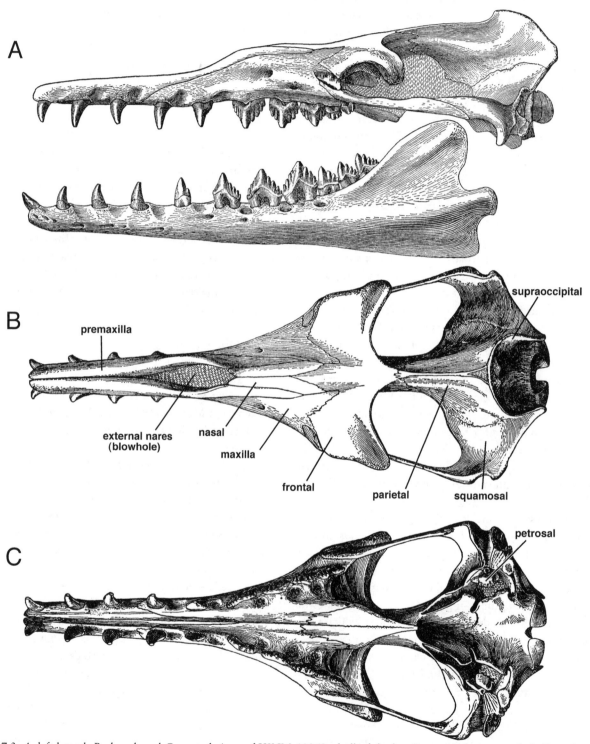

Fig. 17.3. *A*, left lateral, *B*, dorsal, and *C*, ventral views of USNM 11962, skull of the late Eocene archaeocete whale *Zygorhiza kochii* from Alabama. Contrast the location of the external nares, the configuration of skull bones, and the heterodont dentition in this specimen with those of a modern cetacean in figure 17.2. The auditory bullae have been removed to show internal ear structures. About 0.17×. After Kellogg (1936); reproduced by permission of the Carnegie Institution of Washington.

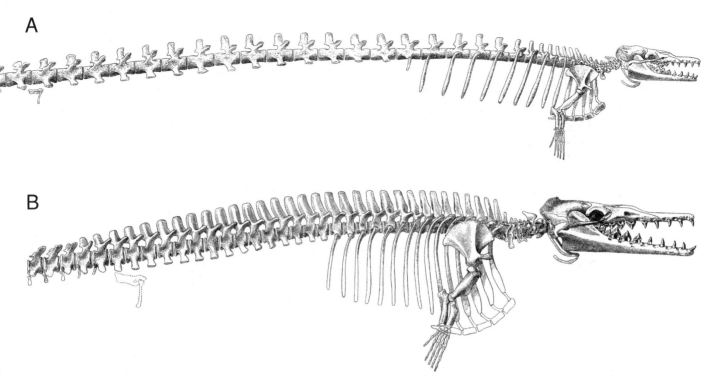

Fig. 17.4. Anterior portion of the skeletons of *A, Basilosaurus cetoides,* and *B, Zygorhiza kochii,* both late Eocene archaeocete whales. *A* about 0.017×; *B* about 0.045×. After Kellogg (1936); reproduced by permission of the Carnegie Institution of Washington.

Basilosaurus cetoides, was the giant of the group. It reached a total length of about 20 meters. Its skull alone was 1.5 meters long. *Basilosaurus* was characterized by its large size and extreme elongation of the centra in the trunk and tail vertebrae (figs. 17.4–17.5B). Vertebrae of *Basilosaurus* may be as much as 45 centimeters in length

and 20 centimeters in diameter. *Zygorhiza kochii* was a considerably smaller archaeocete with a maximum length of 6 meters (figs. 17.3, 17.4B). *Zygorhiza* also differed from *Basilosaurus* in that the vertebrae of *Zygorhiza* did not have elongated centra (fig. 17.5A). The teeth of *Zygorhiza* were similar to those of *Basilosaurus,*

Fig. 17.5. Comparison of the lumbar vertebrae of the Eocene whales *Basilosaurus* and *Zygorhiza* in lateral view. *A,* USNM 12063, fourth lumbar vertebra of *Z. kochii; B,* USNM 12261, seventh lumbar vertebra of *B. cetoides. A* about 0.33×; *B* about 0.17×. After Kellogg (1936); reproduced by permission of the Carnegie Institution of Washington.

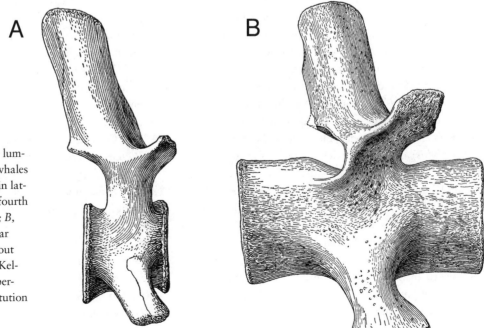

although smaller (fig. 17.3). Although more common elsewhere in the South, *Basilosaurus* is rare in Florida, while *Zygorhiza* is the most common archaeocete in the state. Although all small archaeocete fossils from Florida tend to be identified as *Zygorhiza,* a second genus and species of small basilosaurid was described long ago from South Carolina, *Dorudon serratus.* The presence of a cingulum on the upper premolars of *Zygorhiza* and its absence in *Dorudon* is the key distinguishing feature. Diagnostic cranial material representing *Dorudon* has yet to be found in Florida.

Pontogeneus brachyspondylus is the third Florida basilosaurid. It is known mainly from vertebrae that are of large size (as large as those of *Basilosaurus* in diameter) but are of normal proportions and not greatly elongated. Unfortunately, no cranial material directly associated with these large vertebrae has ever been described, so no one knows exactly what its skull and teeth were like. Fossils of *Pontogeneus* have been found much more often in Florida than those of *Basilosaurus.*

No whale fossils have yet been discovered in Florida's Oligocene rocks. This is unfortunate, as it was during the Oligocene that both the toothed and baleen whales first became common. Paleontological evidence for their evolutionary origins is improving annually but is far from complete, and therefore any find of an Oligocene cetacean from Florida potentially would be of noteworthy scientific importance.

Cetacean fossils are relatively common in Florida's marine sediments of Neogene and Quaternary age. Probably the most productive area for Miocene and Pliocene fossil whales is the Bone Valley phosphate mining region of central Florida. A close second is the Miocene Hawthorn Group of north-central Florida. Fossil cetaceans are also locally common in marine sediments exposed along the southwestern coast of Florida from Hillsborough to Lee County. All of the more complete fossil whales from Florida have come from these regions or from Quaternary beach sands along the eastern coast.

The earliest cetaceans of Miocene age from Florida are fragmentary specimens of long-beaked dolphins (*Pomatodelphis*) and small baleen whales (Cetotheriidae) from the eastern panhandle (Gadsden County). They are about 17 million years old (late early Miocene). These two kinds of cetaceans are so common in Florida's marine faunas for the following 10 million years (through the late Miocene) that Morgan (1994) coined the term "*Metaxytherium-Pomatodelphis*-cetothere assemblage" to refer to them. *Metaxytherium* was the common sea cow of this interval and is usually found in the same sediments as the other two (chap. 16).

Pomatodelphis is currently classified as a member of the Platanistidae, an extant family represented today only by the Asiatic Ganges and Indus River dolphins (genus *Platanista*). Current interpretations of the evolutionary relationships of fossil odontocetes are still in a state of flux, but several workers believe that a large, diverse, globally distributed group of fossil dolphins, including *Pomatodelphis,* are more closely related to *Platanista* than they are to any other living cetacean (Fordyce and Barnes, 1994). This is based on characters of the scapula and skull. Most of these taxa have long snouts or rostra (fig. 17.6), some extremely so, that has lead to their common name of long-beaked dolphins. However, this morphology also appears in some unrelated odontocetes (Eurhinodelphoidea, Iniidae, Pontoporiidae), so this term needs to be used with caution. All long-beaked dolphins appear to be specialized fish eaters.

Two species of *Pomatodelphis* are presently recognized from Florida, a smaller *P. inaequalis* and a larger *P. bobengi.* The latter was formerly allocated to the European genus *Schizodelphis,* a member of the family Eurhinodelphidae. Platanistids like *Pomatodelphis* differed from eurhinodelphids in having flattened rostra and mandibles that were the same length and in having teeth at the anterior end of the snout (fig. 17.7A). A third species named from the Bone Valley region, *Schizodelphis depressus* is now regarded as a junior synonym of *P. inaequalis* (Morgan, 1994). The long, narrow, dorsally-ventrally flattened rostra and mandibles of *Pomatodelphis* are fragile and most commonly found as toothless fragments one to two inches long. However, a few complete or nearly complete skulls are known from the Bone Valley Region, and a partial cranium was found in a creek bed in Gainesville. Rostral and mandibular fragments are actually quite similar and difficult to distinguish. They almost always lack teeth (the specimen in fig. 17.7A is a rare find) as the alveoli are shallow. The alveolii form two straight rows and are closely spaced to each other (fig. 17.6). The teeth are elongate, slightly curved, and have conical enamel crowns (figs. 1.1B, 17.7A, C). The distinctive petrosals and auditory bullae of *Pomatodelphis* are also common finds (fig. 17.8).

A much larger variety of long-beaked dolphin is also known from middle Miocene sediments of the Bone Valley region (fig. 17.9A). It was almost twice the size of *Pomatodelphis* and also differed in its broader rostrum and relatively larger teeth. It was once called *Megalodelphis magnidens,* but unfortunately the holotype specimen of this species turned out to be a jaw of the crocodile *Gavialosuchus* (Morgan, 1986), making this scientific name invalid for a cetacean. This large species

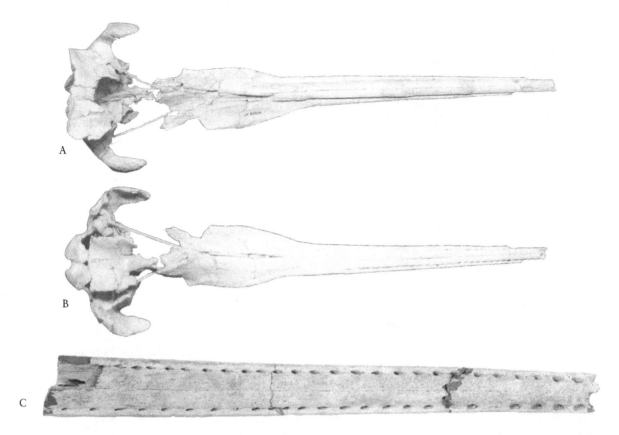

Fig. 17.6. The common long-snouted dolphin *Pomatodelphis inaequalis* from the middle Miocene of the Bone Valley District. *A*, dorsal, and *B*, ventral views of UF 50000, a nearly complete skull from the Phosphoria Mine, Polk County; *C*, dorsal view of UF 132953, partial fused dentaries from the Hookers Prairie Mine, Polk County. As is almost always the case, the loosely attached teeth have all fallen out of the alveoli and have been separated from these specimens (see fig. 17.7A for a rare exception). *A–B* about 0.2×; *C* about 0.5×.

will remain without a formal name until a skull or at least a braincase is found.

Although the vast majority of Miocene odontocete fossils from Florida are the platanistid *Pomatodelphis,* a limited number of specimens represent other families (fig. 17.7B, D, E). Kentriodontids are an extinct family from the Oligocene and Miocene that are related to the true dolphins (the family Delphinidae). Isolated teeth resembling three different kentriodontid genera, *Hadro-*

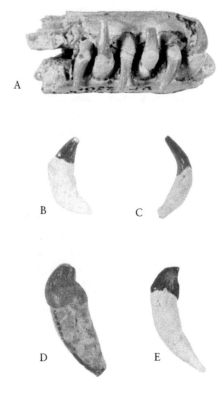

Fig. 17.7. Small odontocete teeth from the Miocene of Florida. *A*, lateral view of UF 32307, anterior rostral fragment with in-place teeth of *Pomatodelphis inaequalis* from the Phosphoria Mine, Polk County; *B*, UF 91196, an isolated tooth of an undetermined genus and species from Mineco Mine, Polk County; *C*, UF 28744, an isolated tooth of *P. inaequalis* from the Phosphoria Mine; *D*, UF 58517, an isolated tooth of *Hadrodelphis* sp. from the Gainesville High School Creek, Alachua County; *E*, UF 131984, an isolated tooth of *Delphinodon* sp. from West Palmetto Mine, Polk County. All about 1×.

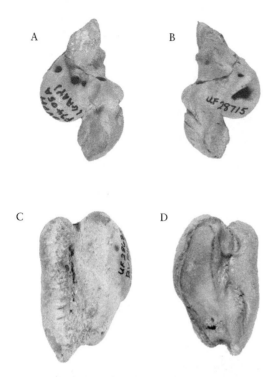

Fig. 17.8. Ear bones of the middle Miocene long-snouted dolphin *Pomatodelphis inaequalis* from the Phosphoria Mine, Polk County. *A*, ventral, and *B*, dorsal views of UF 28715, a petrosal. *C*, ventral (external), and *D*, dorsal (internal) views of UF 28685, an auditory bulla. Both about 1×.

delphis, Delphinodon, and *Lophocetus,* have been recovered from central Florida (Morgan, 1994). Also, a tooth from Gadsden County referred to a squalodontid by Bryant (1991) is now regarded as belonging to the kentriodontid *Hadrodelphis* (Morgan, 1994).

The two other types of odontocetes reported from the Miocene belong to extant families, the sperm whales (Physeteridae) and the beaked whales (Ziphiidae). Beaked whales have a long rostrum, but not as elongated as that of platanistids, which is formed by the premaxillae and maxillae and made of very dense bone. It is durable and fossilizes well, unlike the rest of the skull. Only a single Miocene ziphiid rostrum is known from Florida, and it was dredged from the ocean floor 35 km east of Miami. Whitmore et al. (1986) referred it to the extinct species *Mesoplodon longirostris*. A ziphiid petrosal of uncertain generic affinities is known from a middle Miocene Bone Valley site. The rarity of fossils of this family is probably due in large part to their preference for far offshore, deepwater habitats.

Isolated teeth of sperm whales are not uncommon in marine Miocene and Pliocene deposits. Such finds are difficult to identify accurately to species or even genus.

The most complete specimen known to date is a mandible with sixteen teeth found at the Fort Meade Mine in the Bone Valley region. One commonly found type of sperm whale tooth is about 12 to 20 centimeters long, 4 to 5 centimeters wide, and has a distinct enamel-covered crown (fig. 17.10B). The small crown is striated, may show wear facets, and is separated from the root by a constriction. Below the crown, the bulbous root first widens and then tapers toward its base. Such teeth are assigned to the genus *Scaldicetus* and are apparently limited to the Miocene.

Pliocene odontocetes from Florida are quite distinct from those of the Miocene, with only a single genus held in common. There are a minimum of seven taxa in the early Pliocene Palmetto Fauna of the Bone Valley, and most likely more will eventually be found (Morgan, 1994). *Goniodelphis hudsoni* is an extinct relative of the modern Amazon River dolphin *Inia*. Aside from two fragmentary skulls and a nearly complete set of mandibles, the only known fossils of *G. hudsoni* are isolated teeth and jaw fragments (fig. 17.9B, C). The teeth are readily identified on the basis of their laterally compressed, triangular shaped roots. The elongated rostrum and mandible were compressed laterally, not dorsally and ventrally as in platanistids. Whether *Goniodelphis* lived in fresh or salt water is uncertain, since it is found in fossil assemblages that contain both terrestrial and marine animals.

The second common type of odontocete in the Palmetto Fauna are sperm whales. Two taxa are recognized, *Kogiopsis floridana* and *Physeterula* sp. (fig. 17.10A). The former is definitely known only from the holotype specimen, a portion of the mandibular symphysis with eleven teeth collected in 1924. It is likely that some of the isolated sperm whale teeth and petrosals from this fauna belong to *Kogiopsis*. Teeth of both genera are generally cylindrical in shape, slightly curved, and lack enamel crowns. The outer surface of the tooth is composed of cement and may bear coarse longitudinal grooves or wear facets. If broken to expose the inner core of dentine, it will display a definite layered structure.

The remaining Palmetto Fauna odontocetes are each known only from one or a few specimens. The family Pontoporiidae is represented only by two petrosals and one possible mandibular fragment. The only living member of this family is the South American La Plata River dolphin *Pontoporia*. A single partial mandible (fig. 17.9D) with anterior-posteriorly compressed teeth appears to represent the modern dolphin family Delphinidae, although it does not match any described genus. Beaked whales (Ziphiidae) are represented by a single

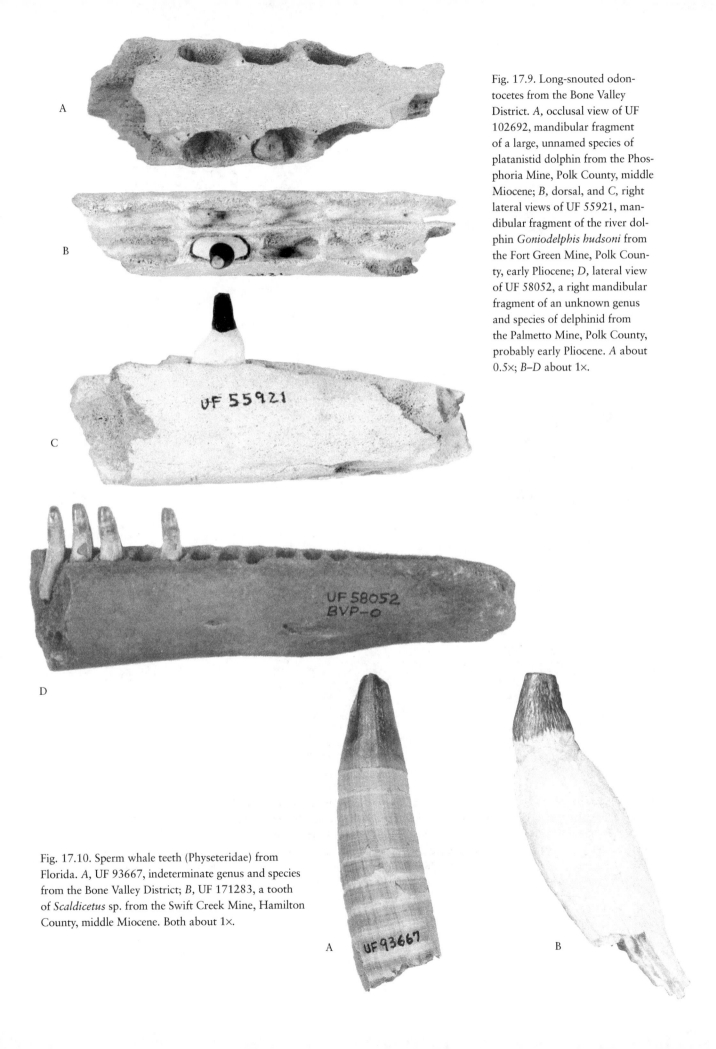

Fig. 17.9. Long-snouted odontocetes from the Bone Valley District. *A,* occlusal view of UF 102692, mandibular fragment of a large, unnamed species of platanistid dolphin from the Phosphoria Mine, Polk County, middle Miocene; *B,* dorsal, and *C,* right lateral views of UF 55921, mandibular fragment of the river dolphin *Goniodelphis hudsoni* from the Fort Green Mine, Polk County, early Pliocene; *D,* lateral view of UF 58052, a right mandibular fragment of an unknown genus and species of delphinid from the Palmetto Mine, Polk County, probably early Pliocene. *A* about 0.5×; *B–D* about 1×.

Fig. 17.10. Sperm whale teeth (Physeteridae) from Florida. *A,* UF 93667, indeterminate genus and species from the Bone Valley District; *B,* UF 171283, a tooth of *Scaldicetus* sp. from the Swift Creek Mine, Hamilton County, middle Miocene. Both about 1×.

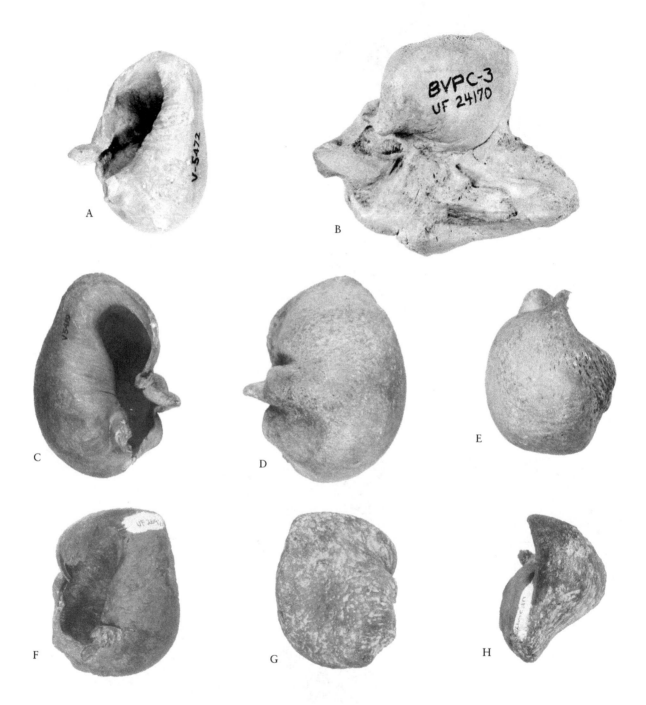

Fig. 17.11. Ear bones of fossil baleen whales (Mysticeti) from Florida. *A,* dorsal view of UF/FGS 5472, a right auditory bulla of a cetothere from the Agrico Mine, Polk County, middle Miocene; *B,* dorsal view of UF 24170, a petrosal of *Balaenoptera floridana* from the Payne Creek Mine, Polk County, early Pliocene; *C,* dorsal, *D,* ventral, and *E,* anterior views of UF/FGS 5419, a left auditory bulla of a balaenopterid, probably *Megaptera novaeangliae,* from the Fox Cut site, Flagler County, Pliocene or Pleistocene; *F,* dorsal, *G,* ventral, and *H,* anterior views of UF 26971, a right auditory bulla of a balaenid (right whale), probably *Eubalaena,* from the APAC (Macasphalt) Shell Pit, Sarasota County, late Pliocene. *A–B* about 1×; *C–H* about 0.5×.

rostrum referable to *Mesoplodon* and three partial mandibles of *Ninoziphius platyrostris.* Unlike most beaked whales, *Ninoziphius* has numerous teeth. *N. platyrostris* was originally described from the early Pliocene of Peru, but is also known from North Carolina as well as Florida.

Middle Pliocene to early Pleistocene odontocetes are uncommon. Sperm whale teeth of the *Physeterula* type have been recovered from the Tamiami Formation (Bee Ridge Fauna) along with the petrosals and teeth of at least two types of small odontocetes. *Pseudorca crassidens,* the false killer whale, first appears in the early Pleistocene as do the living dolphin genera *Stenella* and *Tursiops.*

The fossil record of mysticetes in Florida, like that of odontocetes, begins in the late early Miocene. All Miocene mysticetes found so far are cetotheres, an assemblage of archaic, toothless baleen whales. Cetotheres are small relative to modern mysticetes, but they are certainly still large mammals by most standards. The majority of Florida cetothere fossils are isolated auditory bulla (fig. 17.11A), petrosals, or partial mandibles, but a few skulls are known. In the past, the generic names *Mesocetus* and *Isocetus* were applied to some of these specimens, but in his recent review Morgan (1994) preferred not to attach any names to them. The richest hunting ground for Florida cetotheres is the Bone Valley region, where four different types of middle Miocene age are known.

Florida's mysticete fauna took on a much more modern appearance in the Pliocene. Two species of *Balaenoptera,* the living genus that includes the blue and finback whales, are known from the Palmetto Fauna. One of these, *Balaenoptera floridana,* is a small species by the standards of its family, with an overall body length of about 11 meters. Its mandibles are more than 2 meters in length all by themselves. A second species of this genus is slightly larger and has a more elongate auditory bulla. The Tamiami Formation of the Gulf Coast has produced a number of mysticete skeletons over the past three decades. At least four taxa are represented, two species of *Balaenoptera* (one of which is *B. floridana*), a small right whale (family Balaenidae, fig. 17.11F–H), and a humpback whale, *Megaptera.* The best specimen was collected in Hickey Creek a few miles east of Fort Meyers. This specimen consists of much of the skull, parts of the mandibles, and many associated vertebrae and ribs (fig. 17.12). Based on proportions of skull and body length in living whales of the genus *Balaenoptera,* the Hickey Creek whale would have been about 15 meters long. It is similar in size to medium-sized species of *Balaenoptera,* such as the fin whale (*B. physalus*) and the sei whale (*B. borealis*). Several differences, including the smaller size of the ear bone relative to the skull, indicate that the Hickey Creek whale belongs to an extinct species of *Balaenoptera* rather than a living species. Over on the Atlantic side of the state, a small *Balaenoptera, Megaptera* (fig. 17.11C–E), and a right whale (*Eubalaena*) have all been collected from the late Pliocene Nashua Formation.

Florida's whale fauna was essentially modern by the Pleistocene. Almost any species of whale or dolphin found in Florida's waters today might be found as a fossil in marine Pleistocene sediments. Pleistocene whales are found most commonly in beach deposits, coastal shell pits, and on spoil piles around dredging operations. For

Fig. 17.12. Posterior view of one of the lumbar vertebrae of UF 40200, an associated skeleton of *Balaenoptera* sp. from Hickey Creek, Lee County, early Pliocene. The large centrum with very flat ends and the reduced neural arch are characteristic of modern whale vertebrae. About 0.2×.

example, a partial skull of a right whale (*Eubalaena glacialis*) was collected from beach sands near Jacksonville. Numerous auditory bullae and other cranial fragments of humpback (*Megaptera novaeangliae*) and finback (*Balaenoptera physalus*) whales have been found from dredging sites in the Jacksonville area. Partial skulls of the gray whale, *Eschrichtius robustus*, have been found on Jacksonville Beach and Jupiter Island. It is unclear whether their age is Pleistocene or Holocene. The skull of a late Pleistocene pilot whale from Deland in Volusia County was described as a new species, *Globicephala baekeckeii*. However, it is likely that it represents the modern short-finned pilot whale, *Globicephala macrorhyncus*, which attains a length of 4 to 5 meters.

REFERENCES

Allen, G. S. 1941. A fossil river dolphin from Florida. Bulletin of the Museum of Comparative Zoology, 89:1–8.

Barnes, L. G., D. P. Domning, and C. E. Ray. 1985. Status of studies on fossil marine mammals. Marine Mammal Science, 1:15–53.

Bryant, J. D. 1991. New early Barstovian (middle Miocene) vertebrates from the upper Torreya Formation, eastern Florida panhandle. Journal of Vertebrate Paleontology, 11:472–89.

Fordyce, R. E., and L. G. Barnes. 1994. The evolutionary history of whales and dolphins. Annual Review of Earth and Planetary Sciences, 22:419–55.

Howell, A. B. 1930. *Aquatic Mammals: Their Adaptations to Life in the Water*. Charles Thomas, Springfield, Illinois, 338 p.

Hulbert, R. C., R. M. Petkewich, G. A. Bishop, D. Bukry, and D. P. Aleshire. 1998. A new middle Eocene protocetid whale (Mammalia: Cetacea: Archaeoceti) and associated biota from Georgia. Journal of Paleontology, 72:905–25.

Kellogg, R. 1936. A review of the Archaeoceti. Carnegie Institution of Washington Publication, 482:1–366.

———. 1944. Fossil cetaceans from the Florida Tertiary. Bulletin of the Museum of Comparative Zoology, 94:433–71.

———. 1959. Description of the skull of *Pomatodelphis inaequalis* Allen. Bulletin of the Museum of Comparative Zoology, 121:1–26.

McKenna, M. C., and S. K. Bell. 1997. *Classification of Mammals above the Species Level*. Columbia University Press, New York, 631 p.

Morgan, G. S. 1986. The so-called giant Miocene dolphin *Megalodelphis magnidens* Kellogg (Mammalia: Cetacea) is actually a crocodile (Reptilia: Crocodilia). Journal of Paleontology, 60:411–17.

———. 1994. Miocene and Pliocene marine mammal faunas from the Bone Valley Formation of central Florida. Pp. 239–68 in A. Berta and T. A. Deméré (eds.), Contributions in Marine Mammal Paleontology Honoring Frank C. Whitmore, Jr. San Diego Natural History Society, San Diego.

O'Leary, M. A., and J. H. Geisler. 1991. The position of Cetacea within Mammalia: phylogenetic analysis of morphological data from extinct and extant taxa. Systematic Biology, 48:455–90.

O'Leary, M. A., and M. D. Uhen. 1999. The time of origin of whales and the role of behavioral changes in the terrestrial-aquatic transition. Paleobiology, 25:534–56.

Thewissen, J. G. M. 1994. Phylogenetic aspects of cetacean origins: a morphological perspective. Journal of Mammalian Evolution, 2:157–84.

———. 1998. *The Emergence of Whales*. Plenum Press, New York.

Vaughn, T. A. 1972. *Mammology*. W. B. Saunders, Philadelphis, 463 p.

Whitmore, F. C., G. V. Morejohn, and H. T. Mullins. 1986. Fossil beaked whales—*Mesoplodon longirostris* dredged from the ocean bottom. National Geographic Research, 2:47–56.

Glossary

This glossary provides the definitions for scientific terms used in this book which may not be found in a standard dictionary. Some words have multiple meanings, such as *radiation* or *generic,* but the definitions provided here for such words are limited to a paleontologic, geologic, or anatomical context. Plural forms are shown for irregular nouns only. Abbreviations used are the following: adj., adjective; alt. sp., alternative spelling; n., noun; pl., plural; v., verb.

alveolus (alveoli, pl.) *n.* The depressions or sockets in the tooth-bearing skeletal elements, such as maxillae and dentaries, into which the roots of the teeth are inserted. Especially well developed in mammals and crocodilians.

amniote *n.* A member of the taxonomic group Amniota, such as reptiles, mammals, and birds, characterized by possession of the amnion, a specialized embryonic membrane.

arboreal *adj.* Descriptive term for an animal that primarily lives in trees.

archaeology (archeology, alt. sp.) *n.* The scientific study of human life and culture in the past. Frequently confused with paleontology, but the two are distinctly different fields of study.

atlas *n.* The first or anteriormost vertebra of a tetrapod, the one that articulates with the skull.

auditory bulla (auditory bullae, pl.) *n.* The bone that encloses the inner ear region of the mammalian skull and is found on the ventral surface of the posterior region of the skull. Typically fragile and thus rarely preserved in fossils, but inflated and composed of very solid bone in whales. For that reason, whale auditory bullae are commonly recovered as fossils. Also called the **tympanic bulla.**

axis (axes, pl.) *n.* The second vertebra in a tetrapod; it follows the atlas.

basicranium *n.* The posterior and ventral portion of the skull comprised of the basioccipital, basisphenoid, petrosal, and auditory bulla and portions of the squamosal and pterygoid bones. The evolutionary relationships of mammals can often be deduced by careful study of the arrangement of these bones and the foramina through which pass blood vessels and nerves. **basicranial** *adj.*

bilophodont *adj.* A tooth primarily comprised of two transverse ridges, for example, the lower molars of tapirs. Usually used only in conjunction with mammalian teeth.

brachyodont (brachydont, alt. sp.) *adj.* A tooth with a relatively short crown, especially one in which the crown is shorter than the roots; an animal with such teeth. Usually used only in conjunction with mammalian teeth.

bone *n.* 1. Any single component of the vertebrate skeletal system, regardless of its composition. For example, the maxilla, humerus, and ribs are all bones. 2. A particular type of vertebrate skeletal tissue comprised of a mixture of inorganic hydroxyapatite crystals and proteins.

browser *n.* An animal (usually a mammal) whose diet is entirely or mostly made up of the leaves and stems of shrubs, bushes, and trees.

bunodont *adj.* Descriptive term for a tooth comprised of low, rounded, isolated cusps or cones that are not connected by ridges or lophs. Usually used only in conjunction with mammalian teeth.

canine *n.* The usually large, pointed, single-rooted tooth between the incisor and premolar series in the mammalian dentition.

caniniform *adj.* Any tooth having the sharp, conical appearance of a canine; usually used to describe an incisor or premolar that functionally resembles a typical canine.

carapace *n.* The bony dorsal shell of a turtle or cingulate edentate (armadillos).

carnassial *n.* The specialized bladelike cheek teeth of carnivorous mammals, which are used to cut meat. Usually consists of a pair of teeth (one upper, one lower) and are best developed in the families Felidae and Nimravidae. The carnassial pair in the mammalian order Carnivora consists of the fourth upper premolar and the first lower molar.

carnivore *n.* 1. Any animal whose diet principally includes vertebrate flesh and tissues (**carnivorous** *adj.*). 2. Common name for a member of the mammalian order Carnivora, regardless of diet. To avoid ambiguity, the term **carnivoran** is sometimes used instead in this context.

carpal *n.* Any one of the skeletal elements making up the wrist, such as the unciform or trapezoid. Distally they articulate with the metacarpals, proximally with the radius and ulna. Collectively, they are referred to as the **carpus**.

cartilage *n.* A flexible, organic vertebrate tissue found in the skeletal system. Softer than bone, it is found in places of the skeleton where flexibility is a greater asset than strength, such as on movable joint surfaces, the external ears, and the tip of the nose. Because of its ability to reshape itself (unlike bone), the skeletons of fetal and very young juveniles consist primarily of cartilage, which is gradually replaced by bone tissue as the animal matures. Some animals, such as sharks, retain an internal skeleton made of cartilage as adults. Unless it contains calcium-rich minerals, cartilage rarely preserves in the fossil record.

caudal *adj.* Of or relating to the tail; especially one of the vertebrae of the tail.

cement *n.* A relatively soft, fibrous tissue of the vertebrate skeletal system. It surrounds the roots of teeth in the alveoli of mammals and some reptiles, adhering the teeth to the bone. Also found on the crown of the teeth in some hypsodont or hypselodont mammals where it surrounds the enamel and fills in deep valleys and depressions, thus providing structural support.

centrum (**centra**, pl.) *n.* The spool-shaped or cylindrical body of a vertebra.

cervical *adj.* Of or relating to the neck; especially one of the vertebrae of the neck.

character *n.* Any observable or measurable feature of the anatomy, biochemistry, ecology, physiology, or behavior of a biological organism. For example, "the number of digits on the manus (hand)" is a character, as is "builds a nest using sticks."

character state *n.* The condition of a character found in any particular individual or, more generally, in a species or higher taxonomic group. For the character "the number of digits on the manus," given above, some possible character states are two, three, or five. For the character "builds a nest using sticks," given above, there are only two possible character states, yes or no.

cheek tooth *n.* Any mammalian tooth found posterior to the canine—that is, a premolar or molar.

chondrocranium *n.* The structure comprised of the internal skeletal elements of the skull that surround and support the brain and major sense organs. Bones that form from the cartilaginous chondrocranium include the exoccipital, basioccipital, basisphenoid, and petrosal.

chordate *n.* Any member of the Chordata, a taxon of animals distinguished from other animals by having a notochord, pharyngeal gill slits, a dorsal main nerve chord, V-shaped segmented muscles, and a post-anal tail. Vertebrates are the best-known group of chordates, but the taxon also includes several marine groups, most notably the tunicates and the cephalochordates. Some of the chordate character states are found in vertebrates only during embryonic development and are either lost or highly modified in adults.

cingulum (**cingula**, pl.) *n.* A ridge or shelf found on the cheek teeth of some mammals, located on the outer margin of the tooth. Cusps protruding from the cingulum are referred to as styles (or stylids on lower teeth), such as a parastyle or metastyle. In certain mammalian groups, the cusp referred to as the hypocone originally evolved on the posterolingual region of the cingulum.

collagen *n.* The major type of protein found in vertebrate bone and other hard skeletal tissues.

conspecific *adj.* Of or relating to a single species. For example, two specimens a paleontologist identifies as belong to the same species would be described as conspecific.

cone *n.* The major cusps or hill-like structures found on mammalian teeth. A prefix is added to indicate a specific cone, such as metacone or paracone (see fig. 1.11). The term **conid** is used with lower teeth.

coprolite *n.* Fossilized feces or dung.

crown *n.* The enamel-covered portion of a tooth that usually extends beyond the gums.

crystal *n.* Solid material, such as a mineral, with a three-dimensional atomic structure consisting of a repetitious, regular pattern.

cursorial *adj.* Descriptive term for an animal well adapted for running, usually at high speed or over long distances. Or, any anatomical feature found in such an animal, such as long, slender limb bones.

deciduous *adj.* An anatomical structure, such as a tooth or antler, that is shed or dropped off the body. Especially refers to the deciduous tooth series of juvenile mammals, which consists of a set of incisors, canine, and premolars that erupt early in life (sometimes prior to birth) and which are usually shed sequentially as the animal approaches maturity and is replaced by members of the adult or permanent tooth series.

dentine (**dentin**, alt. sp.) *n.* A hard vertebrate tissue similar in composition to bone but found only in teeth and dermal tissue. Dentine exists in a number of varieties with differing internal structures and degrees of hardness.

dentition *n.* A collective term for the teeth of an individual vertebrate or a taxonomic group of vertebrates.

derived character state *n.* When comparisons are made between differing character states found on two taxa, if one has evolved from (or is thought to have evolved from) the other, then the former is considered the derived character state.

dermal *adj.* Of or relating to the skin.

diaphysis *n.* The shaft of a mammalian limb bone, the portion that excludes the epiphyses.

diastema (**diastemata**, pl.) *n.* The gap between consecutive teeth found in some mammals, especially herbivorous groups such as rodents and ungulates.

digitigrade *adj.* The condition of supporting the body weight and walking on the digits or phalanges, as opposed to the metapodials (plantigrade). Dogs and birds are digitigrade animals.

dignathic *adj.* Relating to both the upper and lower jaws.

edentulous *adj.* Lacking teeth.

enamel *n.* The hardest vertebrate tissue, composed primarily of the mineral hydroxyapatite. A shiny, glossy substance, it forms the outer covering of the teeth in tetrapods and some fishes.

enameloid *n.* A variety of very hard dentine that compositionally resembles true enamel and is found on the teeth and scales of some fish. Can only be distinguished from true enamel microscopically.

epiphysis (epiphyses, pl.) *n.* Separate centers of bone development found at the ends of some bones in juvenile mammals and a few other vertebrates.

epoch *n.* One of the formal units of the geologic time scale, epochs are subdivisions of periods. The term is capitalized when it follows the formal name of such a time unit, for example, the Miocene Epoch.

era *n.* One of the formal units of the geologic time scale, eras are made up of two or more contiguous periods. The term is capitalized when it follows the formal name of such a time unit, for example, the Cenozoic Era.

evolution *n.* The process of morphologic and genetic change in biologic organisms over time that occurs because of natural selection and genetic drift.

extant *adj.* A species or taxon that still exists and has living members; the opposite of extinct.

family *n.* The basic rank in the taxonomic hierarchy between the genus and the order. The formal name of a family must end in the suffix "-idae," which is preceded by the root form of the generic name of the type genus of the family. A common name for most family names can be made by changing the terminal "ae" to an "s" and starting it with a lowercase letter. For example, Felidae is the formal scientific name of a mammalian family, while felids is its informal equivalent. The species making up a family can be further subdivided into subfamilies, tribes, and subtribes.

feral *adj.* A once captive or domesticated species of animal or plant that is now living and breeding in the wild, usually in a region where it is not a native species. For example, the feral burros and horses living in the American Southwest.

foramen (foramina, pl.) *n.* A hole or opening in a bone, usually for the passage of a nerve or blood vessel.

formation *n.* The basic unit of rock in the geologic discipline of stratigraphy, it consists of one or more beds of same type of rock (limestone, sandstone, etc. [or a distinctive combination of rock types]) with a lateral extent of at least several miles. Formal names of formations are capitalized and based on geographic names. Examples from Florida are the Anastasia Formation and the Suwannee Limestone.

fossa (fossae, pl.) *n.* A prominent depression or concavity in a bone, often to house a muscle or gland.

fossorial *adj.* Descriptive term for an animal well adapted for digging; any anatomical feature found in such an animal.

generic *adj.* Of or relating to a genus.

genus (genera, pl.) *n.* The basic taxonomic rank immediately above the species level. A genus consists of a group of species more closely related to each other than any is to a species not included in the genus, or it may contain only a single species. A formal genus name is capitalized and written in italics; for example, *Canis* or *Triceratops*. Optionally, the species making up a genus may be grouped into two or more subgenera. In that case, one of the subgenera will have the same name as the genus, while the other will have different names. They are formally written like generic names but follow the generic name and are enclosed in parentheses. For example, *Hesperotestudo (Hesperotestudo)* and *Hesperotestudo (Caudochelys)* are two subgenera of tortoises found as fossils in Florida.

Gondwana (Gondwanaland, alt. sp.) *n.* Name given by geologists to an ancient landmass or continent that included much of the crust now making up South America, Africa, Australia, Antarctica, parts of Asia including India, and portions of southern Europe. During the Paleozoic Era, Florida was part of Gondwana. Late in the Paleozoic, Gondwana collided with Laurussia to form **Pangaea**.

granivore *n.* An herbivorous animal whose diet is entirely or mostly made up of seeds. **granivorous** *adj.*

grazer *n.* An herbivorous animal (usually a mammal) whose diet is entirely or mostly made up of grasses.

group *n.* 1. A basic unit of rock in the geologic discipline of stratigraphy consisting of two or more contiguous formations. A group is therefore typically a larger and more widespread stratigraphic unit than a formation. An example from Florida is the Hawthorn Group. 2. An informal assemblage of species, more or less synonymous with the term **taxon**.

herbivore *n.* An animal whose diet is entirely or mostly made up of plant matter of any type. **herbivorous** *adj.*

heterodont *adj.* An animal whose teeth have different shapes and functions. The opposite of homodont.

holotype *n.* The single specimen upon which the scientific name of a species is based, and which has explicitly been designated as such in the original published description. All other specimens are only referred to the species.

homodont *adj.* An animal whose teeth all have the same general appearance and function.

homonymy *n.* In biological taxonomy, when two different taxonomic groups have identically spelled scientific names.

hypselodont *adj.* Descriptive term for a high-crowned, ever-growing tooth lacking roots; an animal with such teeth. Usually used only in conjunction with mammalian teeth. Found in xenarthrans, rabbits, some rodents, and a few ungulates.

hypsodont *adj.* Descriptive term for a tooth with a relatively tall or high crown, especially one in which the crown is taller than the roots; an animal with such teeth. Usually used only in conjunction with mammalian teeth.

incisiform *adj.* Descriptive term for a tooth having the appearance of a typical mammalian incisor, especially in reference to the lower canine in some artiodactyls.

incisor *n.* One of the anterior-most teeth in the mammalian dentition, usually single-rooted and with either a chisel-shaped or conical crown.

insectivore *n.* 1. Any animal whose diet principally includes invertebrate tissues, especially insects. 2. Common name for a member of the mammalian grandorder Lipotyphla (=Insectivora), regardless of diet. **insectivorous** *adj.*

interspecific *adj.* Relating to characteristics or features found in two or more species.

intraspecific *adj.* Relating to characteristics or features found within a single species.

keratin *n.* Hard, organic dermal tissue that makes up fingernails and the sheaths of horns; on occasion it preserves in fossils. **keratinous** *adj.*

labial *adj.* Refers to the external or lateral side of a tooth, toward the lips or cheek. An alternate, synonymous term is *buccal*.

lingual *adj.* Refers to the internal side of a tooth, toward the tongue. The opposite of the labial side.

lophodont *adj.* A tooth in which the cusps or cones are connected by ridges (called **lophs**). Usually used only in conjunction with mammalian teeth.

matrix *n.* Samples of fossil-bearing sediment to be processed by screen-washing.

member *n.* A basic unit of rock in the geologic discipline of stratigraphy consisting of a portion of a formation. The subdivision of a formation into two or more members is optional. As an example, in southern Florida the Arcadia Formation includes beds placed in the Tampa Member and the Nocatee Member.

mesic *adj.* A descriptive term for a region or habitat with ample rainfall and vegetation.

mineral *n.* The inorganic, crystalline compounds that make up rocks. Common examples are quartz, calcite, gypsum, and feldspar. The mineral found in the vertebrate skeletal system is hydroxyapatite, a phosphatic mineral that is synthesized by special bone-producing cells.

mixed-feeder *n.* A herbivorous animal (usually a mammal) whose diet is made up of both grasses and leaves.

molar *n.* One of the posterior-most series of mammalian teeth, usually with more than one root and a complicated occlusal surface with numerous cusps and cones. By definition, a molar does not have a deciduous precursor.

molariform *adj.* Descriptive term for a tooth having the appearance of a typical mammalian molar, especially in reference to the posterior premolars in many herbivorous groups.

monophyletic *adj.* A taxonomic group that includes the closest common ancestor of the group and all of the descendants of that species.

monotypic *adj.* A genus or any higher taxonomic category that contains only one species.

neotenic *adj.* Refers to an adult animal that resembles a juvenile or a juvenile feature found in an adult animal.

neotype *n.* A specimen formally designated in a published study to replace a missing or destroyed holotype. After a neotype has been designated, it has the same importance and duties as a holotype.

Neotropics *n.* The region in North and South America characterized by a tropical climate. It stretches from Mexico through Central America to Brazil and Bolivia, as well as the Caribbean islands and southernmost Florida.

notochord *n.* Flexible supporting rod that runs down the dorsal side of embryonic vertebrates. The original vertebrate skeletal system, it is functionally replaced by the vertebrae in most vertebrates.

orbit *n.* The depression or deep concavity in the skull that houses the eye.

osteoderm *n.* A piece of bone found in the skin of some vertebrates (also informally called a scute). Especially well developed in crocodilians, tortoises, and armadillos.

otolith *n.* Small, oval structures made of calcium carbonate that grow in the inner ear region of some fishes. Fossil otoliths are typically only studied by specialists.

paleoecology *n.* The scientific study of the ecology of fossil organisms and ancient environments; a subdiscipline of paleontology.

paleontology *n.* The scientific study of ancient life and fossils.

Pangaea *n.* The name given by geologists to the ancient "supercontinent" that formed late in the Paleozoic Era and broke apart in the Mesozoic Era.

parallelism *n.* The independent acquisition of the same derived character state in two or more taxonomic groups not through a common ancestor. Also sometimes referred to as *convergence*.

paraphyletic *adj.* A taxonomic group that includes the closest common ancestor of the group but not all of the descendants of that species. Paraphyletic groups are not considered valid by some scientists.

paratype *n.* One or more referred specimens specifically designated in the original description of a new species that provide additional information beyond that provided by the holotype.

parsimony *n.* In systematics, the acceptance of one possible phylogenetic arrangement in deference to other arrangements because it requires the fewest number of character state reversals and parallelisms. **parsimonious** *adj.*

pathology *n.* An individual or anatomical part of an individual with an unusual morphology caused by illness, injury, malnourishment, or genetic defect. **pathological** *adj.*

pectoral *adj.* Relating to the anterior paired lateral fin in chondrichthyan and osteichthyan fish, the forelimb in tetrapods, or the region where the forelimb attaches to the torso.

pectoral girdle *n.* The bone(s) in the shoulder region in tetrapods that articulate with the proximal end of the humerus and serve to structurally anchor the forelimb to the torso and provide attachment sites for many of the muscles that move the forelimb. Individual elements of the pectoral girdle include the scapula, coracoid, and clavicle.

pelvic girdle *n.* The bone(s) in tetrapods that articulate with the proximal end of the femur and serve to structurally anchor the hind limb to the torso and provide attachment sites for many of the muscles that move the hind limb. Individual elements of the pelvic girdle are the ilium, ischium, and pubis; in adult mammals these three elements are fused together to form a single bone called the pelvis or innominate.

period *n.* One of the formal units of the geologic time scale, a subdivision of time between the ranks of era and epoch. The term is capitalized when it follows the formal name of such a time unit, for example, the Cretaceous Period.

petrosal *n.* The mammalian skull bone that houses the inner ear. Detailed analysis of these and other features of the ear region of the skull often reveal important evolutionary information.

pharynx *n.* The region of the digestive system between the mouth and throat. **pharyngeal** *adj.*

phylogeny *n.* The pattern of ancestor-descendant relationships among a group of biologic species that is deduced by the study of character states.

plantigrade *adj.* The condition of supporting the body weight and walking on the metapodials, as opposed to the phalanges (digitigrade, op cit.). Humans and bears are examples.

plastron *n.* The ventral portion of the shell of a turtle.

polyphyletic *adj.* A taxonomic group that does not include the closest common ancestor of the group. Polyphyletic groups are not considered valid by scientists.

postcranial *adj.* Of or related to that portion of the skeletal system other than the skull and jaws.

premolar *n.* One of the mammalian teeth found posterior to the canine and anterior to the molars, usually with more than one root. They vary in morphology from simple to complicated and molariform, depending on the type of mammal. Premolars usually have a deciduous precursor, but in some cases the deciduous premolar is retained in the adult.

preorbital *adj.* Anatomically located anterior to the orbit.

primitive character state *n.* When comparisons are made between two differing character states, if one has evolved from (or is thought to have evolved from) the other, then the latter is consider the primitive character state.

radiation *n.* When one or a small number of species evolve rapidly and give rise to many new species over a relatively short period of time.

radiometric dating *n.* The process of determining the geologic age of a rock or fossil by analysis of the relative amounts of radioactive isotopes and their daughter products. Over time, the amount of radioactive isotopes decreases at a known rate, and the amount of daughter products increase. Also called **radio-isotopic dating.**

ramus (rami, pl.) *n.* One side of a mammalian mandible. Technically, *mandible* refers to the combined right and left rami, but it is often used informally to indicate only one side.

reversal *n.* 1. In the context of systematics, when a character state evolves from the derived condition back to the primitive state. 2. In the context of magnetostratigraphy, when the earth's magnetic field changes its polarity, either from normal to reversed or reversed to normal.

reworked *adj.* Descriptive term for a fossil that eroded out of the sediments or rock enclosing it and was subsequently deposited in younger sediments. By this manner, fossils of different ages can be found in the same rock unit. Reworked fossils can often be recognized by their worn appearance.

root *n.* The basal portion of a tooth (as opposed to the enamel-covered crown) where it attaches to the jaw.

rostrum (rostra, pl.) *n.* The anterior portion of the skull, usually consisting of the maxillary, premaxillary, and nasal bones. **rostral** *adj.* Of or relating to the rostrum.

sacral *adj.* Of or relating to the vertebra or vertebrae that articulate with the pelvis. In some vertebrates, these vertebrae are fused to form a **sacrum.**

scansorial *adj.* Descriptive term for an animal well adapted for climbing; any anatomical feature found in such an animal.

screen-wash *v.* The process of separating fossils from matrix by the use of running water and one or more sets of boxes with bottoms made of different-sized screens. Sedimentary particles smaller than the screen openings wash through, leaving a residue of concentrated grains and fossils that is sorted after it has dried. Screen-washing is a widely used method to collect fossils of small rodents, birds, lizards, snakes, salamanders, frogs, and fish.

selenodont *adj.* A special type of lophodont tooth in which the cusps or cones are connected by crescentic ridges. Characteristic of ruminant and tylopod artiodactyls.

serrated *adj.* Having a notched or jagged edge similar to the cutting edge of a saw. In paleontology the term is most often used to describe teeth. For example, the canine teeth of some saber-toothed cats had serrated edges.

species *n.* The lowest commonly used rank in the taxonomic hierarchy. A species is a group of morphologically similar individuals with one or more diagnostic character states that distinguish it from all other species. Some taxonomists define a species as a group of interbreeding individuals, although this is difficult to test in modern animals and impossible in fossils. A formal species name consists of two words written in italics, for example, *Homo sapiens* or *Tyrannosaurus rex*.

subfossil *n*. A fossil less than 10,000 years old.

subspecies *n*. The lowest taxonomic rank, subdivisions of a species, used to formally separate slightly different populations within a species usually on criteria such as color patterns or size. A formal subspecies name consists of three words written in italics, for example, *Lynx rufus koakudsi*. Subspecies are infrequently used in vertebrate paleontology.

supraorbital *adj*. Anatomically located dorsal to the orbit.

symphysis (**symphyses**, pl.) *n*. Immovable (or nearly so) joint where the right and left sides of a skeletal element fuse together. Examples are the mandibular symphysis and the pelvic symphysis.

synonymy *n*. In biological taxonomy, when a single taxonomic group has two or more scientific names.

systematics *n*. The scientific study of the phylogeny of biological organisms and the methods used to reconstruct their evolutionary relationships.

tarsal *n*. Any one of the skeletal elements making up the ankle, such as the cuboid or astragalus. Distally they articulate with the metatarsals, proximally with the tibia and fibula. Collectively they are referred to as the **tarsus**.

taxon (**taxa**, pl.) *n*. Any group of biologic organisms that together make up one of the scientific hierarchies of classification. For example, the genus *Canis*, the family Bovidae, and the class Actinopterygii are all taxa.

taxonomy *n*. The scientific study of classifying and naming biologic organisms.

tetrapod *n*. Common name applied to the taxon Tetrapoda, which includes amphibians, reptiles, birds, and mammals. Most tetrapods live on land, although some are secondarily aquatic.

thoracic *adj*. Of or relating to the chest region (thorax); especially one of the vertebrae in a mammal that articulates with a rib.

trace fossil *n*. A fossil that consists of something made by ancient organism while it was alive. Fossilized footprints, burrows, nests, stomach stones, and coprolites are examples of trace fossils.

trilophodont *n*. A tooth primarily composed of three transverse ridges. Usually used only in conjunction with mammalian teeth, especially proboscideans (chap. 15).

ungual *adj*. Of, or relating to, a hoof, claw, or nail.

variation *n*. The differences in size, color, morphology, behavior, and other biological characteristics between two or more individuals.

vertebrate *n*. Common name for a member of the monophyletic taxon Vertebrata, a group which includes the jawless fishes, bony fishes, sharks and rays, and tetrapods. Among chordates, vertebrates uniquely possess a brain that coordinates the nerves and senses of the body and is protected and supported by a chondrocranium. Almost all vertebrates have a series of bones known as vertebrae (hence the name) and the capacity to produce calcium phosphate crystals.

volant *adj*. Descriptive term for an organism that is capable of powered flight, such as most birds and bats.

xeric *adj*. Used to describe a dry habitat characterized by low levels of rainfall, such as an arid or semiarid region.

Index

Page numbers for figures are in italics.